VOLUME ONE HUNDRED AND THIRTEEN

CURRENT TOPICS IN DEVELOPMENTAL BIOLOGY

The Maternal-to-Zygotic Transition

CURRENT TOPICS IN DEVELOPMENTAL BIOLOGY

"A meeting-ground for critical review and discussion of developmental processes"
A.A. Moscona and Alberto Monroy (Volume 1, 1966)

VOLUME ONE HUNDRED AND THIRTEEN

CURRENT TOPICS IN DEVELOPMENTAL BIOLOGY

The Maternal-to-Zygotic Transition

Edited by

HOWARD D. LIPSHITZ

Department of Molecular Genetics,
University of Toronto,
Toronto, Ontario, Canada

ELSEVIER

AMSTERDAM • BOSTON • HEIDELBERG • LONDON
NEW YORK • OXFORD • PARIS • SAN DIEGO
SAN FRANCISCO • SINGAPORE • SYDNEY • TOKYO
Academic Press is an imprint of Elsevier

Academic Press is an imprint of Elsevier
125 London Wall, London, EC2Y 5AS, UK
525 B Street, Suite 1800, San Diego, CA 92101-4495, USA
225 Wyman Street, Waltham, MA 02451, USA
The Boulevard, Langford Lane, Kidlington, Oxford OX5 1GB, UK

First edition 2015

Notices

Knowledge and best practice in this field are constantly changing. As new research and experience broaden our understanding, changes in research methods, professional practices, or medical treatment may become necessary.

Practitioners and researchers must always rely on their own experience and knowledge in evaluating and using any information, methods, compounds, or experiments described herein. In using such information or methods they should be mindful of their own safety and the safety of others, including parties for whom they have a professional responsibility.

To the fullest extent of the law, neither the Publisher nor the authors, contributors, or editors, assume any liability for any injury and/or damage to persons or property as a matter of products liability, negligence or otherwise, or from any use or operation of any methods, products, instructions, or ideas contained in the material herein.

ISBN: 978-0-12-409523-6
ISSN: 0070-2153

For information on all Academic Press publications visit our website at store.elsevier.com

CONTENTS

CONTRIBUTORS

Tristan Aguero
Department of Cell Biology, University of Miami, Miller School of Medicine, Miami, Florida, USA

Célia Baroux
Institute of Plant Biology & Zurich-Basel Plant Science Center, University of Zurich, Zurich, Switzerland

Shelby A. Blythe
Department of Molecular Biology and Howard Hughes Medical Institute, Princeton University, Princeton, New Jersey, USA

Michael B. Eisen
Howard Hughes Medical Institute; Department of Molecular and Cell Biology; Department of Integrative Biology, and QB3 Institute, University of California, Berkeley, Berkeley, California, USA

Vedran Franke
Bioinformatics Group, Division of Biology, Faculty of Science, Zagreb University, Zagreb, Croatia

Antonio J. Giraldez
Department of Genetics, and Yale Stem Cell Center, Yale University School of Medicine, New Haven, Connecticut, USA

Ueli Grossniklaus
Institute of Plant Biology & Zurich-Basel Plant Science Center, University of Zurich, Zurich, Switzerland

Melissa M. Harrison
Department of Biomolecular Chemistry, School of Medicine and Public Health, University of Wisconsin-Madison, Madison, Wisconsin, USA

Mary Lou King
Department of Cell Biology, University of Miami, Miller School of Medicine, Miami, Florida, USA

John D. Laver
Department of Molecular Genetics, University of Toronto, Toronto, Ontario, Canada

Rueyling Lin
Department of Molecular Biology, UT Southwestern Medical Center, Dallas, Texas, USA

Howard D. Lipshitz
Department of Molecular Genetics, University of Toronto, Toronto, Ontario, Canada

Alexander J. Marsolais
Department of Biochemistry, University of Toronto, Toronto, Ontario, Canada

Scott Robertson
Department of Molecular Biology, UT Southwestern Medical Center, Dallas, Texas, USA

Richard M. Schultz
Department of Biology, University of Pennsylvania, Philadelphia, Pennsylvania, USA

Michael D. Sheets
Department of Biomolecular Chemistry, School of Medicine and Public Health, University of Wisconsin, Madison, Wisconsin, USA

Craig A. Smibert
Department of Molecular Genetics, and Department of Biochemistry, University of Toronto, Toronto, Ontario, Canada

Meng-Xiang Sun
Department of Cell and Developmental Biology, College of Life Sciences, State Key Laboratory of Hybrid Rice, Wuhan University, Wuhan, China

Petr Svoboda
Institute of Molecular Genetics, Academy of Sciences of the Czech Republic, Prague, Czech Republic

S. Zachary Swartz
Department of Molecular and Cellular Biology, Brown University, Providence, Rhode Island, USA

Gary M. Wessel
Department of Molecular and Cellular Biology, Brown University, Providence, Rhode Island, USA

Eric F. Wieschaus
Department of Molecular Biology and Howard Hughes Medical Institute, Princeton University, Princeton, New Jersey, USA

Jing Yang
Department of Comparative Biosciences, College of Veterinary Medicine, University of Illinois at Urbana-Champaign, Urbana, Illinois, USA

Valeria Yartseva
Department of Genetics, Yale University School of Medicine, New Haven, Connecticut, USA

Peng Zhao
Department of Cell and Developmental Biology, College of Life Sciences, State Key Laboratory of Hybrid Rice, Wuhan University, Wuhan, China

PREFACE

Since the early years of experimental embryology, it has been known that maternal factors loaded into the egg are capable of directing early embryogenesis. The most definitive experiments involved enucleation of amphibian or echinoderm embryos, which were shown to be capable of undergoing cleavage divisions and even forming partial blastulae (Briggs, Green, & King, 1951; Fankhauser, 1934; Harvey, 1936). These experiments complemented earlier ones on interspecific hybrid sea urchins that led to the conclusion that the embryo's own nuclei were essential for subsequent development (Boveri, 1893; Tennent, 1914).

The 125 years since those initial experiments have seen the advent of developmental genetics, molecular biology, recombinant DNA, high-resolution imaging methods, and, most recently, techniques that permit global analyses of the genome, transcriptome, proteome, and many other "-omes." Thus, for almost any plant or animal, it is now possible to define almost all of the changes in cellular components and behavior that occur from the production of the gametes, through fertilization and early embryogenesis, into later developmental stages. When combined with genetic, reverse-genetic, or surrogate-genetic strategies, it has been possible not only to define the changes in "-omes" but to identify key regulatory processes and the factors that control them.

Studies on a wide range of plants and animals have revealed shared and distinct features during the crucial, early developmental stages. Perhaps most striking has been the discovery that all plants and animals undergo a maternal-to-zygotic transition, during which two temporally overlapping processes occur: phased elimination of a subset of the maternal gene products that had been loaded into the female gamete and phased activation of the embryo's own genome. The deep conservation of the two components of the maternal-to-zygotic transition reflects the fact that sexually reproducing, multicellular organisms evolved similar overall strategies to hand developmental control from the female parent to its offspring.

Given the enormous advances in elucidating the content, scale, dynamics, regulation, and functions of the maternal-to-zygotic transition, it seemed timely to assemble a book that describes our current understanding of this process. While there have been review articles that have covered animals (Tadros & Lipshitz, 2009) or both animals and plants

(Baroux, Autran, Gillmor, Grimanelli, & Grossniklaus, 2008), many exciting discoveries have been made since these were published. Furthermore, a single review article cannot capture a field with the same depth and breadth as a multichapter book. I was therefore delighted with but not at all surprised by the enthusiasm with which this project was received by the contributors to this volume of *Current Topics in Developmental Biology*.

How does the maternal-to-zygotic transition relate to other developmental events, such as the mid-blastula transition? We previously defined the maternal-to-zygotic transition as "the period that begins with the elimination of maternal transcripts, continues through the production of zygotic transcripts and ends with the first major morphological requirement for zygotic transcripts in embryonic development" (Tadros & Lipshitz, 2009). Importantly, this defines the maternal-to-zygotic transition as spanning a period of development rather than a specific time point, thus distinguishing it from the mid-blastula transition, which is the first developmental event that depends upon the embryo's own genome rather than that of its female parent. Several of the chapters in this book take an even broader view by considering the oocyte-to-embryo transition, which includes events that occur well before fertilization (e.g., oocyte maturation), during which there are also global changes in RNA and protein populations that are crucial for egg activation and early embryogenesis. Furthermore, it is now clear that regulated and large-scale clearance of maternal mRNAs continues after the mid-blastula transition, at least in *Xenopus* and *Drosophila*. Thus, large-scale changes to the transcriptome and proteome brought about by posttranscriptional and posttranslational processes occur well before fertilization and extend well beyond the mid-blastula transition *per se*. While the biological role of several specific transcripts and proteins during the maternal-to-zygotic transition is understood, the role of global changes in the transcriptome and proteome remains to be determined.

The contributors were given flexibility in their choice of topic. Several chose to consider the MZT in both soma and germline, while others focused on only one of these. In addition, while some chose to review mechanistic aspects of maternal product clearance and/or zygotic genome activation, others focused more on the timing and function of these processes. I encouraged the contributors to provide personal perspectives and to speculate. Some chapters are more philosophical than others, reflecting I suspect more the personality of the author(s) than the state of understanding in their particular developmental model.

Together, the chapters in this volume provide a diversity of perspectives written by top researchers on essentially all of the animals and plants in which we now have a deep understanding of the maternal-to-zygotic transition. No doubt a decade from now we will have even deeper functional and mechanistic insights, but I hope that readers will view the current volume as both timely and interesting.

Howard D. Lipshitz

REFERENCES

Baroux, C., Autran, D., Gillmor, C. S., Grimanelli, D., & Grossniklaus, U. (2008). The maternal to zygotic transition in animals and plants. *Cold Spring Harbor Symposia on Quantitative Biology*, *73*, 89–100.

Boveri, T. (1893). An organism produced sexually without characteristics of the mother. *American Naturalist*, *27*, 222–233.

Briggs, R., Green, E. U., & King, T. J. (1951). An investigation of the capacity for cleavage and differentiation in Rana pipiens eggs lacking "functional" chromosomes. *The Journal of Experimental Zoology*, *116*(3), 455–499.

Fankhauser, G. (1934). Cytological studies on egg fragments of the salamander Triton. III. The early development of the sperm nuclei in egg fragments without the egg nucleus. *Journal of Experimental Zoology*, *67*, 159–215.

Harvey, E. B. (1936). Parthenogenetic merogony or cleavage without nuclei in *Arbacia punctulata*. *Biological Bulletin*, *71*, 101–121.

Tadros, W., & Lipshitz, H. D. (2009). The maternal-to-zygotic transition: A play in two acts. *Development*, *136*(18), 3033–3042.

Tennent, D. H. (1914). The early influence of the spermatozoon upon the characters of echinoid larvae. *Carnegie Institution of Washington Publication*, *182*, 127–138.

CHAPTER ONE

The Maternal-to-Zygotic Transition in *C. elegans*

Scott Robertson[1], Rueyling Lin
Department of Molecular Biology, UT Southwestern Medical Center, Dallas, Texas, USA
[1]Corresponding author: e-mail address: scott.robertson@utsouthwestern.edu

Contents

Abstract

In *Caenorhabditis elegans*, the first zygotic transcription can be detected in the 4-cell stage *C. elegans* embryo, a little over 2 h after fertilization. However, early development until the onset of gastrulation at approximately the 28-cell stage takes place normally even in the absence of zygotic transcription. Therefore, posttranslational and posttranscriptional regulation of the maternal proteins and mRNAs, respectively, that are loaded into the developing oocytes is sufficient to direct development prior to gastrulation. Protein phosphorylation is extensively used throughout the *C. elegans* maternal-to-zygotic transition (MZT): (1) for maternal protein activation, (2) for coordination of the meiotic and mitotic cell cycle, (3) to mark specific proteins for degradation, and/or (4) to switch the biochemical activity of specific proteins. Maternally loaded mRNAs are regulated primarily by a set of maternal RNA-binding proteins (RBPs), each of which binds to sometimes overlapping target sequences within the mRNA 3′UTRs and either promotes or inhibits translation. Most maternal transcripts are uniformly distributed throughout the embryo but specific transcripts are translated only in certain blastomeres. This control is achieved by the asymmetric distribution of the maternal RBPs,

Current Topics in Developmental Biology, Volume 113
ISSN 0070-2153
http://dx.doi.org/10.1016/bs.ctdb.2015.06.001

such that the blastomere-specific constellation of RBPs present, and their relative levels, determines the translational readout for their target transcripts. In certain well-studied cases, such as the specification of the sole endodermal precursor in the 8-cell embryo, the maternal transcripts and proteins along with their directly targeted zygotic genes have been identified.

ABBREVIATIONS

APC anaphase-promoting complex
GPCR G protein-coupled receptor
MSP major sperm protein
MZT maternal-to-zygotic transition
NEBD nuclear envelope breakdown
OET oocyte-to-embryo transition
ZGA zygotic genome activation

1. INTRODUCTION

In mammalian embryos, primary oocytes initiate meiosis, then arrest at prophase of meiosis I, and remain so until puberty. Cyclical hormone surges lead to a small group of oocytes resuming meiosis and arresting a second time at metaphase of meiosis II. Following ovulation, meiosis completes if the egg is fertilized (Fig. 1A). This cycle continues in mammals over a period of months, years, or even decades (reviewed in Von Stetina & Orr-Weaver, 2011). The *Drosophila* oocyte differentiates as one member of an interconnected 16-cell cyst, with the other 15 cells forming polyploid nurse cells that support oocyte growth (Von Stetina & Orr-Weaver, 2011). The oocyte arrests at prophase I and then is induced to undergo meiotic maturation by an unknown extrinsic signal, although prostaglandin hormones or ecdysone are likely candidates. On meiotic maturation, the *Drosophila* oocyte arrests for a second time at metaphase I (Fig. 1A). The signal to resume meiosis in *Drosophila* is, surprisingly, independent of fertilization, requiring instead the mechanical stress associated with passage through the oviduct (Mahowald, Goralski, & Caulton, 1983). It is thought that this represents a holdover from an ancestral form that reproduced asexually, as some extant *Drosophila* species, such as *Drosophila mercatorum*, are capable of parthenogenic development (Eisman & Kaufman, 2007). However, it can lead to the seemingly wasteful situation where eggs can be laid that have not been fertilized.

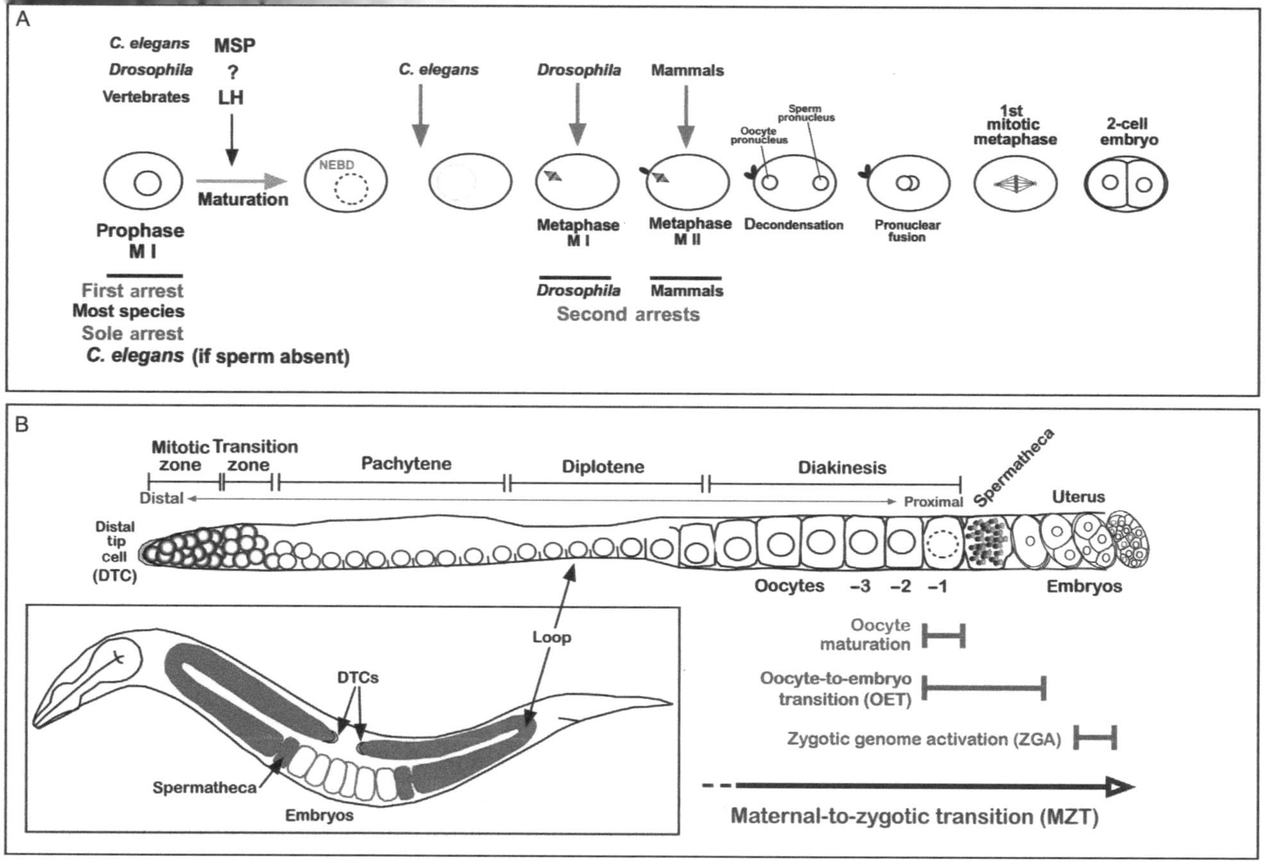

Figure 1 (A) Oocyte development and fertilization. General schematic of oocyte maturation and fertilization, highlighting the meiotic arrest(s) (red text), inducers of oocyte maturation (orange arrow), and timing of fertilization (green arrows) for *C. elegans*, *Drosophila*, and mammals. MI, meiosis I; MII, meiosis II; NEBD, nuclear envelope breakdown; MSP, major sperm protein; LH, luteinizing hormone. (B) Schematic of *C. elegans* gonad. One arm of the adult *C. elegans* hermaphrodite gonad is represented. Specific mitotic/meiotic stages are indicated above the gonad by the bars. Below the gonad, bars and arrow indicate oocyte maturation, OET, ZGA, and MZT, respectively. Inset: Schematic of an adult hermaphrodite, highlighting the gonad (red fill) and developing embryos (red outline). DTCs, distal tip cells. (See the color plate.)

Free-living, solitary *Caenorhabditis elegans* are self-fertilizing hermaphrodites and offer several distinct advantages for the study of oocyte development, maturation, fertilization, and the maternal-to-zygotic transition (MZT). The adult body, including the germline, is transparent and the germline develops in a highly spatially ordered, linear manner. A *C. elegans* hermaphrodite generates sperm in the L3 larval stage, which are stored in the spermatheca, and then switches completely to the generation of oocytes in the L4 larval stage, continuing to do so for the remainder of its reproductive life (Hirsh, Oppenheim, & Klass, 1976; Schedl, 1997). Hermaphrodites contain two tube-like syncytial gonad arms (Fig. 1B). In each gonad, nuclei in the distal region divide mitotically to provide continuing supplies of new germ nuclei. As nuclei move away from the distal region, they initiate and progress through different stages of meiosis, cellularize, and grow in volume in an assembly line-like fashion. The fully grown oocytes undergo maturation, are ovulated through the spermatheca, and are fertilized. Meiosis is completed upon fertilization and embryonic development initiates within the uterus, where the embryos again line up in developmental order before expulsion through the vulva during gastrulation (Fig. 1B).

The MZT includes both initiation within the embryo of transcription of select genes (zygote genome activation, ZGA), as well as inactivation of many maternal mRNAs and proteins that functioned to regulate developmental events following fertilization. The actual timing of the onset of the MZT, as well as its duration, can vary considerably between species. For *C. elegans*, no zygotic transcription occurs in late oocytes or early embryos prior to the 4-cell stage. In fact, zygotic transcription is not needed for embryos to develop normally—including the stereotypic asymmetric early cleavages, orientation of cleavage planes, and lineage-specific timing of early divisions—until the time when gastrulation should occur, at the 28-cell stage. However, a number of processes are already underway in the developing and maturing oocytes that directly or indirectly affect the MZT. For the purpose of this review, we will describe *C. elegans* MZT as from late-stage oocytes to approximately 28-cell embryos. An overriding theme of *C. elegans* MZT regulation is that it is controlled primarily posttranscriptionally and posttranslationally. We aim to show how the combination of asymmetric partitioning of maternal factors, protein modification-mediated changes in function, protein degradation, and highly regulated translational repression ensures a smooth transition.

We will divide the MZT into four key components: (1) oocyte maturation, ovulation, and fertilization; (2) the transition from meiosis to mitosis;

(3) the special case of the *C. elegans* 1-cell embryo; and (4) the transition from a single-cell embryo to a multicell embryo, including the initiation of zygotic transcription. We will review our current understanding of these processes and discuss how they are coordinated one to another and to the cell cycle. We will also highlight four important regulators/events that play key roles in coordinating this transition.

The MZT in *C. elegans* demonstrates nicely how genetics, cell biology, and biochemistry can be brought to bear upon a highly complex developmental process in a model organism. While genetic screens led to the isolation of mutations defective in individual processes and, thereby, key genes in each process, cell biological and biochemical analyses allowed us to determine how these separate events are precisely timed and coordinated.

2. COMPONENTS OF THE MZT

2.1 Oocyte Maturation, Ovulation, and Fertilization

The details of oocyte maturation, ovulation, and fertilization have been reviewed recently elsewhere (Kim, Spike, & Greenstein, 2013; Marcello, Singaravelu, & Singson, 2013; Marcello & Singson, 2010), and we direct readers there for a more complete discussion on these topics.

The solitary worm has only a very fixed number of sperm available for self-fertilization (~150 in each gonad), and mechanisms have evolved to ensure that oocyte development and maturation are closely controlled by, and coordinated with, the availability of sperm (reviewed in Han, Cottee, & Miller, 2010; Kim et al., 2013; Kuwabara, 2003). The presence of sperm drives these processes, and every sperm results in a successful fertilization, with exhaustion of available sperm leading to cessation of oocyte development. These mechanisms ensure that no sperm is wasted and that only oocytes that will result in fertilization undergo maturation. In addition, the mechanisms that terminate maternal control of development actually begin at oocyte maturation. For example, activation of the MBK-2 kinase that occurs as part of oocyte maturation leads to phosphorylation of several maternal proteins that directly participate in the oocyte-to-embryo transition (OET) (Stitzel, Pellettieri, & Seydoux, 2006) (see further below).

The distal-most region of the gonad consists of continuously dividing mitotic germ nuclei under the influence of a Notch signal originating from the somatic distal tip cell (DTC) that either represses meiosis, promotes mitosis, or both (Kimble & Crittenden, 2007; Kimble & Simpson, 1997; Kimble & White, 1981). As the germ nuclei move away from the DTC and the Notch signal, they transition from mitosis to meiosis and enter an extended period of pachytene. The germ nuclei that will form oocytes (approximately half of the germ nuclei will undergo apoptosis and be engulfed by the overlying somatic sheath cells; reviewed in Gartner, Boag, & Blackwell, 2008) exit pachytene as they approach the loop, cellularize as they leave the loop, and continue to grow in size. Fully grown oocytes line up in single file according to age in the proximal region of the gonad, in prophase of meiosis I (Fig. 1B). In young healthy hermaphrodites, sperm is always present and oocyte maturation continues without an apparent arrest. However, in the absence of sperm, the oocytes arrest at prophase of meiosis I, and pile up in the proximal gonad due to the lack of maturation and ovulation. When sperm is reintroduced via mating, oocytes exit arrest and resume the cell cycle (reviewed in Yamamoto, Kosinski, & Greenstein, 2006). In the presence of sperm, the oocyte immediately adjacent to the spermatheca (termed the −1 oocyte) is induced to undergo maturation, and approximately 5–6 min later is ovulated through the spermatheca where it is fertilized (Fig. 2). This cycle repeats itself approximately every 23 min in young hermaphrodites (McCarter, Bartlett, Dang, & Schedl, 1999). The maturation-inducing signal is the major sperm protein (MSP), a structural protein essential for sperm motility (Miller et al., 2001). It was found that MSP is also secreted in vesicles to generate an MSP concentration gradient from the spermatheca distally toward the oocytes (Kosinski, McDonald, Schwartz, Yamamoto, & Greenstein, 2005). Remarkably, MSP also functions in a hormone-like manner, engaging receptors on neighboring oocytes and the somatic gonadal sheath cells that surround the proximal gonad. The somatic sheath cells express muscle components such as actomyosin filaments organized into dense networks, forming a myoepithelial sheath consistent with their contractile function (Kosinski et al., 2005; Miller et al., 2001; Yamamoto et al., 2006). The gonadal sheath cells play a central role in MSP signal reception and response (Fig. 2). Indeed, protein kinase A (PKA) signaling in the sheath cells is required for all MSP-dependent responses in the germline (Govindan, Cheng, Harris, & Greenstein, 2006; Govindan, Nadarajan, Kim, Starich, & Greenstein, 2009; Kim, Govindan, Tu, & Greenstein, 2012). In response to the MSP signal, nuclear envelope breakdown (NEBD) occurs; the MAPK, MBK-2, is phosphorylated by Cyclin-dependent kinase 1 (CDK-1), a step required for MBK-2 activity (Arur et al., 2009; Lee et al., 2007; Miller et al., 2001); cortical and cytoskeletal rearrangements occur (Harris et al., 2006); the meiosis I spindle begins assembly concurrent with NEBD; and the Aurora

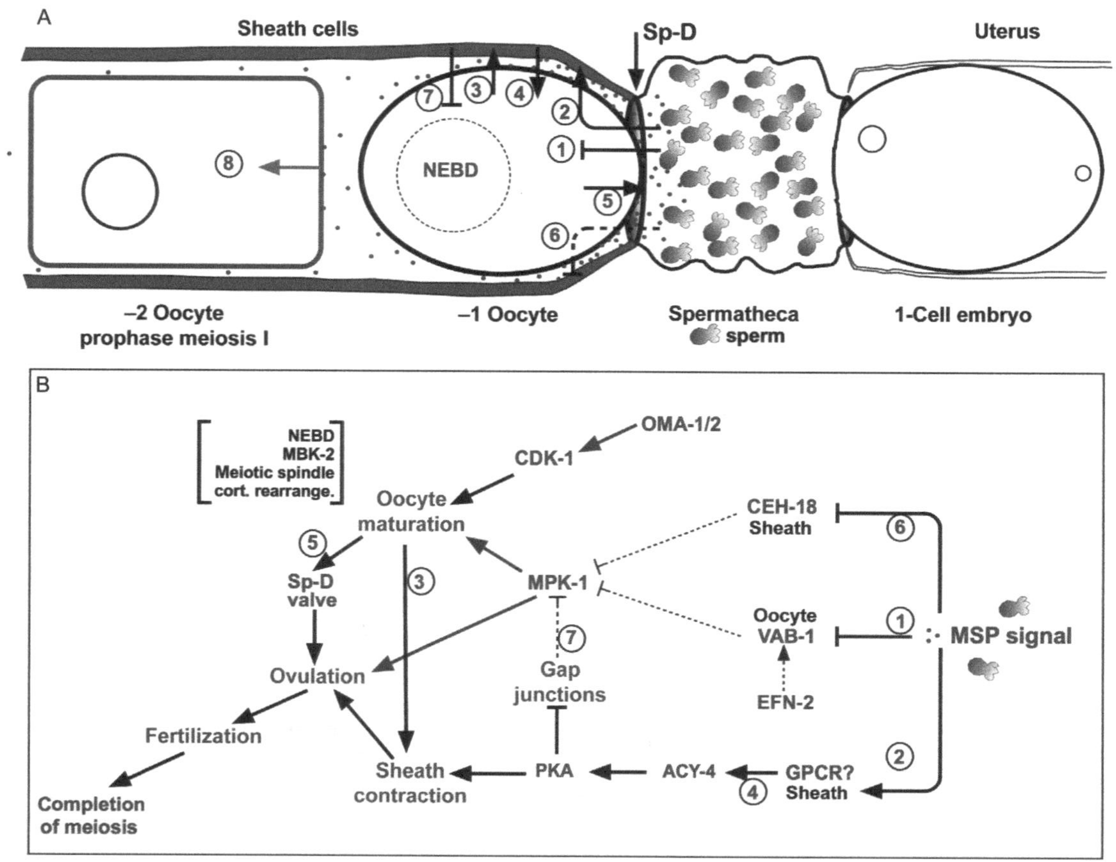

Figure 2 See legend on next page.

B kinase, AIR-2, relocates to chromatin (Govindan et al., 2009; Schumacher, Golden, & Donovan, 1998). Although the −2 and −3 oocytes also demonstrate changes indicating a response to the MSP signal, they do not initiate NEBD, and it is only the −1 oocyte that completes maturation (Harris et al., 2006; Lee et al., 2007; Miller et al., 2001) (Fig. 2). Exactly how this selectivity of regulation is achieved is unclear.

What is clear, however, is that control of ovulation (and therefore fertilization) involves a complex interplay of signals between sperm, the −1 oocyte, and the somatic gonad (Govindan et al., 2006, 2009; Kim et al., 2012; Starich, Hall, & Greenstein, 2014) (Fig. 2). MSP displaces ephrin bound to the oocyte VAB-1 receptors, resulting in MAPK activation and maturation. MSP also binds other non-VAB receptors on the oocytes and gonadal sheath cells, promoting ovulation of the −1 oocyte into the spermatheca. A large portion of the gonad is enclosed by five pairs of sheath cells. The MSP signal causes the proximal three pairs of gonadal sheath cells to contract, and this effect is amplified by signals from the maturing oocyte to the overlying sheath cells (Govindan et al., 2006; Greenstein, 2005; Greenstein et al., 1994; Han et al., 2010; McCarter, Bartlett, Dang, & Schedl, 1997, 1999; Miller, Ruest, Kosinski, Hanks, & Greenstein, 2003). These oocyte-derived signals, which include the worm homolog

Figure 2 Oocyte maturation, ovulation, and fertilization in *C. elegans*. (A) Schematic of the −2 and −1 oocytes, spermatheca, and 1-cell embryo, along with the somatic sheath cells (green; dark gray in the print version) and the distal spermatheca valve (Sp-D). Arrows and bars indicate the various positive and negative interactions between MSP (small brown (gray in the print version) ovals), the maturing oocyte, the sheath cells, and the distal spermatheca valve (Sp-D). Indicated interactions: (1) ephrin binding to VAB-1 receptor on −1 oocyte inhibits maturation, MSP displaces EFN-2 thereby promoting maturation; (2) MSP signals to receptors (likely GPCRs) on the sheath cells; (3) activating adenyl cyclase (ACY-4) which increases cAMP levels leading to activation of protein kinase A (PKA) and induction of sheath cell contraction; (4) oocyte LIN-3 (EGF) signals the sheath cells via the LET-23 receptor with downstream IP_3 signaling promoting sheath cell contractions; (5) similar signaling to the spermathecal–distal valve to induce dilation and ovulation; (6) in the absence of sperm, the POU-class homeodomain transcription factor CEH-18 in the sheath cells inhibits oocyte maturation; (7) PKA activation in the sheath cells leads to inhibition of the gap junction negative signal; (8) several oocytes distal to the −1 oocyte show indications of receiving an MSP signal (gray arrow; presumably weak due to the diffusion gradient of MSP). (B) Positive and inhibitory pathways functioning during oocyte maturation, ovulation, and fertilization, highlighting key components as described above. Solid lines indicate processes that are active as a result of MSP signal (i.e., sperm present). Dashed line indicates pathways that are inhibited as a result of the MSP signal. Numbering is as above.

of epidermal growth factor (EGF), also activate dilation of the adjacent distal spermatheca valve (Clandinin, DeModena, & Sternberg, 1998; Iwasaki, McCarter, Francis, & Schedl, 1996). The combination of sheath cell contractions along with dilation of the distal spermatheca valve results in the spermatheca expanding around the −1 oocyte and engulfing it. Fertilization rapidly occurs in the spermatheca, the spermatheca–uterus valve dilates, and the 1-cell zygote is ejected into the uterus. The sheath cell contractions and the spermathecal valve dilations require intracellular calcium release. The worm EGF homolog, LIN-3, induces inositol triphosphate (IP_3) signaling and calcium release in the sheath cells and spermathecal valves (Bui & Sternberg, 2002; Clandinin et al., 1998; Yin, Gower, Baylis, & Strange, 2004). Sperm signaling through MSP antagonizes the inhibitory VAB-I and CEH-18 pathways to promote oocyte maturation, MAPK activation, and spermathecal valve dilation (Clandinin et al., 1998; Govindan et al., 2006; Greenstein, 2005; Iwasaki et al., 1996). VAB-1 receptors also positively regulate the sheath cell contractions (Miller et al., 2003). The somatic sheath cells also interact with the oocyte via innexin14/22-based gap junctions (Whitten & Miller, 2007). This is intriguing, as it parallels the gap junction-based communication between mammalian follicle cells and the oocyte. The sheath cell–oocyte gap junctions appear to negatively regulate oocyte maturation and ovulation, and genetic epistasis experiments suggest that MSP-dependent Gα(s) signaling in the sheath cells antagonizes the gap junctions, thereby promoting maturation (Cheng, Govindan, & Greenstein, 2008; Govindan et al., 2006; Harris et al., 2006; Whitten & Miller, 2007) (Fig. 2). The types of molecules that are being exchanged between sheath cells and oocytes (and even the direction of exchange) through the gap junctions, as well as the exact mechanism by which Gα(s) antagonizes innexin function, are not known. Given the probable involvement of multiple classes of Gα proteins within the sheath cells in regulating oocyte maturation and ovulation, it is likely that the MSP receptors on the sheath cells include various G protein-coupled receptors (GPCRs), although these have not yet been identified. Consistent with this notion, mammalian follicle cells express the LH receptor, which is a GPCR (reviewed in Dufau, 1998; Mehlmann, 2005).

The closely related proteins, OMA-1 and OMA-2, are redundantly required for oocyte maturation (Detwiler, Reuben, Li, Rogers, & Lin, 2001; Shimada, Kawahara, & Doi, 2002). OMA protein levels increase in growing oocytes and reach their highest levels in the −1 oocyte and the 1-cell embryo. The proteins are rapidly degraded following the first mitotic division (Detwiler et al., 2001; Lin, 2003; Nishi & Lin, 2005; Shimada et al., 2002; Shirayama et al., 2006). The single *oma-1* or *oma-2* mutants show no phenotype, but *oma-1;oma-2* double-mutant worms are viable but sterile due to the inability of the −1 oocyte to complete maturation (Detwiler et al., 2001). The −1 oocyte in *oma-1;oma-2* double mutants can be shown to initiate maturation, but fails to complete it, arresting at a defined point in meiotic prophase I. In the absence of completion of maturation, ovulation is blocked and oocytes compress and back up within the proximal gonad, sometimes to the loop and beyond. The *oma-1;oma-2* meiotic arrest is partially suppressed by *wee-1.3(RNAi)*, suggesting that successful progression through meiotic prophase I requires inhibition of WEE-1.3 by either OMA-1 or OMA-2.

The OMA proteins contain tandem CCCH Tis-11-like zinc fingers, a motif that functions in RNA binding (Detwiler et al., 2001; Lai et al., 1999; Pagano, Farley, McCoig, & Ryder, 2007; Shimada et al., 2002), and are associated with ribonucleoprotein (RNP) complexes in the oocyte cytoplasm. These RNP complexes have been purified and components of the complex identified in an attempt to molecularly characterize other proteins that function in oocyte maturation (Spike, Coetzee, Eichten, et al., 2014; Spike, Coetzee, Nishi, et al., 2014). OMA RNPs were found to contain LIN-41 (TRIM-NHL protein), GLD-2, MEX-1, MEX-3, POS-1, and SPN-4. LIN-41 functions along with the OMA proteins to control and coordinate oocyte growth and meiotic maturation. Interestingly, it was found that whereas the OMA proteins promote oocyte cellularization and M-phase entry, LIN-41 inhibits both of these processes (Spike, Coetzee, Eichten, et al., 2014; Spike, Coetzee, Nishi, et al., 2014). LIN-41 protein is translated in the germline before the OMA proteins (at mid-pachytene vs. late pachytene just before the loop, respectively), and LIN-41 levels decline rapidly following the induction of oocyte maturation. The antagonistic functions of OMA and LIN-41 appear to coordinate the opposing processes of oocyte growth and maturation, suggesting that the OMA proteins may function in LIN-41 degradation. The OMA-containing RNP complexes contain both translational repressors, such as LIN-41, as well as translational activators such as GLD-2 (Ivshina, Lasko, & Richter, 2014; Nousch, Yeroslaviz, Habermann, & Eckmann, 2014). These findings raise the possibility that different versions of these complexes could either activate translation of a target mRNA (via the GLD-2-containing OMA complex) or repress translation of the same mRNA (via the LIN-41-containing OMA complex), depending upon developmental program or cellular requirements. Such a switchable target mRNA has not yet been identified.

Fertilization serves multiple functions beyond deposition of the haploid sperm nucleus into the oocyte cytoplasm (reviewed in Marcello et al., 2013). First, it is the signal for the resumption of meiosis, via activation of the anaphase-promoting complex/cyclosome (APC/C), triggering progression past anaphase (reviewed in Yeong, 2004). Meiotic progression itself has its own downstream ramifications for the MZT, including MBK-2 activation (Cheng, Klancer, Singson, & Seydoux, 2009) (see Section 2.2). Second, dilation of the spermathecal–uterine valve, and deposition of the 1-cell zygote into the uterus, requires fertilization (Marcello et al., 2013). Third, fertilization signals the rapid assembly of the chitinous eggshell that surrounds the developing embryo until hatching (Zhang, Foster, Nelson, Ma, & Carlow, 2005). Fourth, *C. elegans* oocytes do not have inherent polarity, and it is the point of sperm entry that provides the first polarity cue, defining the posterior end of the embryo (Goldstein & Hird, 1996). Sperm entry destabilizes the actomyosin network in the surrounding cortex and initiates a flow of cortically localized nonmuscle myosin and actin. This cortical flow carries other cortical proteins, including particular abnormal embryonic partitioning of cytoplasm (PAR) proteins, to the opposite cortex (Munro, Nance, & Priess, 2004), cementing the anterior–posterior polarity of the developing embryo (see Motegi & Seydoux, 2013; Rose & Gonczy, 2014 for reviews). It has been shown that it is, in fact, the centrosome that accompanies the sperm nucleus that actually initiates the changes that define the anterior–posterior axis of the embryo (Bienkowska & Cowan, 2012; Cowan & Hyman, 2004).

Recent RNA-seq analyses of oocytes, and 1-cell and 2-cell *C. elegans* embryos have identified an early, sharp wave of maternal transcript clearance between the oocyte and the 1-cell embryo (the OET) (Stoeckius et al., 2014). Although these maternal transcripts include the expected meiosis-specific factors, such as *mei-1*, and egg-shell factors, such as *chs-1* (chitin synthetase), the scale of the clearance, involving approximately 25% (just over 2000 transcripts) of the maternal mRNAs present in the oocyte, was unexpected. These authors identified a C-rich motif in the 3′UTRs associated with about half of these cleared maternal transcripts and demonstrated the expression of several *C. elegans* homologs of the mammalian polyC-binding proteins that are capable of binding this sequence, consistent with these RNA-binding proteins (RBPs) playing a role in regulating the stability of these transcripts following fertilization (Stoeckius et al., 2014). A significant proportion, however, of these maternal transcripts that are cleared immediately following fertilization did not contain an identifiable C-rich motif. How these maternal transcripts are rapidly cleared after fertilization in the 1-cell embryo is unclear.

2.2 Transition from Meiosis to Mitosis

Following fertilization, the oocyte-derived nucleus completes both meiotic divisions, setting up short acentrosomal meiotic spindles very close to, and originally parallel to but then rotating to be orthogonal to, the overlying plasma membrane (Fig. 3). This conformation insures that the polar bodies can be extruded with minimal cytoplasm at each reductive division (Fabritius, Ellefson, & McNally, 2011; Muller-Reichert, Greenan, O'Toole, & Srayko, 2010; Yamamoto et al., 2006). Three *C. elegans* genes are known to contribute to oocyte meiotic spindle assembly—*aspm-1*, which encodes a protein with one N-terminal calponin homology domain and two calmodulin-binding domains, which functions in rotation of the meiotic spindle (Connolly et al., 2014; van der Voet et al., 2009); *mei-1*, which encodes the catalytic subunit of the meiosis-specific microtubule-severing complex katanin (Clandinin & Mains, 1993; Clark-Maguire & Mains, 1994a; McNally, Audhya, Oegema, & McNally, 2006; McNally & McNally, 2011; Srayko, Buster, Bazirgan, McNally, & Mains, 2000; Yang, McNally, & McNally, 2003); and *klp-18*, which encodes a kinesin motor protein required for the assembly of a disordered array of acentrosomal microtubules into an ordered, functional bipolar spindle (Segbert et al., 2003). Both *aspm-1* and *klp-18* function in mitosis, whereas *mei-1* expression is meiosis restricted (Clandinin & Mains, 1993; Segbert et al., 2003; van der Voet et al., 2009).

At the completion of the two meiotic divisions, the oocyte-derived pronucleus decondenses and migrates to meet and fuse with the decondensed sperm-derived pronucleus. The 1-cell zygote then immediately enters metaphase and establishes the first mitotic spindle. In comparison to the meiotic spindle, mitotic spindles are larger and longer, requiring the generation of long arrays of astral microtubules from the sperm-supplied centrosomes. The time from completion of meiosis II to the first mitotic division is approximately 20 min, which requires very tight regulation (McCarter et al., 1999). *Meiosis defective* proteins (MEI-1 and MEI-2) form a heterodimeric, meiosis-specific katanin whose microtubule-severing activity is important for the formation of the meiotic spindle (Clark-Maguire &

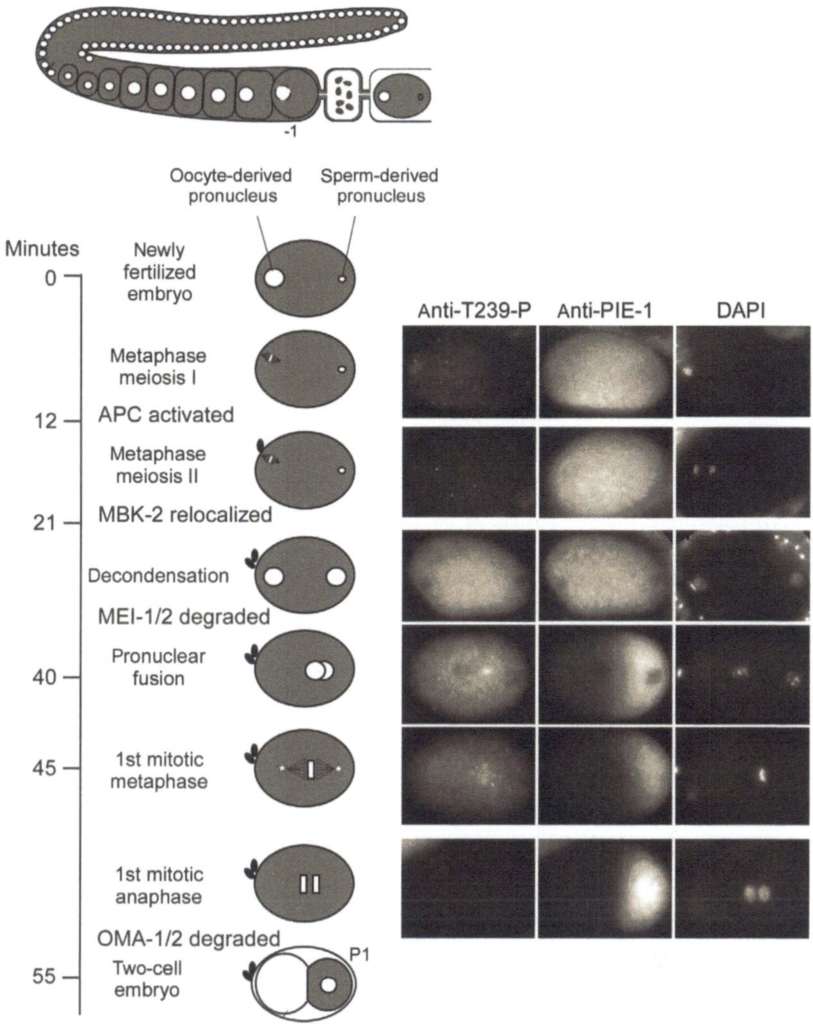

Figure 3 MBK-2 activation. (Left) Schematic of *C. elegans* development from newly fertilized embryo to first mitotic division, below a drawing of a gonad arm. Timescale is in minutes postfertilization. Black text: morphological mileposts. Blue (dark gray in the print version) text: key molecular events. Red (gray in the print version): germline. Small black ovals: polar bodies. (Right) Fixed embryos corresponding to the stages indicated in the schematic, showing OMA-1 anti-T239-P fluorescence (left column), anti-PIE-1 fluorescence (as positive control; central column), and DNA by DAPI-staining (right column). Note the narrow window during which phosphorylated OMA-1 is present: from the pronuclear decondensation stage to the first meiotic metaphase.

Mains, 1994a, 1994b; Mains, Kemphues, Sprunger, Sulston, & Wood, 1990; Srayko et al., 2000). However, continued presence of MEI-1 or MEI-2 in the embryo greatly interferes with formation of the mitotic spindle (Clandinin & Mains, 1993; Clark-Maguire & Mains, 1994a; Mains et al., 1990), and both proteins must be degraded rapidly following completion of meiosis II to allow formation of the first mitotic spindle (Fig. 3).

The rapid and coordinated degradation of MEI-1 and -2 at the completion of meiosis II is regulated in three different ways. First, the E3 ligase that degrades MEI-1 is not activated until both meiosis I and II are completed. MEI-1 is degraded by a CUL-3-containing E3 ubiquitin ligase that contains the MATH and BTB/POZ domain-containing protein MEL-26 as the substrate-binding component (Dow & Mains, 1998; Pintard et al., 2003; Xu et al., 2003). MEL-26 levels are low during meiosis I and II, but increase rapidly following completion of meiosis and remain high throughout the early mitotic divisions (Johnson et al., 2009). Second, MEI-1 is marked for degradation by phosphorylation at Serine 92 by the MBK-2 kinase in meiosis II (see next two paragraphs) (Cheng et al., 2009; Pang et al., 2004; Pellettieri, Reinke, Kim, & Seydoux, 2003; Quintin, Mains, Zinke, & Hyman, 2003; Stitzel et al., 2006). Third, maternal *mei-1* mRNA is still present in the early embryo, and therefore, its translation is actively repressed (see Section 3.2).

MBK-2 is translated in oocytes well before its activity is required, and like all DYRK kinases, it undergoes cotranslational self-phosphorylation on tyrosine within a YTY motif present in the kinase domain (Cheng et al., 2009; Lochhead, Sibbet, Morrice, & Cleghon, 2005). This self-phosphorylation is necessary for MBK-2 activity. Once made, mature DYRKs are no longer capable of tyrosine phosphorylation (Lochhead et al., 2005) and phosphorylate only serine or threonine residues in their targets. Therefore, it is important that phosphorylated tyrosine is shielded from any endogenous tyrosine phosphatase activity present in the oocyte before MBK-2 reaches its substrates. In addition, all known substrates of MBK-2 are present in the oocytes in addition to embryos (Clark-Maguire & Mains, 1994a; Detwiler et al., 2001; Schubert, Lin, de Vries, Plasterk, & Priess, 2000). Active MBK-2 in oocytes would result in precocious phosphorylation of these substrates, with severely detrimental effects. Elegant studies have shown how MBK-2 is subjected to multiple layers of regulation, which ensure that its activity is both preserved and "held in check" until the right place and the right time (Cheng et al., 2009).

In oocytes, two tyrosine pseudophosphatases, EGG-4 and EGG-5, bind to MBK-2 by interacting with the tyrosine-phosphorylated YTY motif, thereby inactivating MBK-2 enzymatic activity (Cheng et al., 2009; Parry et al., 2009). Pseudophosphatases are proteins that resemble phosphatases and bind specific phosphorylated motifs found on their targets, but lack enzymatic activity (reviewed in Langeberg & Scott, 2015). It is believed that one of the functions of pseudophosphatases is to shield phosphorylated proteins from the action of *bona fide* phosphatases. A third tyrosine pseudophosphatase, EGG-3, functions to tether the MBK-2/EGG-4/EGG-5 complex to the oocyte cortex (Cheng et al., 2009; Maruyama et al., 2007; Parry et al., 2009), preventing MBK-2 from accessing its substrates, all of which are cytoplasmic. Final activation of MBK-2 and phosphorylation of its *in vivo* substrates require two events: First, phosphorylation of MBK-2 by CDK-1 at Serine 68, which enhances MBK-2 activity *in vivo* (Cheng et al., 2009). This occurs during oocyte maturation when CDK-1 becomes active to drive progression through M phase at both meiosis I and II (reviewed in Archambault & Glover, 2009; Boxem, 2006). The second event is the release of MBK-2 from the pseudophosphatases and relocalization to the cytoplasm. The release is dependent on the APC, which is a Cyclin-specific E3 RING ubiquitin ligase. The APC regulates progression from metaphase to anaphase of both meiosis I and II, and is dependent on fertilization (McNally & McNally, 2005). The APC also promotes the subsequent proteasome-mediated degradation of EGG-3, EGG-4, and EGG-5 (Cheng et al., 2009; Maruyama et al., 2007; Parry & Singson, 2011; Parry et al., 2009). Coinciding with the APC activity in anaphase I, active MBK-2 relocates from the cortex, coalescing into cytoplasmic puncta (Pellettieri et al., 2003; Stitzel et al., 2006). These studies highlight the importance of MBK-2 activity and its spatiotemporal regulation, as well as the important role the cell cycle plays in coordinating MZT component processes.

2.3 The 1-Cell Embryo: A Special Case

The first embryonic division gives rise to two daughters of different size, maternal factor composition, and developmental fates (Rose & Gonczy, 2014) (Fig. 4). How the mitotic spindle aligns determines how asymmetrically arranged maternal developmental determinants are segregated. As mentioned earlier, the polarity of the first mitotic spindle is established by a series of events triggered by the sperm centrosome. The centrosome generates a localized inactivation of actomyosin cortical contractility, which leads to the establishment of an asymmetric distribution of cortical PAR proteins along the anterior–posterior axis of the embryo

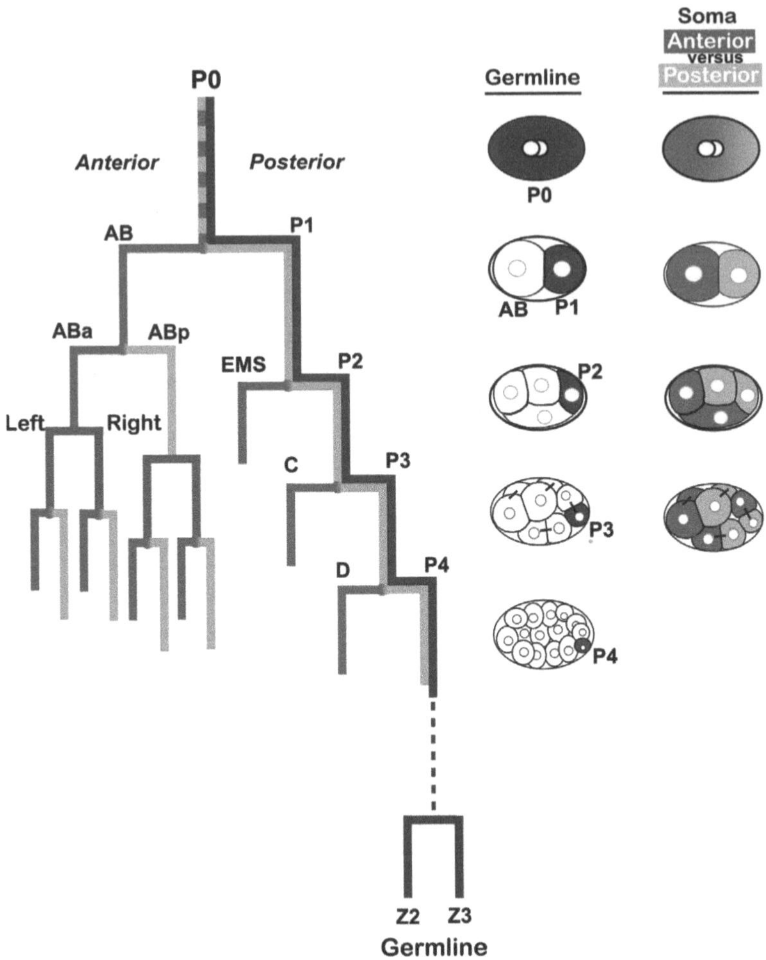

Figure 4 Segregation of germline versus anterior or posterior developmental potential in the early *C. elegans* embryo. The cytoplasm of the 1-cell embryo (P0) contains multiple maternal factors, both protein and mRNA, that direct lineage-specific differentiation. Segregation of these developmental determinants begins with the first mitosis and occurs at each following mitosis to generate the correct blastomere-specific constellation of proteins and maternal transcripts to effect both early development in the absence of zygotic transcription as well as an orderly, blastomere-specific transition from maternal-to-zygotic control of embryogenesis. (See the color plate.)

(Munro et al., 2004). This polarized distribution of specific PAR proteins to either the anterior or posterior cortical domains establishes an A–P polarity that is maintained by mutual negative interactions between the components of the two domains (reviewed in Nance, 2005; Rose & Gonczy, 2014). Critical for the asymmetric division is precise placement of centrioles and the mitotic spindle, as this will determine the position along the A–P axis of the cleavage furrow. The oocyte pronucleus and sperm pronucleus (with associated centrosomes) move toward each other, meeting and fusing in the posterior half of the embryo. The movement of the pronuclei toward each other is thought to involve the microtubule-associated motor protein, dynein, located on the surface of the female pronucleus, capturing the plus ends of astral microtubules originating from the sperm centrioles (Payne, Rawe, Ramalho-Santos, Simerly, & Schatten, 2003). Following pronuclear fusion, the zygotic nucleus and centrosomes move to the center of the embryo, during which time the centrosomes undergo a 90° rotation and become aligned with the A–P axis of the embryo. Although the spindle initially sets up in the center of the embryo aligned with the A–P axis, it is displaced toward the posterior during metaphase and anaphase, resulting in the larger, anterior AB blastomere and the smaller, P1 posterior blastomere. These processes have been reviewed in great detail recently (Rose & Gonczy, 2014) and we will not consider them further here.

Due to its complete dependence upon maternally supplied factors, with many developmental regulators already present as proteins in oocytes, the *C. elegans* 1-cell embryo is faced with something of a developmental conundrum. It is a precursor for both somatic cells and germline blastomeres. In addition, it is a precursor for both anterior blastomeres and posterior blastomeres. The 1-cell embryo contains key maternal regulators for the specification of somatic lineages, germline blastomeres, anterior blastomeres, and posterior blastomeres, all within a common cytoplasm (Fig. 4). Only after the first mitotic division are many key regulators for AB-derived cells versus P1-derived cells, and somatic cells versus germline blastomeres, asymmetrically segregated. The unique dilemma the *C. elegans* 1-cell embryo faces is how to keep these potent maternal regulators in check.

While many details are still missing, recent studies have shed light on two strategies that *C. elegans* employs to restrict or exclude certain activities to, or from, the 1-cell embryo, respectively. The first one is the 1-cell embryo-restricted expression of OMA-1 and OMA-2 (Detwiler et al., 2001; Nishi & Lin, 2005; Shimada et al., 2002). This restricted embryonic expression to only the 1-cell embryo is unique, not being observed for any other *C. elegans* protein to date. We have shown that OMA-1 and OMA-2 bind to an essential component of the transcription initiation complex, TAF-4, in 1-cell embryos, sequestering TAF-4 in the cytoplasm and globally repressing transcription (Fig. 5) (Guven-Ozkan, Nishi, Robertson, & Lin, 2008).

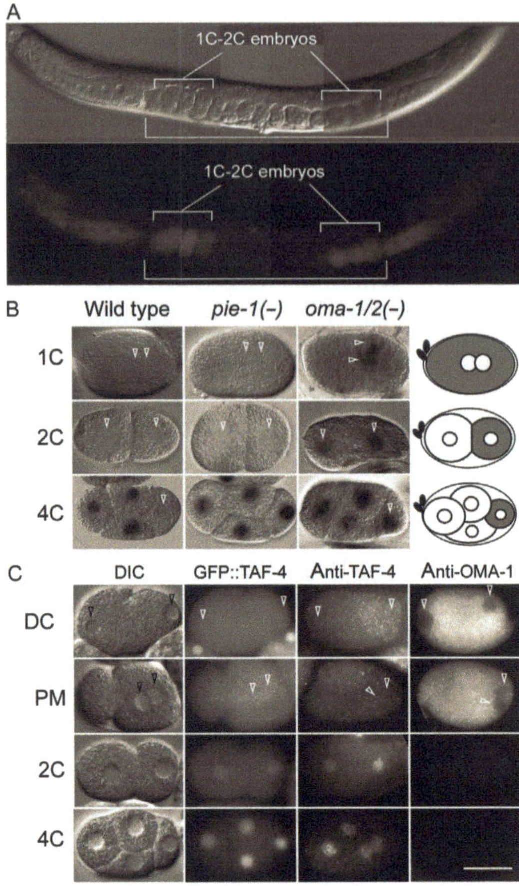

Figure 5 Control of ZGA in *C. elegans*. (A) OMA-1::GFP expression. Matching DIC and fluorescence images of an adult gravid hermaphrodite expressing OMA-1::GFP. Note the increasing fluorescence in the developing oocytes, and high expression in 1-cell and early 2-cell embryos in the uterus, which then rapidly disappears. (B) Timing of OMA-1/2 versus PIE-1 transcriptional repression. *In situ* hybridization for *vet-5* (very early transcript 5) nuclear-restricted transcript (Schauer & Wood, 1990) in wild type, *pie-1(−)*, and *oma-1(−);oma-2(−)* embryos at 1-cell, 2-cell, and 4-cell stages. *vet-5* nuclear transcripts are first detectable at the 4-cell stage in the three somatic blastomeres in wild-type embryos. Note that no transcription occurs in P2 due to repression by PIE-1. In *pie-1 (−)* embryos, 4-cell stage somatic expression of *vet-5* is detected, as well as derepressed expression in P2. In the *oma-1/2(−)* embryos, precocious *vet-5* expression can be detected as early as the 1-cell stage, with expression at the 4-cell stage in all blastomeres due to lack of protection by OMA-1/2 (repressing *zif-1* translation) and therefore precocious degradation of PIE-1 in the oocytes. (C) TAF-4 sequestration by OMA-1. Column 1 (left, top to bottom): DIC images of wild-type embryos at pronuclear decondensation (DC), meeting of pronuclei (PM), 2-cell and 4-cell stages. Arrowheads indicate pronuclei. Column 2: GFP::TAF-4 fluorescence. Note the diffuse cytoplasmic signal at DC, and the beginning of nuclear accumulation at PM, with continuing nuclear accumulation as development continues from 2-cell to 4-cell stage embryos. Column 3: Staining with an anti-TAF-4 antibody. Localization pattern same as for GFP::TAF-4. Column 4 (right): Staining with an anti-OMA-1 antibody. Note the early diffuse cytoplasmic signal with nuclear exclusion, and the rapid reduction in OMA-1 levels (but still detectable) after the first mitotic division.

OMA proteins bind to TAF-4 via a small domain (~35 aa) at the N-terminus distinct from their RNA-binding zinc fingers. The two OMA protein functions, one in oocytes, where they repress translation of specific maternal mRNAs, and the other in the 1-cell embryo, where they bind and sequester TAF-4 protein in the cytoplasm, are separable spatiotemporally and can be genetically uncoupled (Guven-Ozkan, Robertson, Nishi, & Lin, 2010). In addition, the two functions may also be regulated by the MBK-2 phosphorylation of OMA proteins (Nishi & Lin, 2005; Shirayama et al., 2006).

OMA-1 is phosphorylated on T239 by MBK-2 *in vitro* (Nishi & Lin, 2005; Stitzel et al., 2006). Using an antibody to OMA-1 phospho-T239, we showed that OMA-1 is phosphorylated at meiosis II, immediately after the time when MBK-2 is activated (Fig. 3) (Nishi & Lin, 2005; Pellettieri et al., 2003). Phosphorylation by MBK-2, therefore, creates two populations of OMA proteins that are separated temporally during the MZT: the first population, restricted to oocytes, is not phosphorylated at T239, whereas the second population, restricted to the 1-cell embryo, is phosphorylated at T239. We showed that OMA-1 phosphorylated at T239 by MBK-2 is better at binding to and sequestering TAF-4 in the cytoplasm, whereas the unphosphorylated OMA-1 is better at repressing translation of at least one of its *in vivo* target mRNAs (Guven-Ozkan et al., 2008, 2010). Therefore, phosphorylation of OMA-1 at T239 by MBK-2 serves as a molecular "switch" that converts OMA function from translational repressor in oocytes to global transcriptional repressor in the 1-cell embryo. Interestingly, the same site of MBK-2 phosphorylation that facilitates OMA-1 binding to TAF-4 also serves to earmark OMA-1 for degradation (Nishi & Lin, 2005; Shirayama et al., 2006).

The second strategy that *C. elegans* employs to restrict certain activities to the 1-cell embryo is to have the activity controlled by multiple factors that colocalize only in the 1-cell embryo. Translation of *zif-1* mRNA in oocytes is repressed by the OMA proteins (Guven-Ozkan et al., 2010) but, following phosphorylation by MBK-2, the OMA proteins lose their translation repression activity, and yet *zif-1* remains translationally repressed in the 1-cell embryo. We have recently shown that translational repression of *zif-1* in 1-cell embryos requires the combined action of two other RBPs, MEX-3 and SPN-4, with neither protein alone being sufficient to repress translation (Oldenbroek et al., 2012). These two RBPs are differentially segregated after the first mitotic division and therefore no longer colocalize within the same cell (Draper, Mello, Bowerman, Hardin, & Priess, 1996;

Ogura, Kishimoto, Mitani, Gengyo-Ando, & Kohara, 2003), effectively restricting their *zif-1* translational repression to only the 1-cell embryo. This strategy is effective in delaying the translation of maternally deposited mRNAs until after the first mitotic division. However, how general this strategy is in delaying translation of maternal mRNA remains to be determined.

2.4 Transition from a 1-Cell Embryo to a Multicell Embryo

We highlight three key elements that are especially important for the transition from a 1-cell embryo to a multicell embryo.

2.4.1 Posttranscriptional Regulation

The first few embryonic divisions have very short cell-cycle timing (10–15 min) with only alternating S and M phases (no G1 or G2) (Rose & Gonczy, 2014). These divisions occur in a highly reproducible fashion, with cell-cycle timing for each cell, the position of each cell within the embryo, relative sizes of each cell, and the developmental fate of each cell being basically invariant from an embryo to another. This transition is driven almost entirely by posttranscriptional events. Inhibition of zygotic transcription in embryos by RNAi against *ama-1*, which encodes the large subunit of RNA Pol II, leads to no observable phenotype until the 28-cell stage, a stage when gastrulation begins in wild-type embryos (Powell-Coffman, Knight, & Wood, 1996). Gastrulation does not occur without zygotic transcription (Powell-Coffman et al., 1996). Therefore, the timing and axis of all early divisions, the sizes and positions of blastomeres, and the localization of all maternally supplied proteins up to gastrulation all occur independent of transcription in the zygote.

Many genes identified genetically as having essential functions in specification of early blastomere fates encode RBPs (Detwiler et al., 2001; Draper et al., 1996; Guedes & Priess, 1997; Li, DeBella, Guven-Ozkan, Lin, & Rose, 2009; Mello et al., 1996; Ogura et al., 2003; Schubert et al., 2000; Tabara, Hill, Mello, Priess, & Kohara, 1999). In the rapidly dividing *C. elegans* embryo, these maternally supplied RNA-binding regulators are segregated in a spatially and temporally dynamic fashion unique for each protein and each blastomere. Our recent studies suggest that the 3′UTRs of many maternally supplied mRNAs contain a combinatorial code that determines the precise topography of RNA-binding regulators, both positive and negative, that can bind (Guven-Ozkan et al., 2010; Oldenbroek et al., 2013). Therefore, the spatiotemporal expression pattern of any particular maternal mRNA is determined primarily by the precise combination of RNA-binding regulators present in the cell that are capable of binding to its 3′UTR (see also Jadhav, Rana, & Subramaniam, 2008; Jens & Rajewsky, 2015; Wright et al., 2011). This is in many ways analogous to transcriptional enhancers providing binding sites for transcription regulators, which orchestrate the composition and localization of transcriptional machinery at the transcription start site (see Maston, Landt, Snyder, & Green, 2012; Ong & Corces, 2011; Shlyueva, Stampfel, & Stark, 2014) for reviews). Whether a particular gene is transcribed in a cell or tissue is determined primarily by the precise combination of transcription factors that are present and are able to bind to the enhancer sequence in that specific cell.

2.4.2 Maintenance of Transcriptional Repression in Germline Blastomeres

This first asymmetric division initiates a series of events by which *C. elegans* specifies its primordial germ cells. Primordial germ cells, the eventual source of gametes in the adult animal and the genetic link between two generations, are specified early during embryogenesis in most animals. In *C. elegans*, the primordial germ cells are specified through a series of four asymmetric divisions, beginning with the zygote, P0 (Strome, 2005). Each of these divisions results in a smaller, germline precursor (P1–P4, termed the P lineage) and a larger, somatic sister cell (Fig. 3). P4 divides symmetrically to produce the primordial germ cells, Z2 and Z3 (Fig. 4). A hallmark of primordial germ cells in all animals examined is that they are transcriptionally silent for a period after they form (Lesch & Page, 2012; Nakamura & Seydoux, 2008). Similarly, all P cells (P0–P4) in *C. elegans* embryos are transcriptionally repressed, whereas their somatic sisters undergo rapid transcriptional activation and lineage-specific differentiation (Seydoux, 1996; Seydoux & Dunn, 1997) (Fig. 5). Therefore, dynamic and readily reversible mechanisms of transcriptional repression are crucial in *C. elegans* germline blastomeres (Schaner, Deshpande, Schedl, & Kelly, 2003; Seydoux, 1996; Seydoux & Dunn, 1997; Seydoux et al., 1996). Epigenetic marks characteristic of transcriptionally competent chromatin are found in *C. elegans* germline blastomeres (Schaner et al., 2003), consistent with a genome "poised" for transcription. *C. elegans* has been shown to employ two mechanisms to maintain transcriptional repression in germline blastomeres. Both provide dynamic and easily reversed ways to repress transcriptionally competent chromatin. We mentioned above that OMA-1/2 repress transcription in 1-cell embryos by binding to and sequestering TAF-4 in the cytoplasm (Guven-Ozkan et al., 2008) (Fig. 5). After the first mitosis, OMA proteins are degraded. Continued repression of transcription in the subsequent germline blastomeres (P2–P4) is dependent upon another tandem CCCH zinc-finger-containing protein, PIE-1 (Seydoux et al., 1996) (Fig. 5). PIE-1 repression of transcription, similar to the OMA proteins, is also independent of RNA binding (Batchelder et al., 1999). However, PIE-1 functions unlike the OMA proteins by inhibiting the posttranslational modification of the RNA Pol II C-terminal tail required for both transcription initiation and elongation (Batchelder et al., 1999; Ghosh & Seydoux, 2008; Zhang, Barboric, Blackwell, & Peterlin, 2003). One intriguing aspect of PIE-1 function is that although PIE-1 protein is made in oocytes and is present at a high level in P0 and P1 (Fig. 4) (Mello et al., 1996), it is nonetheless insufficient to repress transcription in P0 or P1 (Guven-Ozkan et al., 2008). While it is not clear why PIE-1 is not utilized in P0 or why OMA proteins are employed to repress transcription in 1-cell embryos, this observation nonetheless highlights the uniqueness of the 1-cell embryo.

PIE-1 and OMA-1/2 provide an effective means for repressing transcriptionally competent chromatin in a way that is both dynamic and easily reversed. Asymmetric segregation of PIE-1 to the germline blastomere facilitates rapid resumption of transcription in the somatic cell, presumably from preexisting Pol II preinitiation complexes (Reese, Dunn, Waddle, & Seydoux, 2000; Seydoux & Dunn, 1997; Seydoux et al., 1996; Zhang et al., 2003). Rapid degradation of OMA-1/2 allows for TAF-4 enrichment in nuclei and transcription to occur within the next two, very short cell cycles (Guven-Ozkan et al., 2008). However, the asymmetric segregation of PIE-1 is not 100% effective. At each germline blastomere division, whereas the majority of PIE-1 is segregated to the new germline blastomere, there is a portion of PIE-1 that remains in its somatic sister (Reese et al., 2000). PIE-1 could function as a transcriptional repressor in these somatic blastomeres and therefore needs to be rapidly degraded so that lineage-appropriate transcription and translation can initiate. Degradation of PIE-1 in somatic cells is carried out by a CUL-2-containing E3 ligase (DeRenzo, Reese, & Seydoux, 2003). The substrate-binding subunit of this E3 ligase, ZIF-1, binds to the first CCCH zinc finger of PIE-1 (DeRenzo et al., 2003). Although *zif-1* RNA is provided maternally to the oocytes and is present throughout the early embryo, ZIF-1 protein is not made until the 4-cell stage, and then only in somatic cells (Fig. 4) (Guven-Ozkan et al., 2010). This then ensures PIE-1 stability in oocytes and germline blastomeres, and PIE-1 degradation in the somatic sisters of the germline blastomeres.

Translational repression of *zif-1* in oocytes protects PIE-1 (along with other maternally supplied zinc-finger-containing ZIF-1 targets) from premature degradation, thereby indirectly promoting transcriptional repression in the later germline lineage blastomeres. We want to point out that the two functions of the OMAs, one in oocytes, where they repress translation of *zif-1* mRNA, and the other in the 1-cell embryo, where they bind and sequester TAF-4 protein in the cytoplasm, both work toward the same goal: maintaining the germline blastomeres in a transcriptionally repressed and totipotent state.

2.4.3 Activation of Zygotic Transcription and Clearance of Maternal Products

The *C. elegans* ZGA begins in the somatic blastomeres of 4-cell embryos (Edgar, Wolf, & Wood, 1994; Seydoux & Fire, 1994). As has been described in Section 2.3, the spatiotemporal regulation of the ZGA is likely regulated by the absence of the global repressors OMA-1/2 and PIE-1. Degradation of OMA-1/2 after the first mitosis will allow nuclear enrichment of TAF-4, which, combined with exclusion of PIE-1 from the somatic blastomeres, enables stage/blastomere-specific zygotic transcription to occur in the somatic blastomeres starting at the 4-cell stage.

Maternally supplied proteins include transcription factors, components in signal transduction or cell–cell communication pathways, and RBPs. These proteins, or their activities, are often restricted spatiotemporally in order to achieve lineage-specific transcription of key zygotic regulators (reviewed in Robertson & Lin, 2013) (Fig. 6). A well-characterized example of the direct handoff from maternally supplied developmental factors to transcription of a specific zygotic gene (or genes) is the transcription of two GATA-family transcription factor encoding genes, *end-1* and *end-3*, in the E blastomere of an 8-cell stage embryo. END-1 and END-3 are redundant master endoderm-specifying transcription factors that activate a transcriptional cascade that leads to endoderm development (reviewed in McGhee, 2013). Transcription of *end-1* and *end-3* is both necessary and sufficient for the specification of E as the endoderm precursor. Indeed, ectopic expression of either gene elsewhere in the early embryo results in conversion of the expressing cell(s) to endoderm fate (Maduro et al., 2005). Transcription of *end-1/3* in the E blastomeres requires the lineage-specific, maternally provided transcription factor, SKN-1. SKN-1 activity is restricted to EMS, the mother cell of E (Bowerman, Draper, Mello, & Priess, 1993; Bowerman, Eaton, & Priess, 1992) where it activates transcription of *end-1* and *end-3*

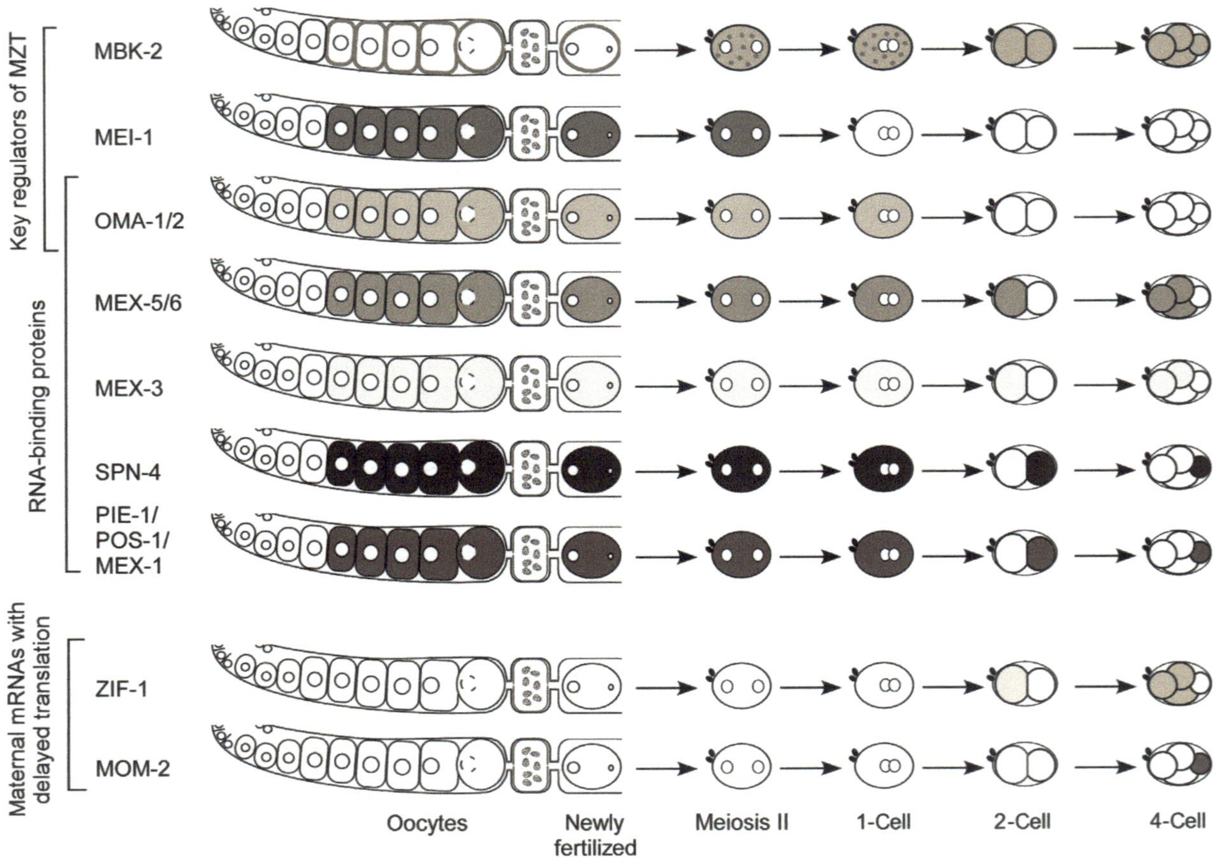

Figure 6 See legend on opposite page.

indirectly via transcriptional activation of *med-1* and *med-2* (Maduro, Meneghini, Bowerman, Broitman-Maduro, & Rothman, 2001). MED-1 and MED-2 are also GATA transcription factors. They are expressed and active in both daughters of the EMS blastomere (Maduro et al., 2001). A signal from P2 to EMS is required to restrict the expression of *end-1* and *end-3* to only the E blastomeres (Maduro et al., 2005; Shetty, Lo, Robertson, & Lin, 2005). Genetic studies identified components in the Wnt, MAP kinase, and SRC-1 signaling pathways, all maternally supplied, as having redundant functions in this P2-EMS signaling (Bei et al., 2002; Ishitani et al., 1999; Meneghini et al., 1999; Rocheleau et al., 1997, 1999; Shin et al., 1999; Thorpe, Schlesinger, Carter, & Bowerman, 1997). This signaling event enhances direct transcriptional activation of *end-1* in E by the sole TCF protein, POP-1, and its coactivating β-catenin, SYS-1 (Huang, Shetty, Robertson, & Lin, 2007; Phillips, Kidd, King, Hardin, & Kimble, 2007; Shetty et al., 2005). In the other EMS daughter, MS, *end-1* and *end-3* are repressed, also by POP-1, which is complexed with the corepressors, Groucho (UNC-37) and HDAC (HDA-1) (Calvo et al., 2001), despite the presence of MED-1 and MED-2.

Approximately coincident with the ZGA, a large proportion (estimated at approximately 30% of the entire transcriptome) of the maternal factors, both protein and mRNA, is degraded (Baugh, Hill, Slonim, Brown, & Hunter, 2003; see Tadros & Lipshitz, 2009 for review). Not all proteins that are degraded during this transition have been tested to determine whether their degradation is critical for embryogenesis to proceed properly. However, for those where genetic tools are available, failure to degrade with wild-type kinetics resulted in detrimental effects in embryogenesis (Lin, 2003; Pintard et al., 2003). The example of strictly timed MEI-1 degradation has already been discussed (Section 2.2). The OMA proteins present a

Figure 6 Distribution of key maternal proteins in early *C. elegans* embryos. The top 7 rows indicate the spatiotemporal localization patterns for 11 maternally supplied proteins, including two redundant pairs (OMA-1/2 and MEX-5/6) and one pattern shown for the three germline-localized proteins—PIE-1, POS-1, and MEX-1. Note that, although not indicated, PIE-1, unlike POS-1 and MEX-1, concentrates in the nuclei of the germline blastomeres. Many of these patterns result from asymmetric protein segregation at mitotic divisions (asymmetry usually precedes division) and/or selective degradation. The bottom two rows are examples of regulated translation of maternally supplied mRNAs, primarily through the highly spatiotemporally regulated binding of a cohort of positive- and negative-acting maternally supplied RNA-binding proteins that produce a blastomere-specific functional readout for what appears to be a 3′UTR "code" of binding sites (see text for details). (See the color plate.)

second example. *oma-1(zu405)* is a missense mutation (a proline to leucine change at amino acid residue 240) that interferes with MBK-2 phosphorylation at the neighboring T239 residue (Lin, 2003; Nishi & Lin, 2005). OMA-1(P240L) is not degraded in early embryos, resulting in temperature-sensitive embryonic lethality (Lin, 2003). Knockdown of the ectopic OMA-1 protein by *oma-1(RNAi)* in *oma-1(zu405)* worms rescues the lethality (Detwiler et al., 2001), showing that perdurance of OMA-1 is responsible for the embryonic lethality.

Early *in situ* hybridization studies with *C. elegans* embryos showed that some maternal mRNAs appeared to be maintained in all cells in the embryo (termed Class I transcripts), whereas others were cleared from somatic cells but were maintained in germline cells (Class II) (Seydoux & Fire, 1994). Those initial observations of clearance of certain maternal mRNAs have been extended by microarray analysis (Baugh et al., 2003) and, more recently, by RNA-seq (Stoeckius et al., 2014), which demonstrate a significant clearance of maternal mRNAs (Class II) that begins about the 4-cell stage and ends around the time of gastrulation initiation, although the kinetics for specific transcripts can differ. Whereas deadenylation of both transcripts and small RNAs has been implicated in similar maternal transcript clearance observed in other systems (Bushati, Stark, Brennecke, & Cohen, 2008; Giraldez et al., 2006; Tadros et al., 2007), the mechanism for the clearance of Class II maternal transcripts in *C. elegans* remains unclear.

3. MAJOR REGULATORS/EVENTS THAT COORDINATE THE MZT

3.1 MBK-2

MBK-2 kinase plays a critical role in the OET, coordinating directly or indirectly a number of MZT component processes with successful completion of meiosis (Pellettieri et al., 2003). Phosphorylation by MBK-2 results in either degradation of maternal proteins that could disrupt development after completion of meiosis (for example, meiosis-specific proteins like MEI-1 that would disrupt mitosis), or a phosphorylation-dependent change in function, or both. OMA-1 (and likely OMA-2) and MEI-1 have been identified as *in vivo* substrates of MBK-2 (Nishi & Lin, 2005; Stitzel et al., 2006). Phosphorylation by MBK-2 is required for the proper developmentally timed degradation of MEI-1 and OMA-1 (Nishi & Lin, 2005; Shirayama et al., 2006; Stitzel et al., 2006). As mentioned earlier, failure to degrade either MEI-1 or OMA-1 results in embryonic lethality (Clandinin &

Mains, 1993; Lin, 2003; Mains et al., 1990). In addition, phosphorylation of OMA-1 at T239 by MBK-2 serves to switch OMA function from translational repressor in oocytes to TAF-4-binding transcriptional repressor in the 1-cell embryo (Guven-Ozkan et al., 2010). Immunofluorescence using antibodies to OMA-1 phosphorylated at T239 and MEI-1 phosphorylated at S92 shows that MBK-2 phosphorylation of both proteins occurs immediately following meiotic anaphase I (Nishi & Lin, 2005; Stitzel et al., 2006). Phosphorylation of the OMA proteins by MBK-2 is followed by a second phosphorylation by GSK-3 (at T339 for OMA-1), and both phosphorylations are required for correctly timed OMA protein degradation (Nishi & Lin, 2005; Shirayama et al., 2006).

Other likely MBK-2 substrates include MEX-5 and MEX-6, two closely related and redundant maternal CCCH zinc finger proteins (Fig. 5), although there is no direct *in vivo* evidence as yet (Nishi, Rogers, Robertson, & Lin, 2008). MEX-5/6 localize toward the anterior cytoplasm of the 1-cell embryo in response to PAR asymmetry, and preferentially localize to the somatic daughters following germline blastomere (P lineage) divisions (Schubert et al., 2000). Degradation of germline-specific proteins, such as PIE-1, in somatic blastomeres is dependent on MEX-5/6 (DeRenzo et al., 2003). Therefore, MEX-5/6 function to restrict these proteins to the germline blastomeres, playing key roles in both germline development and maintenance of embryonic polarity. MEX-5/6 function in zygote polarity is activated following phosphorylation by the polo-family kinases, PLK-1 and PLK-2 (Nishi et al., 2008). For polo kinases to bind to their substrates requires prior phosphorylation of a polo docking site by another kinase (Archambault & Glover, 2009). MBK-2 phosphorylates MEX-5 at T186, thereby creating a docking site for the *C. elegans* polo kinase, PLK-1.

In the absence of MBK-2, oocyte maturation and fertilization occur normally but 100% of embryos from homozygous *mbk-2(−)* mothers exhibit maternal-effect lethality. These embryos present a complex phenotype including lack of a normal mitotic spindle due to perdurance of MEI-1; transcription in the 1-cell embryo due to the inability of OMA proteins to bind to and sequester the essential RNA Pol II component TAF-4; continued translational repression of specific maternal mRNAs due to perdurance of nonphosphorylated OMA proteins; and a lack of embryo polarization as MEX-5/6 are not activated. The very complexity of this mutant phenotype attests to the number of critical processes in the very early embryo regulated by MBK-2.

3.2 OMA Proteins: Distinct and Temporally Independent Functions

One of the major biochemical functions for OMA-1/2 is RNA binding via the Tis-11-like CCCH zinc fingers. Several loss- or reduction-of-function mutations in *oma-1* are missense mutations in the zinc fingers (Detwiler et al., 2001). At least three direct targets have been demonstrated *in vivo*: mRNAs for *zif-1* (Guven-Ozkan et al., 2010), the Wnt ligand *mom-2* (Oldenbroek et al., 2013), and a *nanos*-related gene *nos-2* (Jadhav et al., 2008). In all three cases, *oma-1* (and likely OMA-2) binds to the 3′UTR of the target mRNA, repressing its translation in oocytes and 1- to 2-cell embryos (Guven-Ozkan et al., 2010; Jadhav et al., 2008; Oldenbroek et al., 2013). Defective translational repression of *zif-1* or *mom-2* does not account for the oocyte maturation defective phenotype associated with *oma-1;oma-2* mutant animals because depletion of *zif-1* or *mom-2* in *oma-1;oma-2* animals does not suppress the Oma phenotype. Similar experiments have not been performed with *nos-2*. However, it is possible that aberrant translational regulation of a yet to be identified protein, or combination of proteins, may underlie the Oma phenotype.

OMA-1 binds to SPN-2, which also binds the 5′ Cap-binding protein, eIF4E (Li et al., 2009). Depletion of *spn-2* results in derepression of translation for the OMA-1 substrates, *zif-1* and *mom-2* (Guven-Ozkan et al., 2010; Oldenbroek et al., 2013). Therefore, the mechanism by which OMA-1/2 repress translation is likely through the interaction of 3′UTR-bound OMA-1 and 5′ cap-bound SPN-2, creating a circularized, translationally repressed transcript. We have shown that SPN-2 binding to the *zif-1* 3′UTR is not compatible with MBK-2 phosphorylation of OMA protein (and possibly additional proteins), which provides a mechanistic explanation for MBK-2-dependent inhibition of *zif-1* translation (Guven-Ozkan et al., 2010).

OMA-1 protein also binds to the 3′UTR of *mei-1* mRNA *in vitro*. In embryos mutant for *spn-2*, excess MEI-1 protein was detected in embryos, resulting in mitotic spindle defects (Li et al., 2009). Experiments suggest that the excess MEI-1 is not the result of defective degradation, but rather is due to derepressed translation of *mei-1* mRNA. Therefore, it appears that SPN-2 and OMA-1 also function to negatively regulate translation of maternal *mei-1* mRNA. However, it is unclear how translational repression of *mei-1* by OMA-1 and SPN-2 is regulated, as it does not occur in oocytes, where OMA-1 represses other maternal mRNAs, but does occur in the 1-cell embryo, when MBK-2 phosphorylation of OMA-1 should have rendered it no longer functional in translational repression via 3′UTR binding.

The importance of the OMA proteins in the OET is underscored by the fact that they have critical functions in three key components of this transition: oocyte maturation, degradation of the meiotic spindle, and proper transition from a 1-cell to a multicell embryo. The switch from one OMA-1 function in oocytes to another mechanistically unrelated function in the 1-cell embryo by MBK-2 phosphorylation at OMA-1 T239 is particularly intriguing. MBK-2 phosphorylation soon after fertilization not only promotes OMA-1 binding to TAF-4 (Guven-Ozkan et al., 2008) but also simultaneously inactivates the *zif-1* translational repression exhibited by OMA-1 in oocytes (Guven-Ozkan et al., 2010). The two OMA functions operate through completely different biochemical mechanisms, and switching between these two functions is effected by a single posttranslational modification. The fact that the two OMA functions exhibit strict mutual exclusivity suggests that both functions must not overlap within the organism, but why this should be the case is unclear. There might be a developmental requirement that the first function be completed before the second function initiates, or that the second function initiate without delay upon termination of the first function. A robust functional switch is readily achieved via a single dual-function protein whose modification, which is also stringently timed, terminates the first function while simultaneously activating the second function. Furthermore, phosphorylation of the OMA proteins by MBK-2 not only results in their switch in function but also marks the proteins for proteasomal degradation (Nishi & Lin, 2005). This very effectively delimits the second OMA protein function, sequestration of TAF-4, only to the 1-cell embryo, without requiring additional regulation by, for example, a phosphatase to inactivate OMA-1 function in the early embryo. This coupling of a posttranslational modification that both modifies activity as well as marking the protein for degradation has recently been shown to be the case for the Notch ICD (intracellular domain) following methylation by CARM1, which appears to make the ICD a more potent, but short-lived, transcriptional activator (Hein et al., 2015). It is clear that OMA proteins, in collaboration with MBK-2, play key roles in driving and coordinating MZT, and specifically the OET in *C. elegans*.

3.3 Translational Regulation

Translation of the majority of maternally supplied mRNAs during *C. elegans* embryogenesis is regulated by the corresponding 3′UTR (Merritt, Rasoloson, Ko, & Seydoux, 2008). The 3′UTR, in most cases, is sufficient to confer the correct temporal and spatial expression pattern to a reporter protein. Importantly, many of the key maternal proteins, identified either molecularly or genetically, that regulate this transition contain an RNA-binding motif (Detwiler et al., 2001; Draper et al., 1996; Guedes & Priess, 1997; Li et al., 2009; Mello et al., 1996; Ogura et al., 2003; Schubert et al., 2000; Tabara et al., 1999). Translational repression appears to be a widely used and important posttranscriptional regulatory mechanism in the *C. elegans* germline. Regulation of translational repression during the mitosis-to-meiosis transition in the adult germline and during the various stages of meiotic progression has been reviewed elsewhere recently (Nousch & Eckmann, 2013). We highlight below some key instances of translational repression during the MZT. *oma-1/2* translation is repressed by GLD-1, a STAR domain RBP, in the distal arm of the gonad (Hansen & Schedl, 2013; Lee & Schedl, 2001). As oocytes progress from pachytene to diplotene, GLD-1 levels drop sharply and are undetectable in diakinesis oocytes in the germline-proximal region (Jones, Francis, & Schedl, 1996). Since *gld-1* mRNA remains abundant, the dramatic disappearance of GLD-1 is likely a result of translational repression combined with rapid protein degradation. *oma-1* and *oma-2* transcripts, along with several other transcripts repressed by GLD-1, are relieved of GLD-1 translational repression and their protein levels increase (Lee & Schedl, 2001). Translation of OMA-1 and -2 in proximal oocytes results in the repression of certain maternal transcripts, including *zif-1*, *mom-2*, and *nos-2*, which should not be translated until after fertilization (Guven-Ozkan et al., 2010; Jadhav et al., 2008; Oldenbroek et al., 2013).

Genetic screens have identified many cell-fate-controlling RBPs. These proteins are translated in oocytes and localize to one or a few blastomeres in a spatially and temporally dynamic fashion unique for each protein and each blastomere (Fig. 6). Our recent studies of the two *in vivo* target mRNAs, *zif-1* and *mom-2*, have shed light on the mechanism(s) by which they regulate these targets (Oldenbroek et al., 2012, 2013). The maternally supplied *zif-1* and *mom-2* mRNAs have reciprocal translational patterns, with *zif-1* translated only in somatic and *mom-2* translated only in germline blastomeres. However, the spatiotemporal translation patterns of *mom-2* and *zif-1* are regulated by surprisingly similar sets of maternally supplied RBPs. Seven RBPs all directly bind to the *zif-1* mRNA and function as positive or negative regulators of its translation *in vivo*. A similar observation was made for *mom-2* mRNA, where nine RBPs, including the same seven that regulate *zif-1* mRNA, regulate *mom-2* translation. The net translational readout for the target mRNAs is determined by competitive binding between positively acting and negatively acting RNA-binding regulators for the 3′UTR, along with the distinct spatiotemporal localization patterns of these regulators. These analyses suggest that the 3′UTR of maternal mRNAs contains a combinatorial code that determines the topography of RNA-binding regulators, integrating positive and negative translational inputs in a fashion analogous to transcriptional enhancer elements.

3.4 Protein Degradation

The role that protein degradation plays in the regulation of OMA1/2 and MEI-1 function has already been discussed. In addition, protein degradation plays a critical role in several other processes during the OET. As in mitotic cell cycles, protein degradation plays a crucial role during meiotic divisions (reviewed in Bowerman & Kurz, 2006; DeRenzo & Seydoux, 2004; Pesin & Orr-Weaver, 2008; Stitzel & Seydoux, 2007; Verlhac, Terret, & Pintard, 2010). CDK activities, determined by association with regulatory Cyclin subunits and modified by positive and negative regulation, drive the cell cycle. Entry into and progression through meiosis requires high CDK activity associated with interaction with A- and B-type Cyclins. CDK activity peaks at metaphase with all chromosomes attached to the spindle and aligned at the metaphase plate. The transition from metaphase to anaphase requires the E3 ubiquitin ligase APC. APC initiates (1) chromosome segregation through proteolytic degradation of Securin, which leads to activation of Separase resulting in cleavage of cohesin, and (2) degradation of B-type Cyclins, which leads to downregulation of CDK activity and eventual exit from meiosis (reviewed in Pesin & Orr-Weaver, 2008). In addition, APC regulates degradation of EGG-3, EGG-4, and EGG-5, which is critical for correct timing of MBK-2 activation (Cheng et al., 2009; Stitzel & Seydoux, 2007). The asymmetric segregation of PIE-1, MEX-1, and POS-1, three germline blastomere-specific CCCH finger proteins, is, in part, regulated by the degradation of these proteins in nongerm cell precursors (Reese et al., 2000). Degradation of all three proteins is dependent upon ZIF-1 and MEX-5/6 (DeRenzo et al., 2003).

4. CONCLUDING REMARKS

It is impressive the degree to which the component processes of the MZT are coordinately regulated and timed relative to one another (Fig. 7). For example, MBK-2 activation is tightly coordinated with the completion

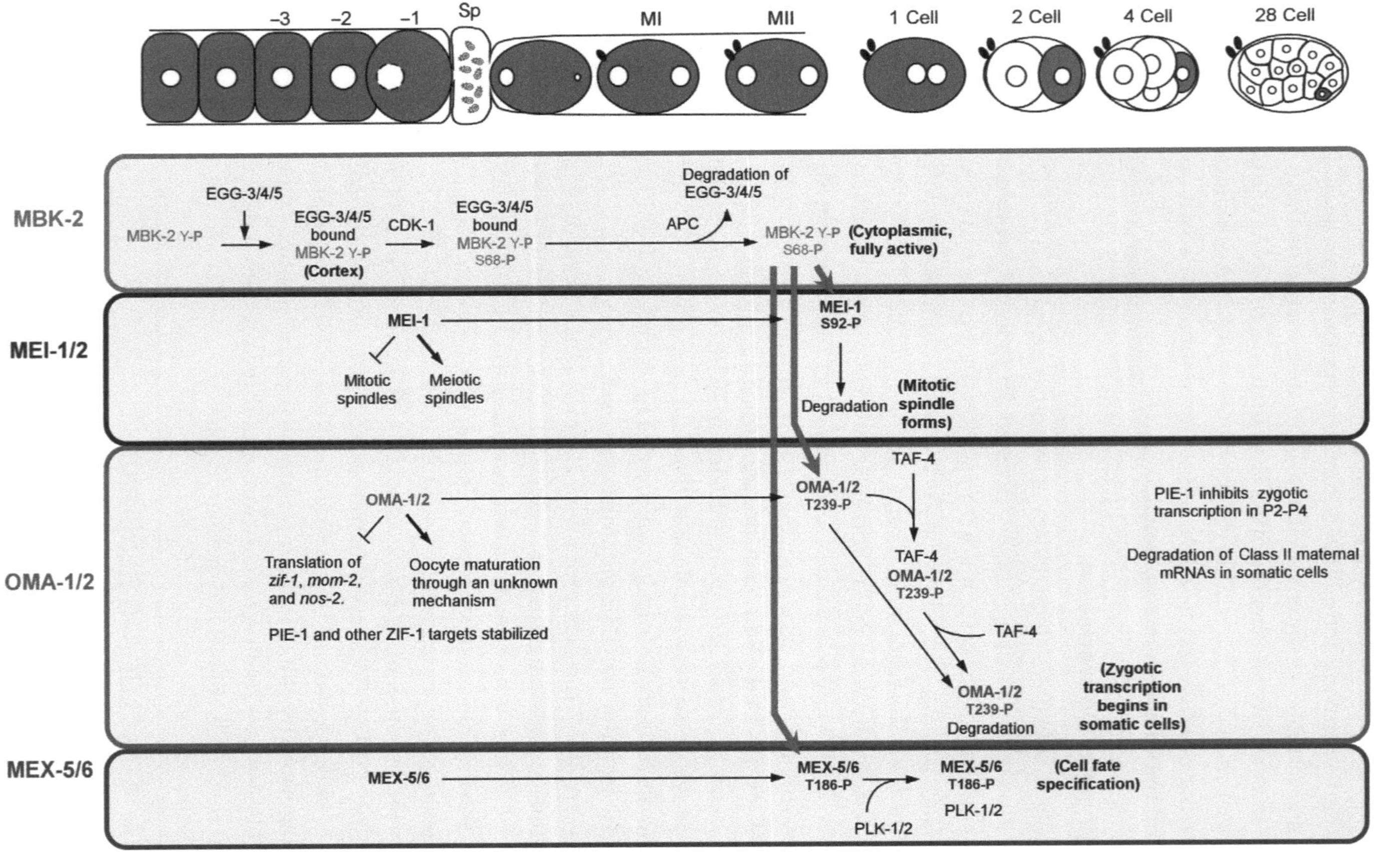

Figure 7 Summary of relationships between key regulators of the MZT in *C. elegans*. Schematic of major temporal processes during the MZT for MBK-2 (pink; light gray in the print version), MEI-1/2 (blue; dark gray in the print version), OMA-1/2 (green; light gray in the print version), and MEX-5/6 (brown; gray in the print version), highlighting the interrelationships between them and the key role played by MBK-2 (see text for details).

of meiosis, and MBK-2 phosphorylation sets in motion several pathways critical to the OET. The OMA proteins are dramatically switched in their function as a result of MBK-2 phosphorylation, which also marks them for degradation. Maternal regulatory factors exhibit a complicated interplay over both space and time. In one example, activated MBK-2 kinase is held in check until precisely the right moment in the OET, and in another example, one protein, OMA-1, functions in the oocyte to ensure that another maternally supplied protein, PIE-1, is protected from degradation, and then both proteins function sequentially later in the embryo to maintain transcriptional silencing of germline precursors. The OET and MZT are currently the subject of intense investigation, and there can be little doubt that this critical first developmental transition in the zygote will be dissected in ever-greater detail in the near future.

How the MZT is coordinated in mammals is less well understood. Molecular events that take place in individual processes, such as oocyte maturation, fertilization, and meiotic divisions, are very similar in outline, with minor differences between *C. elegans* and mammals. However, there are three significant differences between worms and mammals regarding the MZT. *First*, the time span it takes for the transition to occur differs greatly. In mammals, oocytes remain arrested in prophase of meiosis I for extended periods (months, years, even decades), eventually being triggered to resume meiosis and be ovulated by a hormonal signal. The ovulated oocytes arrest again at metaphase of meiosis II and complete meiosis only if fertilized (every oocyte in *C. elegans* is fertilized). Because of the arrest before fertilization of mammalian eggs, precisely timed tight coordination prior to fertilization is not as critical. *Second*, the dependence on maternal proteins prepackaged into the oocytes differs greatly. The *C. elegans* oocyte is huge in comparison to the mammalian oocyte, as development all the way to the first larval stage occurs external to the mother within an eggshell not much larger than the oocyte itself. In mammals, soon after fertilization the developing zygote implants into the wall of the uterus and, thereafter, receives oxygen and nutrients from the mother. In mammals, zygotic transcription can start as early as the 1-cell embryo, and cell-fate specification begins significantly later than in the *C. elegans* embryo. As a result, early embryonic development in mammals is not as dependent on maternal factors in the oocyte as compared to *C. elegans*. *Third*, the early cleavage patterns and the mechanisms by which primordial germ cells are specified differ greatly. The first embryonic divisions in mammals are symmetrical and blastomeres remain totipotent up to the blastocyst stage. In the mouse, for example, primordial germ cells are not specified until much later, around embryonic day 6.5, when pluripotent epiblast cells respond to signals from neighboring extra-embryonic tissue and a pool of primordial germ cells migrates to the genital ridge. By contrast, in *C. elegans* the 1-cell embryo is a germline precursor, and only four divisions later (a little over 2 h following fertilization) the single precursor for the entire germline of the adult (P4) has been specified. These differences suggest that coordination of certain aspects of the MZT is likely to be regulated quite differently between mammals and *C. elegans*, even if component processes share considerable similarity. For example, the outline of oocyte maturation, and in particular the communication via gap junctions between somatic follicle cells (mammals) or sheath cells (*C. elegans*) with their respective oocyte, and the role of maturation signal-receiving GPCRs in these somatic cells, is remarkably similar. The worm will certainly help to bring a better understanding of these processes, as well as a deeper appreciation for similarities between worms and man.

ACKNOWLEDGMENTS

Work conducted in the authors' laboratory was supported by NIH grants HD37933 and GM84198.

REFERENCES

Archambault, V., & Glover, D. M. (2009). Polo-like kinases: Conservation and divergence in their functions and regulation. *Nature Reviews. Molecular Cell Biology*, *10*, 265–275.

Arur, S., Ohmachi, M., Nayak, S., Hayes, M., Miranda, A., Hay, A., et al. (2009). Multiple ERK substrates execute single biological processes in *Caenorhabditis elegans* germ-line development. *Proceedings of the National Academy of Sciences of the United States of America*, *106*, 4776–4781.

Batchelder, C., Dunn, M. A., Choy, B., Suh, Y., Cassie, C., Shim, E. Y., et al. (1999). Transcriptional repression by the *Caenorhabditis elegans* germ-line protein PIE-1. *Genes & Development*, *13*, 202–212.

Baugh, L. R., Hill, A. A., Slonim, D. K., Brown, E. L., & Hunter, C. P. (2003). Composition and dynamics of the *Caenorhabditis elegans* early embryonic transcriptome. *Development*, *130*, 889–900.

Bei, Y., Hogan, J., Berkowitz, L. A., Soto, M., Rocheleau, C. E., Pang, K. M., et al. (2002). SRC-1 and Wnt signaling act together to specify endoderm and to control cleavage orientation in early *C. elegans* embryos. *Developmental Cell*, *3*, 113–125.

Bienkowska, D., & Cowan, C. R. (2012). Centrosomes can initiate a polarity axis from any position within one-cell *C. elegans* embryos. *Current Biology: CB*, *22*, 583–589.

Bowerman, B., Draper, B. W., Mello, C. C., & Priess, J. R. (1993). The maternal gene skn-1 encodes a protein that is distributed unequally in early *C. elegans* embryos. *Cell*, *74*, 443–452.

Bowerman, B., Eaton, B. A., & Priess, J. R. (1992). skn-1, a maternally expressed gene required to specify the fate of ventral blastomeres in the early *C. elegans* embryo. *Cell*, *68*, 1061–1075.

Bowerman, B., & Kurz, T. (2006). Degrade to create: Developmental requirements for ubiquitin-mediated proteolysis during early *C. elegans* embryogenesis. *Development, 133*, 773–784.

Boxem, M. (2006). Cyclin-dependent kinases in *C. elegans*. *Cell Division, 1*, 6.

Bui, Y. K., & Sternberg, P. W. (2002). *Caenorhabditis elegans* inositol 5-phosphatase homolog negatively regulates inositol 1,4,5-triphosphate signaling in ovulation. *Molecular Biology of the Cell, 13*, 1641–1651.

Bushati, N., Stark, A., Brennecke, J., & Cohen, S. M. (2008). Temporal reciprocity of miRNAs and their targets during the maternal-to-zygotic transition in Drosophila. *Current Biology: CB, 18*, 501–506.

Calvo, D., Victor, M., Gay, F., Sui, G., Luke, M. P., Dufourcq, P., et al. (2001). A POP-1 repressor complex restricts inappropriate cell type-specific gene transcription during *Caenorhabditis elegans* embryogenesis. *The EMBO Journal, 20*, 7197–7208.

Cheng, H., Govindan, J. A., & Greenstein, D. (2008). Regulated trafficking of the MSP/Eph receptor during oocyte meiotic maturation in *C. elegans*. *Current Biology: CB, 18*, 705–714.

Cheng, K. C., Klancer, R., Singson, A., & Seydoux, G. (2009). Regulation of MBK-2/DYRK by CDK-1 and the pseudophosphatases EGG-4 and EGG-5 during the oocyte-to-embryo transition. *Cell, 139*, 560–572.

Clandinin, T. R., DeModena, J. A., & Sternberg, P. W. (1998). Inositol trisphosphate mediates a RAS-independent response to LET-23 receptor tyrosine kinase activation in *C. elegans*. *Cell, 92*, 523–533.

Clandinin, T. R., & Mains, P. E. (1993). Genetic studies of mei-1 gene activity during the transition from meiosis to mitosis in *Caenorhabditis elegans*. *Genetics, 134*, 199–210.

Clark-Maguire, S., & Mains, P. E. (1994a). Localization of the mei-1 gene product of *Caenorhabditis elegans*, a meiotic-specific spindle component. *The Journal of Cell Biology, 126*, 199–209.

Clark-Maguire, S., & Mains, P. E. (1994b). mei-1, a gene required for meiotic spindle formation in *Caenorhabditis elegans*, is a member of a family of ATPases. *Genetics, 136*, 533–546.

Connolly, A. A., Osterberg, V., Christensen, S., Price, M., Lu, C., Chicas-Cruz, K., et al. (2014). *Caenorhabditis elegans* oocyte meiotic spindle pole assembly requires microtubule severing and the calponin homology domain protein ASPM-1. *Molecular Biology of the Cell, 25*, 1298–1311.

Cowan, C. R., & Hyman, A. A. (2004). Centrosomes direct cell polarity independently of microtubule assembly in *C. elegans* embryos. *Nature, 431*, 92–96.

DeRenzo, C., Reese, K. J., & Seydoux, G. (2003). Exclusion of germ plasm proteins from somatic lineages by cullin-dependent degradation. *Nature, 424*, 685–689.

DeRenzo, C., & Seydoux, G. (2004). A clean start: Degradation of maternal proteins at the oocyte-to-embryo transition. *Trends in Cell Biology, 14*, 420–426.

Detwiler, M. R., Reuben, M., Li, X., Rogers, E., & Lin, R. (2001). Two zinc finger proteins, OMA-1 and OMA-2, are redundantly required for oocyte maturation in *C. elegans*. *Developmental Cell, 1*, 187–199.

Dow, M. R., & Mains, P. E. (1998). Genetic and molecular characterization of the *Caenorhabditis elegans* gene, mel-26, a postmeiotic negative regulator of mei-1, a meiotic-specific spindle component. *Genetics, 150*, 119–128.

Draper, B. W., Mello, C. C., Bowerman, B., Hardin, J., & Priess, J. R. (1996). MEX-3 is a KH domain protein that regulates blastomere identity in early *C. elegans* embryos. *Cell, 87*, 205–216.

Dufau, M. L. (1998). The luteinizing hormone receptor. *Annual Review of Physiology, 60*, 461–496.

Edgar, L. G., Wolf, N., & Wood, W. B. (1994). Early transcription in *Caenorhabditis elegans* embryos. *Development, 120*, 443–451.

Eisman, R., & Kaufman, T. C. (2007). Cytological investigation of the mechanism of parthenogenesis in *Drosophila mercatorum*. *Fly*, *1*, 317–329.

Fabritius, A. S., Ellefson, M. L., & McNally, F. J. (2011). Nuclear and spindle positioning during oocyte meiosis. *Current Opinion in Cell Biology*, *23*, 78–84.

Gartner, A., Boag, P. R., & Blackwell, T. K. (2008). Germline survival and apoptosis. *WormBook*, ed. The *C. elegans* Research Community, WormBook. http://dx.doi.org/10.1895/wormbook.1.145.1. http://www.wormbook.org.

Ghosh, D., & Seydoux, G. (2008). Inhibition of transcription by the *Caenorhabditis elegans* germline protein PIE-1: Genetic evidence for distinct mechanisms targeting initiation and elongation. *Genetics*, *178*, 235–243.

Giraldez, A. J., Mishima, Y., Rihel, J., Grocock, R. J., Van Dongen, S., Inoue, K., et al. (2006). Zebrafish MiR-430 promotes deadenylation and clearance of maternal mRNAs. *Science*, *312*, 75–79.

Goldstein, B., & Hird, S. N. (1996). Specification of the anteroposterior axis in *Caenorhabditis elegans*. *Development*, *122*, 1467–1474.

Govindan, J. A., Cheng, H., Harris, J. E., & Greenstein, D. (2006). $G\alpha_{o/i}$ and $G\alpha_s$ signaling function in parallel with the MSP/Eph receptor to control meiotic diapause in *C. elegans*. *Current Biology*, *16*, 1257–1268.

Govindan, J. A., Nadarajan, S., Kim, S., Starich, T. A., & Greenstein, D. (2009). Somatic cAMP signaling regulates MSP-dependent oocyte growth and meiotic maturation in *C. elegans*. *Development*, *136*, 2211–2221.

Greenstein, D. (2005). Control of oocyte meiotic maturation and fertilization. *WormBook*, ed. The *C. elegans* Research Community, WormBook. http://dx.doi.org/10.1895/wormbook.1.53.1. http://www.wormbook.org.

Greenstein, D., Hird, S., Plasterk, R. H., Andachi, Y., Kohara, Y., Wang, B., et al. (1994). Targeted mutations in the *Caenorhabditis elegans* POU homeo box gene ceh-18 cause defects in oocyte cell cycle arrest, gonad migration, and epidermal differentiation. *Genes & Development*, *8*, 1935–1948.

Guedes, S., & Priess, J. R. (1997). The *C. elegans* MEX-1 protein is present in germline blastomeres and is a P granule component. *Development*, *124*, 731–739.

Guven-Ozkan, T., Nishi, Y., Robertson, S. M., & Lin, R. (2008). Global transcriptional repression in *C. elegans* germline precursors by regulated sequestration of TAF-4. *Cell*, *135*, 149–160.

Guven-Ozkan, T., Robertson, S. M., Nishi, Y., & Lin, R. (2010). zif-1 translational repression defines a second, mutually exclusive OMA function in germline transcriptional repression. *Development*, *137*, 3373–3382.

Han, S. M., Cottee, P. A., & Miller, M. A. (2010). Sperm and oocyte communication mechanisms controlling *C. elegans* fertility. *Developmental Dynamics: An Official Publication of the American Association of Anatomists*, *239*, 1265–1281.

Hansen, D., & Schedl, T. (2013). Stem cell proliferation versus meiotic fate decision in *Caenorhabditis elegans*. *Advances in Experimental Medicine and Biology*, *757*, 71–99.

Harris, J. E., Govindan, J. A., Yamamoto, I., Schwartz, J., Kaverina, I., & Greenstein, D. (2006). Major sperm protein signaling promotes oocyte microtubule reorganization prior to fertilization in *Caenorhabditis elegans*. *Developmental Biology*, *299*, 105–121.

Hein, K., Mittler, G., Cizelsky, W., Kuhl, M., Ferrante, F., Liefke, R., et al. (2015). Site-specific methylation of Notch1 controls the amplitude and duration of the Notch1 response. *Science Signaling*, *8*, ra30.

Hirsh, D., Oppenheim, D., & Klass, M. (1976). Development of the reproductive system of *Caenorhabditis elegans*. *Developmental Biology*, *49*, 200–219.

Huang, S., Shetty, P., Robertson, S. M., & Lin, R. (2007). Binary cell fate specification during *C. elegans* embryogenesis driven by reiterated reciprocal asymmetry of TCF POP-1 and its coactivator beta-catenin SYS-1. *Development*, *134*, 2685–2695.

Ishitani, T., Ninomiya-Tsuji, J., Nagai, S., Nishita, M., Meneghini, M., Barker, N., et al. (1999). The TAK1-NLK-MAPK-related pathway antagonizes signalling between beta-catenin and transcription factor TCF. *Nature*, *399*, 798–802.

Ivshina, M., Lasko, P., & Richter, J. D. (2014). Cytoplasmic polyadenylation element binding proteins in development, health, and disease. *Annual Review of Cell and Developmental Biology*, *30*, 393–415.

Iwasaki, K., McCarter, J., Francis, R., & Schedl, T. (1996). emo-1, a *Caenorhabditis elegans* Sec61p gamma homologue, is required for oocyte development and ovulation. *The Journal of Cell Biology*, *134*, 699–714.

Jadhav, S., Rana, M., & Subramaniam, K. (2008). Multiple maternal proteins coordinate to restrict the translation of *C. elegans* nanos-2 to primordial germ cells. *Development*, *135*, 1803–1812.

Jens, M., & Rajewsky, N. (2015). Competition between target sites of regulators shapes post-transcriptional gene regulation. *Nature Reviews. Genetics*, *16*, 113–126.

Johnson, J. L., Lu, C., Raharjo, E., McNally, K., McNally, F. J., & Mains, P. E. (2009). Levels of the ubiquitin ligase substrate adaptor MEL-26 are inversely correlated with MEI-1/katanin microtubule-severing activity during both meiosis and mitosis. *Developmental Biology*, *330*, 349–357.

Jones, A. R., Francis, R., & Schedl, T. (1996). GLD-1, a cytoplasmic protein essential for oocyte differentiation, shows stage- and sex-specific expression during *Caenorhabditis elegans* germline development. *Developmental Biology*, *180*, 165–183.

Kim, S., Govindan, J. A., Tu, Z. J., & Greenstein, D. (2012). SACY-1 DEAD-Box helicase links the somatic control of oocyte meiotic maturation to the sperm-to-oocyte switch and gamete maintenance in *Caenorhabditis elegans*. *Genetics*, *192*, 905–928.

Kim, S., Spike, C., & Greenstein, D. (2013). Control of oocyte growth and meiotic maturation in *Caenorhabditis elegans*. *Advances in Experimental Medicine and Biology*, *757*, 277–320.

Kimble, J., & Crittenden, S. L. (2007). Controls of germline stem cells, entry into meiosis, and the sperm/oocyte decision in *Caenorhabditis elegans*. *Annual Review of Cell and Developmental Biology*, *23*, 405–433.

Kimble, J., & Simpson, P. (1997). The LIN-12/Notch signaling pathway and its regulation. *Annual Review of Cell and Developmental Biology*, *13*, 333–361.

Kimble, J. E., & White, J. G. (1981). On the control of germ cell development in *Caenorhabditis elegans*. *Developmental Biology*, *81*, 208–219.

Kosinski, M., McDonald, K., Schwartz, J., Yamamoto, I., & Greenstein, D. (2005). *C. elegans* sperm bud vesicles to deliver a meiotic maturation signal to distant oocytes. *Development*, *132*, 3357–3369.

Kuwabara, P. E. (2003). The multifaceted *C. elegans* major sperm protein: An ephrin signaling antagonist in oocyte maturation. *Genes & Development*, *17*, 155–161.

Lai, W. S., Carballo, E., Strum, J. R., Kennington, E. A., Phillips, R. S., & Blackshear, P. J. (1999). Evidence that tristetraprolin binds to AU-rich elements and promotes the deadenylation and destabilization of tumor necrosis factor alpha mRNA. *Molecular and Cellular Biology*, *19*, 4311–4323.

Langeberg, L. K., & Scott, J. D. (2015). Signalling scaffolds and local organization of cellular behaviour. *Nature Reviews. Molecular Cell Biology*, *16*, 232–244.

Lee, M. H., Ohmachi, M., Arur, S., Nayak, S., Francis, R., Church, D., et al. (2007). Multiple functions and dynamic activation of MPK-1 extracellular signal-regulated kinase signaling in *Caenorhabditis elegans* germline development. *Genetics*, *177*, 2039–2062.

Lee, M. H., & Schedl, T. (2001). Identification of in vivo mRNA targets of GLD-1, a maxi-KH motif containing protein required for *C. elegans* germ cell development. *Genes & Development*, *15*, 2408–2420.

Lesch, B. J., & Page, D. C. (2012). Genetics of germ cell development. *Nature Reviews. Genetics*, *13*, 781–794.

Li, W., DeBella, L. R., Guven-Ozkan, T., Lin, R., & Rose, L. S. (2009). An eIF4E-binding protein regulates katanin protein levels in *C. elegans* embryos. *The Journal of Cell Biology*, *187*, 33–42.

Lin, R. (2003). A gain-of-function mutation in oma-1, a *C. elegans* gene required for oocyte maturation, results in delayed degradation of maternal proteins and embryonic lethality. *Developmental Biology*, *258*, 226–239.

Lochhead, P. A., Sibbet, G., Morrice, N., & Cleghon, V. (2005). Activation-loop autophosphorylation is mediated by a novel transitional intermediate form of DYRKs. *Cell*, *121*, 925–936.

Maduro, M. F., Hill, R. J., Heid, P. J., Newman-Smith, E. D., Zhu, J., Priess, J. R., et al. (2005). Genetic redundancy in endoderm specification within the genus Caenorhabditis. *Developmental Biology*, *284*, 509–522.

Maduro, M. F., Meneghini, M. D., Bowerman, B., Broitman-Maduro, G., & Rothman, J. H. (2001). Restriction of mesendoderm to a single blastomere by the combined action of SKN-1 and a GSK-3beta homolog is mediated by MED-1 and -2 in *C. elegans*. *Molecular Cell*, *7*, 475–485.

Mahowald, A. P., Goralski, T. J., & Caulton, J. H. (1983). In vitro activation of Drosophila eggs. *Developmental Biology*, *98*, 437–445.

Mains, P. E., Kemphues, K. J., Sprunger, S. A., Sulston, I. A., & Wood, W. B. (1990). Mutations affecting the meiotic and mitotic divisions of the early *Caenorhabditis elegans* embryo. *Genetics*, *126*, 593–605.

Marcello, M. R., Singaravelu, G., & Singson, A. (2013). Fertilization. *Advances in Experimental Medicine and Biology*, *757*, 321–350.

Marcello, M. R., & Singson, A. (2010). Fertilization and the oocyte-to-embryo transition in *C. elegans*. *BMB Reports*, *43*, 389–399.

Maruyama, R., Velarde, N. V., Klancer, R., Gordon, S., Kadandale, P., Parry, J. M., et al. (2007). EGG-3 regulates cell-surface and cortex rearrangements during egg activation in *Caenorhabditis elegans*. *Current Biology: CB*, *17*, 1555–1560.

Maston, G. A., Landt, S. G., Snyder, M., & Green, M. R. (2012). Characterization of enhancer function from genome-wide analyses. *Annual Review of Genomics and Human Genetics*, *13*, 29–57.

McCarter, J., Bartlett, B., Dang, T., & Schedl, T. (1997). Soma-germ cell interactions in *Caenorhabditis elegans*: Multiple events of hermaphrodite germline development require the somatic sheath and spermathecal lineages. *Developmental Biology*, *181*, 121–143.

McCarter, J., Bartlett, B., Dang, T., & Schedl, T. (1999). On the control of oocyte meiotic maturation and ovulation in *Caenorhabditis elegans*. *Developmental Biology*, *205*, 111–128.

McGhee, J. D. (2013). The *Caenorhabditis elegans* intestine. *Wiley Interdisciplinary Reviews: Developmental Biology*, *2*, 347–367.

McNally, K., Audhya, A., Oegema, K., & McNally, F. J. (2006). Katanin controls mitotic and meiotic spindle length. *The Journal of Cell Biology*, *175*, 881–891.

McNally, K. L., & McNally, F. J. (2005). Fertilization initiates the transition from anaphase I to metaphase II during female meiosis in *C. elegans*. *Developmental Biology*, *282*, 218–230.

McNally, K. P., & McNally, F. J. (2011). The spindle assembly function of *Caenorhabditis elegans* katanin does not require microtubule-severing activity. *Molecular Biology of the Cell*, *22*, 1550–1560.

Mehlmann, L. M. (2005). Stops and starts in mammalian oocytes: Recent advances in understanding the regulation of meiotic arrest and oocyte maturation. *Reproduction*, *130*, 791–799.

Mello, C. C., Schubert, C., Draper, B., Zhang, W., Lobel, R., & Priess, J. R. (1996). The PIE-1 protein and germline specification in *C. elegans* embryos. *Nature*, *382*, 710–712.

Meneghini, M. D., Ishitani, T., Carter, J. C., Hisamoto, N., Ninomiya-Tsuji, J., Thorpe, C. J., et al. (1999). MAP kinase and Wnt pathways converge to downregulate an HMG-domain repressor in *Caenorhabditis elegans*. *Nature*, *399*, 793–797.

Merritt, C., Rasoloson, D., Ko, D., & Seydoux, G. (2008). 3′ UTRs are the primary regulators of gene expression in the *C. elegans* germline. *Current Biology: CB*, *18*, 1476–1482.
Miller, M. A., Nguyen, V. Q., Lee, M. H., Kosinski, M., Schedl, T., Caprioli, R. M., et al. (2001). A sperm cytoskeletal protein that signals oocyte meiotic maturation and ovulation. *Science*, *291*, 2144–2147.
Miller, M. A., Ruest, P. J., Kosinski, M., Hanks, S. K., & Greenstein, D. (2003). An Eph receptor sperm-sensing control mechanism for oocyte meiotic maturation in *Caenorhabditis elegans*. *Genes & Development*, *17*, 187–200.
Motegi, F., & Seydoux, G. (2013). The PAR network: Redundancy and robustness in a symmetry-breaking system. *Philosophical Transactions of the Royal Society of London. Series B, Biological Sciences*, *368*, 20130010.
Muller-Reichert, T., Greenan, G., O'Toole, E., & Srayko, M. (2010). The elegans of spindle assembly. *Cellular and Molecular Life Sciences: CMLS*, *67*, 2195–2213.
Munro, E., Nance, J., & Priess, J. R. (2004). Cortical flows powered by asymmetrical contraction transport PAR proteins to establish and maintain anterior-posterior polarity in the early *C. elegans* embryo. *Developmental Cell*, *7*, 413–424.
Nakamura, A., & Seydoux, G. (2008). Less is more: Specification of the germline by transcriptional repression. *Development*, *135*, 3817–3827.
Nance, J. (2005). PAR proteins and the establishment of cell polarity during *C. elegans* development. *BioEssays: News and Reviews in Molecular, Cellular and Developmental Biology*, *27*, 126–135.
Nishi, Y., & Lin, R. (2005). DYRK2 and GSK-3 phosphorylate and promote the timely degradation of OMA-1, a key regulator of the oocyte-to-embryo transition in *C. elegans*. *Developmental Biology*, *288*, 139–149.
Nishi, Y., Rogers, E., Robertson, S. M., & Lin, R. (2008). Polo kinases regulate *C. elegans* embryonic polarity via binding to DYRK2-primed MEX-5 and MEX-6. *Development*, *135*, 687–697.
Nousch, M., & Eckmann, C. R. (2013). Translational control in the *Caenorhabditis elegans* germ line. *Advances in Experimental Medicine and Biology*, *757*, 205–247.
Nousch, M., Yeroslaviz, A., Habermann, B., & Eckmann, C. R. (2014). The cytoplasmic poly(A) polymerases GLD-2 and GLD-4 promote general gene expression via distinct mechanisms. *Nucleic Acids Research*, *42*, 11622–11633.
Ogura, K., Kishimoto, N., Mitani, S., Gengyo-Ando, K., & Kohara, Y. (2003). Translational control of maternal glp-1 mRNA by POS-1 and its interacting protein SPN-4 in *Caenorhabditis elegans*. *Development*, *130*, 2495–2503.
Oldenbroek, M., Robertson, S. M., Guven-Ozkan, T., Gore, S., Nishi, Y., & Lin, R. (2012). Multiple RNA-binding proteins function combinatorially to control the soma-restricted expression pattern of the E3 ligase subunit ZIF-1. *Developmental Biology*, *363*, 388–398.
Oldenbroek, M., Robertson, S. M., Guven-Ozkan, T., Spike, C., Greenstein, D., & Lin, R. (2013). Regulation of maternal Wnt mRNA translation in *C. elegans* embryos. *Development*, *140*, 4614–4623.
Ong, C. T., & Corces, V. G. (2011). Enhancer function: New insights into the regulation of tissue-specific gene expression. *Nature Reviews. Genetics*, *12*, 283–293.
Pagano, J. M., Farley, B. M., McCoig, L. M., & Ryder, S. P. (2007). Molecular basis of RNA recognition by the embryonic polarity determinant MEX-5. *The Journal of Biological Chemistry*, *282*, 8883–8894.
Pang, K. M., Ishidate, T., Nakamura, K., Shirayama, M., Trzepacz, C., Schubert, C. M., et al. (2004). The minibrain kinase homolog, mbk-2, is required for spindle positioning and asymmetric cell division in early *C. elegans* embryos. *Developmental Biology*, *265*, 127–139.
Parry, J. M., & Singson, A. (2011). EGG molecules couple the oocyte-to-embryo transition with cell cycle progression. *Results and Problems in Cell Differentiation*, *53*, 135–151.

Parry, J. M., Velarde, N. V., Lefkovith, A. J., Zegarek, M. H., Hang, J. S., Ohm, J., et al. (2009). EGG-4 and EGG-5 link events of the oocyte-to-embryo transition with meiotic progression in *C. elegans*. *Current Biology: CB, 19*, 1752–1757.

Payne, C., Rawe, V., Ramalho-Santos, J., Simerly, C., & Schatten, G. (2003). Preferentially localized dynein and perinuclear dynactin associate with nuclear pore complex proteins to mediate genomic union during mammalian fertilization. *Journal of Cell Science, 116*, 4727–4738.

Pellettieri, J., Reinke, V., Kim, S. K., & Seydoux, G. (2003). Coordinate activation of maternal protein degradation during the egg-to-embryo transition in *C. elegans*. *Developmental Cell, 5*, 451–462.

Pesin, J. A., & Orr-Weaver, T. L. (2008). Regulation of APC/C activators in mitosis and meiosis. *Annual Review of Cell and Developmental Biology, 24*, 475–499.

Phillips, B. T., Kidd, A. R., 3rd., King, R., Hardin, J., & Kimble, J. (2007). Reciprocal asymmetry of SYS-1/beta-catenin and POP-1/TCF controls asymmetric divisions in *Caenorhabditis elegans*. *Proceedings of the National Academy of Sciences of the United States of America, 104*, 3231–3236.

Pintard, L., Willis, J. H., Willems, A., Johnson, J. L., Srayko, M., Kurz, T., et al. (2003). The BTB protein MEL-26 is a substrate-specific adaptor of the CUL-3 ubiquitin-ligase. *Nature, 425*, 311–316.

Powell-Coffman, J. A., Knight, J., & Wood, W. B. (1996). Onset of *C. elegans* gastrulation is blocked by inhibition of embryonic transcription with an RNA polymerase antisense RNA. *Developmental Biology, 178*, 472–483.

Quintin, S., Mains, P. E., Zinke, A., & Hyman, A. A. (2003). The mbk-2 kinase is required for inactivation of MEI-1/katanin in the one-cell *Caenorhabditis elegans* embryo. *EMBO Reports, 4*, 1175–1181.

Reese, K. J., Dunn, M. A., Waddle, J. A., & Seydoux, G. (2000). Asymmetric segregation of PIE-1 in *C. elegans* is mediated by two complementary mechanisms that act through separate PIE-1 protein domains. *Molecular Cell, 6*, 445–455.

Robertson, S., & Lin, R. (2013). The oocyte-to-embryo transition. *Advances in Experimental Medicine and Biology, 757*, 351–372.

Rocheleau, C. E., Downs, W. D., Lin, R., Wittmann, C., Bei, Y., Cha, Y. H., et al. (1997). Wnt signaling and an APC-related gene specify endoderm in early *C. elegans* embryos. *Cell, 90*, 707–716.

Rocheleau, C. E., Yasuda, J., Shin, T. H., Lin, R., Sawa, H., Okano, H., et al. (1999). WRM-1 activates the LIT-1 protein kinase to transduce anterior/posterior polarity signals in *C. elegans*. *Cell, 97*, 717–726.

Rose, L., & Gönczy, P. (2014). Polarity establishment, asymmetric division and segregation of fate determinants in early *C. elegans* embryos. *WormBook,* ed. The *C. elegans* Research Community, WormBook. http://dx.doi.org/10.1895/wormbook.1.30.2. http://www.wormbook.org.

Schaner, C. E., Deshpande, G., Schedl, P. D., & Kelly, W. G. (2003). A conserved chromatin architecture marks and maintains the restricted germ cell lineage in worms and flies. *Developmental Cell, 5*, 747–757.

Schauer, I. E., & Wood, W. B. (1990). Early *C. elegans* embryos are transcriptionally active. *Development, 110*, 1303–1317.

Schedl, T. (1997). Developmental genetics of the germ line. In D. L. Riddle, T. Blumenthal, B. J. Meyer, & J. R. Priess (Eds.), C. elegans *II* (2nd ed., pp. 241–269). New York: Cold Spring Harbor.

Schubert, C. M., Lin, R., de Vries, C. J., Plasterk, R. H., & Priess, J. R. (2000). MEX-5 and MEX-6 function to establish soma/germline asymmetry in early *C. elegans* embryos. *Molecular Cell, 5*, 671–682.

Schumacher, J. M., Golden, A., & Donovan, P. J. (1998). AIR-2: An Aurora/Ipl1-related protein kinase associated with chromosomes and midbody microtubules is required for polar body extrusion and cytokinesis in *Caenorhabditis elegans* embryos. *The Journal of Cell Biology*, *143*, 1635–1646.

Segbert, C., Barkus, R., Powers, J., Strome, S., Saxton, W. M., & Bossinger, O. (2003). KLP-18, a Klp2 kinesin, is required for assembly of acentrosomal meiotic spindles in *Caenorhabditis elegans*. *Molecular Biology of the Cell*, *14*, 4458–4469.

Seydoux, G. (1996). Mechanisms of translational control in early development. *Current Opinion in Genetics & Development*, *6*, 555–561.

Seydoux, G., & Dunn, M. A. (1997). Transcriptionally repressed germ cells lack a subpopulation of phosphorylated RNA polymerase II in early embryos of *Caenorhabditis elegans* and *Drosophila melanogaster*. *Development*, *124*, 2191–2201.

Seydoux, G., & Fire, A. (1994). Soma-germline asymmetry in the distributions of embryonic RNAs in *Caenorhabditis elegans*. *Development*, *120*, 2823–2834.

Seydoux, G., Mello, C. C., Pettitt, J., Wood, W. B., Priess, J. R., & Fire, A. (1996). Repression of gene expression in the embryonic germ lineage of *C. elegans*. *Nature*, *382*, 713–716.

Shetty, P., Lo, M. C., Robertson, S. M., & Lin, R. (2005). *C. elegans* TCF protein, POP-1, converts from repressor to activator as a result of Wnt-induced lowering of nuclear levels. *Developmental Biology*, *285*, 584–592.

Shimada, M., Kawahara, H., & Doi, H. (2002). Novel family of CCCH-type zinc-finger proteins, MOE-1, -2 and -3, participates in *C. elegans* oocyte maturation. *Genes to Cells: Devoted to Molecular & Cellular Mechanisms*, 7, 933–947.

Shin, T. H., Yasuda, J., Rocheleau, C. E., Lin, R., Soto, M., Bei, Y., et al. (1999). MOM-4, a MAP kinase kinase kinase-related protein, activates WRM-1/LIT-1 kinase to transduce anterior/posterior polarity signals in *C. elegans*. *Molecular Cell*, *4*, 275–280.

Shirayama, M., Soto, M. C., Ishidate, T., Kim, S., Nakamura, K., Bei, Y., et al. (2006). The conserved kinases CDK-1, GSK-3, KIN-19, and MBK-2 promote OMA-1 destruction to regulate the oocyte-to-embryo transition in *C. elegans*. *Current Biology: CB*, *16*, 47–55.

Shlyueva, D., Stampfel, G., & Stark, A. (2014). Transcriptional enhancers: From properties to genome-wide predictions. *Nature Reviews. Genetics*, *15*, 272–286.

Spike, C. A., Coetzee, D., Eichten, C., Wang, X., Hansen, D., & Greenstein, D. (2014). The TRIM-NHL protein LIN-41 and the OMA RNA-binding proteins antagonistically control the prophase-to-metaphase transition and growth of *Caenorhabditis elegans* oocytes. *Genetics*, *198*, 1535–1558.

Spike, C. A., Coetzee, D., Nishi, Y., Guven-Ozkan, T., Oldenbroek, M., Yamamoto, I., et al. (2014). Translational control of the oogenic program by components of OMA ribonucleoprotein particles in *Caenorhabditis elegans*. *Genetics*, *198*, 1513–1533.

Srayko, M., Buster, D. W., Bazirgan, O. A., McNally, F. J., & Mains, P. E. (2000). MEI-1/MEI-2 katanin-like microtubule severing activity is required for *Caenorhabditis elegans* meiosis. *Genes & Development*, *14*, 1072–1084.

Starich, T. A., Hall, D. H., & Greenstein, D. (2014). Two classes of gap junction channels mediate soma-germline interactions essential for germline proliferation and gametogenesis in *Caenorhabditis elegans*. *Genetics*, *198*, 1127–1153.

Stitzel, M. L., Pellettieri, J., & Seydoux, G. (2006). The *C. elegans* DYRK kinase MBK-2 marks oocyte proteins for degradation in response to meiotic maturation. *Current Biology: CB*, *16*, 56–62.

Stitzel, M. L., & Seydoux, G. (2007). Regulation of the oocyte-to-zygote transition. *Science*, *316*, 407–408.

Stoeckius, M., Grun, D., Kirchner, M., Ayoub, S., Torti, F., Piano, F., et al. (2014). Global characterization of the oocyte-to-embryo transition in *Caenorhabditis elegans* uncovers a novel mRNA clearance mechanism. *The EMBO Journal*, *33*, 1751–1766.

Strome, S. (2005). Specification of the germ line. *WormBook,* ed. The *C. elegans* Research Community, WormBook. http://dx.doi.org/10.1895/wormbook.1.9.1. http://www.wormbook.org.

Tabara, H., Hill, R. J., Mello, C. C., Priess, J. R., & Kohara, Y. (1999). pos-1 encodes a cytoplasmic zinc-finger protein essential for germline specification in *C. elegans*. *Development, 126,* 1–11.

Tadros, W., Goldman, A. L., Babak, T., Menzies, F., Vardy, L., Orr-Weaver, T., et al. (2007). SMAUG is a major regulator of maternal mRNA destabilization in Drosophila and its translation is activated by the PAN GU kinase. *Developmental Cell, 12,* 143–155.

Tadros, W., & Lipshitz, H. D. (2009). The maternal-to-zygotic transition: A play in two acts. *Development, 136,* 3033–3042.

Thorpe, C. J., Schlesinger, A., Carter, J. C., & Bowerman, B. (1997). Wnt signaling polarizes an early *C. elegans* blastomere to distinguish endoderm from mesoderm. *Cell, 90,* 695–705.

van der Voet, M., Berends, C. W., Perreault, A., Nguyen-Ngoc, T., Gonczy, P., Vidal, M., et al. (2009). NuMA-related LIN-5, ASPM-1, calmodulin and dynein promote meiotic spindle rotation independently of cortical LIN-5/GPR/Galpha. *Nature Cell Biology, 11,* 269–277.

Verlhac, M. H., Terret, M. E., & Pintard, L. (2010). Control of the oocyte-to-embryo transition by the ubiquitin-proteolytic system in mouse and *C. elegans*. *Current Opinion in Cell Biology, 22,* 758–763.

Von Stetina, J. R., & Orr-Weaver, T. L. (2011). Developmental control of oocyte maturation and egg activation in metazoan models. *Cold Spring Harbor Perspectives in Biology, 3,* a005553.

Whitten, S. J., & Miller, M. A. (2007). The role of gap junctions in *Caenorhabditis elegans* oocyte maturation and fertilization. *Developmental Biology, 301,* 432–446.

Wright, J. E., Gaidatzis, D., Senften, M., Farley, B. M., Westhof, E., Ryder, S. P., et al. (2011). A quantitative RNA code for mRNA target selection by the germline fate determinant GLD-1. *The EMBO Journal, 30,* 533–545.

Xu, L., Wei, Y., Reboul, J., Vaglio, P., Shin, T. H., Vidal, M., et al. (2003). BTB proteins are substrate-specific adaptors in an SCF-like modular ubiquitin ligase containing CUL-3. *Nature, 425,* 316–321.

Yamamoto, I., Kosinski, M. E., & Greenstein, D. (2006). Start me up: Cell signaling and the journey from oocyte to embryo in *C. elegans*. *Developmental Dynamics: An Official Publication of the American Association of Anatomists, 235,* 571–585.

Yang, H. Y., McNally, K., & McNally, F. J. (2003). MEI-1/katanin is required for translocation of the meiosis I spindle to the oocyte cortex in C elegans. *Developmental Biology, 260,* 245–259.

Yeong, F. M. (2004). Anaphase-promoting complex in *Caenorhabditis elegans*. *Molecular and Cellular Biology, 24,* 2215–2225.

Yin, X., Gower, N. J., Baylis, H. A., & Strange, K. (2004). Inositol 1,4,5-trisphosphate signaling regulates rhythmic contractile activity of myoepithelial sheath cells in *Caenorhabditis elegans*. *Molecular Biology of the Cell, 15,* 3938–3949.

Zhang, F., Barboric, M., Blackwell, T. K., & Peterlin, B. M. (2003). A model of repression: CTD analogs and PIE-1 inhibit transcriptional elongation by P-TEFb. *Genes & Development, 17,* 748–758.

Zhang, Y., Foster, J. M., Nelson, L. S., Ma, D., & Carlow, C. K. (2005). The chitin synthase genes chs-1 and chs-2 are essential for *C. elegans* development and responsible for chitin deposition in the eggshell and pharynx, respectively. *Developmental Biology, 285,* 330–339.

CHAPTER TWO

Regulation and Function of Maternal Gene Products During the Maternal-to-Zygotic Transition in *Drosophila*

John D. Laver*, Alexander J. Marsolais†, Craig A. Smibert*,†, Howard D. Lipshitz*,[1]

*Department of Molecular Genetics, University of Toronto, Toronto, Ontario, Canada
†Department of Biochemistry, University of Toronto, Toronto, Ontario, Canada
[1]Corresponding author: e-mail address: howard.lipshitz@utoronto.ca

Contents

Abstract

Drosophila late-stage oocytes and early embryos are transcriptionally silent. Thus, control of gene expression during these developmental periods is posttranscriptional and posttranslational. Global changes in the transcriptome and proteome occur during oocyte maturation, after egg activation and fertilization, and upon zygotic genome activation. We review the scale, content, and dynamics of these global changes; the factors that regulate these changes; and the mechanisms by which they are accomplished. We highlight the intimate relationship between the clearance of maternal gene

Current Topics in Developmental Biology, Volume 113
ISSN 0070-2153
http://dx.doi.org/10.1016/bs.ctdb.2015.06.007

products and the activation of the embryo's own genome, and discuss the fact that each of these complementary components of the maternal-to-zygotic transition can be subdivided into several phases that serve different biological roles and are regulated by distinct factors.

ABBREVIATIONS

ARE AU-rich element
ARE-BP ARE-binding protein
GO gene ontology
MBT midblastula transition
miRNA microRNA
MPF maturation-promoting factor
MZT maternal-to-zygotic transition
N:C nuclear:cytoplasmic
NC nuclear cycle
PGC primordial germ cell
RBP RNA-binding protein
SRE Smaug recognition element
UTR untranslated region (of an mRNA)
ZGA zygotic genome activation

1. INTRODUCTION

In plants and animals, the earliest stages of embryogenesis are under maternal genetic control, directed by gene products produced from the female genome during gametogenesis. In insects such as *Drosophila*, the oocyte nucleus and the nuclei of the early embryo are transcriptionally silent; thus, maturation and activation of the oocyte, as well as early embryogenesis, are regulated posttranscriptionally and posttranslationally. In this chapter, we discuss the developmental events and regulation of gene expression during this maternally controlled phase with a particular emphasis on the clearance of a subset of maternal mRNAs and proteins from early embryos and concomitant transcriptional activation of the zygotic genome, processes together referred to as the maternal-to-zygotic transition (MZT).

We begin with a brief overview of oogenesis and early embryogenesis. This is followed by more detailed discussions of oocyte maturation, egg activation, and early embryogenesis, with a focus on how the progression of developmental events is regulated, descriptions of important developmental processes taking place during each time period, and examination of the changes in gene expression that occur during these developmental transitions. We highlight the role of maternally provided factors: both how they regulate various developmental processes and how they are, in turn, regulated by other maternal factors and, later, by zygotic factors. The roles and the regulation of both maternally supplied mRNA and protein are discussed, and we examine how they accomplish the tasks to which they are dedicated and, in many cases, how they are ultimately eliminated.

2. OVERVIEW OF OOGENESIS AND EARLY EMBRYOGENESIS

During *Drosophila* oogenesis, a germ line stem cell divides asymmetrically, giving another germ line stem cell and a cystoblast, the latter of which then divides four times with incomplete cytokinesis at each division, resulting in a 16-cell cyst connected by cytoplasmic bridges referred to as ring canals (Spradling, 1993). One of these 16 cells develops into the oocyte, while the other 15 become the so-called nurse cells, which become polyploid, expressing large quantities of mRNA and protein. Over a period of about 3 days, these gene products are transported from the nurse cells into the oocyte, where mechanisms that control mRNA localization and spatially regulated translation play central roles in specification of the anteroposterior and dorsoventral axes of the oocyte and future embryo (reviewed in Lasko, 2012). While the transport of mRNA and protein from the nurse cells to the oocyte during the early stages of oogenesis is an active, regulated process, towards the end of oogenesis the nurse cells dump their contents into the oocyte and then degenerate.

In addition to the initial stages of axis specification, the oocyte must negotiate the process of meiosis, in which the diploid maternal genome is recombined and reduced to haploid status in preparation for fusion with the male pronucleus after fertilization (Page & Hawley, 2003). In *Drosophila*, meiosis initiates shortly after the oocyte is specified, but stalls at Prophase I for several days until maturation directs progression to a second pause at Metaphase I. The mature oocyte passes from the ovary into the oviduct and then the uterus, during which time it undergoes a process referred to as egg activation, which triggers the resumption and completion of meiosis. After activation, and while still in the uterus, the egg is fertilized. It is then deposited into the external environment, where embryogenesis occurs. The stages of oogenesis and early embryogenesis that are considered in this chapter are schematized in Fig. 1.

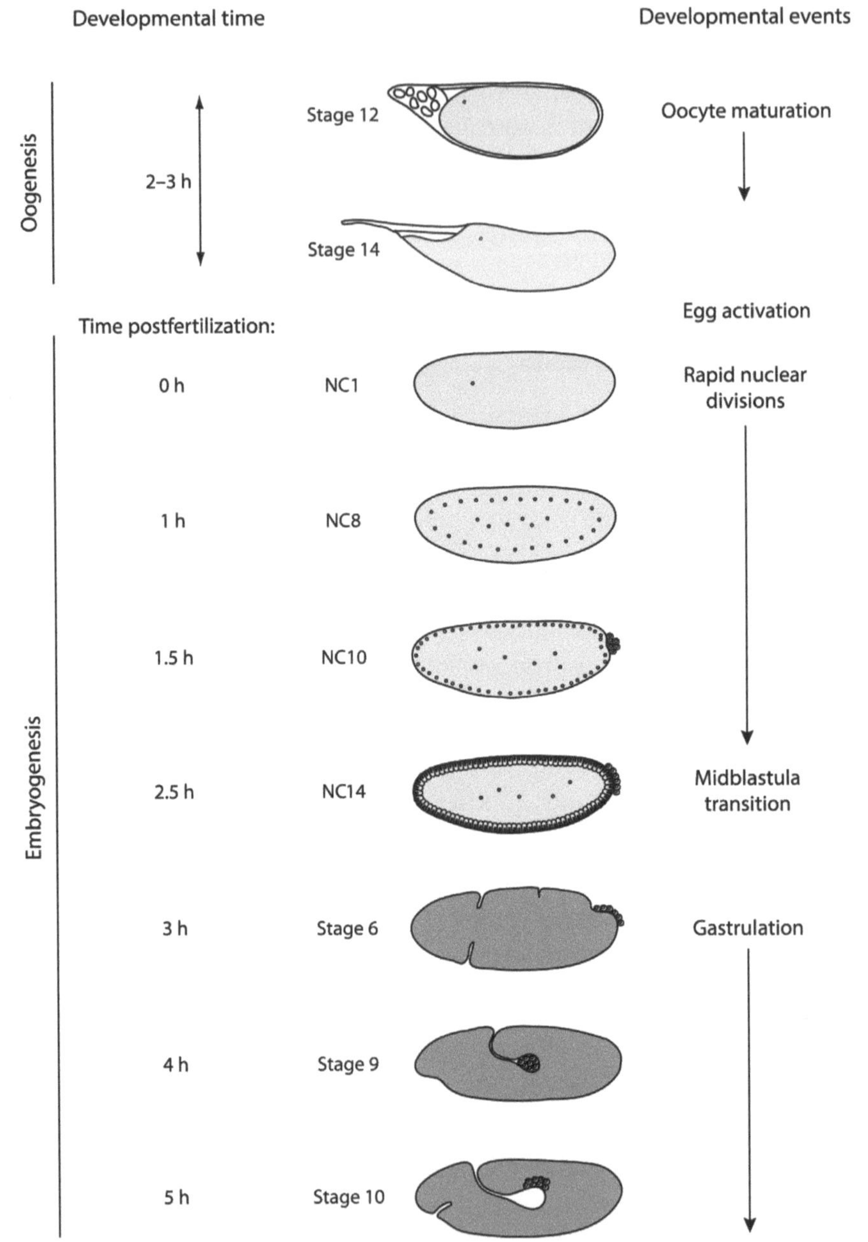

Figure 1 Overview of late oogenesis and early embryogenesis in *Drosophila*. Illustrations of the stages of late oogenesis and early embryogenesis that are discussed in this chapter. The oocyte and embryo are indicated by gray shading. During Stage 12 of oogenesis, nurse cells are degenerating after having dumped their contents into the oocyte. Oocyte maturation occurs during Stages 12 and/or 13. Embryogenesis begins

During early embryogenesis, after completion of meiosis and fusion of the male and female pronuclei, the *Drosophila* embryo initiates rapid and nearly synchronous (metasynchronous) nuclear divisions involving only DNA replication (S phase) and mitosis (M phase), which occur in the absence of cytokinesis. Thirteen such divisions occur, thereby producing, in a period of a little over 2 h, a large multinucleated syncytium containing approximately 6000 nuclei. The first five nuclear cycles (NCs) occur within the yolk, after which the majority of nuclei begin to migrate to the periphery of the embryo such that, by the end of NC8, there are about 200 nuclei distributed around the circumference of the embryo. At this point, nuclei that arrive at the posterior bud to form the primordial germ cells (PGCs) while the remaining, somatic, nuclei continue the rapid S–M divisions. As these divisions proceed, they gradually lengthen in duration, most notably during NC10–NC13, in a manner dependent on the DNA replication checkpoint. During the interphase of the fourteenth division, gap phases are introduced, and plasma membrane invaginates between the peripherally located somatic nuclei in a process referred to as cellularization. Cellularization is the first developmental process that depends on transcription from the zygotic genome and, thus, marks the midblastula transition (MBT; Tadros & Lipshitz, 2009).

As mentioned earlier, during maturation, activation, and the first seven NCs of embryogenesis, the genome is transcriptionally silent. Thus, maternally deposited factors direct these developmental processes. Despite the absence of transcription, these stages of development are characterized by large-scale changes in the transcriptome and proteome, which occur by posttranscriptional and posttranslational mechanisms. At NC8, an early

Figure 1—Cont'd with egg activation and fertilization, followed by 13 rapid nuclear divisions that occur in the absence of cytokinesis, producing a syncytium of ~6000 nuclei. By the end of nuclear cycle (NC) 8, the majority of nuclei have migrated to the periphery of the embryo. Immediately after NC 10, primordial germ cells (PGCs) form at the posterior of the embryo. During interphase of NC14, cellularization occurs as plasma membrane invaginates around the remaining peripherally located somatic nuclei, marking the midblastula transition (MBT). Gastrulation begins after cellularization, at embryonic Stage 6. During embryonic Stages 6, 9, and 10, the darker gray shading represents the presence of gastrulating cells and tissues; for simplicity, only the PGCs are shown during these stages. At Stage 6, the PGCs begin moving dorsally as the underlying blastoderm cells shift their position; by Stage 9 the PGCs are located in the posterior midgut pocket; during Stage 10 the PGCs migrate through the dorsal wall of the midgut pocket. Anterior is to the left, dorsal toward the top of the page. See main text for further descriptions of the developmental events depicted.

phase of zygotic transcription initiates. Large-scale recruitment of poised RNA Polymerase II is observed at NC13 followed by high-level zygotic genome activation (ZGA), and the MBT. ZGA and the MBT are discussed in detail in the chapters "Transcriptional activation of the zygotic genome in *Drosophila*" by Harrison and Eisen and "Coordinating cell cycle remodeling with transcriptional activation at the *Drosophila* MBT" by Blythe and Wieschaus. This chapter, therefore, focuses largely on maternal gene products and the dynamic changes that they undergo during the MZT in both the soma and the PGCs; we discuss ZGA in these cell types largely in the context of regulation by maternal factors.

3. MATERNAL FACTORS DURING OOCYTE MATURATION

3.1 Regulation of Oocyte Maturation

During oogenesis, arrest in Prophase I of meiosis is relieved in a process referred to as oocyte maturation. In *Drosophila*, maturation occurs during stages 12 and/or 13 of oogenesis, which span a length of 2–3 h. In most species in which it has been studied, meiotic maturation is initiated by extrinsic signals, typically in the form of hormones or other signaling molecules. While maturation may be directed by such signals in *Drosophila* as well, such molecule(s) have not yet been identified.

Oocyte maturation is best characterized in other species, particularly in *Xenopus*; indeed, of the key factors that have been shown to be required for relief of meiotic arrest in *Drosophila*, several are homologous to factors identified in amphibians. In amphibians and other animals, for example, oocyte maturation and the resumption of meiosis depend on the activity of Cdk1 and Cyclin B, which, when complexed together, form "maturation-promoting factor" (MPF). The activity of MPF promotes reentry into the meiotic cell cycle, beginning with breakdown of the nuclear envelope in preparation for metaphase. In *Drosophila*, mutation of Cdk1 delays maturation and the resumption of meiosis, suggesting that MPF plays a similar role (Von Stetina et al., 2008). In addition, several factors that regulate the activity of MPF have also been shown to have a role in oocyte maturation. For example, Twine, a Cdc25 homolog, activates Cdk1 and is required for meiotic maturation to occur correctly (Alphey et al., 1992; Courtot, Fankhauser, Simanis, & Lehner, 1992; Von Stetina et al., 2008; White-Cooper, Alphey, & Glover, 1993; Xiang et al., 2007). Upstream of Twine is its activator, the Polo kinase, whose repression by Matrimony (Xiang et al., 2007) as well as the Greatwall kinase (Archambault, Zhao,

White-Cooper, Carpenter, & Glover, 2007) is relieved at the onset of maturation. The position of the Polo kinase atop a signaling cascade required for maturation is conserved in other species (Abrieu et al., 1998; Chase et al., 2000; Roshak et al., 2000). In *Drosophila*, Polo kinase itself is regulated by α-Endosulfine, a predicted protein phosphatase inhibitor, which is required for meiotic maturation due to its ability to upregulate Polo and Twine protein levels; Early girl, a predicted E3 ubiquitin ligase which interacts with α-Endosulfine, is also critical for the timing of meiotic maturation (Von Stetina et al., 2008).

Once maturation is completed, oocytes arrest again in Metaphase I of meiosis until egg activation. This secondary arrest is maintained by the formation of chiasmata during early metaphase, which put tension on kinetochores, resulting in meiotic arrest (Jang, Messina, Erdman, Arbel, & Hawley, 1995; McKim, Jang, Theurkauf, & Hawley, 1993). In addition, heterochromatin in nonexchange chromosomes is required for meiotic arrest at this stage (Dernburg, Sedat, & Hawley, 1996; Hawley et al., 1992; Hughes et al., 2009).

3.2 Changes to the Oocyte Proteome During Oocyte Maturation

Oocyte maturation is marked by large-scale changes in mRNA translation and, thus, in protein levels. For instance, cytoplasmic polyadenylation mediated by oo18 RNA-binding protein, the *Drosophila* homolog of CPEB, and by Wispy, a noncanonical poly(A) polymerase, leads to increased levels of proteins such as Cyclin B upon oocyte maturation (Benoit et al., 2005; Benoit, Papin, Kwak, Wickens, & Simonelig, 2008). Indeed, it has been shown that Wispy is required for the polyadenylation of transcripts from more than 2000 genes in late-stage oocytes, although the relevance of this polyadenylation to changes in mRNA translation is not yet clear (Cui, Sartain, Pleiss, & Wolfner, 2013).

In addition to well-studied regulators such as Cyclin B, hundreds of other proteins, representing 30% of the detected proteome, are either up- or downregulated during oocyte maturation (over 500 detected in each category; Kronja, Whitfield, et al., 2014). Interestingly, the set of upregulated proteins is enriched for functions crucial to the resumption of meiosis, including the cell cycle, nuclear division, and spindle and microtubule organization (Kronja, Whitfield, et al., 2014). Notably, an inhibitor of the APC, Rca1, is among the set of upregulated proteins, possibly facilitating the secondary arrest at Metaphase I (Kronja, Whitfield, et al., 2014).

Intriguingly, 66 proteins have been identified that are upregulated at oocyte maturation but then downregulated upon egg activation, suggesting a highly specific role for these factors in the processes occurring at maturation (Kronja, Whitfield, et al., 2014). Further investigation of these proteins may yield insights into mechanisms underlying the regulation of maturation.

4. MATERNAL FACTORS DURING EGG ACTIVATION AND EARLY EMBRYOGENESIS

4.1 Triggers and Developmental Events During Egg Activation and Early Embryogenesis

In *Drosophila*, the transition from a mature oocyte to one that has been "activated" and is poised to undergo fertilization and embryogenesis involves the resumption and completion of meiosis, cross-linking of vitelline membrane and chorion proteins leading to an impermeable eggshell, and a variety of changes to the maternally supplied mRNA and protein complements of the oocyte (reviewed in Horner & Wolfner, 2008b). In most species, egg activation is triggered by fertilization but, in many insects, this process occurs independently of fertilization. In *Drosophila*, activation occurs over a 20-min period concurrently with ovulation, as the mature oocyte passes from the ovary into the oviduct (Heifetz, Yu, & Wolfner, 2001), and is triggered by fluid uptake and mechanical stress in the oviduct (Heifetz et al., 2001; Horner & Wolfner, 2008a).

Egg activation in all species, regardless of the trigger, is associated with a rise of intracellular calcium. In *Drosophila*, it appears that mechanical stress triggers calcium uptake from the extracellular environment through mechanosensitive calcium channels, leading to waves of calcium that begin at one or both poles and then spread through the egg in a manner dependent upon internal calcium stores and the IP_3 system (Kaneuchi et al., 2015; York-Andersen et al., 2015). Although the details of how this rise in intracellular calcium is transduced into the subsequent cellular and molecular changes associated with egg activation is not well understood in *Drosophila*, genetic evidence suggests that it involves, at least in part, the calmodulin-dependent phosphatase, Calcineurin, and its partner, Sarah (*Drosophila* Calcipressin); mutations in either gene lead to an inability to complete meiosis and, in the case of *sarah* mutants, defects in other processes associated with egg activation (Horner et al., 2006; Takeo, Hawley, & Aigaki, 2010; Takeo, Tsuda, Akahori, Matsuo, & Aigaki, 2006).

Once egg activation has occurred, and after the completion of meiosis and pronuclear fusion, the *Drosophila* embryo initiates rapid, metasynchronous mitotic divisions, consisting only of S and M phases, which occur in the absence of cytokinesis, as described in Section 2 (also see Fig. 1). A key regulator of the onset of these mitotic divisions at egg activation is the Pan gu kinase, mutations in which disrupt these divisions (Fenger et al., 2000; Freeman & Glover, 1987; Freeman, Nusslein-Volhard, & Glover, 1986; Shamanski & Orr-Weaver, 1991), although the mechanism by which Pan gu activity is linked to egg activation is not understood.

4.2 Changes to the Proteome upon Egg Activation

Immediately following egg activation and until approximately 1 h postfertilization, there are minimal changes in mRNA levels (Tadros et al., 2007; Thomsen, Anders, Janga, Huber, & Alonso, 2010). The earliest events following egg activation, therefore, rely on translational and posttranslational controls to regulate the protein complement of the early embryo. For example, with regard to posttranslational regulation, the resumption and completion of meiosis upon egg activation requires the activity of Cortex, a meiosis-specific activator of the APC/C E3 ubiquitin ligase, to degrade Cyclins B and B3 as well as the Polo kinase inhibitor, Matrimony (Pesin & Orr-Weaver, 2007; Swan & Schupbach, 2007; Whitfield, Chisholm, Hawley, & Orr-Weaver, 2013). An example of translational regulation upon egg activation involves the derepression of translation of the mRNA encoding Smaug (Tadros et al., 2007), an RNA-binding protein (RBP) with an essential role in the decay of maternally expressed transcripts during the MZT (see Section 4.3.2 below). Translational derepression of the *smaug* mRNA requires the Pan gu kinase complex (Tadros et al., 2007). Pan gu also regulates the early embryonic cell cycle (Fenger et al., 2000; Vardy & Orr-Weaver, 2007); however, Pan gu's role in regulation of *smaug* mRNA translation and of maternal transcript destabilization is independent of its role in cell cycle control (Tadros et al., 2003, 2007).

In addition to these specific examples of regulated translation and protein degradation upon egg activation, recent work has provided a more global view of these processes. Indeed, it has been found that, of the mRNAs produced by a total of over 5000 genes, hundreds are translationally upregulated (over 800) or downregulated (over 400) upon egg activation, the majority of which are dependent on Pan gu (60% of up- and 70% of downregulated mRNAs; Kronja, Yuan, et al., 2014). Similarly, the levels of hundreds of proteins are altered upon egg activation: of over 3600 proteins that were quantified, about 10% decreased and 8% increased significantly in abundance (Kronja, Yuan, et al., 2014). As would be expected, in many cases, increases in protein levels were reflective of translationally upregulated mRNAs, whereas decreases in protein levels were largely posttranslational: while 40% of proteins whose levels increased significantly were encoded by mRNAs that underwent upregulation of their translation, only 16% of the mRNAs encoding proteins that decreased significantly were translationally downregulated (Kronja, Yuan, et al., 2014). Surprisingly, however, approximately 75% of the translationally upregulated mRNAs encode proteins whose levels do not increase upon egg activation. Evidence suggests that these may represent degradation of the oocyte-supplied protein and replacement with newly translated protein. This might be required in cases where the oocyte-supplied protein is in some way different from the newly translated version (e.g., differentially modified or localized; Kronja, Yuan, et al., 2014). Additional work will be required to further investigate the scale of this turnover and its potential functions.

Notably, it was also found that, in mature oocytes but not in activated eggs, a large fraction of mRNAs that run in the "polysome" region of sucrose gradients do so in a manner that is not sensitive to the translational inhibitor puromycin, which disrupts polysomes (Kronja, Yuan, et al., 2014); these transcripts may therefore represent maternal mRNAs that are stored in ribonucleoprotein particles with similar sedimentation characteristics to polysomes but which are, in fact, distinct.

While changes in protein levels and the translational control of mRNAs upon egg activation are clearly widespread, the mechanisms underlying this regulation are not fully understood. Decreases in protein levels could result from the activity of the APC/C, or other ubiquitin ligases, a number of which are upregulated upon egg activation (Kronja, Yuan, et al., 2014). With regard to translational control, one possible mechanism might be regulation of the polyadenylation status of mRNAs. For example, it has been estimated that mRNAs produced by at least 1800 genes are polyadenylated upon egg activation in a manner dependent on the cytoplasmic poly(A) polymerase, Wispy (Cui et al., 2013). While it is likely that this is important for the translational activation of many mRNAs, the exact contribution of this polyadenylation to translational control is unclear. Indeed, in many instances, additional mechanisms may be required for translational activation. For example, in the case of translational activation of the *smaug* mRNA, it is polyadenylated, but this polyadenylation is not sufficient for translation, which also requires Pan gu-dependent release of translational repression by the Pumilio RBP and additional unidentified repressors (Tadros et al., 2007).

Finally, in addition to changes in mRNA translation and protein stability, egg activation is likely marked by widespread changes in posttranslational modifications, which can have major effects on protein function. For instance, extrapolating from the approximately 300 proteins whose phosphorylation status was shown to change upon egg activation, about 30% of the oocyte proteome is phosphomodified during this developmental process, representing a potentially globally important regulatory mechanism (Krauchunas, Horner, & Wolfner, 2012). Interestingly, that study found that more proteins are dephosphorylated than phosphorylated upon egg activation. A subset of these changes in phosphorylation status depends on the activity of Sarah and Calcineurin, as well as the Cortex APC/C, although whether these factors are directly responsible for the changes in phosphorylation or, instead, represent upstream regulators, remains to be determined (Krauchunas, Sackton, & Wolfner, 2013).

4.3 mRNA Decay During the Maternal-to-Zygotic Transition

4.3.1 Global Analyses of Maternal mRNAs

While the initial regulation of gene expression post-egg-activation largely involves translational and posttranslational control with minimal changes in mRNA levels, after approximately 1 h of embryogenesis a large-scale remodeling of the transcriptome of the embryo begins as developmental control is gradually transferred from maternal gene products to the zygotic genome (Fig. 2). The first part of this remodeling involves the widespread degradation of maternally supplied mRNAs. Genome-wide studies have shown that transcripts representing half to three-quarters of the *Drosophila* protein-coding genome (i.e., mRNAs encoded by ~7000–10,000 genes) are maternally expressed and loaded into the early embryo (Lecuyer et al., 2007; Tadros et al., 2007; Thomsen et al., 2010) and, of these, one- to two-thirds are either eliminated or undergo a significant reduction in levels during the first 3 h of embryogenesis (i.e., clearance of mRNAs representing ~2300–6700 genes; De Renzis, Elemento, Tavazoie, & Wieschaus, 2007; Thomsen et al., 2010). This therefore represents a rapid and drastic change to the transcriptome of the developing embryo.

mRNAs that are stable versus degraded during this period encode proteins enriched for different functions. For example, stable transcripts are enriched for those encoding ribosomal constituents and mitochondrial

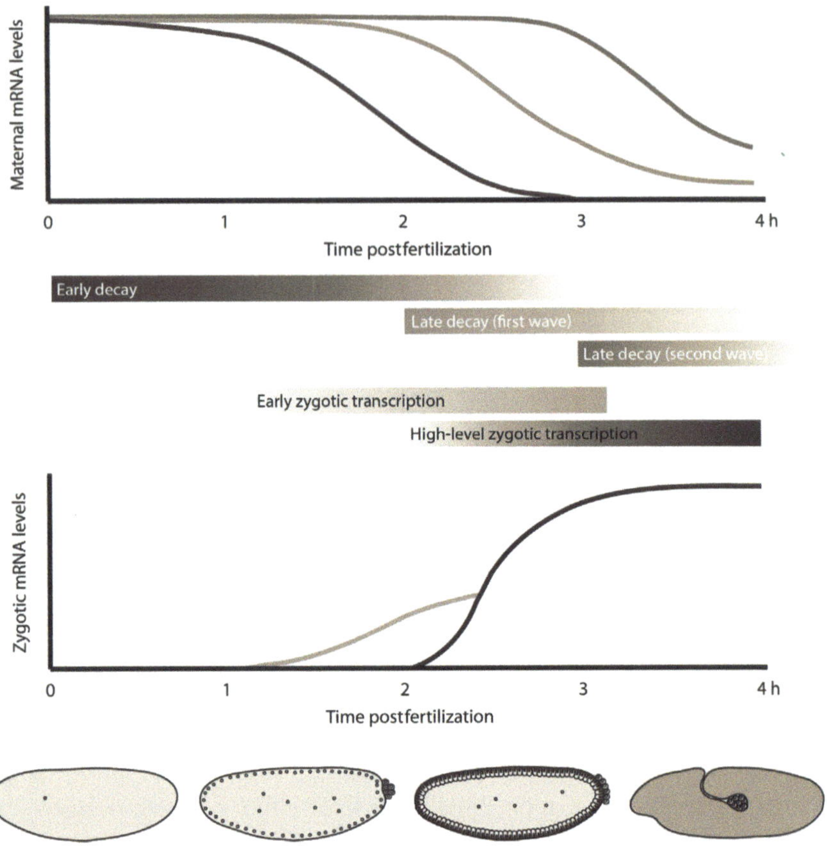

Figure 2 The maternal-to-zygotic transition (MZT) in *Drosophila*. The MZT comprises two processes: degradation of a subset of maternally expressed mRNAs (top), and activation of transcription from the zygotic genome (bottom). Maternal mRNA degradation is mediated by two different types of machineries: first, early-acting machineries which are dependent solely on maternally provided factors, and second, late-acting machineries which depend on zygotically transcribed factors. The early-acting decay machinery (red, black in the print version) is triggered by egg activation, although the major effect of this machinery is not observed until approximately 1 h postfertilization. The activity of the early-acting machinery is largely complete by approximately 3 h postfertilization. The late-acting decay machinery appears to trigger degradation in two waves, the first being observed 2–3 h postfertilization (light orange, light gray in the print version), and the second observed after 3 h postfertilization (dark orange, gray in the print version). The onset of zygotic transcription, referred to as zygotic genome activation (ZGA), also occurs in two waves. An early wave of transcription of a subset of genes occurs gradually (light blue, light gray in the print version), with the earliest signs of transcription observed at nuclear cycle 8 at approximately 1 h postfertilization. The major onset of zygotic transcription (dark blue, black in the print version) occurs 2–3 h postfertilization, concurrent with the midblastula transition.

respiratory chain proteins, whereas functions enriched among proteins encoded by unstable mRNAs include cell cycle control and DNA metabolism (Tadros et al., 2007; Thomsen et al., 2010). These different functions presumably reflect the changes occurring during the earliest hours of embryogenesis: whereas the protein synthetic and energy production machineries are needed throughout the MZT, there are changes in cell cycle and transcriptional regulation leading up to, at, and after the MBT that require corresponding changes in the expression of genes that regulate these processes. For example, while the early NCs are metasynchronous throughout the syncytial embryo, once cells form their divisions are highly patterned. Likewise, spatially regulated expression of transcription factors prior to and after cellularization is a hallmark of, and essential for, early *Drosophila* development.

4.3.2 Mechanisms Directing Decay of Maternal RNAs

In *Drosophila*, the mechanisms and pathways responsible for maternal mRNA degradation have been studied in some detail. Decay of maternal mRNAs is both highly regulated and transcript specific and has been shown to be directed by at least two different types of machineries (Fig. 2): one or more early-acting or "maternal" machineries that depend on egg activation (not fertilization) and are directed by maternally expressed factors; and one or more late-acting or "zygotic" machineries, which initiate after the onset of high-level zygotic transcription between 2 and 3 h of embryogenesis, and require zygotic gene products (Bashirullah et al., 1999). Given that, in *Drosophila*, egg activation is not coupled to fertilization, the early and late components of maternal transcript decay were initially discovered by comparing activated, unfertilized eggs—in which only the maternally loaded activity is present but in which there is no zygotic transcription—to fertilized embryos in which both components are active (Bashirullah, Cooperstock, & Lipshitz, 2001; Bashirullah et al., 1999; Tadros et al., 2003, 2007). Using a quite different strategy—genetic removal of chromosomes or parts of chromosomes from early embryos (but not their mothers)—it was possible to define globally the maternal mRNAs that depend on zygotic transcription for elimination, as well as to map components of the late machinery to particular chromosomes or chromosome arms (De Renzis et al., 2007). Genome-wide studies using these different strategies have provided varying estimates with regard to the relative contributions of the early and late machineries. Based on two studies that estimated the total proportion of maternally expressed and degraded transcripts to be about one-third: 20% of maternally expressed mRNAs (i.e., representing ~1400–2000 genes) are degraded, at least in part, by the early-acting maternally encoded machinery, with an additional 15% (representing ~1000–1500 genes) requiring the activity of a late-acting machinery (De Renzis et al., 2007; Tadros et al., 2007). In another study, which estimated the total proportion of maternally expressed and degraded transcripts at close to two-thirds: 14% (representing ~1000–1400 genes) are targeted exclusively by an early-acting machinery, 22% (representing ~1500–2200 genes) exclusively by a late-acting machinery, and 25% (representing ~1750–2500 genes) by the combined activity of both early- and late-acting machineries (Thomsen et al., 2010).

Several *trans*-acting factors have been shown to function in promoting the decay of specific sets of transcripts, as part of either the early- or late-acting machineries (Fig. 3). The first such factor to be discovered was the RBP, Smaug. Smaug has multiple functions in posttranscriptional regulation, acting to repress translation and/or to promote decay of its target mRNAs (Chen et al., 2014; Dahanukar, Walker, & Wharton, 1999; Dahanukar & Wharton, 1996; Nelson, Leidal, & Smibert, 2004; Semotok et al., 2005; Smibert, Lie, Shillinglaw, Henzel, & Macdonald, 1999; Smibert, Wilson, Kerr, & Macdonald, 1996). Strikingly, Smaug has a crucial role in mediating early decay, as up to two-thirds of the mRNAs degraded by the early-acting maternal machinery are dependent, directly or indirectly, on Smaug for their elimination (Tadros et al., 2007). Smaug targets specific transcripts by binding to *cis*-acting stem-loops, referred to as Smaug recognition elements (SREs), and promotes mRNA degradation by recruiting the CCR4/POP2/NOT-deadenylase complex to these transcripts, thereby triggering their deadenylation and decay (Semotok et al., 2005, 2008; Fig. 3). Smaug can also repress the translation of target mRNAs by recruiting the eIF4E-binding protein, Cup, and/or Argonaute 1 (Nelson et al., 2004; Pinder & Smibert, 2013). Indeed, it has recently been shown that Smaug is a global regulator, not only of the degradation but also of the translation of a large number of maternal mRNAs in early embryos, and that the majority of its ~350 identified direct targets undergo Smaug-dependent regulation of both of these posttranscriptional processes (Chen et al., 2014).

The *smaug* mRNA itself is translationally repressed during oogenesis (Smibert et al., 1996, 1999). This repression is essential since ectopic expression of Smaug in the ovarian cyst results in an absence of mid- and late-stage follicles (Semotok et al., 2005). Smaug protein is synthesized upon egg activation by Pan gu-dependent relief of repression (Tadros et al., 2007), as described earlier. The accumulation of Smaug post-egg-activation appears

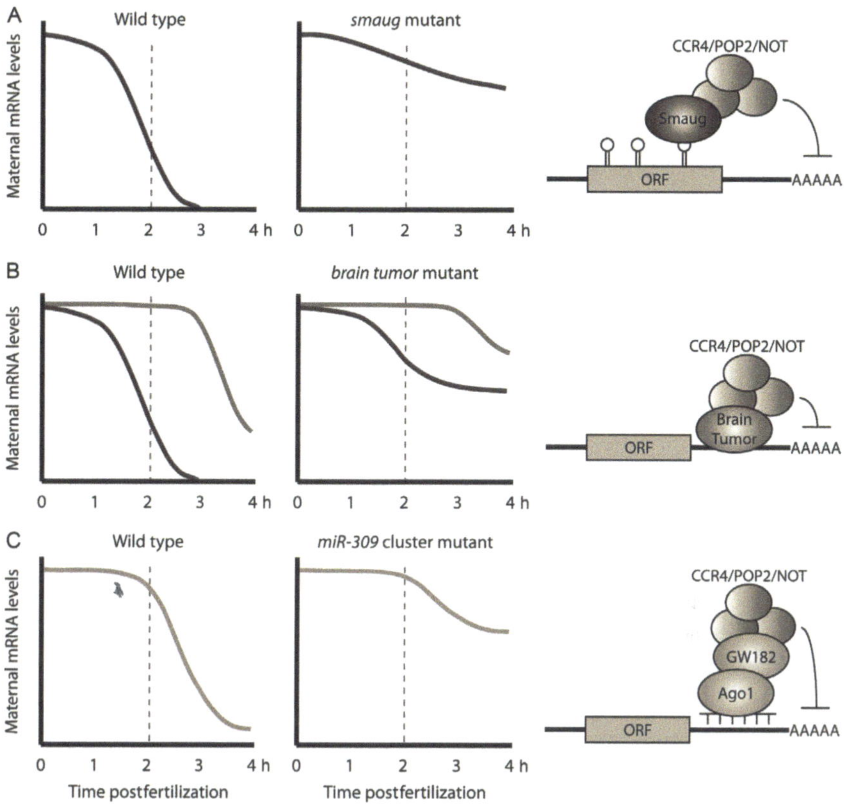

Figure 3 *Trans*-factors that mediate maternal mRNA degradation during the maternal-to-zygotic transition in *Drosophila*. Three *trans*-acting factors have been definitively demonstrated to function in maternal mRNA degradation during the MZT in *Drosophila*. (A) The RNA-binding protein, Smaug, functions to degrade mRNAs as part of the early-acting machinery. In embryos lacking functional Smaug protein, approximately two-thirds of the mRNAs degraded by the early machinery are stabilized (Tadros et al., 2007). Smaug recognizes its target mRNAs by binding to stem-loop structures called Smaug recognition elements (SREs), which are often located in the open-reading frame (ORF), and triggers mRNA degradation through recruitment of the CCR4/POP2/NOT-deadenylase complex, which leads to deadenylation of targeted transcripts, and subsequent decay (Semotok et al., 2005, 2008). (B) The RNA-binding protein, Brain Tumor, functions to degrade mRNAs as part of both the early-acting machinery and the second wave of the late-acting machinery. In embryos lacking functional Brain Tumor protein, a subset of mRNAs degraded by each of these machineries is stabilized (Laver et al., 2015). Whether Brain Tumor also functions as part of the first wave of the late-acting machinery is unclear. Brain Tumor recognizes its target mRNAs by binding to single-stranded motifs that contain a core "UGUU" sequence, and are most often located in the 3′UTR (Laver et al., 2015). While the mechanism by which Brain Tumor mediates mRNA decay has not been elucidated, it has been shown to interact with the CCR4/POP2/NOT-deadenylase complex (Temme et al., 2010), suggesting that it, like Smaug, may trigger

(Continued)

to be an important timer regulating early embryo development, since Smaug is required for activation of the DNA replication checkpoint as well as for ZGA, and an engineered anteroposterior gradient of Smaug levels produces a gradient in the timing of both the MZT (viz., maternal transcript degradation and zygotic transcription) and MBT (viz., cellularization; Benoit et al., 2009). It should be noted, however, that increasing Smaug levels in an otherwise wild-type embryo does not speed up these processes, suggesting that other factor(s) must become limiting (Benoit et al., 2009).

Smaug protein rapidly disappears at the MBT and ZGA is required for Smaug clearance since Smaug protein persists in unfertilized eggs (Benoit et al., 2009; Smibert et al., 1999). Interestingly, many ubiquitous maternal mRNAs are degraded and then reexpressed in specific patterns upon ZGA (De Renzis et al., 2007). Given that these include Smaug targets, one plausible hypothesis as to the reason for the rapid clearance of Smaug is that it must be removed in order to permit stable reexpression of these mRNAs.

The TRIM-NHL family protein, Brain Tumor, has recently been identified as a second RBP with an important role in promoting maternal mRNA degradation (Laver et al., 2015). Brain Tumor was initially shown to be a translational repressor of *hb* mRNA in early embryos, and was long considered to associate with RNA only indirectly, recruited via protein–protein interactions with the RBPs, Pumilio and Nanos (Sonoda & Wharton, 2001). However, it has recently been shown that Brain Tumor directly binds *hb* RNA through its C-terminal NHL domain (Loedige et al., 2014). Furthermore, Brain Tumor has been found to associate, in early embryos, with the mRNAs encoded by almost 1200 genes (Laver et al., 2015). It binds these transcripts in a sequence-specific manner via a motif in their 3′-untranslated regions (3′UTRs) consisting of a "UGUU" core

Figure 3—Cont'd mRNA decay through recruitment of this complex and deadenylation of its target transcripts. (C) The *miR-309* cluster of miRNAs functions to degrade mRNAs as part of the first wave of the late-acting machinery. The *miR-309* cluster is not maternally expressed, but is zygotically transcribed prior to the MBT. In *miR-309*-mutant embryos, approximately 400 mRNAs are stabilized 2–3 h postfertilization (Bushati, Stark, Brennecke, & Cohen, 2008). miRNAs mediate mRNA decay as part of the Argonaute 1 (AGO1)-containing miRNA-induced silencing complex (miRISC), which induces mRNA degradation via recruitment of the CCR4/POP2/NOT-deadenylase, an interaction mediated through the Argonaute 1-interacting protein GW182 (Braun, Huntzinger, Fauser, & Izaurralde, 2011; Chekulaeva et al., 2011; Fabian et al., 2011). Recent work suggests that miRISC also promotes mRNA decay by directly recruiting decapping factors (not shown; Barisic-Jager, Krecioch, Hosiner, Antic, & Dorner, 2013; Nishihara, Zekri, Braun, & Izaurralde, 2013).

sequence (Laver et al., 2015). Brain Tumor-associated transcripts are highly enriched for maternal mRNAs that are translationally repressed and degraded in early embryos, and genome-wide studies of mRNA levels in embryos lacking functional Brain Tumor protein have demonstrated that it has a key role in promoting mRNA decay; almost 600 maternally expressed mRNAs depend on Brain Tumor for their degradation (Laver et al., 2015). Interestingly, prior to the identification of Brain Tumor's RNA-binding activity, computational analysis identified a "UUGUU" motif strikingly similar to the Brain Tumor-binding motif as one of two motifs that are enriched among the entire set of maternally expressed and degraded mRNAs (De Renzis et al., 2007). Indeed, Brain Tumor acts to degrade maternal mRNAs as part of both early and late decay pathways, representing the first RBP that participates in a late-acting machinery (Laver et al., 2015). Why certain Brain Tumor targets are degraded late while others are degraded early is unknown.

Brain Tumor's early role is likely independent of Smaug (i.e., the subsets of mRNAs degraded by these RBPs do not significantly overlap), suggesting that Smaug and Brain Tumor represent two separate pathways for early decay (Laver et al., 2015). While the mechanism by which Brain Tumor mediates mRNA degradation has not yet been elucidated, it, like Smaug, has been found to interact with components of the CCR4/POP2/NOT complex (Temme et al., 2010; Fig. 3), suggesting that both RBPs act as specificity factors, recruiting this deadenylase to subsets of target mRNAs thereby triggering their decay. Homologs of Smaug and Brain Tumor have been shown to regulate mRNA stability and/or translation in cultured mammalian cells (Baez & Boccaccio, 2005; Loedige, Gaidatzis, Sack, Meister, & Filipowicz, 2013); however, it is not yet known whether they act during the MZT in species other than *Drosophila*.

In addition to Smaug and Brain Tumor, microRNAs (miRNAs) have been shown to have a role in mediating maternal mRNA decay during the MZT. The most definitive evidence for this role comes from analysis of a family of miRNAs encoded by the *miR-309* cluster. These miRNAs are not expressed maternally but are transcribed zygotically and have been shown to target a large population of mRNAs after the MBT; embryos mutant for the *miR-309* cluster exhibit stabilization of the mRNA products of several hundred genes (Bushati et al., 2008). Interestingly, the *miR-309* cluster provides a link between early and late decay activities, as transcription of the *miR-309* cluster is dependent on Smaug (Benoit et al., 2009).

In addition to the *miR-309* cluster, various analyses of the set of maternally expressed and degraded mRNAs suggest that other miRNAs may also play a role in mRNA degradation, as the seed sequences for several embryonic miRNAs are enriched in maternally expressed mRNAs in general, as well as in maternally expressed mRNAs degraded by the joint action of both the maternal and zygotic degradation activities (Tadros et al., 2007; Thomsen et al., 2010).

It is worth noting that miRNAs also play a role in maternal transcript decay in other species. Zebrafish *miR-430* and *Xenopus miR-427*, for example, destabilize maternal mRNAs (Giraldez et al., 2006; Lund, Liu, Hartley, Sheets, & Dahlberg, 2009). While *miR-309* cluster miRNAs are not homologous to these vertebrate miRNAs, miRNA-directed degradation of maternal transcripts may represent a conserved mechanism during the MZT in metazoa.

In addition to the demonstrated role of the aforementioned *trans*-acting factors, a number of other *trans*-factors and *cis*-elements have been predicted to function in promoting maternal mRNA decay during the MZT in *Drosophila*. For instance, genome-wide studies of mRNAs associated with the RBP, Pumilio, in early embryos have revealed an enrichment among Pumilio-associated mRNAs for transcripts that are degraded during the MZT by a late-acting, zygotic machinery, suggesting a role for Pumilio in this process (Gerber, Luschnig, Krasnow, Brown, & Herschlag, 2006; Laver et al., 2015; Thomsen et al., 2010). Interestingly, this role may involve cooperation with Brain Tumor: although the sets of mRNAs associated with Pumilio or Brain Tumor in early embryos are largely nonoverlapping, those mRNAs that are cobound by both RBPs are, in fact, enriched for transcripts degraded by a zygotic machinery (Laver et al., 2015).

In addition to Pumilio, a role in maternal transcript decay has been predicted for *cis*-acting AU-rich elements (AREs), which are enriched in the set of destabilized maternal mRNAs (De Renzis et al., 2007; Thomsen et al., 2010). AREs have been shown to cause degradation through recruitment of ARE-binding proteins (ARE-BPs; Brewer & Ross, 1988; Chen et al., 2001; Chen & Shyu, 1994; Gao, Wilusz, Peltz, & Wilusz, 2001; Mukherjee et al., 2002; Shyu, Belasco, & Greenberg, 1991; Wilson & Treisman, 1988). While the involvement of AREs in maternal mRNA decay has yet to be definitively demonstrated in *Drosophila*, AREs are known to function in maternal mRNA decay in *Xenopus* embryos, together with the Embryonic Deadenylation Element

Binding Protein (Detivaud, Pascreau, Karaiskou, Osborne, & Kubiak, 2003; Graindorge et al., 2008; Paillard et al., 1998; Voeltz & Steitz, 1998).

Finally, it should be noted that not only mRNAs but also maternally produced miRNAs are eliminated during the MZT. For example, Wispy, the noncanonical poly(A) polymerase discussed earlier in the context of its role in oocyte maturation and egg activation, has also been implicated in adenylation and clearance of miRNAs inherited from the mother in *Drosophila* (Lee et al., 2014). Since adenylation fails and miRNAs are stabilized in *wispy*-mutant, activated but unfertilized eggs, this represents an early, maternally encoded decay pathway for miRNAs. Wispy interacts physically with Argonaute 1, but the exact role of this interaction in maternal miRNA recognition and adenylation is not yet clear. Notably, Wispy-dependent adenylation of maternal miRNAs is conserved in sea urchin and mouse embryos (Lee et al., 2014).

4.3.3 Timing of Maternal Transcript Clearance

Whereas some maternal mRNAs begin to be degraded soon after egg activation (Bashirullah et al., 1999), the major effect of the maternally encoded, early-acting decay machineries becomes apparent after the first hour of embryogenesis and is largely complete by 3 h (Tadros et al., 2007; Thomsen et al., 2010). In contrast, action of the zygotically encoded, late-acting machineries begins with the major onset of high-level zygotic transcription, at 2–3 h of embryogenesis. For example, studies of the decay of *string* and *Hsp83* mRNAs indicate that the contribution of zygotic machineries to their decay largely occurs at this time (Bashirullah et al., 1999). Furthermore, targets of the *miR-309* cluster, one such late-acting machinery, are stabilized at this time-point (Bushati et al., 2008).

Interestingly, analysis of transcripts dependent on Brain Tumor for their decay suggests that there might be a second wave of late, zygotic transcription-dependent decay since a subset of Brain Tumor-dependent transcripts begins to degrade at about 3 h into embryogenesis (Laver et al., 2015), approximately 1 h after the effects of the *miR-309* cluster are already observable. Consistent with this idea, analysis of transcripts shown to be degraded exclusively by a zygotic-dependent activity prior to 3 h revealed that many of them continue to be degraded beyond this time-point, and it was observed that additional mRNAs classified as stable during the first 3 h of embryogenesis begin to be degraded at 3–4 h (Thomsen et al., 2010). Together, these data suggest that, while the early-acting decay machineries and a first wave of zygotic machineries act in the syncytial embryo, additional zygotic decay machineries act after the MBT is complete and gastrulation is well underway (Fig. 2). These different phases of decay, therefore, occur in profoundly different environments—before versus after cells and tissues form. It will be interesting to determine whether the Brain Tumor-dependent, late decay machinery (and possibly others) displays any cell or tissue specificity, since this might contribute to the determination of cell fates and tissue morphogenesis.

Finally, it should be noted, in light of the preceding data, that the molecular processes that define the MZT (turnover of maternal gene products and activation of the zygotic genome) continue for several hours after the MBT (cellularization, the first developmental process requiring zygotic gene products) is complete. Thus, whereas the MZT was previously proposed to end at the MBT (see, for example, Tadros & Lipshitz, 2009), this concept may need to be revised in *Drosophila* and, likely, also other animals. In fact, as an extreme example, *Xenopus* has a particularly extended MZT since maternal mRNA clearance occurs after rather than before the MBT (see the chapters "Building the future: Posttranscriptional regulation of cell fate decisions prior to the *Xenopus* midblastula transition" by Sheets and "The *Xenopus* maternal-to-zygotic transition from the perspective of the germline" by Yang, Aguero, and King).

4.3.4 Functions of Maternal Transcript Clearance

Despite the impressive scale of maternal transcript clearance and the quite detailed mechanistic insights gained thus far, the biological functions of this process remain unclear. Given the multifunctional nature of the factors involved, it is difficult to extrapolate the function of maternal mRNA decay from studies of mutant, often very pleiotropic, phenotypes. However, a number of potential functions for maternal transcript clearance can be envisioned. Since we have discussed these in some detail elsewhere (Tadros & Lipshitz, 2009; Walser & Lipshitz, 2011), here we will focus on a few of the most plausible functions, which, we note, are not mutually exclusive.

First, the clearance of specific transcripts may play a passive role in early embryos. For example, transcript degradation may function simply to remove factors that were required during oogenesis but which are no longer necessary in the early embryo.

Second, clearance may play a permissive role, being necessary to allow newly synthesized zygotic factors to carry out their functions. For example, degradation of ubiquitous maternal mRNAs may be required to allow for subsequent, spatially regulated zygotic expression and function of those same transcripts (Tadros & Lipshitz, 2009). In support of such a role, genome-wide analysis of zygotically expressed mRNAs that replace ubiquitous maternally expressed transcripts that are degraded during the MZT, revealed that they display patterned expression at NC14 significantly more often than would be expected for the average gene (De Renzis et al., 2007). One purpose of transcript degradation may, thus, be to permit the proper function of zygotically expressed genes in patterning and morphogenesis of the embryo.

Third, transcript decay may play instructive roles, driving and/or dictating the timing of the progression of developmental events in the early embryo. For instance, gradually reducing the levels of mRNAs encoding cell cycle regulators may contribute to the progressive lengthening of the syncytial NCs prior to the MBT. Consistent with this, in *smaug* mutants, mRNAs encoding a number of cell cycle regulators are not degraded, and embryos fail to slow their nuclear divisions (Benoit et al., 2009). Indeed, an engineered anteroposterior gradient of Smaug produces a gradient in NC length (Benoit et al., 2009). As a second example, the clearance of mRNAs encoding particular transcriptional repressors may be required to allow for the onset of zygotic transcription. Support for this idea comes from the fact that *smaug* mutants fail to activate zygotic transcription (Benoit et al., 2009) and that transcripts bound and regulated by Smaug include a set of more than two dozen that encode transcription factors and chromatin regulators (Chen et al., 2014). We note that, in these examples, Smaug would act as a developmental "timer" that is triggered upon egg activation (Benoit et al., 2009). Furthermore, this hypothetical timer would run independent of the nuclear: cytoplasmic (N:C) ratio since production of Smaug and its action in transcript clearance do not depend on fertilization or the cell cycle (Bashirullah et al., 1999; Benoit et al., 2009; Tadros et al., 2003, 2007). However, any conclusion related to the role of Smaug-mediated decay in controlling the cell cycle and/or ZGA must be tempered by the fact that Smaug also functions as a global translational repressor (Chen et al., 2014). The relative roles of these different modes of Smaug function in cell cycle control and transcriptional regulation in the embryo are unclear. Timers and the N:C ratio in early *Drosophila* embryos are discussed in detail in the chapter "Coordinating cell cycle remodeling with transcriptional activation at the *Drosophila* MBT" by Blythe and Wieschaus.

The fact that mRNA clearance appears to occur in at least three waves, as discussed earlier, with different transcripts being targeted for clearance at different stages, suggests that the precise timing of this decay is important. Each wave of decay may be crucial for progression through particular developmental events. For instance, the early, maternally encoded, wave of decay may be important for slowing the cell cycle and allowing the onset of zygotic transcription; the first wave of zygotic decay may contribute to cellularization; and the subsequent wave of zygotic decay could drive events related to gastrulation and cell differentiation.

In addition to potential functions in regulating the timing of developmental events and in permitting spatially regulated zygotic gene expression, in many cases, mRNA decay contributes to the localization of maternal mRNAs in the embryo. For initially ubiquitous maternal transcripts, this can be accomplished by generalized degradation with local protection in a particular subregion of the cytoplasm. This "degradation-protection" mechanism for mRNA localization has a particularly prominent role in localizing mRNAs to the germplasm and the PGCs at the posterior of the embryo. This mechanism was first discovered for posterior localization of the *Hsp83* mRNA (Bashirullah et al., 1999; Ding, Parkhurst, Halsell, & Lipshitz, 1993; Semotok et al., 2005) and subsequent, genome-scale, studies have found that maternal mRNAs encoded by several hundred genes localize to the germplasm and PGCs via this mechanism (Lecuyer et al., 2007; Siddiqui et al., 2012). Also consistent with a link between mRNA decay and posterior localization, maternally expressed transcripts that are degraded in early embryos—both in general and those degraded by the *trans*-acting factors, Smaug and Brain Tumor—are enriched for posterior localization patterns (Chen et al., 2014; Laver et al., 2015; Thomsen et al., 2010). It will be interesting to determine whether the zygotic decay machinery that acts after cellularization has an analogous role in localizing mRNAs to particular somatic tissues by selectively degrading them in others, just as the early activity localizes transcripts to the PGCs but not the soma.

4.3.5 Comparison of Protein Versus mRNA Levels During Maternal Transcript Clearance

In addition to the drastic changes in mRNA levels observed during the first few hours of embryogenesis, this period is characterized by changes in the levels of maternal proteins. Genome-wide comparison of protein levels in 0–1.5 versus 3–4.5 h embryos (Gouw et al., 2009) demonstrated that many proteins are either up- or downregulated during the course of maternal transcript clearance: of over 2200 quantified proteins, about half changed in abundance, split roughly equally between those that increase versus decrease. Since many of these changes are likely dependent on zygotic factors (since the time-points lie on either side of the major onset of zygotic transcription), it is not clear to what extent the observed changes are caused by translational and posttranslational control versus by changes in mRNA levels that occur as a consequence of transcript decay and synthesis. Comparison of protein levels with those of their corresponding mRNA transcripts has provided conflicting results regarding whether these correlate. In one study, the correlation between changes in protein and mRNA levels from the earlier to later time-points was poor (Gouw et al., 2009), suggesting that changes in mRNA abundance do not necessarily lead to corresponding changes in protein abundance, and that translational and posttranslational controls are therefore critical in determining protein levels during early embryogenesis, as may be the case in many cell types (de Sousa Abreu, Penalva, Marcotte, & Vogel, 2009; Vogel & Marcotte, 2012). However, a second study suggested a stronger correlation, comparing the same protein data with different measurements of transcript levels (Thomsen et al., 2010): mRNAs that are degraded in early embryos were found to be enriched for those encoding proteins whose abundance decreases during the same time period. Additional studies will be required to gain further insight into the degree to which changes in transcript levels are reflected by changes in protein abundance during the MZT, as well as other potential mechanisms for the regulation of protein production and stability during this period.

4.4 The Role of Maternal Factors in Zygotic Genome Activation

4.4.1 Scale, Dynamics, and Mechanisms of Zygotic Genome Activation

In addition to the role played by maternal factors in promoting clearance of maternally supplied mRNAs, they also have a critical role in the second component of the MZT, ZGA. ZGA, like maternal mRNA decay, occurs in waves (Fig. 2). Transcription of a subset of genes initiates gradually, prior to large-scale ZGA, with the first signs of transcription detected as early as NC8 (Chen et al., 2013; De Renzis et al., 2007; Pritchard & Schubiger, 1996), and genetic evidence suggests that transcription may begin even earlier (Ali-Murthy, Lott, Eisen, & Kornberg, 2013). Various genome-wide studies place the number of genes transcribed during this early phase of ZGA between a few dozen and a few hundred: about 60 genes have been detected as upregulated in 1–2 versus 0–1 h embryos (De Renzis et al., 2007), and RNA Polymerase II ChIP-Seq has identified RNA Polymerase II bound and engaged at between 100 and 500 promoters at NC12 or earlier (Blythe & Wieschaus, 2015; Chen et al., 2013). Many of these early expressed genes have functions in sex determination, cellularization, and embryonic patterning (Chen et al., 2013), and they have a tendency to lack introns, a feature likely to be important for production of full-length transcripts during the early, rapid NCs (De Renzis et al., 2007). Subsequent to this initial wave of ZGA, the large-scale onset of zygotic transcription occurs at the MBT. Genome-scale studies suggest that about 1000 genes are zygotically transcribed by NC14, representing about 20% of the detectable transcriptome at that stage (De Renzis et al., 2007; Lecuyer et al., 2007), and RNA Polymerase II has been shown to be recruited, in a paused state, to the promoters of 3000–4000 genes, representing approximately 20–30% of the *Drosophila* genome, at NC13–NC14 (Blythe & Wieschaus, 2015; Chen et al., 2013).

While the mechanisms regulating ZGA are still being elucidated, a number of maternally supplied factors have been shown to have a role. Principle among these is Zelda (Vielfaltig), a transcription factor with an essential and widespread role in the activation of transcription from the earliest zygotic genes (Harrison, Li, Kaplan, Botchan, & Eisen, 2011; Liang et al., 2008; Nien et al., 2011). Zelda appears to act by binding specific sites in DNA referred to as "TAGteam" elements, and increasing chromatin accessibility, thus allowing other transcription factors, such as Bicoid and Dorsal, to bind and activate their target genes (Foo et al., 2014; Li, Harrison, Villalta, Kaplan, & Eisen, 2014; Xu et al., 2014). In addition to Zelda, other maternal factors have also been implicated in ZGA. For example, *Drosophila* STAT (STAT92E) has been shown to act together with Zelda in activating transcription of many early expressed zygotic genes (Tsurumi et al., 2011). Moreover, in contrast to Zelda and STAT92E, other maternal factors have been found to act as transcriptional repressors whose activity may have to be relieved to allow ZGA to occur. For instance, Grainyhead is a transcriptional repressor that competes with Zelda for binding to TAGteam elements. Grainyhead may, therefore, regulate the timing of ZGA by silencing transcription until such time as Zelda protein, which gradually increases in levels during the first few hours of embryogenesis, can outcompete Grainyhead for DNA binding (Harrison, Botchan, & Cline, 2010). A second example of a transcriptional repressor is Tramtrack, which has been shown to repress transcription of the segmentation gene *fushi tarazu* during early cleavage cycles; alterations in Tramtrack levels affect the timing of *fushi tarazu* transcription (Brown, Sonoda, Ueda, Scott, & Wu, 1991; Pritchard & Schubiger, 1996). Finally, BigH1, a linker histone, is required for transcriptional repression in early embryos; in mutants premature ZGA occurs (Perez-Montero,

Carbonell, Moran, Vaquero, & Azorin, 2013). The details of the regulation of ZGA by Zelda and other transcription factors are discussed in greater depth in two other chapters in this book ("Transcriptional activation of the zygotic genome in *Drosophila*" by Harrison and Eisen and "Coordinating cell cycle remodeling with transcriptional activation at the *Drosophila* MBT" by Blythe and Wieschaus).

4.4.2 Interplay Between Posttranscriptional Regulation of Maternal mRNA and Zygotic Genome Activation

In addition to direct roles in regulation of zygotic transcription as transcription factors, it is notable that maternal factors also indirectly influence the onset of zygotic transcription. Particularly interesting are the potential roles of factors also involved in maternal transcript clearance (Fig. 4). For instance, as discussed earlier, the major onset of transcription fails to occur in embryos lacking functional Smaug protein, a phenotype potentially explained in part by Smaug's role in mediating decay and/or translational repression of mRNAs encoding transcriptional repressors (Benoit et al., 2009; Fig. 4A). We previously proposed the mRNA encoding Tramtrack, discussed earlier, as one such candidate (Benoit et al., 2009) because it is dependent on Smaug for degradation (Tadros et al., 2007); it has subsequently been shown that *tramtrack* mRNA is also dependent on Smaug for translational repression (Chen et al., 2014). Interestingly, the mRNA encoding BigH1 is also Smaug-dependent for degradation (Tadros et al., 2007). In addition, in the chapter "Coordinating cell cycle remodeling with transcriptional activation at the *Drosophila* MBT," Blythe and Wieschaus speculate that piRNA pathway proteins may be candidates for negative regulation of ZGA; both the *piwi* and *aubergine* mRNAs are Smaug-dependent for degradation and the former is also dependent on Smaug for translational repression (Chen et al., 2014; Tadros et al., 2007). None of the four mRNAs discussed here has been found to be directly bound by Smaug, however (Chen et al., 2014); thus, effects on their translation and/or stability in *smaug* mutants may be indirect. Nonetheless, we note that, with respect to its potential role as a regulator and timer of ZGA, it is irrelevant whether Smaug acts directly or indirectly on maternal mRNAs encoding transcriptional repressors.

Intriguingly, embryos lacking functional Brain Tumor protein also display defects in ZGA (Laver et al., 2015). However, in contrast to *smaug* mutants, transcriptome-wide analysis of *brain tumor*-mutant embryos has revealed, rather than a failure of transcription, precocious activation of a subset of zygotically transcribed genes (Laver et al., 2015). Interestingly, many

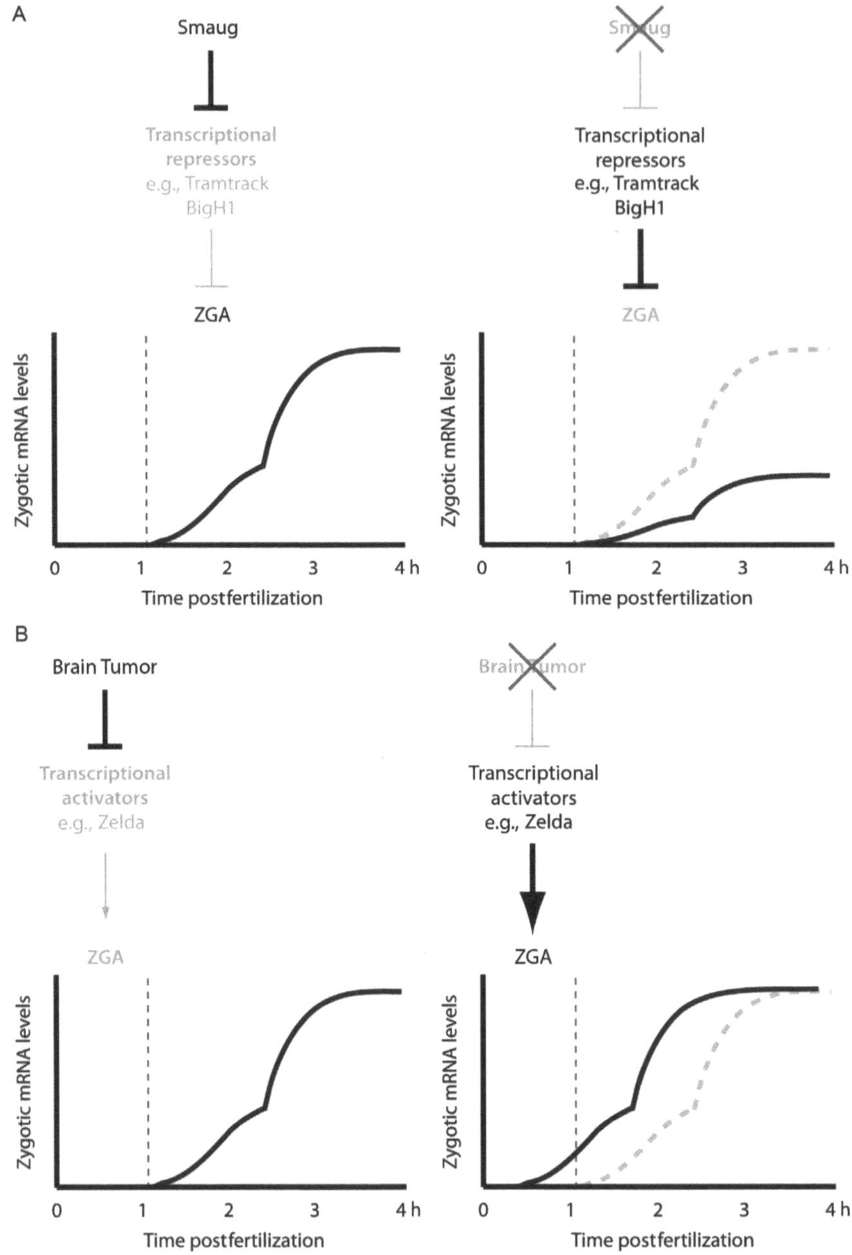

Figure 4 Models for the roles of Smaug and Brain Tumor in regulating the onset of zygotic transcription. (A) Smaug is required for the transcription of the majority of zygotically transcribed genes at 2–3 h postfertilization; in embryos lacking functional Smaug protein, these genes fail to be upregulated (Benoit et al., 2009). While the mechanism underlying the effect of Smaug on zygotic genome activation (ZGA) remains to

of these are direct targets of Zelda, as previously defined by Zelda ChIP-Seq (Harrison et al., 2011). The *zelda* mRNA itself is bound by Brain Tumor, leading to the hypothesis that this RBP normally downregulates the expression of Zelda protein in early embryos to control the timing of the onset of ZGA (Laver et al., 2015; Fig. 4B). This is a particularly intriguing model given that, as described earlier, accumulation of Zelda to a critical level at which it can outcompete Grainyhead for binding to DNA may be important in regulating the timing of the onset of transcription of Zelda's targets (Harrison et al., 2010). However, given the widespread role of Brain Tumor in posttranscriptional regulation, it is possible that, in addition to the *zelda* mRNA, it regulates maternal transcripts encoding other transcription factors required for ZGA, which may contribute to the precocious onset of ZGA observed in *brain tumor*-mutant embryos.

While further investigation will be required to elucidate the exact mechanisms by which Smaug and Brain Tumor regulate zygotic transcription, the defects observed in ZGA in both *smaug* and *brain tumor* mutants highlight the connection between the posttranscriptional regulation of maternal mRNAs and the onset of zygotic transcription.

Figure 4—Cont'd be elucidated, Smaug is required for the posttranscriptional repression—degradation and/or translational repression—of mRNAs encoding transcriptional repressors such as Tramtrack and BigH1. Thus, one possibility is that, in wild-type embryos, Smaug-dependent repression of these factors is required for the onset of zygotic transcription, and the failure of zygotic transcription in embryos lacking functional Smaug is caused by the upregulation of transcriptional repressors. As discussed in the text, we note that Smaug may not directly regulate the mRNAs encoding these repressors. (B) Brain Tumor regulates the timing of the onset of zygotic transcription. In embryos lacking functional Brain Tumor protein, precocious upregulation of a subset of zygotically transcribed genes is observed (Laver et al., 2015). While the mechanism underlying the effect of Brain Tumor on timing ZGA remains to be elucidated, Brain Tumor has been shown to associate with the mRNA encoding the transcription factor Zelda, which has an essential role in mediating the activation of zygotic transcription. Thus, one possibility is that, in wild-type embryos, Brain Tumor represses expression of Zelda, likely via translational repression, whereas in embryos lacking functional Brain Tumor protein, Zelda is upregulated, leading to premature onset of zygotic transcription. In both (A) and (B), the early phase of zygotic transcription and high-level zygotic transcription are indicated by the single, biphasic, blue (black in the print version) curve. The left-hand panels depict events in wild-type embryos, whereas the right-hand panels depict effects observed in embryos lacking functional Smaug or Brain Tumor protein; in the right-hand panels, the normal pattern of transcription in wild-type embryos is indicated by the light blue (light gray in the print version) dotted curves, whereas the observed pattern of expression in embryos lacking Smaug or Brain Tumor is indicated by the dark blue (black in the print version) curves. The normal onset of the early phase of zygotic transcription is indicated by the vertical dotted gray lines.

In addition to the role of posttranscriptional regulation of maternal mRNAs in controlling the onset of ZGA, it should be emphasized that zygotic transcription has an important role in regulating maternal mRNA decay. As discussed earlier, degradation of maternal mRNAs by late-acting machineries requires zygotically transcribed factors. Indeed, just as Smaug and Brain Tumor indirectly regulate the onset of zygotic transcription, Zelda is required indirectly for the degradation of mRNAs by the late-acting machinery, in part via its role in promoting the transcription of the *miR-309* cluster of miRNAs (Fu, Nien, Liang, & Rushlow, 2014; Liang et al., 2008). In addition, it is interesting to speculate that the first and second waves of late, zygotic transcription-dependent decay, described earlier, reflect degradation mediated by factors transcribed via the first and second waves of ZGA, respectively (Fig. 2 shows the relative timing of these different processes). Consistent with this idea, the *miR-309* cluster of miRNAs is known to be transcribed prior to the onset of high-level zygotic transcription at the MBT (Aboobaker, Tomancak, Patel, Rubin, & Lai, 2005; Biemar et al., 2005; Chen et al., 2013), and acts during the first wave of late mRNA decay (Bushati et al., 2008). Further studies will be necessary to understand the precise relationship between the timing of zygotic transcription and mRNA decay via the late-acting machineries.

4.5 Regulation of mRNA Translation and Localization in Early Embryos

In addition to the large-scale and temporally regulated changes in gene expression during egg activation and early embryogenesis discussed thus far, posttranscriptional regulatory mechanisms are essential more generally in the early embryo for regulation of gene expression, particularly given the lack of transcription from the zygotic genome. Genome-wide studies have provided insight into the regulation of both mRNA translation and mRNA localization on a global scale.

With regard to translation, mRNAs encoded by hundreds of genes have been found either to be preferentially translated or to be translationally repressed in 0–2 h embryos (Chen et al., 2014; Qin, Ahn, Speed, & Rubin, 2007), indicating the importance of translational control to the regulation of gene expression at these stages. Interestingly, translationally active mRNAs are enriched for those encoding proteins with roles in transcription and pattern specification (Qin et al., 2007), reflecting the molecular and developmental events underway at this time.

Large-scale analyses of mRNA localization have similarly emphasized the importance of this process during early embryogenesis. An analysis of the localization of mRNAs encoded by ~3000 genes revealed that, strikingly, 70% exhibit subcellular localization in syncytial embryos (Lecuyer et al., 2007). In those cases examined in detail, transcript localization preceded expression of, and correlated with the localization of, the corresponding proteins (Lecuyer et al., 2007), emphasizing the importance of the localization of mRNAs in controlling the spatial expression of their encoded protein products. A particularly prominent localization class utilizes generalized degradation in the bulk cytoplasm together with local protection at the posterior pole to restrict certain maternal mRNAs to the germplasm and PGCs (Section 5).

Regulation of mRNA localization and translation in early embryos has been most studied for factors involved in axis specification. For instance, *bicoid* mRNA, which is localized to the anterior of the oocyte during oogenesis, is anchored there in early embryos by Staufen protein, and locally translated, leading to a gradient of Bicoid protein essential for determining anterior fates (Berleth et al., 1988; Driever & Nusslein-Volhard, 1988a,1988b; St Johnston, Driever, Berleth, Richstein, & Nusslein-Volhard, 1989). A second transcription factor, Hunchback, is also localized to the anterior of the embryo, in part through posttranscriptional regulation of maternal *hunchback* mRNA and in part through Bicoid-dependent transcription of zygotic *hunchback* mRNA in the anterior (Driever & Nusslein-Volhard, 1989; Schroder, Tautz, Seifert, & Jackle, 1988; Tautz, 1988). With respect to the former, in contrast to *bicoid* mRNA, maternal *hunchback* mRNA is present ubiquitously throughout the embryo, and localization of Hunchback protein to the anterior depends on translational repression of maternal *hunchback* mRNA in the posterior by a complex consisting of the RBPs Brain Tumor, Pumilio, and Nanos (Hulskamp, Schroder, Pfeifle, Jackle, & Tautz, 1989; Irish, Lehmann, & Akam, 1989; Murata & Wharton, 1995; Sonoda & Wharton, 1999, 2001; Struhl, 1989).

During both oogenesis and early embryogenesis, mRNA localization and translation are mechanistically coupled to precisely localize protein products. Production of Nanos protein itself provides a well-studied example of this coupling. Nanos protein is an essential determinant of posterior identity, and is localized to the posterior of the oocyte and early embryo (Gavis & Lehmann, 1992, 1994; Wang & Lehmann, 1991). However, in early embryos, only 4% of *nanos* mRNA is present at the posterior and the remaining 96% is distributed throughout the embryo (Bergsten & Gavis, 1999). Posterior localization of Nanos protein therefore depends on a combination of this inefficient mRNA localization and translational repression of unlocalized *nanos* mRNA by Smaug, which binds to SREs in the *nanos* 3′UTR and represses translation by recruiting the proteins Cup and Argonaute 1 (Nelson et al., 2004; Pinder & Smibert, 2013; Smibert et al., 1996, 1999).

5. THE MATERNAL-TO-ZYGOTIC TRANSITION IN PRIMORDIAL GERM CELLS

The germplasm at the posterior of the *Drosophila* oocyte and embryo, which is required for formation and specification of the PGCs, has RNA and protein components that differ from those in the bulk cytoplasm either in relative concentration or by being restricted to the germplasm and, thus, absent from the bulk cytoplasm. Furthermore, the PGCs have been shown to be transcriptionally silent at the time that ZGA initiates in the soma (Hanyu-Nakamura, Sonobe-Nojima, Tanigawa, Lasko, & Nakamura, 2008; Martinho, Kunwar, Casanova, & Lehmann, 2004). RNA Polymerase II C-terminal domain (CTD) Ser2 phosphorylation, a hallmark of transcription elongation, is apparent in the somatic nuclei in 1-h-old embryos but is absent from the PGCs until 3 h of embryogenesis. An important repressor of transcription in the PGCs is the Polar Granule Component protein, which physically and genetically interacts with P-TEFb, the kinase complex responsible for CTD Ser2 phosphorylation, thus preventing P-TEFb recruitment to active promoters (Hanyu-Nakamura et al., 2008).

These data indicate that there are likely to be differences in the MZT in the PGCs compared to the bulk cytoplasm. However, almost all of the genome-wide analyses of the transcriptome and proteome during the *Drosophila* MZT have used whole embryos. Since the PGCs, which bud from the posterior pole 90 min after fertilization, represent less than 1% of the embryo, it is not possible, based on whole-embryo analyses, to assess whether their MZT is of a similar scale and content to that in the soma. However, by sorting GFP-labeled PGCs from early embryos and comparing their transcriptome and proteome to that of the unlabeled, somatic cells, it was possible to define the proteome and transcriptome of these two cell types and to identify proteins and mRNAs highly enriched in the PGCs relative to the somatic cells (and vice versa; Siddiqui et al., 2012).

PGC-specific or -enriched proteins include known components of the germplasm (e.g., Vasa, Oskar, Tudor, Aubergine, Piwi), additional RBPs (e.g., Zn72D), as well as components of the proteasome, DNA replication machinery, and ribosome (Siddiqui et al., 2012). As discussed earlier, certain maternal transcripts are localized to the germplasm and the PGCs; thus, it is not surprising that the transcriptome of PGCs is quite distinct from that of the soma (Siddiqui et al., 2012). Notably, in PGCs, of mRNAs encoded by more than 5600 genes, about a third (over 1700) are enriched relative to the soma while, of the mRNAs in the soma encoded by over 5600 genes, about a quarter (almost 1300) are enriched relative to the PGCs. Whereas soma-enriched mRNAs are enriched for gene ontology (GO) terms related to development, cell fate, and morphogenesis, and include many transcription factors and signaling molecules, PGC-enriched mRNAs are enriched for GO terms related to germ cell fate, the meiotic cell cycle (including stem cell maintenance and proliferation, and DNA damage checkpoints), metabolism, and energy production. Furthermore, the transcriptome of the PGCs does not correlate with the PGC proteome (Siddiqui et al., 2012), as might be expected from the fact that certain maternal proteins are taken up by PGCs but their mRNAs are excluded while other maternal transcripts are loaded into PGCs but kept translationally repressed.

The PGC transcriptome was defined at three time-points (Siddiqui et al., 2012): 1–3 h of embryogenesis, when they bud from and reside at the posterior pole; 3–5 h, when they are internalized within the posterior endodermal pocket; and 5–7 h, when they migrate through the endoderm and come to lie near the mesodermal component of the gonad (Fig. 5). This made it possible to identify the scale, timing, and content of maternal mRNA degradation and ZGA in the PGCs. The scale of mRNA degradation and synthesis in the PGCs is similar to that in the soma, with, in each case, transcripts encoded by 1000–2000 genes (15–35% of the maternal mRNA pool) cleared and those encoded by roughly 1000 genes newly synthesized. However, the nature of the cleared and newly produced transcripts differs substantially between soma and PGCs, presumably reflective of the distinct biological and developmental processes at play in these cell types. Similarly, the timing of the MZT differs in the soma and PGCs with both maternal transcript clearance and ZGA delayed in the PGCs relative to the soma. Nonetheless, despite these differences in content and timing, similar waves of decay and synthesis of transcripts are seen in both cell types: early decay and late decay, early synthesis and later synthesis. Together these results suggest that, while the content and timing of the MZT in the soma and PGCs differ, both cell types may share fundamental regulatory mechanisms.

As discussed earlier, a key player in the somatic MZT is Smaug, which directly regulates maternal transcript clearance and, indirectly, is required for

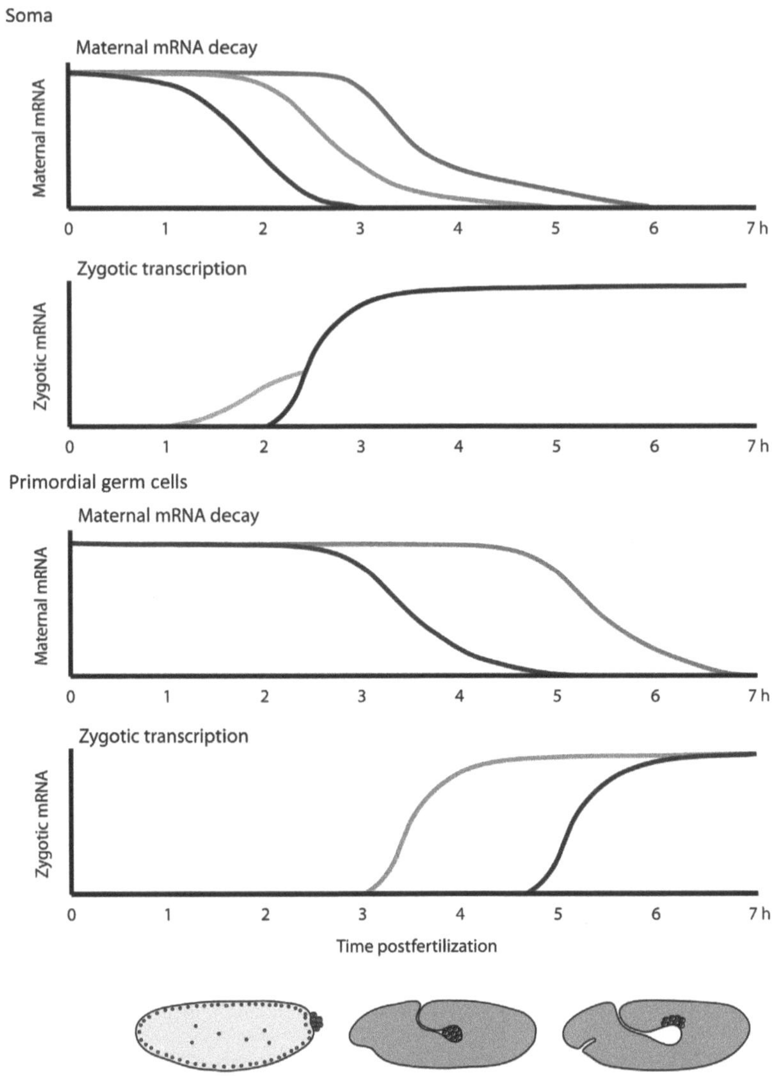

Figure 5 Comparison of the timing of events during the maternal-to-zygotic transition in the soma versus primordial germ cells in *Drosophila*. In the soma (top panels), as depicted in Fig. 2, maternal mRNA degradation occurs via two machineries: early-acting machineries dependent on maternal factors (red, black in the print version) which are triggered by egg activation and whose major effects are first observed at approximately 1 h postfertilization, and late-acting machineries dependent on zygotic factors. Degradation via the late-acting machineries likely occurs in two waves, the first observed 2–3 h postfertilization (light orange, light gray in the print version), and the second observed after 3 h postfertilization (dark orange, gray in the print version). Zygotic transcription in the soma also occurs in two waves: an early phase of zygotic transcription of

ZGA. Whereas Smaug protein rapidly turns over at the somatic MBT (NC14, see Benoit et al., 2009), it persists in the PGCs (Siddiqui et al., 2012; Smibert et al., 1999). To assess a role for Smaug in the PGC MZT, PGCs were purified from *smaug*-mutant embryos at the same three time-points as wild type and the transcriptome of the mutant PGCs defined (Siddiqui et al., 2012). As in the soma, Smaug plays a major role in PGCs in both maternal transcript decay and ZGA albeit to a somewhat lower extent: more than half of the unstable maternal mRNAs in the soma are Smaug-dependent for clearance while a third are Smaug-dependent in the PGCs; over three-quarters of zygotically synthesized transcripts in the soma are Smaug-dependent, while about a third are Smaug-dependent in the PGCs. Smaug-dependent unstable maternal mRNAs in both the soma and PGCs are enriched for SREs, while Smaug-dependent newly synthesized transcripts in both cell types are depleted for SREs (Chen et al., 2014; Siddiqui et al., 2012; Tadros et al., 2007), the latter consistent with the fact that Smaug protein is still present in both cell types at the time of ZGA.

Apart from SREs, binding sites for Pumilio and for ARE-BPs are enriched in degraded PGC transcripts (Siddiqui et al., 2012), suggestive of a possible role for these factors in both the PGC and somatic MZT. miRNA target sites are also enriched in degraded PGC transcripts (Siddiqui et al., 2012), including sites for *miR-309*, which is discussed earlier as a component of a late, ZGA-dependent decay machinery in the soma. However, it is known that *miR-309* production in the soma occurs after the PGCs bud from the posterior (Bushati et al., 2008) and that the

Figure 5—Cont'd a subset of genes begins at approximately 1 h postfertilization (light blue, light gray in the print version), and high-level zygotic genome activation occurs 2–3 h postfertilization (dark blue, black in the print version). In primordial germ cells (bottom panels), both maternal mRNA degradation and zygotic transcription also occur in multiple waves, and, while the transcripts affected differ substantially between the soma and the primordial germ cells, the number of transcripts degraded or transcribed is similar (Siddiqui et al., 2012). However, the timing of both of these processes is delayed in the primordial germ cells relative to the soma. In the primordial germ cells, the earliest observed degradation of maternal mRNAs is seen at 3–5 h postfertilization, with additional degradation observed 5–7 h postfertilization (Siddiqui et al., 2012). The relative contribution of maternal versus zygotic factors to these waves of decay in primordial germ cells is unknown. Zygotic transcription in primordial germ cells is silent until approximately 3 h postfertilization, after which time a first wave of transcription is observed, with additional transcription occurring 5–7 h postfertilization (Siddiqui et al., 2012). Therefore, in primordial germ cells, both maternal mRNA degradation and zygotic transcription are delayed by about 2 h relative to the soma.

miR-309 cluster is not transcribed in the PGCs in embryos that have recently undergone the MBT (Aboobaker et al., 2005). Whether the cluster is transcribed during the late phase of the PGC MZT is not known. In addition, a potential role for Brain Tumor in degrading transcripts in the PGCs has not been investigated, although it is interesting to note that mRNAs dependent on Brain Tumor for their decay are enriched for those localized to PGCs (Laver et al., 2015). As mentioned earlier, this might reflect a role for Brain Tumor in localizing transcripts to the posterior via degradation of transcripts in the bulk cytoplasm, but could also indicate a role for Brain Tumor in promoting decay of these transcripts within the PGCs.

6. CONCLUSIONS AND FUTURE PROSPECTS

Maternally supplied mRNAs and proteins direct the development of the *Drosophila* oocyte and early embryo, including critical processes such as oocyte maturation, egg activation, the rapid early embryonic NCs, clearance of a subset of maternal RNAs and proteins, and activation of transcription from the zygotic genome. Decades of study in *Drosophila* have led to a detailed understanding of the progression of these events and many key factors involved in their regulation. More recently, new technologies have allowed genome-scale descriptions of the maternally supplied transcriptome and proteome of oocytes and embryos—in the latter case separately for the soma and the PGCs—and the profound changes these undergo during the MZT. These new datasets have led to the elucidation of distinct temporal phases of both maternal mRNA decay and ZGA, as well as the identification of novel regulators of these processes. A complete understanding of the functional significance of the MZT will require further elucidation of how these global changes in gene expression are controlled, and how they, in turn, direct the multiple molecular, cellular, and developmental processes that underlie early embryonic development.

ACKNOWLEDGMENTS

Our research on RNA-binding proteins, posttranscriptional regulation, and the maternal-to-zygotic transition is supported by an operating grant from the Canadian Institutes for Health Research (MOP-14409 to H.D.L.) and Discovery Grants from the Natural Sciences and Engineering Research Council of Canada (RGPIN-435985 to C.A.S. and RGPIN-201 to H.D.L.). J.D.L. and A.J.M. were supported in part by Ontario Graduate Scholarships and University of Toronto Open Fellowships.

REFERENCES

Aboobaker, A. A., Tomancak, P., Patel, N., Rubin, G. M., & Lai, E. C. (2005). Drosophila microRNAs exhibit diverse spatial expression patterns during embryonic development. *Proceedings of the National Academy of Sciences of the United States of America, 102*(50), 18017–18022. http://dx.doi.org/10.1073/pnas.0508823102.

Abrieu, A., Brassac, T., Galas, S., Fisher, D., Labbé, J. C., & Dorée, M. (1998). The Polo-like kinase Plx1 is a component of the MPF amplification loop at the G2/M-phase transition of the cell cycle in Xenopus eggs. *Journal of Cell Science, 111*, 1751–1757.

Ali-Murthy, Z., Lott, S. E., Eisen, M. B., & Kornberg, T. B. (2013). An essential role for zygotic expression in the pre-cellular Drosophila embryo. *PLoS Genetics, 9*(4), e1003428. http://dx.doi.org/10.1371/journal.pgen.1003428. PGENETICS-D-13-00016 [pii].

Alphey, L., Jimenez, J., White-Cooper, H., Dawson, I., Nurse, P., & Glover, D. M. (1992). twine, a cdc25 homolog that functions in the male and female germline of Drosophila. *Cell, 69*(6), 977–988. 0092-8674(92)90616-K [pii].

Archambault, V., Zhao, X., White-Cooper, H., Carpenter, A. T., & Glover, D. M. (2007). Mutations in Drosophila Greatwall/Scant reveal its roles in mitosis and meiosis and interdependence with Polo kinase. *PLoS Genetics, 3*(11), e200. http://dx.doi.org/10.1371/journal.pgen.0030200. 07-PLGE-RA-0454 [pii].

Baez, M. V., & Boccaccio, G. L. (2005). Mammalian Smaug is a translational repressor that forms cytoplasmic foci similar to stress granules. *The Journal of Biological Chemistry, 280*(52), 43131–43140. http://dx.doi.org/10.1074/jbc.M508374200.

Barisic-Jager, E., Krecioch, I., Hosiner, S., Antic, S., & Dorner, S. (2013). HPat a decapping activator interacting with the miRNA effector complex. *PLoS One, 8*(8), e71860. http://dx.doi.org/10.1371/journal.pone.0071860.

Bashirullah, A., Cooperstock, R. L., & Lipshitz, H. D. (2001). Spatial and temporal control of RNA stability. *Proceedings of the National Academy of Sciences of the United States of America, 98*(13), 7025–7028. http://dx.doi.org/10.1073/pnas.111145698.

Bashirullah, A., Halsell, S. R., Cooperstock, R. L., Kloc, M., Karaiskakis, A., Fisher, W. W., et al. (1999). Joint action of two RNA degradation pathways controls the timing of maternal transcript elimination at the midblastula transition in Drosophila melanogaster. *The EMBO Journal, 18*(9), 2610–2620. http://dx.doi.org/10.1093/emboj/18.9.2610.

Benoit, B., He, C. H., Zhang, F., Votruba, S. M., Tadros, W., Westwood, J. T., et al. (2009). An essential role for the RNA-binding protein Smaug during the Drosophila maternal-to-zygotic transition. *Development, 136*(6), 923–932. http://dx.doi.org/10.1242/dev.031815. 136/6/923 [pii].

Benoit, B., Mitou, G., Chartier, A., Temme, C., Zaessinger, S., Wahle, E., et al. (2005). An essential cytoplasmic function for the nuclear poly(A) binding protein, PABP2, in poly(A) tail length control and early development in Drosophila. *Developmental Cell, 9*(4), 511–522. http://dx.doi.org/10.1016/j.devcel.2005.09.002. S1534-5807(05)00336-9 [pii].

Benoit, P., Papin, C., Kwak, J. E., Wickens, M., & Simonelig, M. (2008). PAP- and GLD-2-type poly(A) polymerases are required sequentially in cytoplasmic polyadenylation and oogenesis in Drosophila. *Development, 135*(11), 1969–1979. http://dx.doi.org/10.1242/dev.021444. dev.021444 [pii].

Bergsten, S. E., & Gavis, E. R. (1999). Role for mRNA localization in translational activation but not spatial restriction of nanos RNA. *Development, 126*(4), 659–669.

Berleth, T., Burri, M., Thoma, G., Bopp, D., Richstein, S., Frigerio, G., et al. (1988). The role of localization of bicoid RNA in organizing the anterior pattern of the Drosophila embryo. *The EMBO Journal*, 7(6), 1749–1756.

Biemar, F., Zinzen, R., Ronshaugen, M., Sementchenko, V., Manak, J. R., & Levine, M. S. (2005). Spatial regulation of microRNA gene expression in the Drosophila embryo. *Proceedings of the National Academy of Sciences of the United States of America, 102*(44), 15907–15911. http://dx.doi.org/10.1073/pnas.0507817102. 0507817102 [pii].

Blythe, S. A., & Wieschaus, E. F. (2015). Zygotic genome activation triggers the DNA replication checkpoint at the midblastula transition. *Cell, 160*(6), 1169–1181. http://dx.doi.org/10.1016/j.cell.2015.01.050.

Braun, J. E., Huntzinger, E., Fauser, M., & Izaurralde, E. (2011). GW182 proteins directly recruit cytoplasmic deadenylase complexes to miRNA targets. *Molecular Cell, 44*(1), 120–133. http://dx.doi.org/10.1016/j.molcel.2011.09.007.

Brewer, G., & Ross, J. (1988). Poly(A) shortening and degradation of the 3' A+U-rich sequences of human c-myc mRNA in a cell-free system. *Molecular and Cellular Biology, 8*(4), 1697–1708.

Brown, J. L., Sonoda, S., Ueda, H., Scott, M. P., & Wu, C. (1991). Repression of the Drosophila fushi tarazu (ftz) segmentation gene. *The EMBO Journal, 10*(3), 665–674.

Bushati, N., Stark, A., Brennecke, J., & Cohen, S. M. (2008). Temporal reciprocity of miRNAs and their targets during the maternal-to-zygotic transition in Drosophila. *Current Biology, 18*, 501–506. http://dx.doi.org/10.1016/j.cub.2008.02.081. S0960-9822(08)00365-5 [pii].

Chase, D., Serafinas, C., Ashcroft, N., Kosinski, M., Longo, D., Ferris, D. K., et al. (2000). The polo-like kinase PLK-1 is required for nuclear envelope breakdown and the completion of meiosis in Caenorhabditis elegans. *Genesis, 26*(1), 26–41. http://dx.doi.org/10.1002/(SICI)1526-968X(200001)26:1<26::AID-GENE6>3.0.CO;2-O. [pii].

Chekulaeva, M., Mathys, H., Zipprich, J. T., Attig, J., Colic, M., Parker, R., et al. (2011). miRNA repression involves GW182-mediated recruitment of CCR4-NOT through conserved W-containing motifs. *Nature Structural & Molecular Biology, 18*(11), 1218–1226. http://dx.doi.org/10.1038/nsmb.2166.

Chen, L., Dumelie, J. G., Li, X., Cheng, M. H., Yang, Z., Laver, J. D., et al. (2014). Global regulation of mRNA translation and stability in the early Drosophila embryo by the Smaug RNA-binding protein. *Genome Biology, 15*(1), R4. http://dx.doi.org/10.1186/gb-2014-15-1-r4. gb-2014-15-1-r4 [pii].

Chen, C. Y., Gherzi, R., Ong, S. E., Chan, E. L., Raijmakers, R., Pruijn, G. J., et al. (2001). AU binding proteins recruit the exosome to degrade ARE-containing mRNAs. *Cell, 107*, 451–464. S0092-8674(01)00578-5 [pii].

Chen, K., Johnston, J., Shao, W., Meier, S., Staber, C., & Zeitlinger, J. (2013). A global change in RNA polymerase II pausing during the Drosophila midblastula transition. *eLife, 2*, e00861. http://dx.doi.org/10.7554/eLife.00861. 00861 [pii].

Chen, C. Y., & Shyu, A. B. (1994). Selective degradation of early-response-gene mRNAs: Functional analyses of sequence features of the AU-rich elements. *Molecular and Cellular Biology, 14*(12), 8471–8482.

Courtot, C., Fankhauser, C., Simanis, V., & Lehner, C. F. (1992). The Drosophila cdc25 homolog twine is required for meiosis. *Development, 116*(2), 405–416.

Cui, J., Sartain, C. V., Pleiss, J. A., & Wolfner, M. F. (2013). Cytoplasmic polyadenylation is a major mRNA regulator during oogenesis and egg activation in Drosophila. *Developmental Biology, 383*(1), 121–131. http://dx.doi.org/10.1016/j.ydbio.2013.08.013. S0012-1606(13)00438-7 [pii].

Dahanukar, A., Walker, J. A., & Wharton, R. P. (1999). Smaug, a novel RNA-binding protein that operates a translational switch in Drosophila. *Molecular Cell, 4*(2), 209–218.

Dahanukar, A., & Wharton, R. P. (1996). The Nanos gradient in Drosophila embryos is generated by translational regulation. *Genes & Development, 10*(20), 2610–2620.

De Renzis, S., Elemento, O., Tavazoie, S., & Wieschaus, E. F. (2007). Unmasking activation of the zygotic genome using chromosomal deletions in the Drosophila embryo. *PLoS*

Biology, *5*(5), e117. http://dx.doi.org/10.1371/journal.pbio.0050117. 06-PLBI-RA-1961R2 [pii].

Dernburg, A. F., Sedat, J. W., & Hawley, R. S. (1996). Direct evidence of a role for heterochromatin in meiotic chromosome segregation. *Cell*, *86*(1), 135–146. S0092-8674(00)80084-7 [pii].

deSousa Abreu, R., Penalva, L. O., Marcotte, E. M., & Vogel, C. (2009). Global signatures of protein and mRNA expression levels. *Molecular bioSystems*, *5*(12), 1512–1526. http://dx.doi.org/10.1039/b908315d.

Detivaud, L., Pascreau, G., Karaiskou, A., Osborne, H. B., & Kubiak, J. Z. (2003). Regulation of EDEN-dependent deadenylation of Aurora A/Eg2-derived mRNA via phosphorylation and dephosphorylation in Xenopus laevis egg extracts. *Journal of Cell Science*, *116*(Pt. 13), 2697–2705. http://dx.doi.org/10.1242/jcs.00477. jcs.00477 [pii].

Ding, D., Parkhurst, S. M., Halsell, S. R., & Lipshitz, H. D. (1993). Dynamic Hsp83 RNA localization during Drosophila oogenesis and embryogenesis. *Molecular and Cellular Biology*, *13*(6), 3773–3781.

Driever, W., & Nusslein-Volhard, C. (1988a). The bicoid protein determines position in the Drosophila embryo in a concentration-dependent manner. *Cell*, *54*(1), 95–104. 0092-8674(88)90183-3 [pii].

Driever, W., & Nusslein-Volhard, C. (1988b). A gradient of bicoid protein in Drosophila embryos. *Cell*, *54*(1), 83–93. 0092-8674(88)90182-1 [pii].

Driever, W., & Nusslein-Volhard, C. (1989). The bicoid protein is a positive regulator of hunchback transcription in the early Drosophila embryo. *Nature*, *337*(6203), 138–143. http://dx.doi.org/10.1038/337138a0.

Fabian, M. R., Cieplak, M. K., Frank, F., Morita, M., Green, J., Srikumar, T., et al. (2011). miRNA-mediated deadenylation is orchestrated by GW182 through two conserved motifs that interact with CCR4-NOT. *Nature Structural & Molecular Biology*, *18*(11), 1211–1217. http://dx.doi.org/10.1038/nsmb.2149.

Fenger, D. D., Carminati, J. L., Burney-Sigman, D. L., Kashevsky, H., Dines, J. L., Elfring, L. K., et al. (2000). PAN GU: A protein kinase that inhibits S phase and promotes mitosis in early Drosophila development. *Development*, *127*(22), 4763–4774.

Foo, S. M., Sun, Y., Lim, B., Ziukaite, R., O'Brien, K., Nien, C. Y., et al. (2014). Zelda potentiates morphogen activity by increasing chromatin accessibility. *Current Biology*, *24*(12), 1341–1346. http://dx.doi.org/10.1016/j.cub.2014.04.032. S0960-9822(14)00470-9 [pii].

Freeman, M., & Glover, D. M. (1987). The gnu mutation of Drosophila causes inappropriate DNA synthesis in unfertilized and fertilized eggs. *Genes & Development*, *1*(9), 924–930.

Freeman, M., Nusslein-Volhard, C., & Glover, D. M. (1986). The dissociation of nuclear and centrosomal division in gnu, a mutation causing giant nuclei in Drosophila. *Cell*, *46*(3), 457–468. 0092-8674(86)90666-5 [pii].

Fu, S., Nien, C. Y., Liang, H. L., & Rushlow, C. (2014). Co-activation of microRNAs by Zelda is essential for early Drosophila development. *Development*, *141*(10), 2108–2118. http://dx.doi.org/10.1242/dev.108118. dev.108118 [pii].

Gao, M., Wilusz, C. J., Peltz, S. W., & Wilusz, J. (2001). A novel mRNA-decapping activity in HeLa cytoplasmic extracts is regulated by AU-rich elements. *The EMBO Journal*, *20*(5), 1134–1143. http://dx.doi.org/10.1093/emboj/20.5.1134.

Gavis, E. R., & Lehmann, R. (1992). Localization of nanos RNA controls embryonic polarity. *Cell*, *71*(2), 301–313.

Gavis, E. R., & Lehmann, R. (1994). Translational regulation of nanos by RNA localization. *Nature*, *369*(6478), 315–318. http://dx.doi.org/10.1038/369315a0.

Gerber, A. P., Luschnig, S., Krasnow, M. A., Brown, P. O., & Herschlag, D. (2006). Genome-wide identification of mRNAs associated with the translational regulator PUMILIO in Drosophila melanogaster. *Proceedings of the National Academy of Sciences of the United States of America*, *103*(12), 4487–4492. http://dx.doi.org/10.1073/pnas.0509260103. 0509260103 [pii].

Giraldez, A. J., Mishima, Y., Rihel, J., Grocock, R. J., Van Dongen, S., Inoue, K., et al. (2006). Zebrafish MiR-430 promotes deadenylation and clearance of maternal mRNAs. *Science*, *312*(5770), 75–79. http://dx.doi.org/10.1126/science.1122689.

Gouw, J. W., Pinkse, M. W., Vos, H. R., Moshkin, Y., Verrijzer, C. P., Heck, A. J., et al. (2009). In vivo stable isotope labeling of fruit flies reveals post-transcriptional regulation in the maternal-to-zygotic transition. *Molecular & Cellular Proteomics: MCP*, *8*(7), 1566–1578. http://dx.doi.org/10.1074/mcp.M900114-MCP200. M900114-MCP200 [pii].

Graindorge, A., Le Tonqueze, O., Thuret, R., Pollet, N., Osborne, H. B., & Audic, Y. (2008). Identification of CUG-BP1/EDEN-BP target mRNAs in Xenopus tropicalis. *Nucleic Acids Research*, *36*(6), 1861–1870. http://dx.doi.org/10.1093/nar/gkn031. gkn031 [pii].

Hanyu-Nakamura, K., Sonobe-Nojima, H., Tanigawa, A., Lasko, P., & Nakamura, A. (2008). Drosophila Pgc protein inhibits P-TEFb recruitment to chromatin in primordial germ cells. *Nature*, *451*(7179), 730–733. http://dx.doi.org/10.1038/nature06498.

Harrison, M. M., Botchan, M. R., & Cline, T. W. (2010). Grainyhead and Zelda compete for binding to the promoters of the earliest-expressed Drosophila genes. *Developmental Biology*, *345*(2), 248–255. http://dx.doi.org/10.1016/j.ydbio.2010.06.026. S0012-1606(10)00846-8 [pii].

Harrison, M. M., Li, X. Y., Kaplan, T., Botchan, M. R., & Eisen, M. B. (2011). Zelda binding in the early Drosophila melanogaster embryo marks regions subsequently activated at the maternal-to-zygotic transition. *PLoS Genetics*, 7(10), e1002266. http://dx.doi.org/10.1371/journal.pgen.1002266. PGENETICS-D-11-01423 [pii].

Hawley, R. S., Irick, H., Zitron, A. E., Haddox, D. A., Lohe, A., New, C., et al. (1992). There are two mechanisms of achiasmate segregation in Drosophila females, one of which requires heterochromatic homology. *Developmental Genetics*, *13*(6), 440–467. http://dx.doi.org/10.1002/dvg.1020130608.

Heifetz, Y., Yu, J., & Wolfner, M. F. (2001). Ovulation triggers activation of Drosophila oocytes. *Developmental Biology*, *234*(2), 416–424. http://dx.doi.org/10.1006/dbio.2001.0246. S0012-1606(01)90246-5 [pii].

Horner, V. L., Czank, A., Jang, J. K., Singh, N., Williams, B. C., Puro, J., et al. (2006). The Drosophila calcipressin sarah is required for several aspects of egg activation. *Current Biology*, *16*(14), 1441–1446. http://dx.doi.org/10.1016/j.cub.2006.06.024. S0960-9822(06)01751-9 [pii].

Horner, V. L., & Wolfner, M. F. (2008a). Mechanical stimulation by osmotic and hydrostatic pressure activates Drosophila oocytes in vitro in a calcium-dependent manner. *Developmental Biology*, *316*, 100–109. http://dx.doi.org/10.1016/j.ydbio.2008.01.014. S0012-1606(08)00027-4 [pii].

Horner, V. L., & Wolfner, M. F. (2008b). Transitioning from egg to embryo: Triggers and mechanisms of egg activation. *Developmental Dynamics*, *237*(3), 527–544. http://dx.doi.org/10.1002/dvdy.21454.

Hughes, S. E., Gilliland, W. D., Cotitta, J. L., Takeo, S., Collins, K. A., & Hawley, R. S. (2009). Heterochromatic threads connect oscillating chromosomes during prometaphase I in Drosophila oocytes. *PLoS Genetics*, *5*(1), e1000348. http://dx.doi.org/10.1371/journal.pgen.1000348.

Hulskamp, M., Schroder, C., Pfeifle, C., Jackle, H., & Tautz, D. (1989). Posterior segmentation of the Drosophila embryo in the absence of a maternal posterior organizer gene. *Nature*, *338*(6217), 629–632. http://dx.doi.org/10.1038/338629a0.

Irish, V., Lehmann, R., & Akam, M. (1989). The Drosophila posterior-group gene nanos functions by repressing hunchback activity. *Nature*, *338*(6217), 646–648. http://dx.doi.org/10.1038/338646a0.

Jang, J. K., Messina, L., Erdman, M. B., Arbel, T., & Hawley, R. S. (1995). Induction of metaphase arrest in Drosophila oocytes by chiasma-based kinetochore tension. *Science, 268*(5219), 1917–1919.

Kaneuchi, T., Sartain, C. V., Takeo, S., Horner, V. L., Buehner, N. A., Aigaki, T., et al. (2015). Calcium waves occur as Drosophila oocytes activate. *Proceedings of the National Academy of Sciences of the United States of America, 112*(3), 791–796. http://dx.doi.org/10.1073/pnas.1420589112. 1420589112 [pii].

Krauchunas, A. R., Horner, V. L., & Wolfner, M. F. (2012). Protein phosphorylation changes reveal new candidates in the regulation of egg activation and early embryogenesis in D. melanogaster. *Developmental Biology, 370*(1), 125–134. http://dx.doi.org/10.1016/j.ydbio.2012.07.024. S0012-1606(12)00408-3 [pii].

Krauchunas, A. R., Sackton, K. L., & Wolfner, M. F. (2013). Phospho-regulation pathways during egg activation in Drosophila melanogaster. *Genetics, 195*(1), 171–180. http://dx.doi.org/10.1534/genetics.113.150110. genetics.113.150110 [pii].

Kronja, I., Whitfield, Z. J., Yuan, B., Dzeyk, K., Kirkpatrick, J., Krijgsveld, J., et al. (2014). Quantitative proteomics reveals the dynamics of protein changes during Drosophila oocyte maturation and the oocyte-to-embryo transition. *Proceedings of the National Academy of Sciences of the United States of America, 111*(45), 16023–16028. http://dx.doi.org/10.1073/pnas.1418657111. 1418657111 [pii].

Kronja, I., Yuan, B., Eichhorn, S. W., Dzeyk, K., Krijgsveld, J., Bartel, D. P., et al. (2014). Widespread changes in the posttranscriptional landscape at the Drosophila oocyte-to-embryo transition. *Cell Reports*, 7(5), 1495–1508. http://dx.doi.org/10.1016/j.celrep.2014.05.002.

Lasko, P. (2012). mRNA localization and translational control in Drosophila oogenesis. *Cold Spring Harbor Perspectives in Biology. 4*(10). http://dx.doi.org/10.1101/cshperspect.a012294.

Laver, J. D., Li, X., Ray, D., Cook, K. B., Hahn, N. A., Nabeel-Shah, S., et al. (2015). Brain tumor is a sequence-specific RNA-binding protein that directs maternal mRNA clearance during the Drosophila maternal-to-zygotic transition. *Genome Biology, 16*(1), 94. http://dx.doi.org/10.1186/s13059-015-0659-4. [pii].

Lecuyer, E., Yoshida, H., Parthasarathy, N., Alm, C., Babak, T., Cerovina, T., et al. (2007). Global analysis of mRNA localization reveals a prominent role in organizing cellular architecture and function. *Cell, 131*(1), 174–187. http://dx.doi.org/10.1016/j.cell.2007.08.003. S0092-8674(07)01022-7 [pii].

Lee, M., Choi, Y., Kim, K., Jin, H., Lim, J., Nguyen, T. A., et al. (2014). Adenylation of maternally inherited microRNAs by Wispy. *Molecular Cell, 56*(5), 696–707. http://dx.doi.org/10.1016/j.molcel.2014.10.011.

Li, X. Y., Harrison, M. M., Villalta, J. E., Kaplan, T., & Eisen, M. B. (2014). Establishment of regions of genomic activity during the Drosophila maternal to zygotic transition. *eLife, 3*, e03737. http://dx.doi.org/10.7554/eLife.03737.

Liang, H. L., Nien, C. Y., Liu, H. Y., Metzstein, M. M., Kirov, N., & Rushlow, C. (2008). The zinc-finger protein Zelda is a key activator of the early zygotic genome in Drosophila. *Nature, 456*(7220), 400–403. http://dx.doi.org/10.1038/nature07388. nature07388 [pii].

Loedige, I., Gaidatzis, D., Sack, R., Meister, G., & Filipowicz, W. (2013). The mammalian TRIM-NHL protein TRIM71/LIN-41 is a repressor of mRNA function. *Nucleic Acids Research, 41*(1), 518–532. http://dx.doi.org/10.1093/nar/gks1032.

Loedige, I., Stotz, M., Qamar, S., Kramer, K., Hennig, J., Schubert, T., et al. (2014). The NHL domain of BRAT is an RNA-binding domain that directly contacts the hunchback mRNA for regulation. *Genes & Development, 28*(7), 749–764. http://dx.doi.org/10.1101/gad.236513.113. 28/7/749 [pii].

Lund, E., Liu, M., Hartley, R. S., Sheets, M. D., & Dahlberg, J. E. (2009). Deadenylation of maternal mRNAs mediated by miR-427 in Xenopus laevis embryos. *RNA, 15*(12), 2351–2363. http://dx.doi.org/10.1261/rna.1882009.

Martinho, R. G., Kunwar, P. S., Casanova, J., & Lehmann, R. (2004). A noncoding RNA is required for the repression of RNApolII-dependent transcription in primordial germ cells. *Current Biology*, *14*(2), 159–165.

McKim, K. S., Jang, J. K., Theurkauf, W. E., & Hawley, R. S. (1993). Mechanical basis of meiotic metaphase arrest. *Nature*, *362*(6418), 364–366. http://dx.doi.org/10.1038/362364a0.

Mukherjee, D., Gao, M., O'Connor, J. P., Raijmakers, R., Pruijn, G., Lutz, C. S., et al. (2002). The mammalian exosome mediates the efficient degradation of mRNAs that contain AU-rich elements. *The EMBO Journal*, *21*(1–2), 165–174. http://dx.doi.org/10.1093/emboj/21.1.165.

Murata, Y., & Wharton, R. P. (1995). Binding of pumilio to maternal hunchback mRNA is required for posterior patterning in Drosophila embryos. *Cell*, *80*(5), 747–756. 0092-8674(95)90353-4 [pii].

Nelson, M. R., Leidal, A. M., & Smibert, C. A. (2004). Drosophila Cup is an eIF4E-binding protein that functions in Smaug-mediated translational repression. *The EMBO Journal*, *23*(1), 150–159. http://dx.doi.org/10.1038/sj.emboj.7600026. 7600026 [pii].

Nien, C. Y., Liang, H. L., Butcher, S., Sun, Y., Fu, S., Gocha, T., et al. (2011). Temporal coordination of gene networks by Zelda in the early Drosophila embryo. *PLoS Genetics*, 7(10), e1002339. http://dx.doi.org/10.1371/journal.pgen.1002339. PGENETICS-D-11-01777 [pii].

Nishihara, T., Zekri, L., Braun, J. E., & Izaurralde, E. (2013). miRISC recruits decapping factors to miRNA targets to enhance their degradation. *Nucleic Acids Research*, *41*(18), 8692–8705. http://dx.doi.org/10.1093/nar/gkt619.

Page, S. L., & Hawley, R. S. (2003). Chromosome choreography: The meiotic ballet. *Science*, *301*(5634), 785–789. http://dx.doi.org/10.1126/science.1086605.

Paillard, L., Omilli, F., Legagneux, V., Bassez, T., Maniey, D., & Osborne, H. B. (1998). EDEN and EDEN-BP, a cis element and an associated factor that mediate sequence-specific mRNA deadenylation in Xenopus embryos. *The EMBO Journal*, *17*(1), 278–287. http://dx.doi.org/10.1093/emboj/17.1.278.

Perez-Montero, S., Carbonell, A., Moran, T., Vaquero, A., & Azorin, F. (2013). The embryonic linker histone H1 variant of Drosophila, dBigH1, regulates zygotic genome activation. *Developmental Cell*, *26*(6), 578–590. http://dx.doi.org/10.1016/j.devcel.2013.08.011.

Pesin, J. A., & Orr-Weaver, T. L. (2007). Developmental role and regulation of cortex, a meiosis-specific anaphase-promoting complex/cyclosome activator. *PLoS Genetics*, *3*(11), e202. http://dx.doi.org/10.1371/journal.pgen.0030202. 07-PLGE-RA-0630 [pii].

Pinder, B. D., & Smibert, C. A. (2013). microRNA-independent recruitment of Argonaute 1 to nanos mRNA through the Smaug RNA-binding protein. *EMBO Reports*, *14*(1), 80–86. http://dx.doi.org/10.1038/embor.2012.192. embor2012192 [pii].

Pritchard, D. K., & Schubiger, G. (1996). Activation of transcription in Drosophila embryos is a gradual process mediated by the nucleocytoplasmic ratio. *Genes & Development*, *10*(9), 1131–1142.

Qin, X., Ahn, S., Speed, T. P., & Rubin, G. M. (2007). Global analyses of mRNA translational control during early Drosophila embryogenesis. *Genome Biology*, *8*(4), R63. http://dx.doi.org/10.1186/gb-2007-8-4-r63. gb-2007-8-4-r63 [pii].

Roshak, A. K., Capper, E. A., Imburgia, C., Fornwald, J., Scott, G., & Marshall, L. A. (2000). The human polo-like kinase, PLK, regulates cdc2/cyclin B through phosphorylation and activation of the cdc25C phosphatase. *Cellular Signalling*, *12*(6), 405–411. S0898-6568(00)00080-2 [pii].

Schroder, C., Tautz, D., Seifert, E., & Jackle, H. (1988). Differential regulation of the two transcripts from the Drosophila gap segmentation gene hunchback. *The EMBO Journal*, 7(9), 2881–2887.

Semotok, J. L., Cooperstock, R. L., Pinder, B. D., Vari, H. K., Lipshitz, H. D., & Smibert, C. A. (2005). Smaug recruits the CCR4/POP2/NOT deadenylase complex to trigger maternal transcript localization in the early Drosophila embryo. *Current Biology*, *15*(4), 284–294. http://dx.doi.org/10.1016/j.cub.2005.01.048. S0960982205001119 [pii].

Semotok, J. L., Luo, H., Cooperstock, R. L., Karaiskakis, A., Vari, H. K., Smibert, C. A., et al. (2008). Drosophila maternal Hsp83 mRNA destabilization is directed by multiple SMAUG recognition elements in the open reading frame. *Molecular and Cellular Biology*, *28*(22), 6757–6772. http://dx.doi.org/10.1128/MCB.00037-08.

Shamanski, F. L., & Orr-Weaver, T. L. (1991). The Drosophila plutonium and pan gu genes regulate entry into S phase at fertilization. *Cell*, *66*(6), 1289–1300. 0092-8674(91)90050-9 [pii].

Shyu, A. B., Belasco, J. G., & Greenberg, M. E. (1991). Two distinct destabilizing elements in the c-fos message trigger deadenylation as a first step in rapid mRNA decay. *Genes & Development*, *5*(2), 221–231.

Siddiqui, N. U., Li, X., Luo, H., Karaiskakis, A., Hou, H., Kislinger, T., et al. (2012). Genome-wide analysis of the maternal-to-zygotic transition in Drosophila primordial germ cells. *Genome Biology*, *13*(2), R11. http://dx.doi.org/10.1186/gb-2012-13-2-r11.

Smibert, C. A., Lie, Y. S., Shillinglaw, W., Henzel, W. J., & Macdonald, P. M. (1999). Smaug, a novel and conserved protein, contributes to repression of nanos mRNA translation in vitro. *RNA*, *5*(12), 1535–1547.

Smibert, C. A., Wilson, J. E., Kerr, K., & Macdonald, P. M. (1996). Smaug protein represses translation of unlocalized nanos mRNA in the Drosophila embryo. *Genes & Development*, *10*(20), 2600–2609.

Sonoda, J., & Wharton, R. P. (1999). Recruitment of Nanos to hunchback mRNA by Pumilio. *Genes & Development*, *13*(20), 2704–2712.

Sonoda, J., & Wharton, R. P. (2001). Drosophila brain tumor is a translational repressor. *Genes & Development*, *15*(6), 762–773. http://dx.doi.org/10.1101/gad.870801.

Spradling, A. (1993). Developmental genetics of oogenesis. In M. Bate & A. Martinez Arias (Eds.), *The development of Drosophila melanogaster* (pp. 1–70), Plainview, NY: Cold Spring Harbor Laboratory Press.

St Johnston, D., Driever, W., Berleth, T., Richstein, S., & Nusslein-Volhard, C. (1989). Multiple steps in the localization of bicoid RNA to the anterior pole of the Drosophila oocyte. *Development*, *107*(Suppl.), 13–19.

Struhl, G. (1989). Differing strategies for organizing anterior and posterior body pattern in Drosophila embryos. *Nature*, *338*(6218), 741–744. http://dx.doi.org/10.1038/338741a0.

Swan, A., & Schupbach, T. (2007). The Cdc20 (Fzy)/Cdh1-related protein, Cort, cooperates with Fzy in cyclin destruction and anaphase progression in meiosis I and II in Drosophila. *Development*, *134*(5), 891–899. http://dx.doi.org/10.1242/dev.02784. dev.02784 [pii].

Tadros, W., Goldman, A. L., Babak, T., Menzies, F., Vardy, L., Orr-Weaver, T., et al. (2007). SMAUG is a major regulator of maternal mRNA destabilization in Drosophila and its translation is activated by the PAN GU kinase. *Developmental Cell*, *12*(1), 143–155. http://dx.doi.org/10.1016/j.devcel.2006.10.005. S1534-5807(06)00457-6 [pii].

Tadros, W., Houston, S. A., Bashirullah, A., Cooperstock, R. L., Semotok, J. L., Reed, B. H., et al. (2003). Regulation of maternal transcript destabilization during egg activation in Drosophila. *Genetics*, *164*(3), 989–1001.

Tadros, W., & Lipshitz, H. D. (2009). The maternal-to-zygotic transition: A play in two acts. *Development*, *136*(18), 3033–3042. http://dx.doi.org/10.1242/dev.033183.

Takeo, S., Hawley, R. S., & Aigaki, T. (2010). Calcineurin and its regulation by Sra/RCAN is required for completion of meiosis in Drosophila. *Developmental Biology*, *344*(2), 957–967. http://dx.doi.org/10.1016/j.ydbio.2010.06.011. S0012-1606(10)00831-6 [pii].

Takeo, S., Tsuda, M., Akahori, S., Matsuo, T., & Aigaki, T. (2006). The calcineurin regulator sra plays an essential role in female meiosis in Drosophila. *Current Biology*, *16*(14), 1435–1440. http://dx.doi.org/10.1016/j.cub.2006.05.058. S0960-9822(06)01677-0 [pii].

Tautz, D. (1988). Regulation of the Drosophila segmentation gene hunchback by two maternal morphogenetic centres. *Nature, 332*(6161), 281–284. http://dx.doi.org/10.1038/332281a0.

Temme, C., Zhang, L., Kremmer, E., Ihling, C., Chartier, A., Sinz, A., et al. (2010). Subunits of the Drosophila CCR4-NOT complex and their roles in mRNA deadenylation. *RNA, 16*(7), 1356–1370. http://dx.doi.org/10.1261/rna.2145110. rna.2145110 [pii].

Thomsen, S., Anders, S., Janga, S. C., Huber, W., & Alonso, C. R. (2010). Genome-wide analysis of mRNA decay patterns during early Drosophila development. *Genome Biology, 11*(9), R93. http://dx.doi.org/10.1186/gb-2010-11-9-r93. gb-2010-11-9-r93 [pii].

Tsurumi, A., Xia, F., Li, J., Larson, K., LaFrance, R., & Li, W. X. (2011). STAT is an essential activator of the zygotic genome in the early Drosophila embryo. *PLoS Genetics*, 7(5), e1002086. http://dx.doi.org/10.1371/journal.pgen.1002086. PGENETICS-D-10-00149 [pii].

Vardy, L., & Orr-Weaver, T. L. (2007). The Drosophila PNG kinase complex regulates the translation of cyclin B. *Developmental Cell, 12*(1), 157–166. http://dx.doi.org/10.1016/j.devcel.2006.10.017. S1534-5807(06)00469-2 [pii].

Voeltz, G. K., & Steitz, J. A. (1998). AUUUA sequences direct mRNA deadenylation uncoupled from decay during Xenopus early development. *Molecular and Cellular Biology, 18*(12), 7537–7545.

Vogel, C., & Marcotte, E. M. (2012). Insights into the regulation of protein abundance from proteomic and transcriptomic analyses. *Nature Reviews Genetics, 13*(4), 227–232. http://dx.doi.org/10.1038/nrg3185. nrg3185 [pii].

Von Stetina, J. R., Tranguch, S., Dey, S. K., Lee, L. A., Cha, B., & Drummond-Barbosa, D. (2008). alpha-Endosulfine is a conserved protein required for oocyte meiotic maturation in Drosophila. *Development, 135*(22), 3697–3706. http://dx.doi.org/10.1242/dev.025114. dev.025114 [pii].

Walser, C. B., & Lipshitz, H. D. (2011). Transcript clearance during the maternal-to-zygotic transition. *Current Opinion in Genetics & Development, 21*(4), 431–443. http://dx.doi.org/10.1016/j.gde.2011.03.003.

Wang, C., & Lehmann, R. (1991). Nanos is the localized posterior determinant in Drosophila. *Cell, 66*(4), 637–647.

White-Cooper, H., Alphey, L., & Glover, D. M. (1993). The cdc25 homologue twine is required for only some aspects of the entry into meiosis in Drosophila. *Journal of Cell Science, 106*(Pt. 4), 1035–1044.

Whitfield, Z. J., Chisholm, J., Hawley, R. S., & Orr-Weaver, T. L. (2013). A meiosis-specific form of the APC/C promotes the oocyte-to-embryo transition by decreasing levels of the Polo kinase inhibitor matrimony. *PLoS Biology, 11*, e1001648. http://dx.doi.org/10.1371/journal.pbio.1001648. PBIOLOGY-D-12-05056 [pii].

Wilson, T., & Treisman, R. (1988). Removal of poly(A) and consequent degradation of c-fos mRNA facilitated by 3' AU-rich sequences. *Nature, 336*(6197), 396–399. http://dx.doi.org/10.1038/336396a0.

Xiang, Y., Takeo, S., Florens, L., Hughes, S. E., Huo, L. J., Gilliland, W. D., et al. (2007). The inhibition of polo kinase by matrimony maintains G2 arrest in the meiotic cell cycle. *PLoS Biology, 5*(12), e323. http://dx.doi.org/10.1371/journal.pbio.0050323. 07-PLBI-RA-2455 [pii].

Xu, Z., Chen, H., Ling, J., Yu, D., Struffi, P., & Small, S. (2014). Impacts of the ubiquitous factor Zelda on Bicoid-dependent DNA binding and transcription in Drosophila. *Genes & Development, 28*(6), 608–621. http://dx.doi.org/10.1101/gad.234534.113. 28/6/608 [pii].

York-Andersen, A. H., Parton, R. M., Bi, C. J., Bromley, C. L., Davis, I., & Weil, T. T. (2015). A single and rapid calcium wave at egg activation in Drosophila. *Biology Open, 4*(4), 553–560. http://dx.doi.org/10.1242/bio.201411296. bio.201411296 [pii].

CHAPTER THREE

Transcriptional Activation of the Zygotic Genome in *Drosophila*

Melissa M. Harrison[*,1], **Michael B. Eisen**[†,‡,§,¶,1]
[*]Department of Biomolecular Chemistry, School of Medicine and Public Health, University of Wisconsin-Madison, Madison, Wisconsin, USA
[†]Howard Hughes Medical Institute, University of California, Berkeley, Berkeley, California, USA
[‡]Department of Molecular and Cell Biology, University of California, Berkeley, Berkeley, California, USA
[§]Department of Integrative Biology, University of California, Berkeley, Berkeley, California, USA
[¶]QB3 Institute, University of California, Berkeley, Berkeley, California, USA
[1]Corresponding authors: e-mail address: mharrison3@wisc.edu; mbeisen@lbl.gov

Contents

Abstract

During the first stages of metazoan development, the genomes of the highly specified sperm and egg must unite and be reprogrammed to allow for the generation of a new organism. This process is controlled by maternally deposited products. Initially, the zygotic genome is largely transcriptionally quiescent, and it is not until hours later that the zygotic genome takes control of development. The transcriptional activation of the zygotic genome is tightly coordinated with the degradation of the maternal

Current Topics in Developmental Biology, Volume 113
ISSN 0070-2153
http://dx.doi.org/10.1016/bs.ctdb.2015.07.028

products. Here, we review the current understanding of the processes that mediate the reprogramming of the early embryonic genome and facilitate transcriptional activation during the early stages of *Drosophila* development.

1. INTRODUCTION

In nearly all metazoans, the first stages of development are controlled by maternally deposited products. During this time, the zygotic genome is largely transcriptionally quiescent with widespread transcription only initiating hours later. The transcriptional activation of the zygotic genome is coordinated with the degradation of the maternally deposited mRNAs resulting in a monumental shift in the transcriptional profile of the embryo. Thus, during this maternal-to-zygotic transition (MZT), the zygote establishes control of its future development.

In many organisms, these stages of development are characterized by rapid cellular or nuclear divisions resulting in a dramatic increase in nuclear content with little or no increase in cytoplasm. In *Drosophila melanogaster*, early development is characterized by 13 rapid, synchronous nuclear divisions. These early cycles include only a DNA synthesis phase (S) and a division phase (M) with no gap phases and occur approximately every 10 min (Foe & Alberts, 1983; Shermoen, McCleland, & O'Farrell, 2010). The speed of the division cycle is further enhanced by the fact that these divisions happen without cytokinesis such that within 2 h the nuclei have divided 13 times. It is at the 14th nuclear division that the cycle slows and a gap phase (G2) is added (Fig. 1). Furthermore, at this stage of development, the syncytium of now approximately 6000 nuclei undergoes cellularization. Thus, during early development there is a coordinated remodeling of both the division cycle and the transcriptional profile of the embryo.

This transition from maternal to zygotic control over development accompanies an overall change in developmental potential. The processes that govern the MZT provide for the rapid and efficient reprogramming of the specified egg and sperm to generate the totipotent cells of the early embryo. Thus, understanding how these reprogramming events are coordinated within the context of a developing organism will provide insight into how these processes can be facilitated in other contexts.

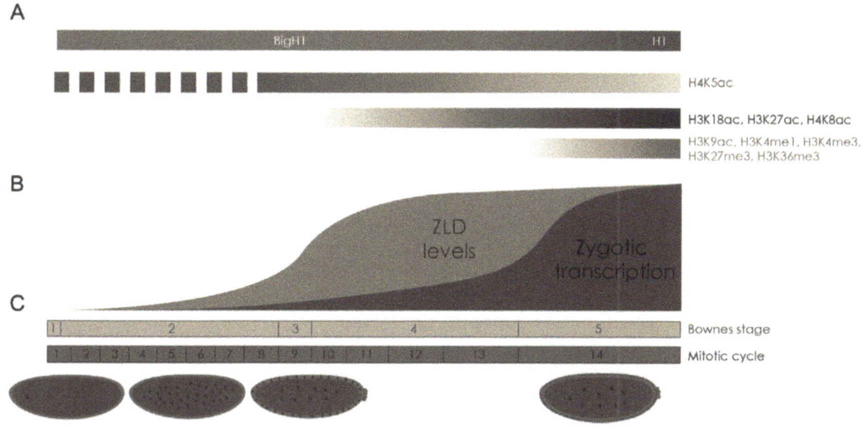

Figure 1 Many dynamic processes shape the maternal-to-zygotic transition. (A) Histone modifications and variants change during the MZT. Linker histone H1 replaces the early embryo-specific variant, dBigH1, at cycle 14. Levels of acetylation of lysine 5 on histone H4 are high at cycle 8 and decrease over the MZT. Boxes indicate unknown levels of H4K5ac prior to cycle 8. Acetylation on H3K18, H3K27, and H4K8 is first detected at cycles 8–10 and increases during the remainder of the MZT. H3K9ac, H3K4me1, H3K4me3, H3K27me3, and H3K36me3 are first detected at cycle 14. (B) ZLD levels increase approximately 1 h after fertilization. Zygotic transcription gradually increases from cycle 2 through cycle 13. At cycle 14, thousands of genes are transcriptionally activated. (C) Mitotic cycles are coordinated with morphological changes demonstrated in the pictures of embryos below and described by the Bownes stages.

2. ACTIVATING TRANSCRIPTION OF THE ZYGOTIC GENOME

Early work demonstrated that RNA production increases dramatically as the embryo develops from fertilization through the blastoderm (Anderson & Lengyel, 1979, 1981; Lamb & Laird, 1976; McKnight & Miller, 1976). Expression of individual genes could be detected as early as nuclear cycle 8 by assaying for nuclear mRNA using *in situ* hybridization (Pritchard & Schubiger, 1996). However, systematic identification of the genes transcribed from the zygotic genome in the blastoderm is complicated by the large store of maternally deposited mRNAs that remain. Recently, multiple high-throughput strategies have been employed to identify a wide array of genes transcribed in the cellular blastoderm. These studies have shown that activation of the zygotic genome is a gradual process that results in the expression of the transcription factors and microRNAs that pattern the embryo. In addition, degradation of maternal mRNAs allows for spatially localized transcription from the zygotic genome. Ultimately, activation of the zygotic genomes provides a mechanism by which developmental control is transferred from the mother to the embryo.

2.1 Identification of Genes Expressed in the Early Embryo

Because of the highly dynamic nature and extensive remodeling of the transcriptional profile of the early embryo, a wide variety of strategies, each possessing their own advantages and disadvantages, have been utilized to identify the transcript profile in the early embryo. Together, these studies have allowed for the identification of hundreds of genes activated over the course of the MZT and laid the groundwork for studies of the mechanisms driving this activation.

By coupling RNA preparations from staged embryos or oocytes with microarrays, initial studies from multiple different groups used changes in expression over developmental time to identify maternally deposited as well as zygotically expressed genes (Arbeitman et al., 2002; Benoit et al., 2009; Pilot, Philippe, Lemmers, Chauvin, & Lecuit, 2006; Tadros et al., 2007). Depending on the time points analyzed and the methods used, there are considerable differences in the estimations of even the percent of the genome that is expressed in mature oocytes and therefore maternally deposited. These estimations range from approximately 30% (Arbeitman et al., 2002) to 55% (Tadros et al., 2007) of the genome being maternally expressed.

Such strategies, based on general changes in expression, are complicated by the fact that maternally deposited products may be degraded concomitantly with expression from the zygotic genome resulting in no significant change in overall expression levels. Therefore to separate transcriptional from posttranscriptional regulatory events, expression levels in embryos lacking specific chromosomal arms were compared to similarly staged wild-type embryos using microarrays (De Renzis, Elemento, Tavazoie, & Wieschaus, 2007). This strategy permitted changes in gene expression to be directly correlated with the presence or absence of the genomic locus and led to the identification of 1158 genes that are zygotically expressed at cycle 14 (~18% of the genes whose transcripts are detected at cycle 14) and 6485 transcripts that are maternally deposited (De Renzis et al., 2007).

The presence of polymorphisms in different wild-type strains of *D. melanogaster* provided another strategy to separate maternally deposited transcripts from zygotically expressed ones. Two different lines of

D. melanogaster (*Canton-S* and *w1*) were crossed, and RNA from single, precisely staged embryos was subjected to high-throughput sequencing (Lott et al., 2011). Detection of polymorphisms specific to either the maternal or paternal chromosome enabled the identification of maternal transcripts from 5598 genes, strictly zygotic transcription from 2210 genes, and 1195 genes expressed both maternally and zygotically (Lott et al., 2011). Furthermore, the high-temporal resolution of these data provides a detailed assessment of exactly when each of these zygotic genes is first transcribed.

Detection of polymerase engagement using global run-on sequencing (GRO-seq) provided a direct assay of transcription in the early embryo (Saunders, Core, Sutcliffe, Lis, & Ashe, 2013). At 2–2.5 h after egg laying, GRO-seq identified 8752 transcriptionally active genes. This is more than double that estimated based on Lott et al. (2011) and included 23 of 27 genes categorized as strictly maternal by Arbeitman et al. (2002). Thus, high-throughput technologies have expanded the number of genes known to be expressed in the zygote, and it is clear that thousands of genes, upward of 62% of the genome, are activated at cycle 14.

High-throughput *in situ* hybridization analysis has been used to determine the expression of approximately 7000 genes over four discrete stages in embryogenesis: stages 1–3, stages 4–5, stages 6–7, and stages 8–9 (Lecuyer et al., 2007). Approximately 5000 genes are expressed in embryos undergoing the MZT (stages 1–3 (5002 genes) and stages 4–5 (4690 genes)) (Fly-FISH website). At these early stages, maternally deposited transcripts are cytoplasmic while actively transcribed genes are detectable by *in situ* hybridization as nuclear dots. Using this distinctive subcellular localization pattern, 56 zygotically expressed genes were identified as being transcribed during stages 1–3 and 736 during stages 4–5. Thus, during these early stages, the large majority of gene expression detected is due to maternal deposition. By this assay, nearly 70% of the genes examined show maternal contribution to the embryo. These data correspond well with microarray and sequencing-based data sets that similarly estimated 55–60% of the genome being maternally contributed (De Renzis et al., 2007; Lott et al., 2011; Tadros et al., 2007).

The high-throughput *in situ* hybridization study made the striking observation that over 70% of mRNAs examined had specific patterns of localization (Lecuyer et al., 2007). More recently, localization of all mRNAs along the anterior–posterior axis has been characterized using the sensitivity of high-throughput sequencing approaches (Combs & Eisen, 2013). Here single, tightly staged embryos were sliced to produce 6–12 sections along the anterior–posterior axis, and RNA from each slice was sequenced individually allowing a quantitative assessment of the amount of a given mRNA along the embryo.

In the past decade, high-throughput strategies have significantly expanded our understanding of transcriptional activation during the MZT. Together, it is the combination of multiple strategies that has provided a detailed understanding of the timing (Lott et al., 2011), extent (Saunders et al., 2013), and pattern (Combs & Eisen, 2013; Lecuyer et al., 2007) of gene expression. These studies cumulatively demonstrated that approximately two-thirds of genes are maternally contributed to the embryo as mRNA and that transcriptional activation of the zygotic genome occurs gradually with a large increase in transcription at cycle 14 (Fig. 1).

2.2 Dynamics of Zygotic Genome Activation

Both genome-wide studies and analyses of individual genes have demonstrated that the zygotic genome does not transition from transcriptionally silent to active at a single discrete time point. Instead, activation of the zygotic genome is a gradual process. Early studies using *in situ* hybridizations to analyze zygotic transcription of 13 different genes demonstrated that *snail* and *sisterless A* were first transcribed starting at cycle 8, whereas transcription of *Toll*, *string*, and *Antennapedia* was not detected until cycle 14 (Pritchard & Schubiger, 1996). High-temporal resolution sequencing analysis enabled this observation to be extended to hundreds of genes (Lott et al., 2011). Eighty-six genes were shown to initiate expression at cycles 10–11, 94 at cycles 12–13, 142 at early cycle 14, and 98 at late cycle 14 (Li, Harrison, Villalta, Kaplan, & Eisen, 2014; Lott et al., 2011). Because the definition of these classes excluded genes that are expressed both maternally and zygotically, the actual numbers of genes transcribed at each developmental time point may be larger. Together, these data demonstrate that the gradual activation of the zygotic genome occurs broadly and is not restricted to just a handful of genes. Furthermore, careful analysis has shown that *engrailed (en)* is first expressed as early as cycle 2 and that zygotic mutations in *en* cause phenotypes prior to formation of the blastoderm (Ali-Murthy, Lott, Eisen, & Kornberg, 2013). High-throughput sequencing identified additional genes that are likely to be expressed prior to blastoderm formation (Ali-Murthy et al., 2013). Thus, gradual activation of the zygotic genome may initiate immediately following fertilization (Fig. 1).

While activation of the genome is a gradual process, transcriptional activation of individual genes can also be stochastic and gradual with individual nuclei initiating transcription at discrete developmental time points (Boettiger & Levine, 2009; Lagha et al., 2013). Nonetheless, other genes are activated in a rapid, synchronous manner, and this is governed in large part by sequences in the promoter, as discussed below (Boettiger & Levine, 2009; Lagha et al., 2013). Thus, transcriptional activation at both the level of the genome as a whole and at individual genes is a gradual process that occurs over the first few hours of development.

2.3 Nature of the Genes Expressed in the Early Embryo

Generally, genes transcriptionally activated early in development are small and lack introns. Of the 59 earliest expressed genes identified by De Renzis et al., 70% lacked introns (De Renzis et al., 2007). The reduced length of these transcription units may be a requirement for early expression because of the rapid nuclear divisions at this time in development. Mitosis disrupts transcription such that only the very short interphase is permissive for transcription in the early embryo (Shermoen & O'Farrell, 1991). This model is supported by the fact that, while the gene *knirps related*, which contains a large 19-kb intron, cannot substitute for loss of *knirps* function early in development, it can substitute if expressed as a short intronless transgene (Rothe, Pehl, Taubert, & Jackle, 1992). These characteristics of early zygotic genes are likely shared among organisms whose early development is characterized by rapid divisions, as early zygotic genes in zebrafish are also generally short and intronless (Heyn et al., 2014).

In addition to a general difference in gene size between maternal and early zygotic transcripts, there is an enrichment of housekeeping genes among those that are maternally deposited and an enrichment for transcription factors among early zygotic genes (De Renzis et al., 2007). Many of these zygotically expressed transcription factors function to pattern the early embryo. Furthermore, mRNAs are not the only RNAs expressed during the MZT. MicroRNAs (miRs) are also expressed (Aboobaker, Tomancak, Patel, Rubin, & Lai, 2005; Biemar et al., 2005; Fu, Nien, Liang, & Rushlow, 2014), and their expression coordinates zygotic genome activation with the degradation of the maternally provided mRNAs (Bushati, Stark, Brennecke, & Cohen, 2008; Giraldez et al., 2006; Lund, Liu, Hartley, Sheets, & Dahlberg, 2009). For example, in *Drosophila* the *mir-309* cluster is instrumental in downregulating a subset of maternally deposited mRNAs (Bushati et al., 2008). Because these miRs target maternal RNAs for degradation, their expression helps to coordinate the transition from maternal to zygotic control of development in flies, frogs, and zebrafish (Bushati et al., 2008; Giraldez et al., 2006; Lund et al., 2009).

While nearly all of the studies identifying zygotically expressed genes have focused on the somatic cells that comprise the majority of the blastoderm embryo, it is important to keep in mind that the primordial germ cells (PGCs) are being specified during these first stages of development. Purification of GFP-labeled PGCs allowed for proteomic and transcriptional analysis of this specialized cell type (Siddiqui et al., 2012). Both degradation of maternal RNAs and transcriptional activation of the zygotic genome are delayed in the PGCs. Furthermore, in contrast to the enrichment for factors important in specifying cell identity found in the somatic cells, early expressed genes in the PGCs are primarily involved in translation and energy production (Siddiqui et al., 2012).

3. MECHANISMS REGULATING TRANSCRIPTIONAL ACTIVATION DURING THE MZT

Studies by Newport and Kirschner in *Xenopus* linked transcriptional activation with changes in the cell cycle and indicated that a key feature regulating the MZT was the increasing DNA content of the zygote (Newport & Kirschner, 1982a, 1982b). They proposed that the increase in DNA content that occurred during the rapid cell cycles of early *Xenopus* development led to the titration of a maternally deposited transcriptional repressor and that this permitted activation of the zygotic genome. In the more than 30 years since these initial studies, it has been shown that this nuclear-to-cytoplasmic ratio is a conserved feature governing early development in *Drosophila* as well as *Xenopus*. However, it has also become clear that additional mechanisms are regulating activation of the zygotic genome.

3.1 Mechanisms Controlling the Timing of Zygotic Genome Activation

Studies in haploid *Drosophila* embryos demonstrated that the nuclear-to-cytoplasmic ratio influences the division cycle, but the effects on transcription were less clear (Edgar, Kiehle, & Schubiger, 1986; Edgar & Schubiger, 1986). Building on these studies, *in situ* hybridization was used to assay the timing of transcriptional initiation when nuclear density was perturbed (Pritchard & Schubiger, 1996). These data showed that local changes in nuclear density drive changes in the timing of gene expression and confirmed the importance of the nuclear-to-cytoplasmic ratio in controlling the initiation of gene expression for two genes, *fushi tarazu (ftz)* and *Kruppel (Kr)*. However, expression of other genes was shown to be independent of the increasing DNA content (Yasuda, Baker, & Schubiger, 1991). Thus, the nuclear-to-cytoplasmic ratio is a key regulator of the timing of the MZT conserved between *Xenopus* and *Drosophila*. Nonetheless, these studies also indicated that other factors, including a timing mechanism, were instrumental in controlling transcriptional activation. Genome-wide microarray studies allowed for a broader comparison of expression profiles between haploid and diploid embryos at specific early cleavage cycles and showed that while 88/290 zygotically expressed genes depended largely on the ratio of nuclear to cytoplasmic content, 127/290 were expressed similarly in age-matched embryos, suggesting that they relied on a timing mechanism to control expression (Lu, Li, Elemento, Tavazoie, & Wieschaus, 2009). Ultimately, these studies demonstrated that neither developmental time nor nuclear-to-cytoplasmic ratio alone could predict gene expression dynamics and that it is likely the coordination of these two processes that enables transcriptional activation of the zygotic genome.

The coordination of the lengthening of the division cycle and transcriptional activation is discussed in detail in another chapter of this book. Nonetheless, it is important to note that there is a clear connection between the two processes. The early, abbreviated division cycles limit the length of transcripts that can be expressed (Shermoen & O'Farrell, 1991). Conversely, recent evidence suggests that transcriptional engagement and the resulting replication stress are instrumental in the slowing of the division cycle (Blythe & Wieschaus, 2015; Sung, Spangenberg, Vogt, & Grosshans, 2013).

3.2 DNA-Sequence Elements Provide Insights into Genome Activation

Key insights into the mechanisms driving transcriptional activation during the MZT were provided by studies of the sex determination pathway as genes that regulate sex determination are among the earliest expressed genes in *Drosophila*. It was initially observed that the heptameric sequence element, CAGGTAG, was located in the upstream regulatory sequences of a number of genes regulating sex determination (*scute*, *outstretched*, *Sex lethal*, and *sisterless B*) and that this sequence element was conserved in other *Drosophila* species (Erickson & Cline, 1998; Sefton, Timmer, Zhang, Beranger, & Cline, 2000). Mutational analysis of sequences upstream of the *scute* promoter demonstrated that both CAGGTAG elements and a related CAGGCAG element were required for driving early embryonic expression of *scute* (ten Bosch, Benavides, & Cline, 2006; Wrischnik, Timmer, Megna, & Cline, 2003). Genome-wide studies identifying genes expressed during the MZT demonstrated the more general importance of the CAGGTAG sequence element and related sequence elements (termed TAGteam elements) in activating early zygotic transcription (De Renzis et al., 2007; Pilot et al., 2006; ten Bosch et al., 2006). Mutation of TAGteam elements delays transcriptional activation, while increasing the number of these heptameric motifs results in precocious gene expression (Foo et al., 2014; Liberman & Stathopoulos, 2009; ten Bosch et al., 2006). Intriguingly, the TAGteam element CAGGTAG was also identified as the most highly enriched sequence in regions bound by the transcription factors that govern anterior–posterior patterning in the early embryo (Li et al., 2008). Together, these studies indicated the presence of a transcriptional activator that could bind to TAGteam sequence elements and drive activation of the zygotic genome.

Multiple approaches, including a DNA-affinity column, biochemical purification following DNase I protection activity, and a yeast one-hybrid assay, identified three different proteins that can bind to TAGteam elements (De Renzis et al., 2007; Harrison, Botchan, & Cline, 2010; Liang et al., 2008). However, only one of these proteins, Zelda (ZLD; also known as Vielfaltig) has been shown to drive transcriptional activation in the early embryo as discussed in depth in the next section (Liang et al., 2008). Nonetheless, the two additional proteins, Bicoid Stability Factor and Grainyhead, may modulate ZLD activity by binding to TAGteam elements at later stages of development (De Renzis et al., 2007; Harrison et al., 2010).

Bioinformatic analysis of sequence elements upstream of 21 zygotically expressed genes also identified enrichment for STAT92E-binding sites, suggesting a role for STAT92E, the sole *Drosophila* homolog of mammalian STAT, in activating gene expression at the MZT (Tsurumi et al., 2011). As in embryos depleted for maternal ZLD, the expression of a subset of zygotically transcribed genes is reduced in embryos depleted of STAT92E. Furthermore, STAT92E and ZLD cooperated to regulate some early zygotic genes. However, unlike embryos depleted for maternal ZLD, these STAT92E-depleted embryos were capable of developing through the MZT (Tsurumi et al., 2011). Together, these data suggest that STAT92E can modulate transcriptional activity in the early embryo, perhaps in conjunction with the essential activator, ZLD.

3.3 Zelda Is a Global Activator of Early Embryonic Transcription

Initial studies of ZLD (Vielfaltig) demonstrated that the gene was widely expressed, predominantly in mitotically active tissues including oocytes and early embryos (Staudt, Fellert, Chung, Jackle, & Vorbruggen, 2006). The gene was named *vielfaltig*, meaning versatile or manifold in German, due to the wide variety of defects in division and cellularization caused by both depletion and overexpression in the early *Drosophila* embryo (Staudt et al., 2006). However, it was the subsequent identification of ZLD as a TAGteam-binding protein that provided an explanation for the many defects observed in the original study (Liang et al., 2008). Gene expression analysis from embryos lacking maternally deposited ZLD demonstrated that, as expected for a TAGteam-binding protein, ZLD was required for the expression of more than 120 genes in the early embryo (Liang et al., 2008). Building on the many studies differentiating zygotically and maternally expressed genes in the early embryo, analysis of transcript abundance in ZLD-depleted embryos showed that maternally deposited mRNAs increased. Thus, ZLD helps to establish a connection between maternal mRNA degradation and zygotic transcriptional activation (Liang et al., 2008). In fact, *in situ* hybridization analysis demonstrated that the *mir-309* cluster, which downregulates maternal mRNAs during the MZT, depended on ZLD for expression, establishing a direct connection between zygotic transcriptional activation and maternal mRNA degradation (Liang et al., 2008). The fundamental role of ZLD in regulating gene expression is illustrated by the fact that lack of maternal ZLD leads to a failure to undergo the MZT, and lack of either maternal or zygotic ZLD leads to embryonic lethality (Liang et al., 2008).

ZLD is ubiquitously expressed throughout the early embryo (Nien et al., 2011; Staudt et al., 2006). While protein can be detected as early as cycle 2, ZLD levels increase approximately 1 h after fertilization coincident with the widespread activation of the zygotic genome (Fig. 1; Harrison et al., 2010; Nien et al., 2011). In embryos, ZLD is a 1596-amino acid protein comprised of six zinc fingers, four of which are clustered in the C-terminal half of the protein (Staudt et al., 2006). This cluster of four zinc fingers constitutes the DNA-binding domain of ZLD, and all of the zinc fingers are required for this activity (Hamm, Bondra, & Harrison, 2015; Struffi et al., 2011).

Apart from the zinc fingers, there are no other identifiable protein domains suggesting function. Nonetheless, a significant portion of the protein is comprised of large stretches of highly repetitive amino acid sequences (low-complexity regions), and at least some of this low-complexity sequence provides an activation domain (Hamm et al., 2015). The lack of any known enzymatic domains along with other *in vivo* and *in vitro* data suggests that the N-terminal region of ZLD interacts with cofactors through these low-complexity regions to mediate transcriptional activation (Hamm et al., 2015).

In vivo analysis of ZLD binding through early development provided the first mechanistic insights into how ZLD might function to drive activation of the zygotic genome (Harrison, Li, Kaplan, Botchan, & Eisen, 2011; Nien et al., 2011). At cycle 14, when widespread transcription initiates, ZLD is bound to thousands of loci, including most if not all of the *cis*-regulatory regions that are known to drive transcription in the early embryo. Depletion of maternal ZLD results in misexpression of a large number of the genes that contain ZLD-binding sites in their *cis*-regulatory regions, suggesting that ZLD is directly regulating many of these genes (Harrison et al., 2011; Liang et al., 2008; Nien et al., 2011). However, interpretation of these results is complicated by the fact that ZLD binds to the promoters of many of the transcription factors that regulate embryonic patterning and drive localized expression at cycle 14. Nonetheless, studies in which individual ZLD-binding sites (TAGteam elements) were mutated or added to the regulatory regions of specific genes have supported the conclusion that at least some of the observed effects on gene expression are direct (Foo et al., 2014; Fu et al., 2014; Liberman & Stathopoulos, 2009; ten Bosch et al., 2006; Xu et al., 2014). These data support a model in which ZLD can activate expression of zygotic genes, and this may be in part through the potentiation of morphogens, such as Dorsal (Kanodia et al., 2012).

Prior to cycle 10, a time in embryonic development when only approximately 100 genes are expressed (De Renzis et al., 2007; Lott et al., 2011), ZLD is already bound to thousands of *cis*-regulatory regions (Harrison et al., 2011). ZLD-bound regions remain remarkably consistent through the stages of development analyzed (cycles 10–14) demonstrating that ZLD binding is established early and is maintained through an hour of development, corresponding to at least four nuclear divisions. These ZLD-bound regions are highly enriched for the canonical TAGteam sequence, CAGGTAG. In fact, ZLD occupies 64% of all CAGGTAG sequences in the genome at cycle 10. This level of occupancy is dramatically larger than that observed for other transcription factors during cycle 14 (Harrison et al., 2011; Li et al., 2008; MacArthur et al., 2009). This widespread ZLD binding suggests that it is the affinity of ZLD for DNA and not interactions with chromatin that drives binding specificity. This may be due to the inherent DNA-binding capacity of ZLD and/or a permissive chromatin structure in the early embryo may allow ZLD access to the underlying DNA. As discussed below, studies of histone modifications in the early embryo suggest that the genome may be in a generally naïve state, and it is possible that this state permits widespread binding by ZLD (Li et al., 2014). Supporting this hypothesis, the limited changes in *in vivo* ZLD binding that occur as the embryo progresses through the MZT lead to a reduction in the percent of CAGGTAG sites occupied by ZLD; at cycle 14, only 39% of CAGGTAG sites are bound by ZLD (Harrison et al., 2011). This suggests that chromatin features may limit the ability of ZLD to access DNA later in development.

ZLD-bound loci are enriched for regions of open chromatin at cycle 14 (Harrison et al., 2011; Nien et al., 2011). Furthermore, these loci are highly occupied by a diversity of transcription factors that drive patterning in the early embryo (Harrison et al., 2011; Li et al., 2008; MacArthur et al., 2009; Nien et al., 2011). Because ZLD binding is established prior to the expression of many of these transcription factors, it was proposed that ZLD was instrumental in establishing regions of open chromatin and thus permitting transcription factor binding (Harrison et al., 2011; Nien et al., 2011). Recent evidence supports this model: histone H3 chromatin immunoprecipitation (ChIP) and DNase I sensitivity have shown that ZLD is important for either the establishment or the maintenance of regions of accessible chromatin (Foo et al., 2014; Li et al., 2014). In addition, ZLD has been shown to be important in determining DNA binding of three distinct transcription factors: Twist, Dorsal, and Bicoid (Foo et al., 2014; Xu et al., 2014; Yanez-Cuna, Dinh, Kvon, Shlyueva, & Stark, 2012).

Together, these data suggest that ZLD binds early in embryonic development to establish or maintain specific regions of open chromatin. Subsequent binding of transcription factors to these areas of chromatin accessibility can then drive specific gene expression at the MZT. This activity is similar to that described for pioneer factors, a unique class of proteins that bind chromatin and establish accessible regions (Fig. 2; Zaret & Carroll, 2011). The concept of pioneer factors is based largely on the FOX family of transcription factors, which drives hepatic development (Cirillo et al., 2002, 1998; Cirillo & Zaret, 1999; Gualdi et al., 1996). However, there is increasing evidence that other factors perform similar functions. For example, the core pluripotency factors, Oct4, Nanog, and SoxB1, are bound to the enhancers of genes required for reprogramming of fibroblasts early in the process of generating induced pluripotent stem cells. The early occupancy by these

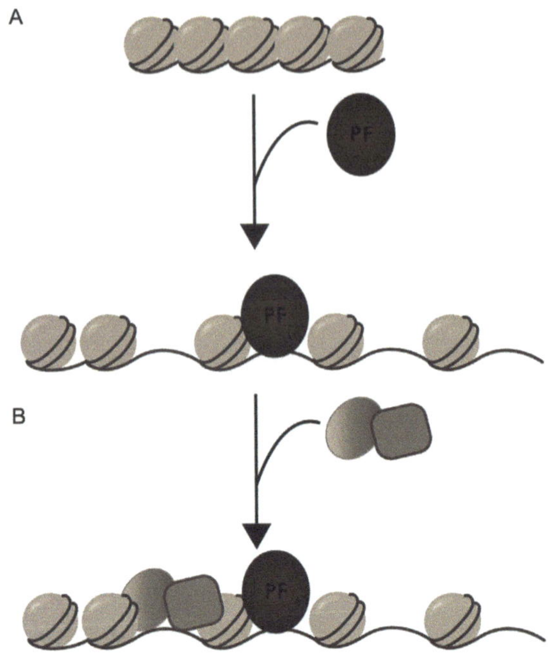

Figure 2 Pioneer factors may regulate chromatin accessibility during the MZT. (A) Pioneer factors (PFs) bind to nucleosomes and open up compacted chromatin. (B) The pioneer factor-mediated accessible DNA is available for binding by additional transcription factors.

factors likely relies on their ability to bind to closed chromatin (Soufi, Donahue, & Zaret, 2012). Furthermore, orthologs of these factors drive genome activation in zebrafish (Lee et al., 2013; Leichsenring, Maes, Mossner, Driever, & Onichtchouk, 2013). Together, this suggests that pioneer factors remodel chromatin in the early embryo, and it is this activity that facilitates the establishment of the pluripotent state.

4. CHROMATIN DYNAMICS DURING THE MZT

Within the nucleus, DNA is organized into chromatin, which regulates accessibility of the encoded genomic information. Thus, changes to chromatin structure can dramatically modulate gene expression. The building blocks of chromatin are the nucleosomes, which are composed of two copies each of the histones H2A, H2B, H3, and H4. Linker histones further compact the nucleosome-wrapped DNA by binding to the DNA between nucleosomes, helping to stabilize chromatin packing. Covalent modifications to the histone proteins, variations in the core histones, modifications to interactions between nucleosomes, as well as the shifting of entire nucleosomes can all cause dramatic changes in gene expression levels. It is, therefore, not surprising that there is dramatic chromatin remodeling during the MZT.

4.1 Histone Modifications and Variants Change During the MZT

Specific covalent modifications to chromatin have been associated with transcriptional activation or repression (Kouzarides, 2007). Changes in these covalent histone modifications through the MZT have been widely studied in mice and zebrafish (Andersen et al., 2013; Fadloun, Eid, & Torres-Padilla, 2013). These studies have demonstrated large changes in the abundance of histone modifications as the embryo progresses through the first stages of development, and it has been proposed that erasure of epigenetic information from the genomes contributed by the egg and sperm may be essential for the establishment of totipotency in the early embryo (Fadloun et al., 2013).

During early *Drosophila* development, the genome similarly undergoes dramatic epigenetic reprogramming (Chen et al., 2013; Li et al., 2014). Trimethylation on H3K4, H3K27, and H3K36, monomethylation on H3K4, and acetylation on H3K9 are first detectable at cycle 14, the developmental time point when widespread transcription initiates (Chen et al., 2013; Li et al., 2014). This is similar to what has been observed in zebrafish, where there is a dramatic increase in methylation at the same time as high levels of transcription initiate (Lindeman et al., 2011; Vastenhouw et al., 2010). Low levels of acetylation on H3K18, H3K27, and H4K8 are evident prior to cycle 14 (Li et al., 2014). Nonetheless, like the other covalent modifications studied, the total levels of these acetylation marks increase dramatically over the MZT. H4K5 acetylation, a mark associated with chromatin following replication, is evident as early as cycle 8 and levels decrease as would be expected from the slowing of the nuclear division cycle (Figs. 1 and 3; Li et al., 2014). Early acetylation marks are enriched at the promoters and enhancers of genes expressed at these very early stages of development (Li et al., 2014). These regions are enriched for ZLD binding, and H3K18 acetylation is decreased in embryos depleted of maternal ZLD (Li et al., 2014). These data suggest that ZLD may mediate transcriptional activation, at least in part, by recruiting a histone acetyltransferase to the *cis*-regulatory regions of genes expressed during the MZT. However, it is also possible that these acetyl marks precede ZLD binding. Given the time resolution of the

ChIP studies, the order of histone acetylation and ZLD binding cannot be determined. Thus as the embryo undergoes the MZT, the genome likely transitions through a relatively naïve state, and this may allow the restoration of developmental plasticity.

During development, variation in the core and linker histones can regulate how the genome is packaged and therefore how it is interpreted. An extreme example is the replacement of histone proteins with protamines during sperm development. Thus, during the first stage of zygotic development, the paternal genome must be repackaged, replacing the protamines with histones provided by the oocyte. An additional role for histone variants during the initial stages of development was shown through the identification of linker histone variants that were specific to oocytes and the early embryo in frogs, fish, and mice (Dimitrov, Almouzni, Dasso, & Wolffe, 1993; Fu et al., 2003; Smith, Dworkin-Rastl, & Dworkin, 1988; Wibrand & Olsen, 2002). The recent identification of a histone H1 variant specific to the *Drosophila* MZT, dBigH1, extends these findings. dBigH1 is expressed during only the first few hours of development (Fig. 1; Perez-Montero, Carbonell, Moran, Vaquero, & Azorin, 2013). Widespread transcription does not occur until dH1 replaces dBigH1, and mutations in dBigH1 lead to premature transcriptional activation of the zygotic genome (Perez-Montero et al., 2013). Together, these data suggest that histone H1 variants specific to the oocyte and early embryo may be instrumental in regulating the transcriptional activity of the zygotic genome during the MZT in a wide variety of metazoans.

4.2 Alterations in Chromatin Accessibility

Histone modifications and variants can affect transcriptional output of the surrounding genes by regulating accessibility to the underlying DNA. For example, acetylation on histone tails neutralizes the positively charged lysines and can result in an opening of the chromatin structure. These alterations in DNA accessibility can subsequently modulate transcription factor binding (Kaplan et al., 2011; Li et al., 2011). Thus, it is possible that local and global changes in accessibility regulate transcriptional activity during the MZT.

Both DNase I sensitivity and Formaldehyde-Assisted Isolation of Regulatory Elements can be used to assess chromatin accessibility. While both techniques have been used to determine chromatin accessibility in cycle 14 embryos, neither technique has been used to analyze changes in chromatin accessibility during the earliest stages of development. This is likely in large part due to the limiting amount of material that can be obtained from earlier stages and the challenges of precisely staging embryos during the rapid division cycles. Analysis of cycle 14 embryos has demonstrated that regions of accessible chromatin are enriched for enhancers and promoters (McKay & Lieb, 2013; Thomas et al., 2011), and this accessibility is likely a driving force in determining where transcription factors bind globally (Kaplan et al., 2011; Li et al., 2011).

Nucleosome density and organization can also govern transcriptional activity and chromatin compaction. Supporting a role of nucleosome density in regulating transcription over the MZT, recent evidence demonstrates that maternally contributed core histones H3 and H4 can act to repress transcriptional activation and cell cycle lengthening in *Xenopus* embryos (Amodeo, Jukam, Straight, & Skotheim, 2015). Global nucleosome density can be estimated from both micrococcal nuclease digestion or antihistone H3 ChIP coupled with high-throughput sequencing. In *Drosophila*, quantitative ChIP-seq against H3 shows a gradual increase ($\sim 1.4\times$) in DNA-associated H3 as the embryo progresses through the MZT, suggesting there may be an overall increase in nucleosome density and compaction (Li et al., 2014). These data in *Drosophila* appear to contradict the *Xenopus* data, which had suggested that titration of core histones by an increasing DNA content can lead to transcriptional activation. It is likely that in addition to global levels of histones, the position of these histones in the genome regulates gene expression. Indeed, well-positioned nucleosome arrays at promoters emerge over the MZT in zebrafish as assayed by micrococcal nuclease (Zhang et al., 2014). Clearly, further studies are necessary to understand the general role of nucleosome density and chromatin accessibility in regulating transcription during the MZT.

5. THE ROLE OF *cis*-REGULATORY ELEMENTS IN REGULATING ZYGOTIC TRANSCRIPTION

To better understand the processes that drive transcriptional activation of the zygotic genome, many studies have focused on the defining features of the *cis*-regulatory regions for those genes that are activated during the MZT. These studies have provided important information not only into the regulation of these early expressed genes but also more generally into the defining features of promoters and enhancers.

5.1 Promoter Sequences Influence Gene Expression Dynamics

Promoters are defined as the DNA sequence immediately surrounding the transcription start site, which is bound by the basal transcription machinery and allows for the initiation of transcription. There are a number of distinct sequence elements identified in promoters. These sequence elements can vary between promoters and can regulate the ability of upstream enhancer elements to drive transcription (Juven-Gershon & Kadonaga, 2010; Rach et al., 2011). For example, while the TATA box sequence upstream of the transcription start site is a core promoter element essential for the expression of a subset of genes, many promoters lack a canonical TATA box and instead possess downstream promoter elements (DPEs) (Juven-Gershon & Kadonaga, 2010). Studies have shown that these differences in promoter sequence permit promoters to respond differentially to enhancer elements (Butler & Kadonaga, 2001; Juven-Gershon, Hsu, & Kadonaga, 2008; Ohtsuki, Levine, & Cai, 1998), and this was recently shown on a genome-wide scale (Zabidi et al., 2015). It has been demonstrated that housekeeping genes and developmentally regulated genes are preferentially enriched for distinct core promoter elements, enabling them to be activated by different enhancers (Gaertner et al., 2012; Zabidi et al., 2015). Because maternally deposited products tend to be enriched for housekeeping genes while the earliest zygotically expressed genes are developmental regulators, these studies of core promoter elements have important implications for gene expression during the MZT.

A defining feature of the promoter is the ability to engage the RNA polymerase II holoenzyme (Pol II) that transcribes the gene. While many studies have focused on the recruitment of the polymerase to the promoter as a key regulatory step in transcriptional activation, recent work has shown that control of elongation by the polymerase is another aspect regulating transcription levels (Adelman & Lis, 2012). Indeed, in the *Drosophila* embryo at cycle 14, Pol II is bound to the promoters of hundreds of genes but is not found in gene bodies, suggesting that it is stalled and not actively transcribing (Zeitlinger et al., 2007). This paused polymerase is instrumental in rapid and synchronous transcriptional activation during the MZT (Boettiger & Levine, 2009; Lagha et al., 2013). Analysis of Pol II binding in cycle 13–14 embryos showed *de novo* recruitment of RNA Pol II to over 4000 promoters, with widespread pausing at the transcription start site (Chen et al., 2013). However, in earlier embryos (cycle 8–12), Pol II was only bound to 117 genes, which showed little evidence of pausing (Chen et al., 2013). The timing of wide-scale polymerase recruitment was further refined to the transition between nuclear cycle 12 and 13 by assaying Pol II binding by ChIP on low numbers of tightly staged embryos (Blythe & Wieschaus, 2015). Furthermore, the results of the ChIP-seq were supported by GRO-seq analysis, which similarly identified paused polymerase at a large number of genes transcribed during cycle 14 (Saunders et al., 2013). These data indicate that, while polymerase pausing may be an essential feature driving synchronous widespread activation at cycle 14, earlier expression during the MZT is likely regulated by polymerase recruitment and not the elongation of poised polymerase (Fig. 3).

Sequence motif analysis has identified elements specifically enriched in different classes of promoters. Early expressed genes (cycle 8–12) are enriched for the TATA element, while zygotically activated genes expressed at cycle 14 are enriched for the downstream promoter element (DPE), the Motif Two Element (MPE), and the pause button (PB) sequences (Fig. 3; Chen et al., 2013). Quantitative studies of the *snail* promoter identified the necessity of promoter sequences for establishing paused Pol II and the importance of this paused

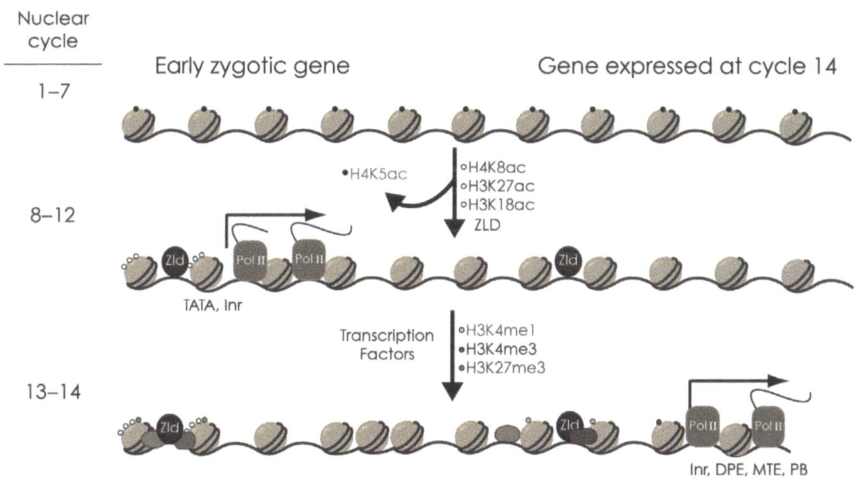

Figure 3 Model of the multiple processes that govern gene expression during the MZT. During nuclear cycles 1–7, the genome is likely reprogrammed to a relatively naïve state. Due to the rapid DNA replication, there are high levels of acetylated histone H4 lysine 5. At cycles 8–12, a number of early zygotic genes are activated. These genes have TATA box-containing promoters. At the *cis*-regulatory regions driving gene expression, ZLD is bound and histones are acetylated. As the embryo transitions through cycle 14, histone methylation is first detected, and ZLD facilitates the binding of additional transcription factors. Widespread RNA polymerase II pausing is established, especially at promoters containing DPE, MTE, and PB elements. (See the color plate.)

polymerase in driving synchronous transcriptional activation (Lagha et al., 2013). Thus, the sequence of the promoter can influence the binding properties of the polymerase, and this varies over even the very initial stages of embryonic development.

5.2 Enhancers

While promoters are essential for transcription, enhancers play a large role in regulating the timing and levels of gene expression; work studying gene expression during the *Drosophila* MZT has provided essential insights into enhancer structure and function starting with the identification of the stripe-specific enhancers for the *even-skipped* gene (Stanojevic, Small, & Levine, 1991). Identification of enhancers has been facilitated by genomic studies mapping the location of transcription factor binding, chromatin accessibility, and histone modifications (Bonn et al., 2012; Li et al., 2014, 2008; MacArthur et al., 2009; Rada-Iglesias et al., 2011; Thomas et al., 2011). Through these studies, regions that are highly occupied by a large number of diverse transcription factors—highly occupied target (HOT) regions—have been identified (Li et al., 2008; MacArthur et al., 2009; Moorman et al., 2006; Roy et al., 2010). These HOT regions are not unique to *Drosophila* embryonic development. They have similarly been identified in worms and humans (Gerstein et al., 2010; The ENCODE Project Consortium et al., 2007). HOT regions are enriched for regions of open chromatin, suggesting that it is chromatin accessibility that enables transcription factor binding in these regions (Roy et al., 2010). In *Drosophila*, many of the early embryonic HOT regions serve as enhancers that drive embryonic gene expression (Kvon, Stampfel, Yanez-Cuna, Dickson, & Stark, 2012). ZLD binding along with the sequence to which it binds, CAGGTAG, is strongly enriched in these HOT regions, suggesting a potential role for ZLD in initiating zygotic genome activation through the establishment of embryonic enhancers (Harrison et al., 2011; Kvon et al., 2012; Li et al., 2008; MacArthur et al., 2009; Negre et al., 2011; Nien et al., 2011; Roy et al., 2010). Thus, ZLD binding early in embryonic development may function to open chromatin and allow additional transcription factors access to the underlying DNA.

5.3 Comparative Studies in Multiple Species Demonstrate a Broad Role for ZLD in Driving the MZT

These detailed studies identifying and characterizing enhancers that drive expression in *Drosophila* embryos during the MZT have provided a foundation for understanding how transcriptional networks evolved. This is enabled by the high degree of morphological and physiological conservation of the MZT among *Drosophila* species (Lin, Deoras, Rasmussen, & Kellis, 2008). Analysis of transcription factor-binding sites across multiple *Drosophila* species identified conservation of between 15% and 38% of sites (He et al., 2011; Paris et al., 2013). When transcription factor binding was compared to gene expression profiles between *D. melanogaster* and three additional *Drosophila* species, mRNA levels were much more highly conserved than transcription factor binding (Paris et al., 2013). This seeming conundrum may be explained in part by the fact that conservation is stronger for those highly occupied sites that are likely to be functionally important in regulating gene expression (He et al., 2011; Paris et al., 2013). If binding strength for each transcription factor, rather than the mere presence or absence of a binding site, was analyzed, it was apparent that changes in binding strength for one factor is often correlated with the changes in the binding profiles for other transcription factors, and this shared divergence can be explained in part by the gain and loss of ZLD-binding sites (Bradley et al., 2010; Paris et al., 2013; Satija & Bradley, 2012). Thus, these comparative studies suggest a role for ZLD in regulating transcription factor-binding profiles and driving the MZT in other *Drosophila* species.

The role of ZLD in activating transcription during the MZT likely extends beyond the *Drosophila* genus. In the mosquito *Aedes aegypti*, sequences similar to the TAGteam element to which ZLD binds (VBRGGTA, V=A/C/G, B=C/G/T, R=A/G) are enriched in the 400 bp upstream of those genes activated during the MZT, and, like TAGteam elements, this sequence can drive zygotic transcription (Biedler, Hu, Tae, & Tu, 2012). Furthermore, ZLD orthologs from *Anopheles gambiae* and *Nasonia vitripennis* can drive TAGteam-dependent transcription in *D. melanogaster* cell culture (Hamm et al., 2015).

6. CONCLUDING REMARKS

The advent of tools for profiling transcription at the genome-wide level has afforded an unprecedented ability to determine the genes that were zygotically activated during the MZT. Building on a series of detailed studies, these analyses demonstrated that genome activation is a gradual process coordinated with the changes in the division cycle and that these processes are controlled by both the nuclear-to-cytoplasmic ratio and a timing mechanism. Given the importance of the nuclear-to-cytoplasmic ratio in regulating both transcription and cell division in *Xenopus*, there are likely to be mechanisms shared between invertebrates and vertebrates that drive the MZT.

Insights into some of the shared mechanisms have been provided by detailed studies of the changing chromatin environment as the embryo passes through the MZT. While the factors involved may be species specific, it is likely that early development of many species requires a transition through a naïve chromatin state. This process may allow for the reprogramming of the genome from that of a specified germ cell to a totipotent state. In the future, it will be essential to determine the proteins that regulate this reprogramming.

Maternally provided pioneer factors are likely to be essential for this reprogramming process and for the initial steps of embryonic development. However, because of the unique chromatin environment of the rapidly cycling embryo, these early embryonic factors may possess distinct properties that differ from canonical pioneer factors that function later in development. Future studies into the mechanisms by which pioneer factors like ZLD and Oct4 drive genome activation in the early embryo are likely to provide key insights into pioneer factor function as well as, more generally, into the process of reprogramming genomes to establish totipotency.

REFERENCES

Aboobaker, A. A., Tomancak, P., Patel, N., Rubin, G. M., & Lai, E. C. (2005). Drosophila microRNAs exhibit diverse spatial expression patterns during embryonic development. *Proceedings of the National Academy of Sciences of the United States of America*, *102*, 18017–18022.

Adelman, K., & Lis, J. T. (2012). Promoter-proximal pausing of RNA polymerase II: Emerging roles in metazoans. *Nature Reviews. Genetics*, *13*, 720–731.

Ali-Murthy, Z., Lott, S. E., Eisen, M. B., & Kornberg, T. B. (2013). An essential role for zygotic expression in the pre-cellular Drosophila embryo. *PLoS Genetics*, *9*, e1003428.

Amodeo, A. A., Jukam, D., Straight, A. F., & Skotheim, J. M. (2015). Histone titration against the genome sets the DNA-to-cytoplasm threshold for the Xenopus midblastula transition. *Proceedings of the National Academy of Sciences of the United States of America*, *112*(10), E1086–E1095.

Andersen, I. S., Lindeman, L. C., Reiner, A. H., Ostrup, O., Aanes, H., Alestrom, P., et al. (2013). Epigenetic marking of the zebrafish developmental program. *Current Topics in Developmental Biology*, *104*, 85–112.

Anderson, K. V., & Lengyel, J. A. (1979). Rates of synthesis of major classes of RNA in Drosophila embryos. *Developmental Biology*, *70*, 217–231.

Anderson, K. V., & Lengyel, J. A. (1981). Changing rates of DNA and RNA synthesis in Drosophila embryos. *Developmental Biology*, *82*, 127–138.

Arbeitman, M. N., Furlong, E. E., Imam, F., Johnson, E., Null, B. H., Baker, B. S., et al. (2002). Gene expression during the life cycle of *Drosophila melanogaster*. *Science*, *297*, 2270–2275.

Benoit, B., He, C. H., Zhang, F., Votruba, S. M., Tadros, W., Westwood, J. T., et al. (2009). An essential role for the RNA-binding protein Smaug during the Drosophila maternal-to-zygotic transition. *Development, 136*, 923–932.

Biedler, J. K., Hu, W., Tae, H., & Tu, Z. (2012). Identification of early zygotic genes in the yellow fever mosquito *Aedes aegypti* and discovery of a motif involved in early zygotic genome activation. *PLoS One, 7*, e33933.

Biemar, F., Zinzen, R., Ronshaugen, M., Sementchenko, V., Manak, J. R., & Levine, M. S. (2005). Spatial regulation of microRNA gene expression in the Drosophila embryo. *Proceedings of the National Academy of Sciences of the United States of America, 102*, 15907–15911.

Blythe, S. A., & Wieschaus, E. F. (2015). Zygotic genome activation triggers the DNA replication checkpoint at the midblastula transition. *Cell, 160*, 1169–1181.

Boettiger, A. N., & Levine, M. (2009). Synchronous and stochastic patterns of gene activation in the Drosophila embryo. *Science, 325*, 471–473.

Bonn, S., Zinzen, R. P., Girardot, C., Gustafson, E. H., Perez-Gonzalez, A., Delhomme, N., et al. (2012). Tissue-specific analysis of chromatin state identifies temporal signatures of enhancer activity during embryonic development. *Nature Genetics, 44*, 148–156.

Bradley, R. K., Li, X. Y., Trapnell, C., Davidson, S., Pachter, L., Chu, H. C., et al. (2010). Binding site turnover produces pervasive quantitative changes in transcription factor binding between closely related Drosophila species. *PLoS Biology, 8*, e1000343.

Bushati, N., Stark, A., Brennecke, J., & Cohen, S. M. (2008). Temporal reciprocity of miRNAs and their targets during the maternal-to-zygotic transition in Drosophila. *Current Biology: CB, 18*, 501–506.

Butler, J. E., & Kadonaga, J. T. (2001). Enhancer-promoter specificity mediated by DPE or TATA core promoter motifs. *Genes & Development, 15*, 2515–2519.

Chen, K., Johnston, J., Shao, W., Meier, S., Staber, C., & Zeitlinger, J. (2013). A global change in RNA polymerase II pausing during the Drosophila midblastula transition. *eLife, 2*, e00861.

Cirillo, L. A., Lin, F. R., Cuesta, I., Friedman, D., Jarnik, M., & Zaret, K. S. (2002). Opening of compacted chromatin by early developmental transcription factors HNF3 (FoxA) and GATA-4. *Molecular Cell, 9*, 279–289.

Cirillo, L. A., McPherson, C. E., Bossard, P., Stevens, K., Cherian, S., Shim, E. Y., et al. (1998). Binding of the winged-helix transcription factor HNF3 to a linker histone site on the nucleosome. *The EMBO Journal, 17*, 244–254.

Cirillo, L. A., & Zaret, K. S. (1999). An early developmental transcription factor complex that is more stable on nucleosome core particles than on free DNA. *Molecular Cell, 4*, 961–969.

Combs, P. A., & Eisen, M. B. (2013). Sequencing mRNA from cryo-sliced Drosophila embryos to determine genome-wide spatial patterns of gene expression. *PLoS One, 8*, e71820.

De Renzis, S., Elemento, O., Tavazoie, S., & Wieschaus, E. F. (2007). Unmasking activation of the zygotic genome using chromosomal deletions in the Drosophila embryo. *PLoS Biology, 5*, e117.

Dimitrov, S., Almouzni, G., Dasso, M., & Wolffe, A. P. (1993). Chromatin transitions during early Xenopus embryogenesis: Changes in histone H4 acetylation and in linker histone type. *Developmental Biology, 160*, 214–227.

Edgar, B. A., Kiehle, C. P., & Schubiger, G. (1986). Cell cycle control by the nucleo-cytoplasmic ratio in early Drosophila development. *Cell, 44*, 365–372.

Edgar, B. A., & Schubiger, G. (1986). Parameters controlling transcriptional activation during early Drosophila development. *Cell, 44*, 871–877.

Erickson, J. W., & Cline, T. W. (1998). Key aspects of the primary sex determination mechanism are conserved across the genus Drosophila. *Development, 125*, 3259–3268.

Fadloun, A., Eid, A., & Torres-Padilla, M. E. (2013). Mechanisms and dynamics of heterochromatin formation during mammalian development: Closed paths and open questions. *Current Topics in Developmental Biology, 104*, 1–45.

Foe, V. E., & Alberts, B. M. (1983). Studies of nuclear and cytoplasmic behaviour during the five mitotic cycles that precede gastrulation in Drosophila embryogenesis. *Journal of Cell Science, 61*, 31–70.

Foo, S. M., Sun, Y., Lim, B., Ziukaite, R., O'Brien, K., Nien, C. Y., et al. (2014). Zelda potentiates morphogen activity by increasing chromatin accessibility. *Current Biology: CB, 24*, 1341–1346.

Fu, G., Ghadam, P., Sirotkin, A., Khochbin, S., Skoultchi, A. I., & Clarke, H. J. (2003). Mouse oocytes and early embryos express multiple histone H1 subtypes. *Biology of Reproduction, 68*, 1569–1576.

Fu, S., Nien, C. Y., Liang, H. L., & Rushlow, C. (2014). Co-activation of microRNAs by Zelda is essential for early Drosophila development. *Development, 141*, 2108–2118.

Gaertner, B., Johnston, J., Chen, K., Wallaschek, N., Paulson, A., Garruss, A. S., et al. (2012). Poised RNA polymerase II changes over developmental time and prepares genes for future expression. *Cell Reports, 2*, 1670–1683.

Gerstein, M. B., Lu, Z. J., Van Nostrand, E. L., Cheng, C., Arshinoff, B. I., Liu, T., et al. (2010). Integrative analysis of the Caenorhabditis elegans genome by the modENCODE project. *Science, 330*, 1775–1787.

Giraldez, A. J., Mishima, Y., Rihel, J., Grocock, R. J., Van Dongen, S., Inoue, K., et al. (2006). Zebrafish MiR-430 promotes deadenylation and clearance of maternal mRNAs. *Science, 312*, 75–79.

Gualdi, R., Bossard, P., Zheng, M., Hamada, Y., Coleman, J. R., & Zaret, K. S. (1996). Hepatic specification of the gut endoderm in vitro: Cell signaling and transcriptional control. *Genes & Development, 10*, 1670–1682.

Hamm, D. C., Bondra, E. R., & Harrison, M. M. (2015). Transcriptional activation is a conserved feature of the early embryonic factor Zelda that requires a cluster of four zinc fingers for DNA binding and a low-complexity activation domain. *The Journal of Biological Chemistry, 290*, 3508–3518.

Harrison, M. M., Botchan, M. R., & Cline, T. W. (2010). Grainyhead and Zelda compete for binding to the promoters of the earliest-expressed Drosophila genes. *Developmental Biology, 345*, 248–255.

Harrison, M. M., Li, X. Y., Kaplan, T., Botchan, M. R., & Eisen, M. B. (2011). Zelda binding in the early *Drosophila melanogaster* embryo marks regions subsequently activated at the maternal-to-zygotic transition. *PLoS Genetics*, 7, e1002266.

He, Q., Bardet, A. F., Patton, B., Purvis, J., Johnston, J., Paulson, A., et al. (2011). High conservation of transcription factor binding and evidence for combinatorial regulation across six Drosophila species. *Nature Genetics, 43*, 414–420.

Heyn, P., Kircher, M., Dahl, A., Kelso, J., Tomancak, P., Kalinka, A. T., et al. (2014). The earliest transcribed zygotic genes are short, newly evolved, and different across species. *Cell Reports, 6*, 285–292.

Juven-Gershon, T., Hsu, J. Y., & Kadonaga, J. T. (2008). Caudal, a key developmental regulator, is a DPE-specific transcriptional factor. *Genes & Development, 22*, 2823–2830.

Juven-Gershon, T., & Kadonaga, J. T. (2010). Regulation of gene expression via the core promoter and the basal transcriptional machinery. *Developmental Biology, 339*, 225–229.

Kanodia, J. S., Liang, H. L., Kim, Y., Lim, B., Zhan, M., Lu, H., et al. (2012). Pattern formation by graded and uniform signals in the early Drosophila embryo. *Biophysical Journal, 102*, 427–433.

Kaplan, T., X-y, Li, Sabo, P. J., Thomas, S., Stamatoyannopoulos, J. A., Biggin, M. D., et al. (2011). Quantitative models of the mechanisms that control genome-wide patterns of transcription factor binding during early *Drosophila* development. *PLoS Genetics*, *7*, e1001290.

Kouzarides, T. (2007). Chromatin modifications and their function. *Cell*, *128*, 693–705.

Kvon, E. Z., Stampfel, G., Yanez-Cuna, J. O., Dickson, B. J., & Stark, A. (2012). HOT regions function as patterned developmental enhancers and have a distinct cis-regulatory signature. *Genes & Development*, *26*, 908–913.

Lagha, M., Bothma, J. P., Esposito, E., Ng, S., Stefanik, L., Tsui, C., et al. (2013). Paused Pol II coordinates tissue morphogenesis in the Drosophila embryo. *Cell*, *153*, 976–987.

Lamb, M. M., & Laird, C. D. (1976). Increase in nuclear poly(A)-containing RNA at syncytial blastoderm in *Drosophila melanogaster* embryos. *Developmental Biology*, *52*, 31–42.

Lecuyer, E., Yoshida, H., Parthasarathy, N., Alm, C., Babak, T., Cerovina, T., et al. (2007). Global analysis of mRNA localization reveals a prominent role in organizing cellular architecture and function. *Cell*, *131*, 174–187.

Lee, M. T., Bonneau, A. R., Takacs, C. M., Bazzini, A. A., Divito, K. R., Fleming, E. S., et al. (2013). Nanog, Pou5f1 and SoxB1 activate zygotic gene expression during the maternal-to-zygotic transition. *Nature*, *503*(7476), 360–364.

Leichsenring, M., Maes, J., Mossner, R., Driever, W., & Onichtchouk, D. (2013). Pou5f1 transcription factor controls zygotic gene activation in vertebrates. *Science*, *341*(6149), 1005–1009.

Li, X. Y., Harrison, M. M., Villalta, J. E., Kaplan, T., & Eisen, M. B. (2014). Establishment of regions of genomic activity during the maternal to zygotic transition. *eLife*, *3*, 3.

Li, X. Y., MacArthur, S., Bourgon, R., Nix, D., Pollard, D. A., Iyer, V. N., et al. (2008). Transcription factors bind thousands of active and inactive regions in the Drosophila blastoderm. *PLoS Biology*, *6*, e27.

Li, X. Y., Thomas, S., Sabo, P. J., Eisen, M. B., Stamatoyannopoulos, J. A., & Biggin, M. D. (2011). The role of chromatin accessibility in directing the widespread, overlapping patterns of Drosophila transcription factor binding. *Genome Biology*, *12*, R34.

Liang, H. L., Nien, C. Y., Liu, H. Y., Metzstein, M. M., Kirov, N., & Rushlow, C. (2008). The zinc-finger protein Zelda is a key activator of the early zygotic genome in Drosophila. *Nature*, *456*, 400–403.

Liberman, L. M., & Stathopoulos, A. (2009). Design flexibility in cis-regulatory control of gene expression: Synthetic and comparative evidence. *Developmental Biology*, *327*, 578–589.

Lin, M. F., Deoras, A. N., Rasmussen, M. D., & Kellis, M. (2008). Performance and scalability of discriminative metrics for comparative gene identification in 12 Drosophila genomes. *PLoS Computational Biology*, *4*, e1000067.

Lindeman, L. C., Andersen, I. S., Reiner, A. H., Li, N., Aanes, H., Ostrup, O., et al. (2011). Prepatterning of developmental gene expression by modified histones before zygotic genome activation. *Developmental Cell*, *21*, 993–1004.

Lott, S. E., Villalta, J. E., Schroth, G. P., Luo, S., Tonkin, L. A., & Eisen, M. B. (2011). Noncanonical compensation of the zygotic X transcription in early *Drosophila melanogaster* development revealed through single-embryo RNA-seq. *PLoS Biology*, *9*, e1000590.

Lu, X., Li, J. M., Elemento, O., Tavazoie, S., & Wieschaus, E. F. (2009). Coupling of zygotic transcription to mitotic control at the Drosophila mid-blastula transition. *Development*, *136*, 2101–2110.

Lund, E., Liu, M., Hartley, R. S., Sheets, M. D., & Dahlberg, J. E. (2009). Deadenylation of maternal mRNAs mediated by miR-427 in Xenopus laevis embryos. *RNA*, *15*, 2351–2363.

MacArthur, S., Li, X. Y., Li, J., Brown, J. B., Chu, H. C., Zeng, L., et al. (2009). Developmental roles of 21 Drosophila transcription factors are determined by quantitative differences in binding to an overlapping set of thousands of genomic regions. *Genome Biology*, *10*, R80.

McKay, D. J., & Lieb, J. D. (2013). A common set of DNA regulatory elements shapes Drosophila appendages. *Developmental Cell*, *27*, 306–318.

McKnight, S. L., & Miller, O. L., Jr. (1976). Ultrastructural patterns of RNA synthesis during early embryogenesis of *Drosophila melanogaster*. *Cell*, *8*, 305–319.

Moorman, C., Sun, L. V., Wang, J., de Wit, E., Talhout, W., Ward, L. D., et al. (2006). Hotspots of transcription factor colocalization in the genome of *Drosophila melanogaster*. *Proceedings of the National Academy of Sciences of the United States of America*, *103*, 12027–12032.

Negre, N., Brown, C. D., Ma, L., Bristow, C. A., Miller, S. W., Wagner, U., et al. (2011). A cis-regulatory map of the Drosophila genome. *Nature*, *471*, 527–531.

Newport, J., & Kirschner, M. (1982a). A major developmental transition in early Xenopus embryos: I. Characterization and timing of cellular changes at the midblastula stage. *Cell*, *30*, 675–686.

Newport, J., & Kirschner, M. (1982b). A major developmental transition in early Xenopus embryos: II. Control of the onset of transcription. *Cell*, *30*, 687–696.

Nien, C. Y., Liang, H. L., Butcher, S., Sun, Y., Fu, S., Gocha, T., et al. (2011). Temporal coordination of gene networks by Zelda in the early Drosophila embryo. *PLoS Genetics*, 7, e1002339.

Ohtsuki, S., Levine, M., & Cai, H. N. (1998). Different core promoters possess distinct regulatory activities in the Drosophila embryo. *Genes & Development*, *12*, 547–556.

Paris, M., Kaplan, T., Li, X. Y., Villalta, J. E., Lott, S. E., & Eisen, M. B. (2013). Extensive divergence of transcription factor binding in Drosophila embryos with highly conserved gene expression. *PLoS Genetics*, *9*, e1003748.

Perez-Montero, S., Carbonell, A., Moran, T., Vaquero, A., & Azorin, F. (2013). The embryonic linker histone H1 variant of Drosophila, dBigH1, regulates zygotic genome activation. *Developmental Cell*, *26*, 578–590.

Pilot, F., Philippe, J. M., Lemmers, C., Chauvin, J. P., & Lecuit, T. (2006). Developmental control of nuclear morphogenesis and anchoring by charleston, identified in a functional genomic screen of Drosophila cellularisation. *Development*, *133*, 711–723.

Pritchard, D. K., & Schubiger, G. (1996). Activation of transcription in Drosophila embryos is a gradual process mediated by the nucleocytoplasmic ratio. *Genes & Development*, *10*, 1131–1142.

Rach, E. A., Winter, D. R., Benjamin, A. M., Corcoran, D. L., Ni, T., Zhu, J., et al. (2011). Transcription initiation patterns indicate divergent strategies for gene regulation at the chromatin level. *PLoS Genetics*, 7, e1001274.

Rada-Iglesias, A., Bajpai, R., Swigut, T., Brugmann, S. A., Flynn, R. A., & Wysocka, J. (2011). A unique chromatin signature uncovers early developmental enhancers in humans. *Nature*, *470*, 279–283.

Rothe, M., Pehl, M., Taubert, H., & Jackle, H. (1992). Loss of gene function through rapid mitotic cycles in the Drosophila embryo. *Nature*, *359*, 156–159.

Roy, S., Ernst, J., Kharchenko, P. V., Kheradpour, P., Negre, N., Eaton, M. L., et al. (2010). Identification of functional elements and regulatory circuits by Drosophila modENCODE. *Science*, *330*, 1787–1797.

Satija, R., & Bradley, R. K. (2012). The TAGteam motif facilitates binding of 21 sequence-specific transcription factors in the Drosophila embryo. *Genome Research*, *22*, 656–665.

Saunders, A., Core, L. J., Sutcliffe, C., Lis, J. T., & Ashe, H. L. (2013). Extensive polymerase pausing during Drosophila axis patterning enables high-level and pliable transcription. *Genes & Development*, *27*, 1146–1158.

Sefton, L., Timmer, J. R., Zhang, Y., Beranger, F., & Cline, T. W. (2000). An extracellular activator of the Drosophila JAK/STAT pathway is a sex-determination signal element. *Nature, 405*, 970–973.

Shermoen, A. W., McCleland, M. L., & O'Farrell, P. H. (2010). Developmental control of late replication and S phase length. *Current Biology: CB, 20*, 2067–2077.

Shermoen, A. W., & O'Farrell, P. H. (1991). Progression of the cell cycle through mitosis leads to abortion of nascent transcripts. *Cell, 67*, 303–310.

Siddiqui, N. U., Li, X., Luo, H., Karaiskakis, A., Hou, H., Kislinger, T., et al. (2012). Genome-wide analysis of the maternal-to-zygotic transition in Drosophila primordial germ cells. *Genome Biology, 13*, R11.

Smith, R. C., Dworkin-Rastl, E., & Dworkin, M. B. (1988). Expression of a histone H1-like protein is restricted to early Xenopus development. *Genes & Development, 2*, 1284–1295.

Soufi, A., Donahue, G., & Zaret, K. S. (2012). Facilitators and impediments of the pluripotency reprogramming factors' initial engagement with the genome. *Cell, 151*, 994–1004.

Stanojevic, D., Small, S., & Levine, M. (1991). Regulation of a segmentation stripe by overlapping activators and repressors in the Drosophila embryo. *Science, 254*, 1385–1387.

Staudt, N., Fellert, S., Chung, H. R., Jackle, H., & Vorbruggen, G. (2006). Mutations of the Drosophila zinc finger-encoding gene vielfaltig impair mitotic cell divisions and cause improper chromosome segregation. *Molecular Biology of the Cell, 17*, 2356–2365.

Struffi, P., Corado, M., Kaplan, L., Yu, D., Rushlow, C., & Small, S. (2011). Combinatorial activation and concentration-dependent repression of the Drosophila even skipped stripe 3+7 enhancer. *Development, 138*, 4291–4299.

Sung, H. W., Spangenberg, S., Vogt, N., & Grosshans, J. (2013). Number of nuclear divisions in the Drosophila blastoderm controlled by onset of zygotic transcription. *Current Biology: CB, 23*, 133–138.

Tadros, W., Goldman, A. L., Babak, T., Menzies, F., Vardy, L., Orr-Weaver, T., et al. (2007). SMAUG is a major regulator of maternal mRNA destabilization in Drosophila and its translation is activated by the PAN GU kinase. *Developmental Cell, 12*, 143–155.

ten Bosch, J. R., Benavides, J. A., & Cline, T. W. (2006). The TAGteam DNA motif controls the timing of Drosophila pre-blastoderm transcription. *Development, 133*, 1967–1977.

The ENCODE Project Consortium, Birney, E., Stamatoyannopoulos, J. A., Dutta, A., Guigo, R., Gingeras, T. R., et al. (2007). Identification and analysis of functional elements in 1% of the human genome by the ENCODE pilot project. *Nature, 447*, 799–816.

Thomas, S., Li, X. Y., Sabo, P. J., Sandstrom, R., Thurman, R. E., Canfield, T. K., et al. (2011). Dynamic reprogramming of chromatin accessibility during Drosophila embryo development. *Genome Biology, 12*, R43.

Tsurumi, A., Xia, F., Li, J., Larson, K., Lafrance, R., & Li, W. X. (2011). STAT is an essential activator of the zygotic genome in the early Drosophila embryo. *PLoS Genetics, 7*, e1002086.

Vastenhouw, N. L., Zhang, Y., Woods, I. G., Imam, F., Regev, A., Liu, X. S., et al. (2010). Chromatin signature of embryonic pluripotency is established during genome activation. *Nature, 464*, 922–926.

Wibrand, K., & Olsen, L. C. (2002). Linker histone H1M transcripts mark the developing germ line in zebrafish. *Mechanisms of Development, 117*, 249–252.

Wrischnik, L. A., Timmer, J. R., Megna, L. A., & Cline, T. W. (2003). Recruitment of the proneural gene scute to the Drosophila sex-determination pathway. *Genetics, 165*, 2007–2027.

Xu, Z., Chen, H., Ling, J., Yu, D., Struffi, P., & Small, S. (2014). Impacts of the ubiquitous factor Zelda on Bicoid-dependent DNA binding and transcription in Drosophila. *Genes & Development, 28*, 608–621.

Yanez-Cuna, J. O., Dinh, H. Q., Kvon, E. Z., Shlyueva, D., & Stark, A. (2012). Uncovering cis-regulatory sequence requirements for context-specific transcription factor binding. *Genome Research, 22*, 2018–2030.

Yasuda, G. K., Baker, J., & Schubiger, G. (1991). Temporal regulation of gene expression in the blastoderm Drosophila embryo. *Genes & Development, 5*, 1800–1812.

Zabidi, M. A., Arnold, C. D., Schernhuber, K., Pagani, M., Rath, M., Frank, O., et al. (2015). Enhancer-core-promoter specificity separates developmental and housekeeping gene regulation. *Nature, 518*, 556–559.

Zaret, K. S., & Carroll, J. S. (2011). Pioneer transcription factors: Establishing competence for gene expression. *Genes & Development, 25*, 2227–2241.

Zeitlinger, J., Stark, A., Kellis, M., Hong, J. W., Nechaev, S., Adelman, K., et al. (2007). RNA polymerase stalling at developmental control genes in the *Drosophila melanogaster* embryo. *Nature Genetics, 39*, 1512–1516.

Zhang, Y., Vastenhouw, N. L., Feng, J., Fu, K., Wang, C., Ge, Y., et al. (2014). Canonical nucleosome organization at promoters forms during genome activation. *Genome Research, 24*, 260–266.

CHAPTER FOUR

Coordinating Cell Cycle Remodeling with Transcriptional Activation at the *Drosophila* MBT

Shelby A. Blythe, Eric F. Wieschaus[1]

Department of Molecular Biology and Howard Hughes Medical Institute, Princeton University, Princeton, New Jersey, USA

[1]Corresponding author: e-mail address: efw@princeton.edu

Contents

Abstract

During the maternal-to-zygotic transition (MZT), major changes in cell cycle regulation coincide with large-scale zygotic genome activation. In this chapter, we discuss the current understanding of how the cell cycle is remodeled over the course of the *Drosophila* MZT, and how the temporal precision of this event is linked to contemporaneous alterations in genome-wide chromatin structure and transcriptional activity. The cell cycle is initially lengthened during the MZT by activation of the DNA replication checkpoint but, subsequently, zygotically supplied factors are essential for establishing lasting modifications to the cell cycle.

Current Topics in Developmental Biology, Volume 113
ISSN 0070-2153
http://dx.doi.org/10.1016/bs.ctdb.2015.06.002

1. INTRODUCTION

Early embryos are invariably concerned with maximizing cellular proliferation in advance of morphogenesis and regionalized functional specialization. The period immediately following fertilization typically consists of a sequence of rapid, near-synchronous mitotic divisions that occur under conditions of minimal cellular differentiation. For many organisms, including *Drosophila*, *Xenopus*, and zebrafish, these early cell cycles are driven by maternally supplied factors that are regulated posttranscriptionally, with no clear requirement for transcription from the zygotic genome. In general, zygotic transcription is constrained during this early period. By delaying transcriptional responses to maternally supplied patterning cues, these early embryonic cells may therefore retain maximal plasticity in their developmental fates. After a precise time interval following fertilization, such embryos undergo a major developmental transition that involves activation of zygotic gene expression and a concomitant remodeling and lengthening of the embryonic cell cycle. This developmental milestone, termed the midblastula transition (MBT), marks the genetic transition between maternal and zygotic control of development, and the initiation of the embryonic programs of morphogenesis and differentiation. Whereas the maternal-to-zygotic transition (MZT) encompasses the entire period during which maternal products are cleared and zygotic genome activation occurs, we shall focus here on the MBT, which defines the particular time-point at which the zygote wrests control of development from its mother.

Although the specific number of mitotic divisions, the overall timescale, and the mechanism for suppressing cellular differentiation will vary considerably across different species, within an individual species, the number of divisions and the timing of accompanying global changes in gene expression are highly reproducible. Three major events occur at the MBT. (1) A subset of maternally supplied mRNAs is degraded. (2) Following major changes in chromatin architecture, the zygotic genome undergoes large-scale transcriptional activation. (3) The rapid, synchronous, maternally controlled cell cycle is remodeled to have a significantly longer period, and falls under control of local, zygotically provided cues. Experiments in a number of different species suggest that the MBT clock consists of at least two types of timers that regulate these three major events. One timer measures elapsed time following egg activation or fertilization, and regulates the onset of maternal mRNA degradation and the activation of a subset of zygotic mRNAs.

Another timer measures the accumulation of zygotic DNA as nuclear content increases, and regulates the timing of cell cycle remodeling and the activation of a distinct subset of zygotic mRNAs. The MBT clock resembles an hourglass, where degradation or accumulation of certain factors serves to count off or demarcate intervals of elapsed time and trigger biological responses at predefined threshold values (Johnson & Day, 2000; Rensing, Meyer-Grahle, & Ruoff, 2001). However, it is currently not understood precisely what factors are measured by this clock, and how these thresholds are mechanistically triggered.

In contrast, the effect of the clock is better understood. Over the course of the MZT, embryos acquire cell biological functions—such as gap phases in the cell cycle or large-scale transcriptional activity—that were missing in pre-MBT stages. Many of these missing functions are acquired at or around the MBT with remarkable temporal precision, and—as is increasingly evident—these functions are often acquired interdependently. Understanding how these systems change and interact over the course of the MBT perhaps gives the best available vantage point to the inner workings of the overall timing mechanism.

In the following review, we present issues central to the temporal coordination of the MBT. In the first part, we will discuss the current understanding of the effects of the clock, i.e., the changes in cell cycle and transcriptional regulation that occur at the MBT. In the second part, we will discuss progress toward understanding how the clock itself functions. We primarily focus on the *Drosophila melanogaster* model system, but draw comparisons and predictions based on work in other models when appropriate.

2. WHAT HAPPENS AT THE MBT?

Following fertilization, *Drosophila* embryos undergo 13 rapid, metasynchronous mitotic divisions prior to undergoing cell cycle lengthening at the MBT (Fig. 1). The early embryo develops as a syncytium, with amplified nuclei occupying a shared cytoplasm without intervening plasma membranes. As such, a cell cycle consists of DNA replication followed by mitosis without cytokinesis. By convention, we refer to a syncytial cell cycle as a "nuclear cycle," abbreviated "NC," as in the sentence, "The pre-MBT period in *Drosophila* typically spans between NC1 and NC13." Following fertilization, the first seven nuclear cycles occur deep within the inner cytoplasm of the embryo. Each subsequent nuclear cycle drives the posteriorward expansion of the mass of amplifying nuclei. From NC8 though

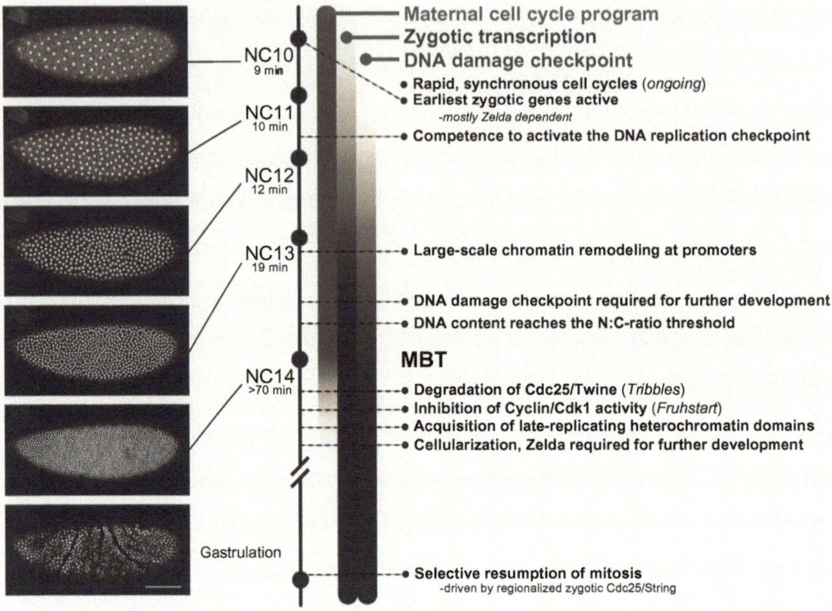

Figure 1 Timeline of early *Drosophila* development. Key events underlying the temporal coordination of cell cycle remodeling and zygotic genome activation over the course of the MBT are highlighted and placed in sequence. Left: frames from a time-lapse micrograph of *Histone H2Av-GFP* transgenic embryos highlight the four cell cycles preceding cell cycle pause at NC14/MBT and gastrulation. Anterior is at left, dorsal is up. Center: a timeline of relative cell cycle lengths is keyed to the micrographs at left. Cell cycle pause occurs following mitosis 13, yielding a period of variable length before any cell undergoes further division. The earliest of these (+70 min after mitosis 13) is pictured in the micrograph labeled "gastrulation," where a small patch of cells is dividing in the dorsal/anterior-most edge of the image. Right: overall durations of the periods of maternal cell cycle control (red, light gray in the print version), zygotic genome activity (blue, dark gray in the print version), and DNA damage checkpoint competence (purple, dark gray in the print version) are indicated. Selected events discussed in the text are highlighted to indicate the relative order of occurrence.

NC10, nuclei migrate outward toward the embryo cortex to form what is called a "syncytial blastoderm." During the syncytial blastoderm stages, from NC10 through NC13, nuclei undergo the final four mitoses of the pre-MBT period on the surface of the embryo. Following mitosis 13, embryos undergo the MBT resulting in a dramatically prolonged NC14 interphase (Fig. 1). Additionally, during NC14, embryos undergo a process called "cellularization" wherein each individual nucleus and its surrounding share of cytoplasm become enveloped in a plasma membrane. This period, between the MBT and gastrulation, is referred to as the "cellular blastoderm" stage of development, and both cell cycle pause and cellularization represent the two major macroscopic indicators that the MZT is well underway.

For any species, the timing of the MBT derives at least some of its temporal precision from a measurement of the nuclear-to-cytoplasmic (N:C) ratio. Eggs from oviparous organisms have cytoplasmic volumes that are several orders of magnitude greater than a typical postembryonic cell, while early embryonic nuclei are similar in size to their adult counterparts. A one-cell embryo with its single nucleus, therefore, has an extremely skewed N:C ratio, owing to the relative overabundance of cytoplasm. Since embryos will not grow during the initial stages of development, their cytoplasmic volume remains constant over the period of early development. Each round of cell cycle activity, however, exponentially amplifies the relative number of nuclei. As the number of nuclei increases over the pre-MBT period, the initially skewed N:C ratio begins to approximate that of a typical postembryonic cell and, at a predetermined threshold of nuclear content, embryos trigger the MBT (Fig. 1).

Some feature of zygotic DNA content contributes to the measurement of the N:C ratio. Similar to other organisms (Kane & Kimmel, 1993; Newport & Kirschner, 1982a), the number of pre-MBT mitoses in *Drosophila* embryos depends on zygotic ploidy. Haploid embryos, with 50% DNA content, will undergo one additional pre-MBT cell cycle than their diploid counterparts (Di Talia et al., 2013; Edgar, Kiehle, & Schubiger, 1986; Lu, Li, Elemento, Tavazoie, & Wieschaus, 2009), whereas triploid embryos, with 150% DNA content, will perform one fewer pre-MBT cell cycle (Erickson & Quintero, 2007). The decision to undergo one more or one less mitosis is quantitatively linked to zygotic DNA content in general. By the use of aneuploid *Drosophila* embryos, it can be demonstrated that 70% DNA content marks the threshold for performing one more or one less pre-MBT mitosis (Lu et al., 2009). The timing system is, therefore, buffered against small changes in genomic content, requiring loss (or gain) of ~½ of the haploid DNA content to alter the number of pre-MBT cell cycles (Lu, Drocco, & Wieschaus, 2010). Yet, despite the demonstrated importance of zygotic DNA content for determining the number of pre-MBT mitoses, the N:C ratio is clearly not the only timer of the MBT. In *Drosophila*, maternally supplied mRNA transcripts are degraded largely on a schedule that is determined by time elapsed after egg activation/fertilization and independent of zygotic nuclear content (Tadros et al., 2003, 2007). Likewise, a significant proportion of zygotic gene activation can take place in *Drosophila* embryos with sub-MBT nuclear content (Edgar et al., 1986; Edgar & Schubiger, 1986; Lu et al., 2009; McCleland & O'Farrell, 2008).

2.1 Large-Scale Cell Cycle Remodeling in *Drosophila* Is Zygotically Controlled and N:C Ratio Dependent

In *Drosophila*, the permanent switch between rapid pre-MBT cell cycles and longer post-MBT cell cycles is a zygotically driven process that involves the acquisition of inhibitory phosphorylation of Cdk1 (Di Talia et al., 2013; Edgar, Sprenger, Duronio, Leopold, & O'Farrell, 1994; Farrell & O'Farrell, 2013). The quick pre-MBT cell cycle consists of sequential rounds of S-phase followed by M-phase with little or no intervening gap phases. Due to high cyclin/Cdk1 (CDK) catalytic activity, these early cell cycles can be as fast as 8 min, during which the entirety of the genome is replicated in as little as 3–5 min (Blumenthal, Kriegstein, & Hogness, 1974; Edgar et al., 1994; Shermoen, McCleland, & O'Farrell, 2010). Two regulatory mechanisms limit CDK activity during these early cell cycles: availability of cyclin protein and regulated inhibitory phosphorylation of Cdk1. Cyclin protein is degraded following each syncytial blastoderm mitotic division and progression through the next mitosis necessitates the synthesis of new cyclin protein from maternally supplied mRNA (Edgar et al., 1994). The kinase that confers inhibitory phosphorylation of CDK, Wee1, is also maternally supplied, but its effect is counteracted early on by the high expression levels of two Cdc25 phosphatases, String and Twine (Price, Rabinovitch, O'Farrell, & Campbell, 2000; Stumpff, Duncan, Homola, Campbell, & Su, 2004). While String seems to be dispensable for these early pre-MBT mitoses, Twine is essential for ensuring the rapid pace of early development (Edgar & Datar, 1996).

Large-scale cell cycle remodeling at NC14 results from the expression of several zygotic transcripts that destabilize Twine protein and thus promote the inhibitory phosphorylation of Cdk1 (Di Talia et al., 2013; Edgar et al., 1994; Farrell & O'Farrell, 2013). Global pharmacological inhibition of RNA Polymerase II (Pol II) demonstrates that embryos require zygotic inputs to halt the maternal cell cycle program (Edgar & Datar, 1996; Farrell & O'Farrell, 2013; Farrell, Shermoen, Yuan, & O'Farrell, 2012; Shermoen et al., 2010). At a high frequency, such inhibited embryos undergo at least one additional synchronous mitotic division, 25–30 min after mitosis 13. Likewise, embryos that prematurely activate at least a subset of zygotic genes can halt cell cycle progression prematurely, after mitosis 12 (Sung, Spangenberg, Vogt, & Großhans, 2013). Before the MBT, Twine protein has a long half-life of 20 min but, following large-scale ZGA at NC14, Twine protein becomes highly unstable and its half-life is reduced to 5 min (Di Talia et al., 2013). Zygotic expression of *tribbles* has been shown to contribute to the process that destabilizes Twine protein: knockdown of *tribbles* by RNAi results in additional synchronous mitotic divisions due to persistence of Twine protein well into NC14 (Farrell & O'Farrell, 2013). However, since the frequency of extra divisions in *tribbles* RNAi embryos is less than that observed with global inhibition of Pol II by α-amanitin, it is presumed that additional, as yet unknown zygotic factors are necessary for destabilization of Twine at the MBT.

A second zygotic factor, *fruhstart*, functions to downregulate Cdk1 activity, independently of Twine degradation. *Fruhstart* is highly transcribed at NC14, and functions as a cyclin-dependent kinase inhibitor to downregulate CDK activity (Gawliński et al., 2007; Grosshans, Müller, & Wieschaus, 2003). Embryos mutant for *fruhstart* undergo extra mitotic divisions at a relatively low rate but, in combination with moderate overexpression of Twine, nearly every *fruhstart* mutant embryo will undergo an extra mitotic division (Grosshans et al., 2003). Translation of *fruhstart* mRNA is positively regulated at NC14 by Fragile X Mental Retardation Protein (Fmr1) and Caprin. Interestingly, Fmr1 and Caprin also negatively regulate translation of *cyclinB* during NC14. Consequently, embryos with attenuated Fmr1 and Caprin function undergo extra mitotic divisions with high phenotypic penetrance (Papoulas et al., 2010). Such cross-regulation of *cyclinB* and *fruhstart* translation may indicate additional levels of control exist to ensure timely and precise remodeling of the early embryonic cell cycle, although the characterization of such additional systems is currently far from complete. Tribbles and Fruhstart therefore represent essential components of a two-step zygotic mechanism for downregulation of CDK activity and to effect large-scale cell cycle remodeling at the MBT.

The onset of these zygotically driven processes is timed by the N:C ratio. Haploid embryos produced by the maternal-effect mutants, *sesame* and *maternal haploid*, develop with a lower overall N:C ratio. Consequently, these embryos undergo one additional pre-MBT mitotic division before activating large-scale cell cycle remodeling (Di Talia et al., 2013; Edgar et al., 1986; Farrell & O'Farrell, 2013). Although not all of ZGA is activated in an N:C ratio-dependent manner, both *tribbles* and *fruhstart* show N:C ratio dependence for their activation (Grosshans et al., 2003; Lu et al., 2009).

Likewise, RNAi knockdown of cyclins A, B, and B3 halts syncytial cell cycle progression, and embryos that arrest with sub-MBT DNA content fail to trigger Twine degradation at a clock-time equivalent to NC14 (Farrell & O'Farrell, 2013; McCleland & O'Farrell, 2008). Given that additional unknown factors are involved in the *tribbles* pathway for Twine destabilization, it may be expected that these will also adhere to an N:C ratio-dependent ZGA timer.

2.2 Transitions in Chromatin: Behind the Scenes During Zygotic Genome Activation

The activation of the zygotic genome represents a series of changes in chromatin structure that occur over the course of the MZT. Arguably, the most critical event of ZGA is its endpoint: the production of mRNA from the zygotic genome. Yet, at the heart of ZGA is a poorly understood transition in chromatin structure that establishes, for the first time, the chromosomal domain architecture and biochemical interactions that realize and underlie these transcriptional regulatory mechanisms. Features of chromatin as fundamental as the distinction between "euchromatin" and "heterochromatin" are arguably first established, in the biochemical sense, at the MBT. In Section 2.2.1, we review some salient features of what is known about the transition of chromatin architecture that underlies ZGA.

2.2.1 Acquisition of an "Active" Chromatin State

Large-scale ZGA commences at NC13 (one cycle prior to large-scale cell cycle remodeling at NC14) with a wave of *de novo* binding of RNA Polymerase II to thousands of promoters across the *Drosophila* genome and a major global transition in chromatin structure (Fig. 1; Blythe & Wieschaus, 2015; Chen et al., 2013; Li, Harrison, Villalta, Kaplan, & Eisen, 2014). To the extent that a "chromatin state" can be defined by patterns of posttranslationally modified histone proteins, pre-MBT chromatin is relatively barren compared to either its post-MBT or its postembryonic counterparts. Chromatin marks typical of active gene expression, including lysine-4 or lysine-36 trimethylated Histone H3 (H3K4me3 and H3K36me3, respectively) are absent (or not detected) from their typical sites of deposition during the pre-MBT stages (Chen et al., 2013; Li et al., 2014). Likewise, chromatin marks associated with Polycomb-mediated transcriptional repression (H3K27me3) also not found prior to the MBT. However, not all chromatin marks are absent from pre-MBT chromatin. For example, acetylated Histone H3 and H4 (H4K8ac, H3K18ac, and H3K27ac) are present to some extent in early embryos, although acetylation of another residue (H3K9ac) is absent (Li et al., 2014). The large-scale activation of zygotic gene expression at the MBT brings with it several changes that typically signify and delineate euchromatic chromatin compartments. Constellations of posttranslational modifications become established throughout chromatin at the MBT, presumably driving the emergence of specialized chromatin states imbued with characteristic regulatory properties (Chen et al., 2013; Li et al., 2014). The nucleosome-free regions that form at promoters of actively expressed genes likewise first become widespread at the MBT (Li et al., 2014). Finally, chromatin architecture corresponding to poised or paused promoters first becomes evident upon initiation of large-scale zygotic gene expression (Blythe & Wieschaus, 2015; Chen et al., 2013).

However, not every early zygotic gene is activated neatly and precisely at the MBT. Direct measurements of zygotic mRNA production or of RNA Pol II occupancy demonstrate that zygotic transcript production is active well before large-scale ZGA (Ali-Murthy, Lott, Eisen, & Kornberg, 2013; Blythe & Wieschaus, 2015; Chen et al., 2013; Garcia, Tikhonov, Lin, & Gregor, 2013; Harrison, Li, Kaplan, Botchan, & Eisen, 2011; Little, Tikhonov, & Gregor, 2013; Pritchard & Schubiger, 1996). This class of pre-MBT transcripts is unique in several ways. Nearly, every pre-MBT gene is short (1–2 kb in length) and transcribed efficiently during the short interphase times of the pre-MBT mitotic divisions, rarely showing evidence of transcriptional pausing or poising (Saunders, Core, Sutcliffe, Lis, & Ashe, 2013). As a population, pre-MBT genes have unique promoter architectures that distinguish them from the majority of their MBT counterparts (Chen et al., 2013). Pre-MBT promoters are highly enriched for consensus sequences for binding Zelda (also known as Vielfaltig), a maternally supplied transcription factor whose function is essential for early events at ZGA (Fig. 1). Zelda functions as a transcriptional activator, binds to thousands of sites in pre-MBT chromatin, and is required for recruitment of RNA Polymerase II to hundreds of genes, which largely fall into the category of N:C ratio-independent transcripts (Blythe & Wieschaus, 2015; De Renzis, Elemento, Tavazoie, & Wieschaus, 2007; Hamm, Bondra, & Harrison, 2015; Harrison et al., 2011; Liang et al., 2008; Lu et al., 2009; Nien et al., 2011; ten Bosch, Benavides, & Cline, 2006).

Zelda appears to be rate limiting for the recruitment of RNA Polymerase II to its targets; *zelda* mutants have a ~50% reduction in overall genome-wide occupancy of RNA Pol II at NC13, accounting for nearly all RNA Pol II binding at Zelda targets. In contrast, loss of *zelda* has little or no effect on Pol II recruitment to *zelda*-independent target genes (Blythe & Wieschaus, 2015). As a population, Zelda target genes also appear to be regulated independently of chromatin-marking mechanisms that accompany transcriptional activity in post-MBT and postembryonic tissues. A modest reduction of protranscriptional chromatin modifications is observed in *zelda* mutant embryos, specifically at enhancers. In contrast, loss of *zelda* function has little or no reported effect on promoter-proximal histone modifications or on nucleosome remodeling around the transcription start sites of its targets (Li et al., 2014).

Perhaps, it is not surprising that loss of function for *zelda* has only a minimal effect on the chromatin modifications that are established at or around the MBT. Zelda target genes are largely expressed early, well before large-scale ZGA, prior to the time when many of these protranscriptional chromatin modifications are established (Chen et al., 2013; Harrison et al., 2011; Li et al., 2014; Liang et al., 2008). Whatever the mechanism through which Zelda functions to activate transcription, it appears to do so independently of regulating the systems for modifying chromatin that have been measured thus far (Li et al., 2014). In sum, with respect to transcriptional activation, there exist at least two classes of early zygotic transcripts that can be delineated by their dependence on Zelda for activation.

2.2.2 Changes in DNA Replication Origin Usage

The chromatin transition may also mark a less well-understood change in DNA replication origin specification. In *Xenopus* and *Drosophila*, DNA replication origins initially appear to be distributed randomly but, after ZGA, origins preferentially localize to promoter regions of actively expressed genes (Hyrien, Maric, & Méchali, 1995; Sasaki, Sawado, Yamaguchi, & Shinomiya, 1999). Unlike in the budding yeast *Saccharomyces cerevisiae*, where origins of replication are associated with distinct, sequence-defined sites of initiation, metazoan origins of replication are DNA sequence independent and rely on structural or biochemical features of chromatin for their specification. One common feature of DNA replication in metazoans, and in *Drosophila* specifically, is a correlation of transcribed regions with DNA replication origins, particularly those that fire early in the course of S-phase (Schübeler et al., 2002; Schwaiger et al., 2009). Experiments in *Xenopus* have shown that the features of ZGA that reorganize replication origins are upstream of entry into productive elongation, but downstream of—and dependent on—transcriptional activation. Using a Gal4-VP16-regulated plasmid reporter assay to measure origin specification in *Xenopus* embryos, it can be demonstrated that origin localization to promoter regions requires binding of transcriptional activators (in this case, Gal4-VP16), and the subsequent maintenance of an activated state. Importantly, origin localization to activated promoter regions occurs efficiently even if transcription is inhibited with α-amanitin (Danis et al., 2004). Although it is difficult to generalize about a genome-wide transition on the basis of these single-gene or single-locus studies, the idea is nonetheless attractive that this feature of euchromatin is established at the MBT in response to the chromatin transition that underlies ZGA. It will be interesting to revisit these concepts with improved temporal resolution using genome-wide approaches as well as using modern optical methods in living embryos. The association of replication origins with actively transcribed regions may underlie a more general feature: that regions of active gene expression typically begin DNA replication early in S-phase compared with repressed or heterochromatic domains.

Indeed, an integral part of the change in chromatin structure of early embryos is the delineation of early- and late-replicating genomic regions, respectively, associated with euchromatic and heterochromatic chromatin domains. Remarkably, much of the post-MBT S-phase lengthening can be attributed to the emergence of late-replicating domains in *Drosophila* heterochromatin (Shermoen et al., 2010). In a typical postembryonic cell, euchromatic regions will initiate DNA replication early in S-phase, whereas heterochromatic regions are replicated later (Schübeler et al., 2002; Schwaiger et al., 2009). These two genomic compartments are not distinguished during the pre-MBT S-phase: genetically "euchromatic" and "heterochromatic" sequences initiate DNA replication simultaneously in the pre-MBT embryo. The distinction between these two genomic neighborhoods as a template for DNA replication is gradually acquired over the course of the MBT, paralleling the onset of large-scale ZGA.

2.2.3 Biochemical Emergence of Heterochromatic States

Heterochromatin is a functionally distinct genomic compartment that is characterized by its relatively low gene density, enrichment for repetitive sequences and transposable elements, highly compact chromatin architecture, and late onset of DNA replication. In a typical postembryonic cell, a large number of chromatin regulatory factors operate to define the physical and functional boundaries between the euchromatic and heterochromatic genomic compartments. These chromatin regulatory factors demarcate these compartments partly by the establishment and maintenance of histone posttranslational modifications. The H3K4me3 histone modification is largely associated with euchromatic compartments, and typically associates with sites of active gene expression. Conversely, H3K9me3 marks heterochromatic compartments and is typically associated with transcriptional repression. Establishment of the H3K9me3 modification at heterochromatin precedes the recruitment of factors, including Heterochromatin Protein 1 (HP1), that promote compact chromatin structure and mediate transcriptional repression. Prior to NC12, histone marks for both compartments are relatively absent from chromatin. From NC12 onward, both H3K4me3 and H3K9me3 appear and begin to demarcate the emergent boundaries between different chromatin neighborhoods (Rudolph et al., 2007). HP1 protein is maternally expressed and nuclear-localized before the MBT. However, before the MBT, HP1 only forms small, transient foci at presumed sites of heterochromatin. In contrast, after the MBT, HP1 forms intense high-concentration assemblies at apically positioned nuclear compartments known to contain the pericentric heterochromatin (Kellum, Raff, & Alberts, 1995; Rudolph et al., 2007; Shermoen et al., 2010). Thus, the emergence of late replication and the establishment of biochemical markers of heterochromatin first become evident at or around the MBT.

Although the mechanism underlying heterochromatin formation in the early embryo is not entirely clear, evidence suggests that the heterochromatic state is dynamically acquired and relies on zygotic transcription. At least part of the requirement for new transcription in the formation of heterochromatin derives from the role of zygotic gene expression during large-scale cell cycle remodeling. Blocking zygotic gene expression with the drug α-amanitin prevents the acquisition of late replication during NC14 (Shermoen et al., 2010). The onset of late replication requires the downregulation of cyclin/Cdk1 activity that accompanies the MBT. Promoting high post-MBT cyclin/Cdk1 activity drives premature replication of late-replicating domains. Since the downregulation of cyclin/Cdk1 activity at the MBT relies on zygotic genome activation (see above), it is likely that loss of late replication in α-amanitin-treated embryos results from preventing cyclin/Cdk1 downregulation. However, transcription may play a more direct role in the acquisition of biochemically distinct heterochromatin. During the MBT in mouse embryos, bursts of dual-strand transcription from heterochromatic satellite repeats are essential for establishing pericentric heterochromatin (Probst & Almouzni, 2011; Probst et al., 2010). This transcription is essential, as depletion of these transcripts results in developmental arrest and failure to fully establish features of pericentric heterochromatin. Interestingly, portions of *Drosophila* heterochromatin are transcribed to some extent before the MBT, although the functional significance of these transcripts has not yet been demonstrated (He et al., 2012). Likewise, the effect of high cyclin/Cdk1 activity on the acquisition of additional biochemical markers of heterochromatin (histone modifications, HP1 localization) has not been tested in *Drosophila*. It remains an interesting possibility that zygotic genome activation influences heterochromatin formation on multiple levels.

2.3 Temporal Coordination of ZGA and the Cell Cycle Via the DNA Replication Checkpoint

Like many organisms, *Drosophila* embryos are initially hypersensitive to genotoxic stress. Following induction of DNA damage, embryos demonstrate a marked failure to fully halt cell cycle progression (Fasulo et al., 2012; Raff & Glover, 1988), and this can likely be attributed to a limited activity of the DNA damage checkpoints in early embryos. Nonetheless, a functional DNA replication checkpoint is necessary for inducing gradual delays in pre-MBT mitosis in response to endogenous developmental cues that cause S-phase lengthening (Figs. 1 and 2). In the context of early embryogenesis, the relevant DNA damage checkpoints are those that respond to damage accrued during the process of DNA replication: the so-called S-phase checkpoints. A checkpoint response requires mechanisms for sensing damaged DNA that function to recruit signal-transducing components, which subsequently modulate cell cycle activity to reversibly halt or slow further progress until the damage can be repaired (Bartek, Lukas, & Lukas, 2004). Damage can come in the form of physical lesions to the DNA polymer (such as double-strand breaks) or in the form of under-replicated DNA. In a typical somatic cell, the type of damage encountered during S-phase determines the molecular mechanism for checkpoint activation. Under conditions where DNA replication capacity is limiting, under-replication of DNA triggers a replication checkpoint that functions to inhibit the firing of new origins of replication, resulting in an overall lengthening of S-phase. Activation of the replication checkpoint also transiently inhibits entry into M-phase to ensure complete copying of the DNA template before cell division. Independently of replication status, another checkpoint, termed the intra-S-phase checkpoint, is triggered in response to double-strand breaks, and functions via a different molecular cascade to slow or halt cell cycle progression.

Despite the critical importance of the DNA damage checkpoint to protect against deleterious genotoxic stress, embryos typically initiate embryonic development without the full benefit of this surveillance mechanism. The

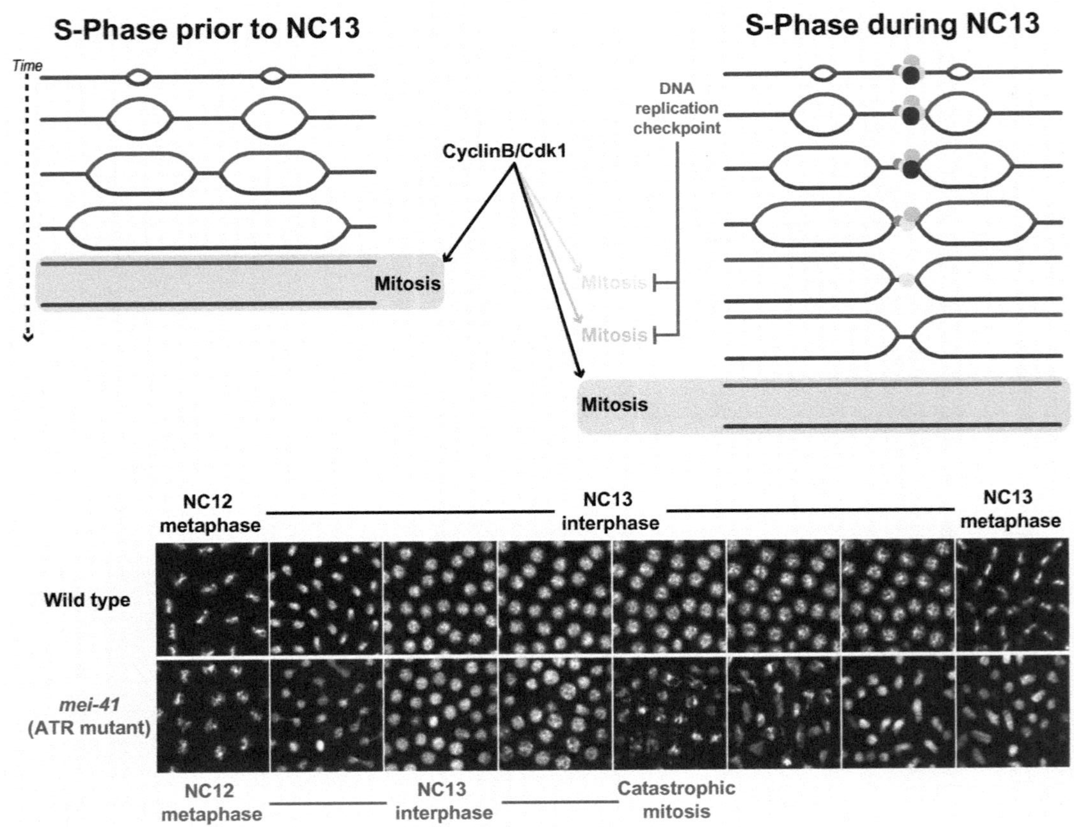

Figure 2 See legend on opposite page.

ability to regulate cell cycle length in response to DNA damage is ultimately acquired over the course of early development. Pre-MBT *Drosophila* embryos lack a functional checkpoint response to double-strand breaks. Inducing double-strand breaks either pharmacologically or with ionizing radiation does not cause a corresponding compensatory lengthening of the cell cycle despite widespread DNA damage (Fasulo et al., 2012). We should stress here the distinction between "DNA damage responses" and "DNA damage checkpoints." Although no major cell cycle checkpoint (that is, cell cycle slowing) is activated in response to double-strand breaks, syncytial blastoderm stage embryos nonetheless use components of the DNA damage response pathway to "detect" damaged nuclei and eliminate them (Iampietro et al., 2014; Takada, Kelkar, & Theurkauf, 2003).

In contrast, embryos have attenuated, but not absent, DNA replication checkpoint functionality. Stalling DNA replication either by inhibition of DNA Polymerase or by depletion of deoxyribonucleotide pools can trigger a transient cell cycle arrest from NC11 onward (Fig. 1). This pre-MBT DNA replication checkpoint is, nonetheless, constrained in its capacity to indefinitely halt cell cycle progression. For example, activating the DNA replication checkpoint with hydroxyurea (HU) treatment of NC12 embryos results in a 2.75-fold increase in interphase time (from 12 to 32 min) before embryos enter a subsequent terminal catastrophic mitosis (Blythe & Wieschaus, 2015; Fasulo et al., 2012). In comparison, mammalian cultured cells can be kept in an HU-induced checkpoint-arrested state for at least 18 h without major deleterious effects (e.g., Adams & Lindsay, 1967). Although the reason for such an anemic replication checkpoint in pre-MBT embryos has not been rigorously determined, it is tempting to speculate that the magnitude of the checkpoint mechanism is dwarfed by the promitotic drive of early embryos.

Figure 2 Cell cycle lengthening by the DNA damage checkpoint. Top panels: prior to NC13 (top left), S-phase proceeds at a maximal rate. During NC13 (top right), chromatin remodeling results in increased promoter occupancy (yellow-green spots) that causes reduced replication fork progression. Such stalling of DNA replication triggers the DNA replication checkpoint, which serves to inhibit transiently mitotic entry as signaled by cyclin/Cdk1 (center). Upon completion of DNA replication, the checkpoint is inactivated and mitosis can proceed. Bottom panels: frames from time-lapse micrographs of wild-type or *mei-41* mutant embryos expressing *Histone H2Av-GFP* demonstrate the relative contribution of DNA replication checkpoint activity on cell cycle slowing during NC13. Note the premature mitotic entry for *mei-41* mutant embryos and the subsequent catastrophic mitosis. Image sequences represent 3-min interval between frames of a 2500 μm^2 area on the embryo surface. (See the color plate.)

2.3.1 N:C Ratio-Dependent Acquisition of DNA Damage Checkpoints

The idea that the pre-MBT mitotic apparatus generally outmatches the DNA replication checkpoint originates from early studies on *Xenopus* embryogenesis and provides a working model for N:C ratio-dependent acquisition of checkpoint function. The pre-MBT silencing of the DNA damage checkpoints could simply result from the great initial imbalance in nuclear content and cytoplasmic volume in the early embryo compared with a typical somatic cell. A newly fertilized *Xenopus* embryo, for example, contains an N:C ratio approximately 1/20,000th that of a somatic cell (Newport & Dasso, 1989). Signals produced by the DNA damage response from a single nucleus would, in this model, be insufficient to halt the progression of a cell cycle oscillator capable of driving mitosis efficiently in a 20,000-fold greater cytoplasmic volume. However, ongoing mitosis increases the N:C ratio, and embryos ultimately reach a threshold where such DNA damage signals could begin to be effective in halting cell cycle progression. Accordingly, *Xenopus* embryos initially fail to halt cell cycle progression in response to DNA replication stress (Kimelman, Kirschner, & Scherson, 1987). After eight mitotic divisions (four divisions before the MBT), *Xenopus* embryos acquire a DNA damage checkpoint—i.e., the ability to halt the cell cycle in response to replication stress—once the embryo contains approximately 256 cells, and the N:C ratio is 1:78 (Newport & Dasso, 1989).

Artificially increasing the DNA-to-cytoplasmic ratio by injecting *Xenopus* embryos with synthetic DNA containing free double-stranded ends triggers premature ATR-dependent checkpoint-mediated cell cycle pause. In these experiments, acquisition of a checkpoint response to double-strand breaks likewise occurs at a DNA-to-cytoplasmic ratio approximating the 256/512-cell stage of development (Conn, Lewellyn, & Maller, 2004). Notably, these experiments also may suggest that the pre-MBT response to double-strand breaks is considerably weaker than in typical somatic cells: acquisition of measurable checkpoint activation occurs at a dosage of as little as 0.6 fmol of free double-strand ends per embryo, or approximately 9.0×10^5 breaks per diploid nucleus (one every 7000 bp or so). For comparison, in postembryonic cells, the DNA damage response can be induced in response to as few as two double-strand breaks per nucleus, with near maximal pathway activation observed with 18 breaks (Bakkenist & Kastan, 2003). Thus, checkpoint silencing in pre-MBT embryos may stem from a deficiency of scale, where early DNA damage signals are simply overpowered by the dominance of CDK activity.

This model implies a conceptually straightforward link between nuclear content and cell cycle progression via the DNA replication checkpoint. Damage encountered early in development *could* trigger a DNA damage response, but this information would be lost in the cavernous cytoplasm. Only upon reaching a critical nuclear threshold would the signal strength of a concerted DNA damage response be strong enough to overcome, albeit temporarily, the drive to enter mitosis. The *Xenopus* model provides a useful framework for thinking about checkpoint activation in *Drosophila* embryos. However, the genetic analyses presented below suggest that the endogenous input phenomena that activate these checkpoints are not just random occurrences of double-strand breaks or stalled replication forks.

2.3.2 A Functional DNA Replication Checkpoint Is Necessary to Complete the MBT in Drosophila

Despite its attenuated nature, the pre-MBT DNA replication checkpoint is nonetheless essential for progression of embryonic development beyond the MBT (Fig. 1). Critical components of the DNA replication checkpoint, including ATR, ATRIP, Chk1, and Claspin, are all necessary for driving gradual Wee1-dependent cell cycle lengthening as embryos approach the MBT (Brodsky, Sekelsky, Tsang, Hawley, & Rubin, 2000; Fasulo et al., 2012; Fogarty, Kalpin, & Sullivan, 1994; Lee et al., 2012; Sibon, Laurençon, Hawley, & Theurkauf, 1999; Sibon, Stevenson, & Theurkauf, 1997; Stumpff et al., 2004). During the late pre-MBT nuclear divisions, more time is required for the completion of S-phase. During these times, the DNA replication checkpoint functions to prevent entry into mitosis before DNA replication is complete. The DNA replication checkpoint is triggered in response to the emergence of additional constraints on the timely completion of S-phase, as we discuss below (Fig. 2). Pharmacologically interfering with ongoing DNA replication triggers a transient ATR/Chk1-dependent block to cell cycle progression due to activation of mechanisms that sense and respond to initiated but stalled DNA forks. In contrast, the DNA replication checkpoint is not activated at all if S-phase is eliminated entirely by blocking preinitiation complex formation (McCleland, Shermoen, & O'Farrell, 2009). Cell cycle time in embryos lacking an S-phase does not reflect the gradual checkpoint-dependent pre-MBT lengthening seen in wild-type embryos. Therefore, the replication checkpoint provides feedback to the pre-MBT cell cycle to drive lengthening in response to signals generated during the process of DNA replication.

Embryos mutant for these checkpoint components all fail to prevent mitotic entry in response to these "damage" signals. Whereas a functional checkpoint is not essential for survival of late- and postembryonic tissues, it is absolutely required for completion of the MBT. Checkpoint mutants undergo a catastrophic mitosis 13, having failed to adequately prevent mitotic entry in the face of increasing S-phase length (Fig. 2). Importantly, this mutant lethality can be largely bypassed by genetically slowing the onset of M-phase, such as by reducing maternal cyclin dosage (Blythe & Wieschaus, 2015; Sibon et al., 1999; Yuan, Farrell, & O'Farrell, 2012). Thus, the DNA replication checkpoint is necessary at the MBT to provide enough time for the completion of DNA replication in response to a developmentally programmed slowing of S-phase.

2.3.3 Factors That Trigger the DNA Damage Response at the MBT

The NC13 S-phase cannot be completed on the same timetable as earlier S-phases. The DNA replication checkpoint is necessary to signal this information to the cell cycle oscillator to keep DNA replication and mitosis in phase. Early models for this process predicted that some maternal component of the DNA replication machinery becomes limiting as embryos approach the MBT, such that incompletely replicated chromatin mimics some features of DNA damage (Newport & Dasso, 1989; Sibon et al., 1999, 1997). Some evidence has been found for this in *Xenopus* (Collart, Allen, Bradshaw, Smith, & Zegerman, 2013). The idea that limiting replication factors could trigger the DNA damage response was particularly attractive, because it nicely accounted for how this process could be tied to the N:C ratio measurement. In this case, the increasing number of nuclei would tie up more and more limiting replication factors, and at some critical concentration, the demand for the initiation of replication would outpace the supply of replication factors and thus necessitate more time to complete S-phase. However, as has been argued elsewhere, the measured changes in the distribution of replication origins between pre- and post-MBT *Drosophila* embryos do not account for the changes seen in S-phase duration over this period (Farrell & O'Farrell, 2014). Likewise, even if limiting replication factors did not affect origin usage but, instead, somehow reduced the overall processivity of the DNA replisome, ectopically induced S-phases should adhere to an imposed "speed limit" of replication. This is not the case. When ectopic S-phases are induced during NC14, either by injection of α-amanitin or by overproduction of active Cdk1, mitosis proceeds without catastrophe on a significantly accelerated pace (Edgar & Datar, 1996; Farrell et al., 2012; Shermoen et al., 2010). Likewise, genetic combinations that suppress the lethality of checkpoint mutants can be found where NC13 is completed quickly without catastrophe (in 13 min, compared with the typical 19 min; Blythe & Wieschaus, 2015). Thus, models for limiting DNA replication capacity in MBT-stage embryos do not account for the perceived lengthening of the cell cycle at the *Drosophila* MBT.

Recent evidence suggests instead that the major trigger for the DNA replication checkpoint is linked to a global change in chromatin structure that accompanies the MBT, independent of the embryo's potential capacity to replicate DNA (Blythe & Wieschaus, 2015). If checkpoint activation did indeed result from limitations on DNA replication capacity, it would be expected that checkpoint activation would scale with DNA content, and that all types of DNA would trigger the checkpoint equally well. Although it can be demonstrated with aneuploid *Drosophila* embryos with known zygotic DNA content that the overall duration of NC13 is roughly proportional to the overall quantity of zygotic DNA, not all types of zygotic DNA are equivalent for activating the replication checkpoint. DNA sequences that can rightly be called "euchromatin" in postembryonic tissues cause greater NC13 lengthening than "heterochromatic" DNA sequences of otherwise equivalent length. These results suggest that checkpoint activation coincides with the less well-understood global changes in chromatin structure that accompany the MBT.

The emergence of euchromatic chromatin architecture is indeed a major trigger of the DNA replication checkpoint at the MBT. As discussed above, the delineation of biochemically distinct euchromatic and heterochromatic compartments is a major feature of the chromatin transition that takes place at the MBT. Establishment of these chromatin compartments coincides with the genetic requirement for a functional checkpoint and, consequently, with the initial gradual changes in cell cycle progression observed at NC13. Besides this temporal coincidence, there are spatial correlations between components of the transcriptional apparatus and of the DNA replication checkpoint. Binding of RpA-70 is a first step in the initiation of the DNA replication checkpoint serving to recruit ATR and its binding partner ATRIP to sites of stalled replication forks to initiate signaling through Chk1 (Zou & Elledge, 2003). When genome-wide RpA-70 distribution is measured in NC13 embryos, it is found in the vicinity of sites that are also bound by RNA Pol II and are engaged in the initial phases of large-scale ZGA, and reducing RNA Pol II recruitment likewise reduces recruitment of RpA-70. These results predict that reducing or eliminating ZGA in an embryo should likewise eliminate the requirement for a functional DNA replication checkpoint at the MBT. Indeed, whereas *mei-41/ATR* mutant embryos undergo a lethal catastrophic mitosis at NC13, a significant proportion of *mei-41/ATR zelda* double-mutant embryos suppresses this catastrophic mitosis. Interestingly, simply blocking transcription from RNA Pol II with α-amanitin is insufficient to suppress the *mei-41/ATR* lethal phenotype, indicating that the trigger of the DNA replication checkpoint lies upstream of entry into productive transcriptional elongation (Blythe & Wieschaus, 2015).

This failure of α-amanitin to suppress the lethal phenotype of checkpoint-deficient embryos is also in agreement with results in *Xenopus* and zebrafish (Ikegami, Rivera-Bennetts, Brooker, & Yager, 1997; Newport & Dasso, 1989; Zhang, Kothari, Mullins, & Lampson, 2014) that showed no effect of α-amanitin on checkpoint-dependent S-phase lengthening at the MBT. Instead, the demonstrated link between checkpoint activation and the chromatin transition at the MBT highlights the need to identify the specific changes that occur in chromatin structure preceding the MBT, especially those that delineate euchromatic and heterochromatic regions and may have a major impact on checkpoint activation.

3. MECHANISMS FOR TIMING ZYGOTIC GENOME ACTIVATION

As discussed in the preceding sections, ZGA plays an integral, causative role in the temporal unfolding of the *Drosophila* MBT. The course of events at the MBT could be considered to result from an initial conflict between the opposing requirements for rapid DNA replication and the emergent chromatin structure required for large-scale gene expression. To mediate this conflict, the DNA replication checkpoint responds to changes in chromatin structure and initiates transient adjustments in the cell cycle to accommodate these new regulatory features. Ultimately, zygotic gene products such as *tribbles* and *fruhstart* are essential for fully instituting the lasting transition to the zygotically controlled, post-MBT cell cycle. Therefore, the key to understanding the MBT clock lies in the understanding of how ZGA is regulated.

It is currently unclear how ZGA is timed. As introduced above, ZGA receives temporal cues from both N:C ratio-dependent and -independent mechanisms. In neither case is it clear how intervals of time are measured by either of these timing mechanisms. In the following sections, we discuss the current models for how these timers function.

3.1 N:C Ratio Dependence and Independence

As discussed above, altering the ploidy of embryos clearly reveals the existence of an N:C ratio-dependent timer whose function is critical for determining when the cell cycle is remodeled and when a subset of zygotic genes is activated. Studies in haploid embryos have also revealed more complex timing mechanisms for zygotic gene activation (Edgar et al., 1986; Edgar & Schubiger, 1986; Lu et al., 2009). Since haploid embryos undergo one additional pre-MBT mitosis, and therefore take at least 10 min longer to reach the MBT than their diploid counterparts, they provide an ideal system to distinguish N:C ratio-dependent from timer-dependent early developmental events. Measuring the onset of zygotic gene expression in *maternal haploid* embryos indicates that a significant fraction of early zygotic transcripts in *Drosophila* is not dependent on the N:C ratio for their activation. ZGA in *maternal haploid* embryos reveals effectively two classes of activated loci that have been brought out of phase by the alteration of embryonic ploidy, one that adheres to the N:C ratio, and a second class that becomes activated in a time- or stage-dependent manner independent of the N:C ratio. Likewise, experiments in which cell cycle progression is halted either by triple RNAi knockdown of cyclin A/B/B3 or by treatment of embryos with cycloheximide during the late-cleavage divisions reveals that ZGA for at least a subset of loci can indeed take place at pre-MBT N:C ratios (Edgar et al., 1986; McCleland & O'Farrell, 2008).

It is not currently known how this additional timer functions nor is it even known whether one or many additional timers exist. Experiments with the translational inhibitor cycloheximide may provide a clue to function. Although, as mentioned above, zygotic transcription becomes active at the right time even when protein synthesis is inhibited during the late-cleavage divisions (at nuclear cycle 12 or greater), cycloheximide treatment earlier in development (prior to nuclear cycle 10) prevents ZGA (Edgar & Schubiger, 1986). Perhaps during the early cleavage divisions, regulated translation results in production of a factor (or factors) permissive for ZGA. In this model, this factor would accumulate to sufficiently high quantities by nuclear cycle 10, but would be constrained in its activity by additional, limiting factors. These additional limiting factors could themselves be linked to the N:C ratio-dependent timer or, instead, could represent maternally supplied factors that are degraded N:C ratio independently.

The zinc-finger transcription factor Zelda is one such candidate for an N:C ratio-independent factor that is permissive for ZGA. As discussed above,

Zelda is essential for controlling the expression of a large subset of early zygotic genes, and Zelda targets are largely N:C ratio independent (Liang et al., 2008; Lu et al., 2009). Measurement of Zelda protein expression during the cleavage divisions suggests that Zelda protein gradually accumulates over the course of early development, and is possibly under some form of translational control (Harrison, Botchan, & Cline, 2010). Another maternally supplied factor generally required for early zygotic gene expression, Smaug, likewise demonstrates evidence of translational control during the early cleavage divisions (Benoit et al., 2009). Smaug protein gradually accumulates following fertilization, climbs to a peak expression level, and then is downregulated at the MBT. Notably, the initial accumulation of Smaug protein is N:C ratio independent, occurring at the right time after fertilization even in activated yet unfertilized eggs. The later downregulation of Smaug, however, does not occur at the right time in unfertilized eggs. This may indicate that Smaug is degraded by an N:C ratio-dependent mechanism and/or by a mechanism that depends on zygotic gene activity. In both of these examples, Zelda and Smaug reach maximal protein expression (1–2 h for Zelda, and after nuclear cycle 10 for Smaug) during the period when embryos are insensitive to cycloheximide treatment for undergoing ZGA. While the requirement for Smaug for large-scale ZGA is clear, its mechanism of action is not fully understood. Smaug is a component of the CCR4/POP4/NOT deadenylase pathway required for the large-scale degradation of maternal mRNAs during the MZT (Benoit et al., 2009; Semotok et al., 2005; Tadros et al., 2007). It is tempting to speculate that Smaug is necessary for degrading the mRNA of a maternally supplied transcriptional inhibitor that could function to suppress ZGA until an appropriate time.

Notably, if protein accumulation is indeed critical for timing N:C ratio-independent events, such effectors of this timer should be especially sensitive to alterations in dose of maternal mRNA. Although experiments testing the dose dependence of ZGA on Zelda expression levels (e.g., by varying maternal gene dosage between 0.5- and 2-fold) have not been reported, similar experiments have been performed with Smaug. Reduction of Smaug expression by 50% results in a slight delay in the onset of cellularization, without changing the number of pre-MBT mitotic divisions. Expression of Smaug mRNA in an anteroposterior gradient will generate embryos that undergo a graded MBT, judged by the criteria of graded maternal transcript clearance and graded onset of zygotic gene expression. However, ~fivefold overexpression of Smaug in the anterior of these graded embryos does not strongly advance the onset of MBT, indicating that additional limiting factors are essential for controlling the timing of the MBT via Smaug (Benoit et al., 2009). Importantly, it has not been tested whether these additional limitations stem from N:C-dependent or -independent mechanisms.

Although an N:C ratio-independent timing function before the MBT has not been strictly defined for other model systems either, evidence for such systems exists and is often overlooked in review articles and textbooks. N:C ratio-independent mechanisms control the proteolytic degradation of maternally supplied Cyclin E protein in early *Xenopus* embryos (Howe & Newport, 1996). Similar to *Drosophila*, treatment of *Xenopus* embryos with cycloheximide three to four cell divisions before the MBT blocks further DNA synthesis such that embryos fail to reach the critical N:C ratio threshold; however, these embryos will nonetheless undergo zygotic activation of at least a limited number of targets, including a number of the most highly expressed transcripts. Embryos treated with cycloheximide six to seven divisions before the MBT fail to undergo this N:C ratio-independent zygotic gene activation, even if exogenous DNA is injected to correct for the deficit in zygotic DNA content (Lund & Dahlberg, 1992). Thus, some set of unknown protranscriptional factors may reach a sufficiently high level of expression by the 256-cell stage in *Xenopus* in order to drive large-scale zygotic gene activation later, at or around the MBT. Interestingly, the earliest known zygotic gene expression events first become detectable at or around this time in *Xenopus* (Collart et al., 2014; Skirkanich, Luxardi, Yang, Kodjabachian, & Klein, 2011; Yang, Tan, Darken, Wilson, & Klein, 2002).

In zebrafish, the transcription factors—Pou5f1, SoxB1, and Nanog—are essential for zygotic gene activation at the MBT. These factors are expected to function much like the Zelda protein in *Drosophila*, as general activators of the zygotic gene expression program. These factors are all maternally deposited and are the three most highly translated sequence-specific transcription factors in pre-MBT zebrafish embryos as estimated by ribosome profiling. Loss of the three factors results in catastrophic loss of zygotic gene expression and developmental arrest similar to that observed upon global inhibition of RNA Polymerase II (Lee et al., 2013; Leichsenring, Maes, Mössner, Driever, & Onichtchouk, 2013). Although Pou5f1 protein appears to be present from fertilization onward (Lippok, Song, & Driever, 2014), it is not known whether Nanog and SoxB1 are under translational control. Notably, cycloheximide treatment of zebrafish embryos shows a similar effect as that seen in *Xenopus* and *Drosophila*, with zygotic gene activation occurring under conditions of presumed sub-MBT nuclear content, and with gradual loss or attenuation of this zygotic gene activation with progressively earlier drug treatment (Lee et al., 2013).

Without additional evidence, however, it is difficult to interpret the effect of cycloheximide on ZGA. On the one hand, if essential factors for ZGA are particularly unstable, then inhibition of protein synthesis could effectively deplete the embryo of these activators. Yet, on the other hand, cycloheximide could prevent the regulated accumulation of one or more of these essential factors. Perhaps comparing the global patterns of zygotic transcript activation in diploid and haploid embryos will provide the most convincing test of whether N:C ratio-independent mechanisms contribute to ZGA in these systems.

3.2 What Suppresses ZGA Before the MBT?

Several models have been proposed for the suppression of zygotic transcription before the MBT. The difficulty with proposing such a model is that it not only must provide for global transcriptional repression, but it must also have a built-in, dose-dependent mechanism for responding to the N:C ratio. The initial, titration model for N:C ratio measurement proposed by Newport and Kirschner remains the most attractive (Newport & Kirschner, 1982a, 1982b). This model predicts that the unfertilized egg contains an excess of a factor (or factors) that when bound to chromatin would suppress transcription. Rounds of DNA replication would gradually titrate this repressor until it was to drop below a threshold for preventing transcription, and zygotic transcription would then commence (Newport & Kirschner, 1982b). From an experimental perspective, it is therefore expected that embryos would be exceptionally sensitive to changes in dosage of such a factor. For example, a twofold increase in a repressor should delay ZGA by one cell cycle, and a twofold decrease should lead to premature ZGA one cell cycle early. For a factor (or a collection of factors) to satisfy the criteria of this model the factor should both show this functional relationship to dosage and it should clearly function as a transcriptional repressor. To date, no set of candidate factors has been found that fully satisfy these criteria in *Drosophila*, although sometimes this is because these criteria have not always been specifically tested for. On the other hand, promising results have recently been reported from the *Xenopus* model system.

The initial evidence for the titration model came from classic experiments in *Xenopus* embryos where precocious transcriptional activation could be induced as a result of altering the N:C ratio by injection of exogenous sources of DNA. In these experiments, the exogenous DNA competes for binding of the maternal repressor, thus allowing endogenous loci to become active ahead of schedule. In a twist on this assay, plasmids containing transcriptional reporters can be used to directly investigate the nature of the repressor. Such plasmids, when injected into a pre-MBT *Xenopus* embryo at high concentrations, will briefly transcribe, but will rapidly fall under control of this repressive mechanism (Almouzni & Wolffe, 1995; Newport & Kirschner, 1982b; Prioleau, Huet, Sentenac, & Méchali, 1994). When supplied as a plasmid, a type III reporter gene (i.e., a gene transcribed by RNA Pol III) will autonomously demonstrate this initial burst of pre-MBT transcription prior to silencing (Newport & Kirschner, 1982b), whereas a type II reporter gene (i.e., a gene transcribed by RNA Pol II) must be preloaded with TATA-binding protein (TBP) to show premature transcription (Prioleau et al., 1994). In either example, the pre-MBT silencing is concurrent with assembly of the plasmid into chromatin, which occurs rapidly following injection. Although such injected plasmids will undergo limited rounds of DNA replication within the timescale of the pre-MBT period, the acquisition of transcriptional silencing is independent of cell cycle progression or of DNA replication (Prioleau et al., 1994). Importantly, silencing by the endogenous, chromatin-dependent pre-MBT repressor could be overcome if, in the case of type II genes, reporters were preloaded with both TBP and an activating transcription factor such as Gal4-VP16 (Almouzni & Wolffe, 1995).

A recent effort to biochemically purify the endogenous pre-MBT repressor identified Histones H3 and H4 as the critical titratable factors that set the N:C ratio in *Xenopus* embryos. In experiments that model pre-MBT transcriptional repression in *Xenopus* egg extracts, it is possible to titrate out a repressive factor by addition of exogenous DNA, and to use this feature as a basis for biochemical purification. Histones H3 and H4 satisfy the dosage criteria of the titration model. In embryos, reducing the dosage of Histone H3 by ~50% advances the onset of cell cycle lengthening by precisely one cell cycle, and likewise induces precocious transcriptional activation. Conversely, overexpressing Histones H3 and H4 delays the onset of cell cycle lengthening, although in this case a strict doubling of H3 and H4 cannot be achieved due to an unspecified early embryonic toxicity (Amodeo, Jukam, Straight, & Skotheim, 2015). Preliminarily, Histones H3 and H4 also satisfy the requirement of the titration model that a factor represses pre-MBT transcription, although the mechanism of action is still sketchy.

Transcriptional control at the MBT is probabilistically determined by access to regulatory domains and promoters (Almouzni & Wolffe, 1995; Amodeo et al., 2015; Prioleau et al., 1994). In this case, histone proteins essentially compete with transcription factors for binding to critical regulatory elements, reducing the probability of transcriptional activation at a specific concentration of histones. When histones are in excess, there are fewer opportunities for transcription factors to gain a foothold in pre-MBT chromatin in order to activate gene expression. Below a threshold concentration, however, limited histone protein would insufficiently prevent transcription factor access and the probability of gene expression would increase markedly (Amodeo et al., 2015).

These results nicely demonstrate that the histone proteins themselves provide a major contribution to the pre-MBT transcriptional repression mechanism, and strongly support the titration model for N:C ratio-dependent MBT timing. These results also raise several additional questions regarding the underlying timing mechanism:

It is not yet clear whether modulating the dosage of histone proteins affects all aspects of ZGA equally well, or whether only certain types of transcripts respond to this repressive mechanism. Notably, in egg extracts, depletion of histones results in the precocious activation of a type III transcript derived from heterochromatic satellite repeats, OAX, which may simply reflect a known requirement for protein synthesis late in the pre-MBT period for the activation of other classes of genes, or may indicate that some types of transcripts are more sensitive to histone depletion than others (Amodeo et al., 2015; Lund & Dahlberg, 1992).

The probabilistic model for the mechanism of histone-based repression could be subject to numerous additional regulatory inputs, whereas the histone proteins themselves could play a passive, yet essential role in the process. For example, additional repressors could further enhance histone-mediated repression, or transcriptional activators could locally counteract histone-mediated repression prior to genome-wide derepression. Coupling these additional factors to N:C ratio-independent timing mechanisms could add additional layers of control. At present, it is unclear to what extent histone-mediated repression is simply the canvas upon which these putative additional regulatory systems function.

Another question is how the probabilistic transcriptional model links loss of repression with cell cycle remodeling. Investigations of the *Xenopus* MBT have adamantly excluded a zygotic contribution to cell cycle remodeling, largely on the basis of insensitivity of cell cycle remodeling to the transcriptional inhibitor α-amanitin (Dasso & Newport, 1990; Newport & Dasso, 1989). In the case of histone titration, α-amanitin almost certainly functions too late in the transcriptional process to prevent this histone timer from functioning. The possibility remains that opening of chromatin at the MBT and early events in transcriptional activation interfere with ongoing DNA replication, as is seen in the *Drosophila* model system (Blythe & Wieschaus, 2015). It remains to be seen whether histone titration is sufficient to activate the replication checkpoint in *Xenopus* embryos, and to what extent this initiates the process of cell cycle remodeling.

Finally, whether such a histone titration mechanism functions in other embryonic model systems remains to be determined, but it seems highly likely that histones will play a critical role in any mechanism for transcriptional control.

3.3 Nuclear Size, Not DNA Content?

One complicating factor to the histone titration model is the observation that nuclear volume, rather than zygotic DNA content, is a major quantified feature of "N" for the determination of the *Xenopus* N:C ratio. As discussed above, classic studies have unambiguously demonstrated that the dosage of DNA is critical for setting the N:C ratio. Yet, in addition to changes in zygotic DNA content, nuclear and cytoplasmic volumes likewise undergo changes in the pre-MBT period. Recent work indicates that the relative ratio between nuclear and cytoplasmic volumes appears to be critical for timing the MBT. From the four-cell stage to the MBT in *Xenopus* embryos, nuclear volume undergoes a modest 3-fold reduction whereas cytoplasmic volume undergoes a more dramatic 70-fold reduction. Although it is not feasible to alter cytoplasmic volume directly, it is possible to alter nuclear volume experimentally to test the effect of skewing the early embryonic volume ratio. Altering rates of nuclear import by overexpression of importin-α, for example, makes nuclei larger in volume, and thereby increases the N:C volume ratio by three- to fourfold. This has a major impact on MBT timing, resulting in premature zygotic expression of at least a subset of genes and early cell cycle lengthening. Likewise, reducing overall nuclear volume has the opposite effects on transcription and the cell cycle. This effect is independent of nuclear import rates *per se*, as similar effects on nuclear size and MBT timing can be achieved by manipulating the rates of endoplasmic reticulum formation (Jevtić & Levy, 2015).

Notably, precocious zygotic transcription induced by alterations in nuclear volume accompanies premature cell cycle lengthening, but on different time-scales. Whereas increasing nuclear volume can drive premature zygotic gene expression (of at least some transcripts) extremely early, at the 16- to 32-cell stage (stages 5–6), an effect on cell cycle lengthening is not observed until the 1024-cell stage (stage 8). Changes in nuclear volume drive increases in zygotic gene expression but not to the same magnitude as is seen at the MBT (Jevtić & Levy, 2015). Additionally, it is not known whether all zygotic genes are equally sensitive to changes in nuclear volume. Curiously, the earliest zygotic gene expression in *Xenopus* embryos occurs well before the MBT in the blastomeres of the vegetal pole, which are significantly larger in cytoplasmic volume than those in the animal pole and marginal region, and would be expected to have a smaller N:C volume ratio than their smaller counterparts (Skirkanich et al., 2011; Takahashi et al., 2000; Yang et al., 2002). How this proposed N:C volume ratio mechanism adapts to the wide range in cell size observed in the *Xenopus* embryo remains to be determined.

These observations are compatible with other observations relating ploidy to the N:C ratio set point, insofar, as nuclear volume should scale with DNA content. However, the experimentally induced increases in nuclear volume would be expected to dilute the nucleoplasmic concentration independently of changes in DNA content and, presumably, histone availability or occupancy, although this has not been rigorously determined. If the histone-dependent transcriptional repression discussed above were further supplemented by additional repressive mechanisms to decrease the probability of pre-MBT zygotic gene expression, it is possible that increases in nuclear volume would dilute these factors and result in a higher probability of premature gene expression. Presumably, however, a counteracting effect could be expected if specific transcriptional activators were required for ZGA and they too were diluted by increased nuclear volume. Indeed, additional work is necessary to understand the nature of the pre-MBT transcriptional repressor in *Xenopus* and how it functions under conditions of altered nuclear volume.

3.4 And What About *Drosophila*?

Compared with the *Xenopus* model system, investigations of the titration hypothesis in other model systems lag far behind. It is clear that, in *Drosophila* embryos, cell cycle regulatory factors indeed are titrated down to near-limiting concentrations at or around the MBT. Increasing or decreasing the dosage of Cyclin B, for example, results in changes in cell cycle timing (Crest, Oxnard, Ji, & Schubiger, 2007; Edgar et al., 1994; Ji, Squirrell, &

Schubiger, 2004). Indeed, early versions of the titration model in *Drosophila* posited that gradual lengthening of the interphase period would become permissive for ZGA, and this placed a special emphasis on cell cycle regulatory factors that would become limiting at the MBT (Edgar & Schubiger, 1986). As has been argued elsewhere (Farrell & O'Farrell, 2014) and in the preceding sections, although increasing interphase time in pre-MBT embryos can result in minor increases in transcriptional activity (Edgar & Schubiger, 1986; Kimelman et al., 1987), this does not fully account for the major changes in zygotic gene activity seen at the MBT. Nevertheless, embryos do not possess the capacity to continue the maternal cell cycle program indefinitely. At present, it is unclear whether such limitations in cell cycle capacity provide direct timing cues to the MBT clock, or—rather—that the MBT occurs when it does in order to avoid encountering such limitations in the maternally supplied components.

With respect to titrated factors that suppress premature *Drosophila* ZGA, much less is known. The transcriptional repressor Tramtrack satisfies the dosage requirement but not the requirement for genome-wide or general repression. Embryos depleted for *tramtrack* display precocious activation of at least one target gene, *fushi tarazu* (*ftz*) and, conversely, embryos show a delay in activation of *ftz* when *tramtrack* is modestly overexpressed. However, Tramtrack fails in the generality department as another early zygotic gene, *Kruppel* does not respond to modulation of *tramtrack* dosage (Pritchard & Schubiger, 1996). It remains possible that Tramtrack is a member of a set of factors that together confer pre-MBT transcriptional repression, but additional members have yet to be identified.

One overlooked feature of the titration model is that if, for example, an embryo were completely depleted of *the* pre-MBT transcriptional repressor, zygotic gene expression would be expected to commence immediately, without delay, provided that this factor was the only limiting factor for ZGA. In light of the observations that ZGA triggers the DNA replication checkpoint at the MBT, and that very early embryos have little to no ability to halt cell cycle progression in the face of induced DNA damage, it would be predicted that complete loss of function for *the* pre-MBT transcriptional repressor would have a phenotype of massive unrepaired DNA damage. Because such phenotypes are difficult to interpret unambiguously as having originated from transcriptional dysregulation, identification of *the* pre-MBT transcriptional repressor by examination of recessive mutant phenotypes may prove impossible. Nonetheless, we highlight some interesting candidates.

The maternally supplied linker histone, BigH1, is highly enriched in pre-MBT chromatin, regulates nucleosomal spacing, and is essential for pre-MBT transcriptional repression. Embryos deficient for *bigH1* display severe defects including mitotic asynchrony, anaphase bridging, DNA damage, and premature zygotic gene expression. The genetics of *bigH1* are curious. Crosses from parents heterozygous for an allele of *bigH1* produce ¼ embryos that fail to survive to the MBT and that are presumed to be homozygous for the *bigH1* mutation. This would suggest a significant zygotic contribution to the mutant phenotype of *bigH1*, which would be unexpected, but not implausible given the prevalence of pre-MBT transcription in the *Drosophila* embryo (Pérez-Montero, Carbonell, Morán, Vaquero, & Azorín, 2013). Since it is predicted that pre-MBT transcriptional repression is driven by a maternally supplied factor, perhaps, the apparent zygotic role of *bigH1* stems from a zygotic requirement to maintain sufficiently high levels of *bigH1* expression, although this has not been formally demonstrated. From the available literature, it is also not clear whether the reduced maternal dosage of *bigH1* additionally has a slight effect on the onset of ZGA that would be expected if it were *the* pre-MBT transcriptional repressor. However, BigH1 is attractive in light of the histone model for pre-MBT transcriptional repression observed in *Xenopus*.

Maternal-effect mutants in transcriptional repressors can give rise to catastrophic pre-MBT mitotic phenotypes. It is tempting to speculate (with appropriate reservations) that these may be involved in regulating the timing of ZGA. Homozygous maternal-effect mutants for components of the Polycomb group (PcG) and the piRNA pathway generate embryos with massive mitotic defects well before the MBT similar to what is seen with *bigH1*. Mutants for *Polycomb*, *polyhomeotic*, *posterior sex combs*, *and additional sex combs* demonstrate mitotic delays, asynchronies, and anaphase bridging indicating pervasive defects in pre-MBT chromatin (O'Dor, Beck, & Brock, 2006). Although these defects are not specifically ascribed to premature zygotic transcription, the PcG represents a major pathway for transcriptional repression at all stages of development. Although the histone methylation mark produced by the PcG (H3K27me3) is exceedingly rare in pre-MBT chromatin (Li et al., 2014), the possibility remains that the PcG functions in a capacity independent of this modification to suppress gene expression. Likewise, mutants in the piRNA pathway (*piwi*, *aubergine*, and *argonaute-3*) display multiple early mitotic phenotypes including abnormal nuclear morphology, asynchronous cell division, abnormal mitosis, and supernumerary mitotic divisions (Mani, Megosh, & Lin, 2014; Megosh, Cox,

Campbell, & Lin, 2006). piRNA pathway mutants additionally show defects in HP1 localization, establishment of H3K9me3, and localization of the DNA origin recognition complex subunit 2. Although a slight increase of a marker of the DNA damage response (γH2Av) is observed as early as the first nuclear cycle, on the basis of a lack of suppression by mutants in *chk2* it is thought that these phenotypes are independent of the DNA damage response (Mani et al., 2014). Similar to the mutants in the PcG, it is unknown whether the piRNA pathway mutants display precocious zygotic gene activation. Determining whether these catastrophic phenotypes result from dysregulation of transcription, as opposed to some other form of structural defect in chromatin, will be an interesting question for future investigation.

4. CONCLUDING REMARKS

Based on the experiments described in the preceding pages, we believe the key to understanding the temporal control of the MBT in *Drosophila* will be to understand the mechanism for controlling the onset of large-scale zygotic genome activation. We have reviewed the current understanding of how alterations in the embryonic cell cycle stem from two consequences of large-scale zygotic genome activation. Via the DNA replication checkpoint, initial alterations in chromatin architecture during large-scale ZGA necessitate transient lengthening of the latter pre-MBT interphases to accommodate delays in S-phase completion. Subsequently, zygotic gene products including—but not limited to—*tribbles* and *fruhstart* act to downregulate cyclin/Cdk1 activity and drive acquisition of a more typical, leisurely cell cycle pace.

The timing of large-scale ZGA remains poorly understood, but recent developments in both the *Drosophila* and *Xenopus* model systems may point the way to developing a comprehensive mechanism that integrates N:C ratio-dependent and -independent timing mechanisms. Notably, factors that contribute to the MBT clock are predicted to have exquisite dosage sensitivity, where twofold increases or decreases in expression should have significant effects on the timing of ZGA. Future work to seek out and rigorously define such dosage-sensitive factors should go a long way toward the mechanistic elucidation of the MBT clock.

ACKNOWLEDGMENT

The authors would like to thank Stefano Di Talia for critical comments on this chapter.

REFERENCES

Adams, R. L., & Lindsay, J. G. (1967). Hydroxyurea reversal of inhibition and use as a cell-synchronizing agent. *The Journal of Biological Chemistry*, *242*, 1314–1317.

Ali-Murthy, Z., Lott, S. E., Eisen, M. B., & Kornberg, T. B. (2013). An essential role for zygotic expression in the pre-cellular Drosophila embryo. *PLoS Genetics*, *9*, e1003428.

Almouzni, G., & Wolffe, A. P. (1995). Constraints on transcriptional activator function contribute to transcriptional quiescence during early Xenopus embryogenesis. *The EMBO Journal*, *14*, 1752–1765.

Amodeo, A. A., Jukam, D., Straight, A. F., & Skotheim, J. M. (2015). Histone titration against the genome sets the DNA-to-cytoplasm threshold for the Xenopus midblastula transition. *Proceedings of the National Academy of Sciences of the United States of America*, *112*(10), E1086–E1095. http://dx.doi.org/10.1073/pnas.1413990112, Epub 2015 Feb 23.

Bakkenist, C. J., & Kastan, M. B. (2003). DNA damage activates ATM through intermolecular autophosphorylation and dimer dissociation. *Nature*, *421*, 499–506.

Bartek, J., Lukas, C., & Lukas, J. (2004). Checking on DNA damage in S phase. *Nature Reviews Molecular Cell Biology*, *5*, 792–804.

Benoit, B., He, C. H., Zhang, F., Votruba, S. M., Tadros, W., Westwood, J. T., et al. (2009). An essential role for the RNA-binding protein Smaug during the Drosophila maternal-to-zygotic transition. *Development*, *136*, 923–932.

Blumenthal, A. B., Kriegstein, H. J., & Hogness, D. S. (1974). The units of DNA replication in Drosophila melanogaster chromosomes. *Cold Spring Harbor Symposia on Quantitative Biology*, *38*, 205–223.

Blythe, S. A., & Wieschaus, E. F. (2015). Zygotic genome activation triggers the DNA replication checkpoint at the midblastula transition. *Cell*, *160*, 1169–1181.

Brodsky, M. H., Sekelsky, J. J., Tsang, G., Hawley, R. S., & Rubin, G. M. (2000). mus304 encodes a novel DNA damage checkpoint protein required during Drosophila development. *Genes & Development*, *14*, 666–678.

Chen, K., Johnston, J., Shao, W., Meier, S., Staber, C., & Zeitlinger, J. (2013). A global change in RNA polymerase II pausing during the Drosophila midblastula transition. *eLife*, *2*, e00861.

Collart, C., Allen, G. E., Bradshaw, C. R., Smith, J. C., & Zegerman, P. (2013). Titration of four replication factors is essential for the Xenopus laevis midblastula transition. *Science*, *341*, 893–896.

Collart, C., Owens, N. D. L., Bhaw-Rosun, L., Cooper, B., De Domenico, E., Patrushev, I., et al. (2014). High-resolution analysis of gene activity during the Xenopus mid-blastula transition. *Development*, *141*, 1927–1939.

Conn, C. W., Lewellyn, A. L., & Maller, J. L. (2004). The DNA damage checkpoint in embryonic cell cycles is dependent on the DNA-to-cytoplasmic ratio. *Developmental Cell*, *7*, 275–281.

Crest, J., Oxnard, N., Ji, J.-Y., & Schubiger, G. (2007). Onset of the DNA replication checkpoint in the early Drosophila embryo. *Genetics*, *175*, 567–584.

Danis, E., Brodolin, K., Menut, S., Maiorano, D., Girard-Reydet, C., & Méchali, M. (2004). Specification of a DNA replication origin by a transcription complex. *Nature Cell Biology*, *6*, 721–730.

Dasso, M., & Newport, J. W. (1990). Completion of DNA replication is monitored by a feedback system that controls the initiation of mitosis in vitro: Studies in Xenopus. *Cell*, *61*, 811–823.

De Renzis, S., Elemento, O., Tavazoie, S., & Wieschaus, E. F. (2007). Unmasking activation of the zygotic genome using chromosomal deletions in the Drosophila embryo. *PLoS Biology*, *5*, e117.

Di Talia, S., She, R., Blythe, S. A., Lu, X., Zhang, Q. F., & Wieschaus, E. F. (2013). Posttranslational control of Cdc25 degradation terminates Drosophila's early cell-cycle program. *Current Biology*, *23*, 127–132.

Edgar, B. A., & Datar, S. A. (1996). Zygotic degradation of two maternal Cdc25 mRNAs terminates Drosophila's early cell cycle program. *Genes & Development*, *10*, 1966–1977.

Edgar, B. A., Kiehle, C. P., & Schubiger, G. (1986). Cell cycle control by the nucleo-cytoplasmic ratio in early Drosophila development. *Cell*, *44*, 365–372.

Edgar, B. A., & Schubiger, G. (1986). Parameters controlling transcriptional activation during early Drosophila development. *Cell*, *44*, 871–877.

Edgar, B. A., Sprenger, F., Duronio, R. J., Leopold, P., & O'Farrell, P. H. (1994). Distinct molecular mechanism regulate cell cycle timing at successive stages of Drosophila embryogenesis. *Genes & Development*, *8*, 440–452.

Erickson, J. W., & Quintero, J. J. (2007). Indirect effects of ploidy suggest X chromosome dose, not the X:A ratio, signals sex in Drosophila. *PLoS Biology*, *5*, e332.

Farrell, J. A., & O'Farrell, P. H. (2013). Mechanism and regulation of Cdc25/Twine protein destruction in embryonic cell-cycle remodeling. *Current Biology*, *23*, 118–126.

Farrell, J. A., & O'Farrell, P. H. (2014). From egg to gastrula: How the cell cycle is remodeled during the Drosophila mid-blastula transition. *Annual Review of Genetics*, *48*, 269–294.

Farrell, J. A., Shermoen, A. W., Yuan, K., & O'Farrell, P. H. (2012). Embryonic onset of late replication requires Cdc25 down-regulation. *Genes & Development*, *26*(7), 714–725. http://dx.doi.org/10.1101/gad.186429.111.

Fasulo, B., Koyama, C., Yu, K. R., Homola, E. M., Hsieh, T. S., Campbell, S. D., et al. (2012). Chk1 and Wee1 kinases coordinate DNA replication, chromosome condensation, and anaphase entry. *Molecular Biology of the Cell*, *23*, 1047–1057.

Fogarty, P., Kalpin, R. F., & Sullivan, W. (1994). The Drosophila maternal-effect mutation grapes causes a metaphase arrest at nuclear cycle 13. *Development*, *120*, 2131–2142.

Garcia, H. G., Tikhonov, M., Lin, A., & Gregor, T. (2013). Quantitative imaging of transcription in living Drosophila embryos links polymerase activity to patterning. *Current Biology*, *23*, 2140–2145.

Gawliński, P., Nikolay, R., Goursot, C., Lawo, S., Chaurasia, B., Herz, H.-M., et al. (2007). The Drosophila mitotic inhibitor Frühstart specifically binds to the hydrophobic patch of cyclins. *EMBO Reports*, *8*, 490–496.

Grosshans, J., Müller, H. A. J., & Wieschaus, E. (2003). Control of cleavage cycles in Drosophila embryos by frühstart. *Developmental Cell*, *5*, 285–294.

Hamm, D. C., Bondra, E. R., & Harrison, M. M. (2015). Transcriptional activation is a conserved feature of the early embryonic factor Zelda that requires a cluster of four zinc fingers for DNA binding and a low-complexity activation domain. *The Journal of Biological Chemistry*, *290*, 3508–3518.

Harrison, M. M., Botchan, M. R., & Cline, T. W. (2010). Grainyhead and Zelda compete for binding to the promoters of the earliest-expressed Drosophila genes. *Developmental Biology*, *345*, 248–255.

Harrison, M. M., Li,, X. Y., Kaplan, T., Botchan, M. R., & Eisen, M. B. (2011). Zelda binding in the early Drosophila melanogaster embryo marks regions subsequently activated at the maternal-to-zygotic transition. *PLoS Genetics*, 7, e1002266.

He, B., Caudy, A., Parsons, L., Rosebrock, A., Pane, A., Raj, S., et al. (2012). Mapping the pericentric heterochromatin by comparative genomic hybridization analysis and chromosome deletions in Drosophila melanogaster. *Genome Research*, *22*, 2507–2519.

Howe, J. A., & Newport, J. W. (1996). A developmental timer regulates degradation of cyclin E1 at the midblastula transition during Xenopus embryogenesis. *Proceedings of the National Academy of Sciences of the United States of America*, *93*, 2060–2064.

Hyrien, O., Maric, C., & Méchali, M. (1995). Transition in specification of embryonic metazoan DNA replication origins. *Science*, *270*, 994–997.
Iampietro, C., Bergalet, J., Wang, X., Cody, N. A. L., Chin, A., Lefebvre, F. A., et al. (2014). Developmentally regulated elimination of damaged nuclei involves a Chk2-dependent mechanism of mRNA nuclear retention. *Developmental Cell*, *29*(4), 468–481. http://dx.doi.org/10.1016/j.devcel.2014.03.025.
Ikegami, R., Rivera-Bennetts, A. K., Brooker, D. L., & Yager, T. D. (1997). Effect of inhibitors of DNA replication on early zebrafish embryos: Evidence for coordinate activation of multiple intrinsic cell-cycle checkpoints at the mid-blastula transition. *Zygote*, *5*, 153–175.
Jevtić, P., & Levy, D. L. (2015). Nuclear size scaling during Xenopus early development contributes to midblastula transition timing. *Current Biology*, *25*, 45–52.
Ji, J.-Y., Squirrell, J. M., & Schubiger, G. (2004). Both cyclin B levels and DNA-replication checkpoint control the early embryonic mitoses in Drosophila. *Development*, *131*, 401–411.
Johnson, M. H., & Day, M. L. (2000). Egg timers: How is developmental time measured in the early vertebrate embryo? *Bioessays*, *22*, 57–63.
Kane, D. A., & Kimmel, C. B. (1993). The zebrafish midblastula transition. *Development*, *119*, 447–456.
Kellum, R., Raff, J. W., & Alberts, B. M. (1995). Heterochromatin protein 1 distribution during development and during the cell cycle in Drosophila embryos. *Journal of Cell Science*, *108*(Pt. 4), 1407–1418.
Kimelman, D., Kirschner, M., & Scherson, T. (1987). The events of the midblastula transition in Xenopus are regulated by changes in the cell cycle. *Cell*, *48*, 399–407.
Lee, M. T., Bonneau, A. R., Takacs, C. M., Bazzini, A. A., DiVito, K. R., Fleming, E. S., et al. (2013). Nanog, Pou5f1 and SoxB1 activate zygotic gene expression during the maternal-to-zygotic transition. *Nature*, *503*, 360–364.
Lee, E.-M., Trinh, T. T. B., Shim, H. J., Park, S.-Y., Nguyen, T. T. T., Kim, M.-J., et al. (2012). Drosophila Claspin is required for the G2 arrest that is induced by DNA replication stress but not by DNA double-strand breaks. *DNA Repair (Amst)*, *11*, 741–752.
Leichsenring, M., Maes, J., Mössner, R., Driever, W., & Onichtchouk, D. (2013). Pou5f1 transcription factor controls zygotic gene activation in vertebrates. *Science*, *341*, 1005–1009.
Li, X. Y., Harrison, M. M., Villalta, J. E., Kaplan, T., & Eisen, M. B. (2014). Establishment of regions of genomic activity during the Drosophila maternal to zygotic transition. *Elife*, *3*. http://dx.doi.org/10.7554/eLife.03737.
Liang, H. L., Nien, C. Y., Liu, H. Y., Metzstein, M. M., Kirov, N., & Rushlow, C. (2008). The zinc-finger protein Zelda is a key activator of the early zygotic genome in Drosophila. *Nature*, *456*, 400–403.
Lippok, B., Song, S., & Driever, W. (2014). Pou5f1 protein expression and posttranslational modification during early zebrafish development. *Developmental Dynamics*, *243*, 468–477.
Little, S. C., Tikhonov, M., & Gregor, T. (2013). Precise developmental gene expression arises from globally stochastic transcriptional activity. *Cell*, *154*, 789–800.
Lu, X., Drocco, J., & Wieschaus, E. F. (2010). Cell cycle regulation via inter-nuclear communication during the early embryonic development of Drosophila melanogaster. *Cell Cycle*, *9*, 2908–2910.
Lu, X., Li, J. M., Elemento, O., Tavazoie, S., & Wieschaus, E. F. (2009). Coupling of zygotic transcription to mitotic control at the Drosophila mid-blastula transition. *Development*, *136*, 2101–2110.
Lund, E., & Dahlberg, J. E. (1992). Control of 4-8S RNA transcription at the midblastula transition in Xenopus laevis embryos. *Genes & Development*, *6*, 1097–1106.
Mani, S. R., Megosh, H., & Lin, H. (2014). PIWI proteins are essential for early Drosophila embryogenesis. *Developmental Biology*, *385*, 340–349.

McCleland, M. L., & O'Farrell, P. H. (2008). RNAi of mitotic cyclins in Drosophila uncouples the nuclear and centrosome cycle. *Current Biology, 18*, 245–254.
McCleland, M. L., Shermoen, A. W., & O'Farrell, P. H. (2009). DNA replication times the cell cycle and contributes to the mid-blastula transition in Drosophila embryos. *The Journal of Cell Biology, 187*, 7–14.
Megosh, H. B., Cox, D. N., Campbell, C., & Lin, H. (2006). The role of PIWI and the miRNA machinery in Drosophila germline determination. *Current Biology, 16*, 1884–1894.
Newport, J., & Dasso, M. (1989). On the coupling between DNA replication and mitosis. *Journal of Cell Science Supplement, 12*, 149–160.
Newport, J., & Kirschner, M. (1982a). A major developmental transition in early Xenopus embryos: I. Characterization and timing of cellular changes at the midblastula stage. *Cell, 30*, 675–686.
Newport, J., & Kirschner, M. (1982b). A major developmental transition in early Xenopus embryos: II. Control of the onset of transcription. *Cell, 30*, 687–696.
Nien, C.-Y., Liang, H.-L., Butcher, S., Sun, Y., Fu, S., Gocha, T., et al. (2011). Temporal coordination of gene networks by Zelda in the early Drosophila embryo. *PLoS Genetics, 7*, e1002339.
O'Dor, E., Beck, S. A., & Brock, H. W. (2006). Polycomb group mutants exhibit mitotic defects in syncytial cell cycles of Drosophila embryos. *Developmental Biology, 290*, 312–322.
Papoulas, O., Monzo, K. F., Cantin, G. T., Ruse, C., Yates, J. R., Ryu, Y. H., et al. (2010). dFMRP and Caprin, translational regulators of synaptic plasticity, control the cell cycle at the Drosophila mid-blastula transition. *Development, 137*, 4201–4209.
Pérez-Montero, S., Carbonell, A., Morán, T., Vaquero, A., & Azorín, F. (2013). The embryonic linker histone H1 variant of Drosophila, dBigH1, regulates zygotic genome activation. *Developmental Cell, 26*, 578–590.
Price, D., Rabinovitch, S., O'Farrell, P. H., & Campbell, S. D. (2000). Drosophila wee1 has an essential role in the nuclear divisions of early embryogenesis. *Genetics, 155*, 159–166.
Prioleau, M. N., Huet, J., Sentenac, A., & Méchali, M. (1994). Competition between chromatin and transcription complex assembly regulates gene expression during early development. *Cell, 77*, 439–449.
Pritchard, D. K., & Schubiger, G. (1996). Activation of transcription in Drosophila embryos is a gradual process mediated by the nucleocytoplasmic ratio. *Genes & Development, 10*, 1131–1142.
Probst, A. V., & Almouzni, G. (2011). Heterochromatin establishment in the context of genome-wide epigenetic reprogramming. *Trends in Genetics, 27*, 177–185.
Probst, A. V., Okamoto, I., Casanova, M., El Marjou, F., Le Baccon, P., & Almouzni, G. (2010). A strand-specific burst in transcription of pericentric satellites is required for chromocenter formation and early mouse development. *Developmental Cell, 19*, 625–638.
Raff, J. W., & Glover, D. M. (1988). Nuclear and cytoplasmic mitotic cycles continue in Drosophila embryos in which DNA synthesis is inhibited with aphidicolin. *The Journal of Cell Biology, 107*, 2009–2019.
Rensing, L., Meyer-Grahle, U., & Ruoff, P. (2001). Biological timing and the clock metaphor: Oscillatory and hourglass mechanisms. *Chronobiology International, 18*, 329–369.
Rudolph, T., Yonezawa, M., Lein, S., Heidrich, K., Kubicek, S., Schäfer, C., et al. (2007). Heterochromatin formation in Drosophila is initiated through active removal of H3K4 methylation by the LSD1 homolog SU(VAR)3-3. *Molecular Cell, 26*, 103–115.
Sasaki, T., Sawado, T., Yamaguchi, M., & Shinomiya, T. (1999). Specification of regions of DNA replication initiation during embryogenesis in the 65-kilobase DNApolalpha-dE2F locus of Drosophila melanogaster. *Molecular and Cellular Biology, 19*, 547–555.
Saunders, A., Core, L. J., Sutcliffe, C., Lis, J. T., & Ashe, H. L. (2013). Extensive polymerase pausing during Drosophila axis patterning enables high-level and pliable transcription. *Genes & Development, 27*, 1146–1158.

Schübeler, D., Scalzo, D., Kooperberg, C., van Steensel, B., Delrow, J., & Groudine, M. (2002). Genome-wide DNA replication profile for Drosophila melanogaster: A link between transcription and replication timing. *Nature Genetics*, *32*, 438–442.

Schwaiger, M., Stadler, M. B., Bell, O., Kohler, H., Oakeley, E. J., & Schübeler, D. (2009). Chromatin state marks cell-type- and gender-specific replication of the Drosophila genome. *Genes & Development*, *23*, 589–601.

Semotok, J. L., Cooperstock, R. L., Pinder, B. D., Vari, H. K., Lipshitz, H. D., & Smibert, C. A. (2005). Smaug recruits the CCR4/POP2/NOT deadenylase complex to trigger maternal transcript localization in the early Drosophila embryo. *Current Biology*, *15*, 284–294.

Shermoen, A. W., McCleland, M. L., & O'Farrell, P. H. (2010). Developmental control of late replication and S phase length. *Current Biology*, *20*, 2067–2077.

Sibon, O. C., Laurençon, A., Hawley, R., & Theurkauf, W. E. (1999). The Drosophila ATM homologue Mei-41 has an essential checkpoint function at the midblastula transition. *Current Biology*, *9*, 302–312.

Sibon, O. C., Stevenson, V. A., & Theurkauf, W. E. (1997). DNA-replication checkpoint control at the Drosophila midblastula transition. *Nature*, *388*, 93–97.

Skirkanich, J., Luxardi, G., Yang, J., Kodjabachian, L., & Klein, P. S. (2011). An essential role for transcription before the MBT in Xenopus laevis. *Developmental Biology*, *357*, 478–491.

Stumpff, J., Duncan, T., Homola, E., Campbell, S. D., & Su, T. T. (2004). Drosophila Wee1 kinase regulates Cdk1 and mitotic entry during embryogenesis. *Current Biology*, *14*, 2143–2148.

Sung, H. W., Spangenberg, S., Vogt, N., & Großhans, J. (2013). Number of nuclear divisions in the Drosophila blastoderm controlled by onset of zygotic transcription. *Current Biology*, *23*, 133–138.

Tadros, W., Goldman, A. L., Babak, T., Menzies, F., Vardy, L., Orr-Weaver, T., et al. (2007). SMAUG is a major regulator of maternal mRNA destabilization in Drosophila and its translation is activated by the PAN GU kinase. *Developmental Cell*, *12*, 143–155.

Tadros, W., Houston, S. A., Bashirullah, A., Cooperstock, R. L., Semotok, J. L., Reed, B. H., et al. (2003). Regulation of maternal transcript destabilization during egg activation in Drosophila. *Genetics*, *164*, 989–1001.

Takada, S., Kelkar, A., & Theurkauf, W. E. (2003). Drosophila checkpoint kinase 2 couples centrosome function and spindle assembly to genomic integrity. *Cell*, *113*, 87–99.

Takahashi, S., Yokota, C., Takano, K., Tanegashima, K., Onuma, Y., Goto, J., et al. (2000). Two novel nodal-related genes initiate early inductive events in Xenopus Nieuwkoop center. *Development*, *127*, 5319–5329.

ten Bosch, J. R., Benavides, J. A., & Cline, T. W. (2006). The TAGteam DNA motif controls the timing of Drosophila pre-blastoderm transcription. *Development*, *133*, 1967–1977.

Yang, J., Tan, C., Darken, R. S., Wilson, P. A., & Klein, P. S. (2002). Beta-catenin/Tcf-regulated transcription prior to the midblastula transition. *Development*, *129*, 5743–5752.

Yuan, K., Farrell, J. A., & O'Farrell, P. H. (2012). Different cyclin types collaborate to reverse the S-phase checkpoint and permit prompt mitosis. *The Journal of Cell Biology*, *198*, 973–980.

Zhang, M., Kothari, P., Mullins, M., & Lampson, M. A. (2014). Regulation of zygotic genome activation and DNA damage checkpoint acquisition at the mid-blastula transition. *Cell Cycle*, *13*, 3828–3838.

Zou, L., & Elledge, S. J. (2003). Sensing DNA damage through ATRIP recognition of RPA-ssDNA complexes. *Science*, *300*, 1542–1548.

CHAPTER FIVE

Germ Line Versus Soma in the Transition from Egg to Embryo

S. Zachary Swartz, Gary M. Wessel[1]
Department of Molecular and Cellular Biology, Brown University, Providence, Rhode Island, USA
[1]Corresponding author: e-mail address: rhet@brown.edu

Contents

Abstract

With few exceptions, all animals acquire the ability to produce eggs or sperm at some point in their life cycle. Despite this near-universal requirement for sexual reproduction, there exists an incredible diversity in germ line development. For example, animals exhibit a vast range of differences in the timing at which the *germ line*, which retains reproductive potential, separates from the *soma*, or terminally differentiated, nonreproductive cells. This separation may occur during embryonic development, after gastrulation, or even in adults, depending on the organism. The molecular mechanisms of germ line segregation are also highly diverse, and intimately intertwined with the overall transition from a fertilized egg to an embryo. The earliest embryonic stages of many species are largely controlled by maternally supplied factors. Later in development,

Current Topics in Developmental Biology, Volume 113
ISSN 0070-2153
http://dx.doi.org/10.1016/bs.ctdb.2015.06.003

patterning control shifts to the embryonic genome and, concomitantly with this transition, the maternally supplied factors are broadly degraded. This chapter attempts to integrate these processes—germ line segregation, and how the divergence of germ line and soma may utilize the egg to embryo transitions differently. In some embryos, this difference is subtle or maybe lacking altogether, whereas in other embryos, this difference in utilization may be a key step in the divergence of the two lineages. Here, we will focus our discussion on the echinoderms, and in particular the sea urchins, in which recent studies have provided mechanistic understanding in germ line determination. We propose that the germ line in sea urchins requires an acquisition of maternal factors from the egg and, when compared to other members of the taxon, this appears to be a derived mechanism. The acquisition is early—at the 32-cell stage—and involves active protection of maternal mRNAs, which are instead degraded in somatic cells with the maternal-to-embryonic transition. We collectively refer to this model as the Time Capsule method for germ line determination.

ABBREVIATIONS

D/N delta–notch
EGA embryonic genome activation
FACS fluorescence-activated cell sorting
GNARLE Global Nanos-associated RNA lability element
GSCs germ line stem cells
MBT midblastula transition
MET maternal-to-embryonic transition
ORF open reading frame
PE posterior enterocoel
PGCs primordial germ cells
PRE pumilio response element
sMics small micromeres

1. THE BIG PICTURE

Multicellularity in an organism allows for a division of labor. Different cellular functions enhance the organism's overall competitiveness in the ecosystem, and expand its range and niche occupancy. Formation of these different functions often results from a progressive loss of developmental potency, as uncommitted cells terminally differentiate into their diverse fates. In most animals, these developmental decisions are controlled by both maternal information and by regulatory decisions engendered during embryogenesis. Differentiation involves not just the acquisition of gene function but also the repression of recent gene activity and the degradation of previously important mRNAs. As development progresses, patterning controls shift increasingly to those newly made in the embryo. This passing of the controls involves several conserved features which occur coordinately, including (1) a change in the *transcriptional* activity of embryonic blastomeres compared to the egg, (2) an extensive *degradation* of maternally supplied mRNA, and (3) remodeling of the *cell cycle* from rapid, often synchronous cleavages to longer, more asynchronous divisions. Here, we will collectively refer to these changes as *maternal-to-embryonic transitions* (*METs*). METs represent a conserved requirement for clearing of the maternal slate, which may facilitate cellular specialization by allowing embryonic gene regulation to rule. Eggs from different organisms exhibit vast differences in the extent that determinants are maternally supplied, and how rapidly these investments are removed in the shift to embryonic developmental control.

The extent of maternal loading of developmentally important information is intimately related to the manner in which embryos accomplish their division of labor. From the yellow crescent material that dictates muscle formation in ascidians, to the anucleate polar lobes that direct mesodermal fates in mollusk embryos, to even the differential yolk accumulation in birds, reptiles, and amphibians that influences cleavage patterns, to the germ plasm important for specifying the germ line in many organisms, differential loading of maternal components greatly influences when and how development proceeds (Conklin, 1905; Crampton & Wilson, 1896). Here, we will specifically explore differences in the segregation of germ line from somatic cells, and how maternal control and the MET may be differentially utilized for this process.

In contrast to the terminal differentiation of the soma, the germ line retains the capacity for totipotency, since it will pass all heritable information to the next generation. From an evolutionary perspective, the establishment of the germ line is therefore of utmost importance. The result of germ line segregation is that committed *primordial germ cells* are developmentally sequestered from the soma and into a reproductive niche that will be solely tasked with creating the *germ line stem cells*, which once in the gonad will produce gametes. Despite the near-universal requirement of the animal germ line, there exists a surprising diversity in the mechanisms by which it is segregated. In some animals, the germ line is formed very early, before the MET, whereas in others, germ line determination occurs after the MET has passed, and may even occur into adulthood. The timing of germ line segregation with respect to the MET necessitates differences in mechanism; that is, does it occur by posttranscriptional control, through induction and gene regulatory networks (GRNs), or by some combination of these mechanisms? The goal of the work presented here is to unravel the relative contributions of maternal information and embryonic gene regulation in the separation of the germ line from the soma. We will explore the interrelationships between METs and germ line development across species, with particular emphasis on the echinoderms. This phylum comprises familiar examples such as the sea lily, the sea star, the brittle star, the sea cucumber, the sea urchin, and the sand dollar (Figs. 1 and 2). Echinoderms offer readily accessible eggs and embryos for studying the mechanisms of developmental transition, and representatives within the phylum exhibit distinct mechanisms of germ line determination, valuable for comparative analysis in developmental function.

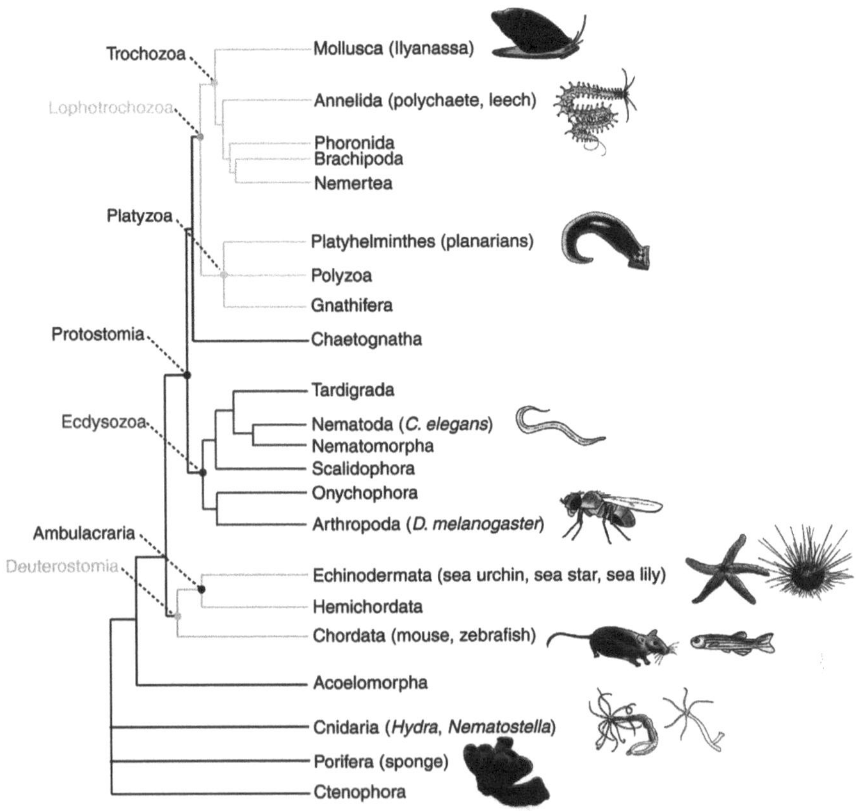

Figure 1 Metazoan phylogenetic tree. Phylogenetic relationships of major animal groups, with representative organisms highlighted within parenthesis and illustrations. *Reproduced from Juliano, Swartz, and Wessel (2010).*

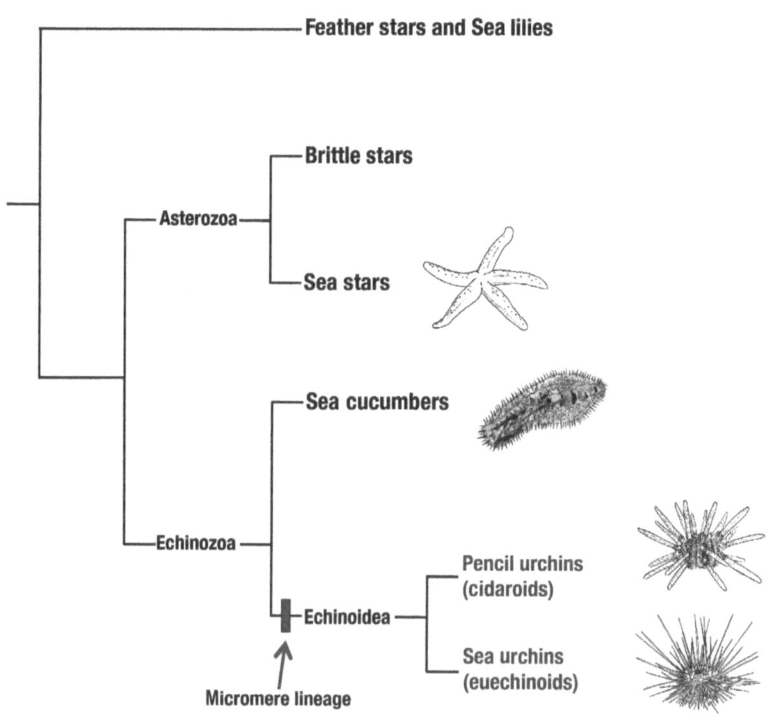

Figure 2 Species relationships of echinoderm classes (Reich, Dunn, Akasaka, & Wessel, 2015). The relationships between major echinoderm groups are indicated, with illustrations provided for species discussed in the text. The inferred stem at which the micromere lineage was acquired is indicated in purple (dark gray in the print version). Extant species with micromeres are indicated in red (light gray in the print version).

2. ARE METs UNIVERSAL?

Analyses of different organisms have taught us that the individual characteristics of METs are indeed widespread. An archetypal example is provided by the *midblastula transition* (*MBT*) in the frog *Xenopus laevis*. In this animal, the rapidly dividing early embryo first undergoes a morphologically apparent change: blastomeres no longer cleave synchronously. Second, based on radioactive nucleotide precursor incorporation experiments as well as more recent genomics, a major activation of embryonic transcription occurs (*embryonic genome activation*; *EGA*). Third, a major portion of the maternally supplied transcriptome is actively degraded (Tadros & Lipshitz, 2009). In animals such as *Drosophila*, *Xenopus*, and the zebrafish *Danio rerio*, whose phylogenetic positions are delineated in Fig. 1, loss of cell cycle synchrony, EGA, and maternal transcript degradation happens simultaneously. Dramatic examples like these have broadly influenced the terminology (e.g., MBT or maternal-to-zygotic transition) in a way that perhaps does not accurately reflect the diversity in development. There exist remarkable differences in when, where, and how EGA and maternal RNA clearance occur.

3. GENERAL ACTIVATION OF THE EMBRYONIC GENOME

Several species of sea urchins, including California's purple *Strongylocentrotus purpuratus*, have been particularly fruitful for investigating transcriptional regulation of embryogenesis. Sea urchin eggs are fertilized externally and adults can produce large cultures of synchronously developing embryos. Early cleavage stages in *S. purpuratus* are rapid (about 1 h per cleavage) and complete. The first three cleavages are equal, but the fourth is asymmetric and produces a 16-cell embryo of three tiers: the animal (top, by convention) most and midsized mesomeres, the large macromeres, and the small vegetal (bottom) micromeres. At this stage, germ layer fates have already begun to be specified. Subsequently, a ciliated blastula forms, which hatches from the fertilization envelope and becomes free swimming. Gastrulation initiates at the vegetal pole, and embryogenesis culminates in a feeding pluteus larva (Fig. 3A).

Several important discoveries in the sea urchin embryo were transformative to the field of development: (1) the early embryo can develop independently of transcription and even independently of a nucleus, (2) the fertilized egg (zygote[1]) begins transcriptional activity as quickly as can be measured, and (3) protein synthesis begins at fertilization and begins independently of new transcriptional activity (Davidson, 1986). Development independent of transcription was shown in a variety of ways. One early indication was by E.B. Harvey, when she stratified the egg into nucleated and nonnucleated fragments and then tested developmental potentials in the resultant pieces. Using a highly pigmented species for this work, the local *Arbacia punctulata* at the Marine Biological Laboratory in Woods Hole, she was able to visualize stratification of major organelles with isopycnic sucrose gradients, and even separate intact halves and quarters of eggs. From this approach, she learned

[1] The term *zygote* (Gr, yoked, or joined together) formally refers to the cell in development following fertilization, when the two gametes have joined into one cell, but before the first cell division. Here, we have retained this definition and refer to events following this as in embryogenesis.

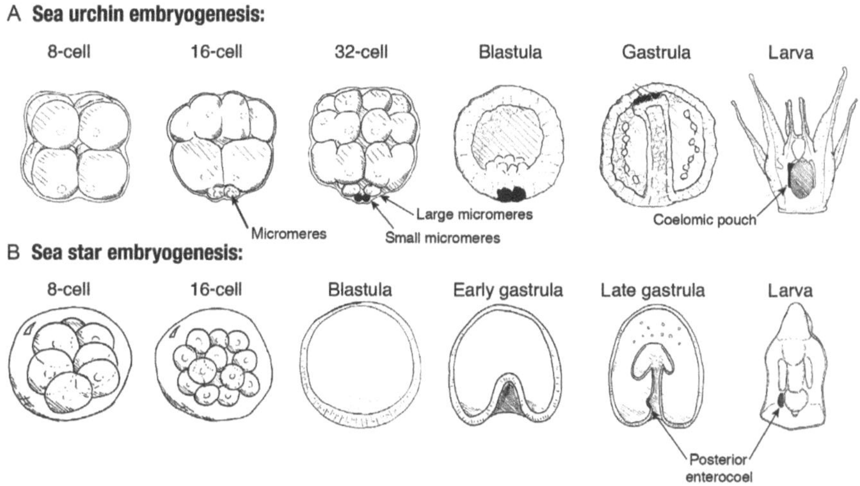

Figure 3 Representative embryonic stages of echinoderms. (A) Embryogenesis in the sea urchin is distinguished by highly regular cleavages, the third of which is asymmetric, yielding a 16-cell embryo of differently sized blastomeres. At the vegetal pole, a quartet of four small blastomeres, the micromeres, is produced. At the next cleavage, the micromeres divide asymmetrically to produce the large and small micromeres (sMics). The large micromeres ingress into the blastocoel and form the skeleton, while the sMics (red, dark gray in the print version) are the PGCs in the sea urchin. The sMics translocate along the tip of the gut during gastrulation, and subsequently assort into the two coelomic pouches. The sMics contained in the left pouch will contribute to the adult germ line. (B) Embryogenesis of the sea star, which may represent the ancestral mode of germ line segregation in echinoderms. The early cleavages are equal, yielding embryos of equally sized blastomeres with no morphologically overt polarity. A large, hollow blastula forms and gastrulation initiates before any mesenchyme enters the blastocoel. In the later gastrula, two coelomic pouches form at the anterior tip of the gut. Subsequently, a smaller posterior pouch called the posterior enterocoel (indicated in red, light gray in the print version) forms on the left of the gut, which is visibly distinct from the gut in the larva. The posterior enterocoel contains the likely PGCs.

that each egg fragment was capable of fertilization and development regardless of whether it contained the egg pronucleus, either as a diploid organism (male and female pronuclear contributions), as a merogone (an enucleated egg fragment that was fertilized), or even as a parthenogenetically activated merogone (with no nucleus). These experiments documented that early cleavage and development can occur in this animal even in the absence of a nucleus and that maternal information was important in early development (Harvey, 1940). Earlier, Theodor Boveri, while working at Stazione Zoologica Anton Dohrn di Napoli, even made use of sea urchin merogones fertilized by the sperm of other species to distinguish between contributions from the maternal stores, relative to the paternal nucleus (Boveri, 1893; Laubichler & Davidson, 2008). Overall, these experiments introduced the nuclear theory of determination but, more importantly for our discussion here, showed that all the RNAs needed for early development in this animal, including mRNAs, rRNAs, tRNAs, and small RNAs, must already be present in the egg prior to fertilization.

Paul Gross made use of the newly identified toxin actinomycin D as an inhibitor of DNA-dependent RNA synthesis (Gross & Cousineau, 1963; Gross, Malkin, & Moyer, 1964). First, he and his colleagues tested how soon newly synthesized RNA was being made by incorporation of a radiolabeled uridine. Although technically limited to global RNA analysis and only by quantitation of radioactive counts, his group was able to detect significant incorporation within the first time point possible in these experiments by 20 min after fertilization. Further, actinomycin D-treated embryos exhibited no detectable transcriptional signal, yet the embryos developed relatively normally. They therefore concluded that new transcription was not necessary for early development, and that protein synthesis was templated by RNAs stored in the egg. It was subsequently demonstrated that the egg contained such information by measuring protein synthesis in the presence of radiolabeled amino acids. David Epel learned that amino acid incorporation occurred following fertilization, also as quickly as could be measured, within 15 min (Epel, 1967). Thus, the egg has substantial stored information that can support early development, and the transition from egg to embryo includes a rapid activation of both transcription and translation. While transcription is not essential for early development, the normal embryo does indeed initiate transcription with fertilization. So while some embryos (frog, fly) may not transcribe significant RNAs immediately after fertilization, others (sea urchins) clearly do, supporting the concept that the orchestrated METs of different species are highly variable.

These early functional investigations into sea urchin embryonic transcription and translation guided much of the thinking in the field for decades. Recently, high-throughput RNA sequencing and other technologies documented these processes in transcriptome-level detail. Sampling multiple time points from 10 to 72 h of sea urchin development onward indicated diverse genome activity (Tu, Cameron, & Davidson, 2014). However, this study did not include time points between fertilization and 10 h of development (approximately the first nine cleavages); thus, the earliest upregulation of transcription was not captured. A further limitation of these approaches is that they do not directly distinguish between maternally supplied and embryonically transcribed RNA; embryonic activation could only be inferred by an increase in relative transcript abundance. In the future, it will be important to test additional and earlier time points with new approaches for capturing nascent RNA, such as groSEQ, or blocking of splicing to detect newly transcribed intronic sequence (Core, Waterfall, & Lis, 2008; Lee et al., 2013). For now, it can at least be deduced that there is an increasing reliance upon embryonic transcriptional activity as development progresses that is perhaps accentuated by important milestones in embryogenesis (summarized in Fig. 4).

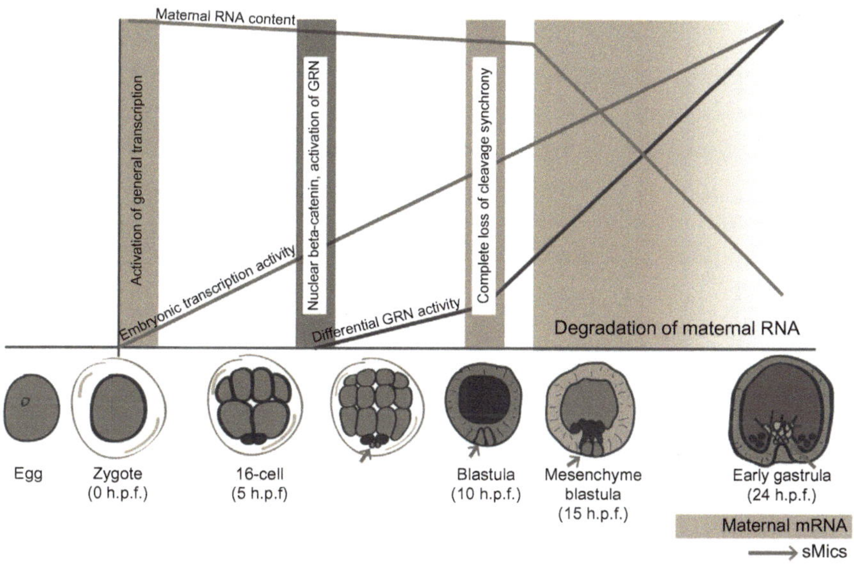

Figure 4 Summary of METs in sea urchin embryogenesis. Important embryonic milestones are denoted with embryo illustrations. General transcription based on incorporation of radioactive nucleotides initiates at fertilization and steadily increases through early development (green line). The regionally specific gene regulatory network and fate specification first occur at the 16-cell stage, when β-catenin is nuclearized in the micromeres (purple line). We hypothesize that the differential GRN further diversifies in activity when the cell cycle is remodeled at approximately 10 h.p.f. and cleavage synchrony is lost. Based on mRNA expression dynamics, we suggest that there is a major degradation of maternal RNA that occurs as the embryo prepares to gastrulate. Maternal mRNA is represented in blue in embryo descriptions, and is lost in somatic cells in the gastrula, but retained in the sMics (green arrows). (See the color plate.)

4. ACTIVATION OF REGIONALLY SPECIFIC GENE REGULATORY NETWORKS

Through gene regulatory analysis, it is clear that lineage-specific gene expression initiates in the sea urchin by the fourth cleavage, when the micromeres express vegetal inductive signals, such as Wnt8 (Cui, Siriwon, Li, Davidson, & Peter, 2014; Wikramanayake et al., 2004). The earliest activation of this localized gene regulatory network (GRN) is downstream of the canonical Wnt/β-catenin pathway (Logan, Miller, Ferkowicz, & McClay, 1999; Wikramanayake, Huang, & Klein, 1998; Fig. 4). Blocking nuclear β-catenin prevents Wnt8 expression, vegetal fate specification, and animalizes the embryo with complete failure in gastrulation. Typically, the canonical Wnt pathway is activated by a secreted Wnt ligand binding to a Frizzled-type receptor. Subsequently, membrane-associated Disheveled protein binds and disassembles a destruction complex, allowing for β-catenin to enter the nucleus and activate transcription. However, the earliest activation of the pathway in the sea urchin embryo may occur independently of a Wnt ligand. Maternally supplied Disheveled protein is enriched in the vegetal cortex of the sea urchin egg and, based on gel mobility, is modified in a form that perhaps indicates constitutive activation (Peng & Wikramanayake, 2013). In effect, the pathway may be hard-wired, implying that the early embryo activates differential axial specification autonomously, and not until later does inductive signaling influence cell fate decisions. Indeed, when cultured in the presence of an inhibitor that blocks all Wnt secretion, embryos still establish animal–vegetal polarity and form a blastopore (Cui et al., 2014). Nuclearization of maternal β-catenin thus may be a primary maternal activator of the embryonic genome. Downstream of this initial anisotropy, regionally specific GRNs pattern the various ectodermal, endodermal, and mesodermal territories of the pluteus larva (Oliveri, Tu, & Davidson, 2008; Peter & Davidson, 2015).

5. LOSS OF CLEAVAGE SYNCHRONY: A CAUSE OR EFFECT OF EMBRYONIC CONTROL?

In *Drosophila* and *Xenopus*, major activations of the embryonic genome coincide with a loss of cleavage synchrony among blastomeres (Farrell & O'Farrell, 2014; Tadros & Lipshitz, 2009). While cleavage patterns exhibit species differences amongst echinoderms, careful investigation of several sea urchin species reveals some commonalities. Only the first four cleavages are completely synchronous—that is, until the birth of the micromere lineage. This timing of synchrony loss would seem to coincide with a major activation of the differential GRN. Subsequently, embryos display "regional synchrony"—that is, particular tiers of related blastomeres divide together, but at different rates than their neighbors. Intriguingly, these regional synchronies follow an animal–vegetal polarity gradient, with vegetal cells (e.g., micromeres) dividing more slowly than the more animal cells. This animal–vegetal wave of cleavage is reminiscent of the metasynchronous cleavage cycles in the early syncytial *Drosophila* embryo (Edgar & O'Farrell, 1989). All cleavage synchrony in the sea urchin is lost by the eighth or ninth cycle at the onset of ciliogenesis and hatching (Fig. 4). Furthermore, the cell cycle is substantially lengthened at this stage (Dan, Tanaka, Yamazaki, & Kato, 1980; Masuda & Sato, 1984). The complete loss of synchrony detected in blastula stages may coincide with the arrests observed when embryos are cultured in transcriptional inhibitors. That is, even though embryos can develop without new transcriptional activity, essential gene activity is present at the time of cell cycle loss of synchrony. More investigation is required, but it is tempting to speculate that there may be an intimate and even causal relationship between the cell cycle and the EGA in the sea urchin. For example, the lagging of vegetal blastomere cleavage (e.g., the micromeres) relative to animal blastomeres may permit the early vegetal activation of GRN expression. Downstream GRNs become activated once other blastomeres in the embryo have divided more, producing enough raw cellular material to create the other territories. The lengthening of the cell cycle at the nineth division may be important to facilitate robust embryonic transcription, as occurs in *Drosophila* (Shermoen & O'Farrell, 1991). The ability of embryos to reach blastula stages with transcriptional inhibitors may simply reflect their maternally endowed ability to divide into many cells, whereas the differentiation of those cells is under embryonic control.

The loss of synchrony, lengthening of the cell cycle, and possible relationship between cell cycle and transcription are consistent MET characters in other organisms. Cell cycle modifications in embryogenesis have been extensively characterized in *Drosophila*, which displays a dramatic extension of cleavage cycle 14—the *Drosophila* MBT. With cycle 14, a typical G2 phase is introduced—in the first 10 cycles, there are no gap phases, and a short G2 only begins to be introduced at cycle 10. One mechanism for this extension is the degradation of the maternally supplied ortholog of the Cdc25 phosphatase, Twine (Farrell & O'Farrell, 2014). The lengthening of the cell cycle may be required for robust transcription, particularly from large genes. Given the extremely rapid early cleavages of the *Drosophila* embryo, transcription is limited by the ability of RNAPII to elongate before being displaced by the DNA replication machinery or by condensation in the beginning of mitosis. Taking advantage of detailed knowledge of cell cycle timing and *in situ* hybridization, it was found that *Ubx* gene transcription is interrupted by mitosis prior to cycle 14, resulting in abortive transcripts (Shermoen & O'Farrell, 1991). Thus, it is possible that the remodeling of the cell cycle observed in the sea urchin and other animals also has direct consequences for transcriptional regulation.

6. DEGRADATION OF THE MATERNALLY SUPPLIED TRANSCRIPTOME

In diverse organisms, such as the zebrafish, frog, and fruit fly, degradation of the maternally supplied transcriptome is essential for developmental progression (Tadros & Lipshitz, 2009). Genomic studies strongly support a major transcriptome remodeling around the blastula to gastrula transition in the sea urchin. Temporal analysis by microarrays suggested that a major fraction of maternally supplied transcripts is degraded by the onset of gastrulation (Wei, Angerer, & Angerer, 2006). Subsequent deep sequencing and cluster analysis indicated that the transcripts of 34% of all genes are maternally deposited into the egg and sharply degraded by the onset of gastrulation (Fig. 4; Tu et al., 2014).

How these RNAs are turned over in the sea urchin remains unknown. In other animals, maternal transcriptome degradation is affected by RNA-binding proteins, as well as by small RNA machinery. *Drosophila* has two phases of degradation: first, a maternal pathway that can occur in an activated, unfertilized eggs (i.e., without a fertilizing sperm) and a second pathway that requires embryonic transcription (Bashirullah et al., 1999; Tadros et al., 2007). The first pathway uses the RNA-binding protein Smaug, which binds stem loop elements in the 3′UTRs of its targets, and can both repress translation and target the mRNA for degradation, typically performing both functions (Chen et al., 2014; Smibert, Wilson, Kerr, & Macdonald, 1996). Translational repression is achieved by recruiting the *Drosophila*-specific factor Cup, which interacts with eIF4E to prevent interaction with eIF4G and recruitment of the 40S ribosomal subunit (Nelson, Leidal, & Smibert, 2004). Transcript degradation is achieved by recruiting the CCR4/POP2/NOT complex, called the *CNOT complex* in vertebrates

(Semotok et al., 2005). This multisubunit complex is a major cytoplasmic deadenylase, which functions by shortening poly-A tails and making the mRNAs substrates for various RNA decay pathways, including the exosome (Collart & Panasenko, 2012). Smaug binding to mRNA in *Drosophila* can recruit the CNOT complex and degrades over 1000 transcripts, at least 339 of which were identified as direct targets (Chen et al., 2014). While most transcripts were both translationally repressed (based on the lack of polysome association) and degraded by Smaug, many transcripts were repressed but not destabilized. This result suggests separable functions for Smaug, and future work may illuminate the combinatorial control that can lead to these distinct outcomes of Smaug binding. The sea urchin contains two putative Smaug paralogs, which should be tested in the future for conservation of function in maternal transcript turnover.

A second pathway mechanism for mRNA decay is via the micro RNA (miRNA) pathway. Together with their associated Argonaute proteins, miRNAs recognize seed sequences in 3′UTRs and promote both translational repression and transcript decay, also by recruiting the CNOT complex (Tritschler, Huntzinger, & Izaurralde, 2010). In *Drosophila*, a second pathway requiring embryonic transcription further reinforces the transcript degradation initiated by Smaug and requires the miR-309 miRNA cluster. This cluster of miRNAs degrades several hundred maternally supplied mRNAs and its deletion results in larval lethality (Bushati, Stark, Brennecke, & Cohen, 2008). The miRNA machinery is used in several organisms for maternal transcript clearance. In the zebrafish, miR-430 is expressed at the MBT and causes the degradation of several hundred maternal transcripts by deadenylation (Giraldez et al., 2006). The *Xenopus* ortholog of miR-430, called miR-427, is required to degrade maternally supplied cyclin A1 and B2 mRNAs; however, whether miR-427 targets a broad subset of transcripts like miR-430 is unknown (Lund, Liu, Hartley, Sheets, & Dahlberg, 2009). The miRNA pathway has also been interrogated in the sea urchin, whose genome contains at least 49 miRNAs present in the egg and embryo. Complete blockage of miRNA biogenesis prevents normal gastrulation and differentiation of the embryo but, strikingly, this defect can be rescued by adding back four abundant miRNAs: miR-1, 31, 2012, and 71 (Song et al., 2012). This result is reminiscent of the rescue obtained when miR-430 is injected into Dicer-mutant zebrafish (Giraldez, 2005). The link has not yet been directly tested in sea urchins; however, it seems likely that these miRNAs would play a large role in the maternal transcript degradation detected by temporal deep sequencing (Tu et al., 2014). Together, these results indicate that maternal transcriptome clearance is an essential transition in embryogenesis. However, since it occurs by different mechanisms in different species, we can infer that convergent evolution has favored multiple acquisitions of this form of MET. There must then exist a strong selective pressure for the process.

We propose that maternal transcriptome degradation is essential for clearing of the slate of the totipotent egg, allowing for specific gene regulatory processes to pattern the different cell fates of the embryo. Differences in developmental strategy that favor rapid development have influenced the evolution of the timing of the event. For example, it occurs late in *Drosophila*, in an embryo of several thousand syncytial nuclei. In contrast, the mouse initiates degradation of maternal transcripts in response to fertilization and is largely complete by the 2-cell stage (Hamatani, Carter, Sharov, & Ko, 2004). The segregation of the germ line is uniquely intertwined with maternal RNA acquisition, and also has evolved multiple developmental mechanisms, which we document below.

7. SEPARATING THE SOMA FROM THE GERM LINE: A CRITICAL FORK IN THE ROAD

The survival of a species depends on a means for the individual to transmit its heritable information to its offspring. In the life cycle of most animals, important reproductive cells called the *germ line* are typically segregated away from the terminally differentiating soma, which lacks reproductive potential and will die with that individual. Species differences in the timing of germ line/soma separation have led to ambiguity in how terminology should be applied. For example, the P blastomeres of the early *Caenorhabditis elegans* embryo are often referred to as a germ line cells because they contain characteristic germ granules (or P-granules) and will give rise to *de facto* germ line stem cells later (Seydoux & Braun, 2006). However, this blastomere will also create a number of somatic cell fates for several cleavage cycles, until finally giving rise to the Z-blastomeres, which are solely germ line competent. In this work, we will designate any cell as being within the *germ line* if it is capable of giving rise to a gamete, whether it is the direct progenitor of that gamete or a general founding blastomere whose later progeny becomes gametogenic. It may be more straightforward to define the germ line by what it is not: a blastomere may be called *somatic* when its fate no longer includes germ line potential. Similarly, we will refer to a blastomere that is solely germ line competent and without somatic potential as a *primordial germ cell* (*PGC*). A *germ line stem cell* (*GSC*) is the mature product of the PGC, which is found in the gonad and undergoes self-renewing divisions to produce meiotically capable cells that will give rise to gametes (eggs or sperm). This discussion also excludes the consideration of cells that *can* make germ line by a variety of experimental or abnormal conditions but do not do so in normal development.

8. DIFFERENT ROADS LEAD TO A CONSERVED GERM LINE PROGRAM

Comparative analyses of both expression pattern and function have illuminated a set of highly conserved germ line regulators, which have been extensively reviewed elsewhere (Ewen-Campen, Schwager, & Extavour, 2010). The most conserved of these factors are posttranscriptional regulators of mRNA stability and translation. Here, we will focus our attention primarily on a translational repressor, *Nanos*, and also an Argonaute family member, *Piwi*, and an RNA helicase, Vasa. In all species examined previously, overlapping expression of these three factors in embryonic blastomeres is predictive of germ line fate, though occasional non-germ line stem cell functions have been identified for them individually (Juliano, Swartz, et al., 2010). Nanos orthologs are zinc finger-containing RNA-binding proteins that act upon 3′UTRs of target mRNAs in the germ line. Nanos itself possesses only nonselective RNA-binding activity, and instead derives its specificity from associating with a partner called Pumilio. Pumilio is the founding member of the PUF-family RNA-binding factors, and recognizes highly conserved *Pumilio Response Elements* (*PREs*) of a "UGUAAAU" consensus (Chen et al., 2012). The primary consequence of Nanos/Pumilio binding a target mRNA is to effect its translational repression and/or destabilization. A conserved mechanism of Nanos action is by recruitment of the CNOT complex, similar to that described for Smaug earlier (Bhandari, Raisch, Weichenrieder, Jonas, & Izaurralde, 2014). Upon recruitment, the nuclease subunits CNOT6 and CNOT7 then degrade the mRNA's poly-A tail resulting in its destabilization. Piwi family members are nucleases guided by small RNAs (piRNAs), which are primarily thought to protect the germ line by opposing transposable element mobilization in a process called the ping-pong cycle (Mani & Juliano, 2013). The precise role of Vasa has been mysterious since its discovery, but has now been identified as an integral component of piRNA biogenesis and transposon suppression in the germ line (Xiol et al., 2014) as well as a general translation factor (Liu, Han, & Lasko, 2009; Yajima & Wessel, 2015).

Despite similarities in the molecular toolkit (e.g., Vasa, Nanos, and Piwi), there exists a surprising amount of diversity in how PGCs are segregated and how germ line specifying genes come to be expressed. The mechanisms of germ line specification across species can be considered within a continuum of what we will call here *preloaded* versus *inductive* mechanisms (sometimes referred to as preformation vs. epigenesis, respectively). In brief, induction involves a conversation of cell–cell signaling between embryonic tissues that instructs select cells to adopt germ line fate (Fig. 5A and B). Conversely, preloaded specification involves maternally supplied factors that are often spatially enriched in one region of the egg, which when acquired early by nascent blastomeres directs them toward a germ line fate. These dense amalgamations of protein and mRNA determinants are collectively referred to as germ plasm (Fig. 5C).

9. PGC SPECIFICATION BY INDUCTION

Comprehensive syntheses of available data across taxa strongly imply that induction represents the ancestral mode of germ line segregation, though the timing of the event can vary (Extavour, 2007; Extavour & Akam, 2003). Perhaps the most intensely studied example of induction is the mouse, which specifies its germ line late, at about 6.5 days postfertilization. The proximal portion of the epiblast, or embryo proper (in closest proximity to the placental attachment), contains a field of mesodermal progenitors. Synergistic signaling from the extraembryonic ectoderm to this field, notably via the secreted molecules Wnt3 and Bmp4, enacts a germ line transcriptional program in approximately eight founding PGCs (Fig. 5A; Lawson et al., 1999; Ohinata et al., 2009; Tam & Zhou, 1996). The exact connections between PGC transcriptional regulators are still being elucidated; however, the mesodermal factor Brachyury/T appears to be one of the earliest activators, and a direct target of canonical signaling from Wnt3. Brachyury is then required for the activation of core mouse PGC regulators Blimp1 and Prdm14 (Aramaki et al., 2013; Magnúsdóttir et al., 2013). Together with AP2γ, these three factors comprise a transcriptional program termed the "tripartite network," which is required to broadly repress somatic cell transcription but enable germ line transcription. Nanos3 is a direct target of AP2γ as determined by ChIP analysis (Magnúsdóttir et al., 2013). It is expressed in early PGCs following their

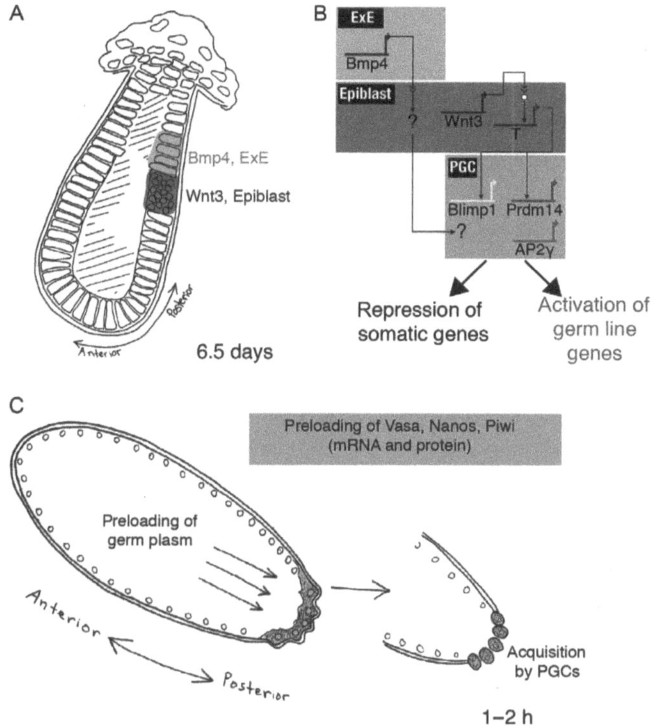

Figure 5 Strategies for germ line segregation by induction or preloading. (A) Germ line specification by induction. Shown is a simplified mouse embryo 6.5 days postfertilization. Bmp4 signaling from the extraembryonic ectoderm (ExE, blue) and Wnt3 expressed in the proximal epiblast (purple) are synergistically required to activate germ line transcriptional regulators in the PGCs (green). (B) Regulatory diagram depicting Wnt3 activating the mesodermal transcription factor T, which in turn activates Blimp1 and Prdm14, which along with AP2γ, form a tripartite network that activates germ line gene expression and represses somatic fate. The activating role of Bmp4 has not been directly delineated, but is required to license the PGCs to respond to the Wnt3 signal. (C) Germ line segregation by preloading. Shown is a syncytial blastoderm-staged *Drosophila* embryo, with maternally localized germ plasm in green. The PGCs are the first to cellularize in this embryo, and acquisition of the germ plasm directs them toward germ line fate. (See the color plate.)

initial segregation and during their migration into the somatic gonad and its knockout leads to loss of PGCs by apoptosis and sterility of both males and females (Suzuki, Tsuda, Kiso, & Saga, 2008; Tsuda et al., 2003).

While induction is mechanistically best understood in the mouse, ample morphological, gene expression, and some functional data indicate its broad usage across species. Some effort has been made into human PGC investigation, but several considerations make their isolation impractical. Recently, *in vitro* schemes have been devised that appear to accurately mimic the developmental course of human PGCs *in vivo*. Already such investigation has revealed some striking differences from the established mouse paradigm—for example, Sox17, which was previously considered an endodermal regulator, is required to activate Blimp1; however, these PGC-like cells also upregulate Brachyury, normally associated with somatic mesoderm formation, and it will be important to test in the future whether Brachyury is required for Blimp1 and Prdm14 activation in human cells, or whether these are competing elements for PGC formation (Irie et al., 2015). These human PGC-like cells also upregulate Nanos3 in a Blimp1- and Sox17-dependent fashion, though it is not yet known whether it is a direct target of these regulators. A third and powerful vertebrate comparison is the axolotl or Mexican salamander, a urodele amphibian. In contrast to mice and humans, axolotl PGC specification seems to require both Fgf and Bmp4 signaling for robust germ line gene activation. Brachyury also enhances germ line induction in the axolotl (Chatfield et al., 2014). Intriguingly, PGC specification appears to be Blimp1 independent in the axolotl, raising the possibility that the upstream Blimp1 expression in mouse PGCs is a derived trait. Further comparisons will be required to more confidently resolve the evolutionary history of vertebrate PGCs. In any case, a conserved theme of vertebrate PGC segregation is intercellular signaling via Bmp signaling and activation of the translational repressor Nanos, whose molecular functions we will return to later.

Accumulating data in invertebrate suggest that induction via Bmp signaling is indeed a deeply conserved mechanism for PGC segregation. The cricket *Gryllus gryllus* has been developed into a functional model organism, and has been suggested to be more representative of ancestral insect development than *Drosophila melanogaster*. In the cricket, PGC clusters arise in posterior segments, and are recognized both by morphology as well as by expression of germ line factors. Knockdown of Bmp pathway components leads to a loss of PGCs, whereas overactivation of the pathway leads to supernumerary germ cells (Donoughe et al., 2014). Whether the Wnt pathway is also required, as observed in the mouse, remains to be tested. Functional data for PGC specification in other taxa are limited; however, the expression patterns of these genes merged onto the embryological considerations broadly imply conservation of induction. For example, morphological and molecular examination indicates a lack of germ plasm or early-forming PGCs in several other arthropods, Lophotrochozoans such as the mollusk *Ilyanassa*, and the cnidarian *Nematostella*, among others (Ewen-Campen, Donoughe, Clarke, & Extavour, 2013; Ewen-Campen, Jones, & Extavour, 2013; Extavour, Pang, Matus, & Martindale, 2005; Rabinowitz, Chan, Kingsley, Duan, & Lambert, 2008; Swartz, Chan, & Lambert, 2008). Highly regenerative animals, such as hydrozoan cnidarians like *Hydra*, demonstrate a capacity to segregate germ cells throughout their life cycles, rather than just in embryogenesis, precluding reliance upon inherited embryonically germ plasm (Juliano et al., 2014). In these animals, a self-renewing multi- or totipotent stem cell called the i-cell continually replenishes somatic lineages, and also contains the germ line, as it will give rise to sperm and eggs upon sexual reproduction (Müller, Teo, & Frank, 2004). The planarian *Schmidtea mediterranea* can regenerate its germ line after being dissected into very small pieces, demonstrating an impressive capacity for induction in the adult (Wang, Zayas, Guo, & Newmark, 2007). Broad comparisons such as these and many others reviewed elsewhere all point to induction being the likely ancestral route for PGC segregation in Metazoa (Extavour, 2007).

10. PGC SPECIFICATION BY PRELOADING

Investigation in diverse animals has yielded multiple examples of early PGC segregation by preloading. In contrast to the aforementioned strategies, PGC segregation here is thought to occur both autonomously and independently of PGC transcriptional activity. Perhaps, the most famous example of germ line specification by preloading is by *Drosophila*, whose PGCs are indeed the very first cells to form, approximately 1.5 h after fertilization (Fig. 5B). In this archetypal example, a cytologically obvious germ plasm is spatially localized to the posterior of the oocyte. Germ plasm localization involves an active process of transport along microtubules, cytoplasmic dumping from the nurse cells followed by streaming in the oocyte, and their anchoring to the posterior cortex (Bergsten & Gavis, 1999; Seydoux & Braun, 2006). Germ plasm assembly follows a distinct assembly hierarchy, with the insect-specific protein, Oskar, serving as an anchor upon which other mRNAs and proteins aggregate. In an intriguing example of evolutionary co-option, the ancestral role of Oskar was likely in the nervous system, while its utility as a scaffold was later employed in the germ plasms of Dipteran species (Ewen-Campen, Srouji, Schwager, & Extavour, 2012). Classic experiments have revealed that germ plasm is both necessary and sufficient to specify PGCs in *Drosophila*: transplant of germ plasm to an irradiated (and infertile) recipient can restore fertility, or even create ectopic PGCs in the anterior region of the embryo (Illmensee & Mahowald, 1974). Revisiting these experiments with genetic approaches has shown that Oskar overexpression results in ectopic, anterior PGCs (Ephrussi & Lehmann, 1992; Smith, Wilson, & Macdonald, 1992).

The nematode worm *C. elegans* also provides a genetically tractable example of preloaded germ line specification. In this organism, the germ plasm comprises a collection of ribonucleoprotein granules called P-granules. The P-granules contain conserved germ line factors, such as the *C. elegans* Vasa orthologs, called GLH-1 and 2 (Gruidl et al., 1996). Before fertilization, P-granules are uniformly distributed in the cytoplasm, but after fertilization and during cleavage, these granules are segregated toward the P blastomere lineage. As development progresses, P-granules become increasingly associated with the nuclei, and are eventually restricted to the Z-blastomeres, which are the *de facto* PGCs (Updike & Strome, 2010). While some analyses have suggested that P-granules are neither strictly required nor sufficient for the embryonic segregation of PGCs in *C. elegans*, their intimate association with the germ line is clear, and they likely perform diverse tasks pertaining to RNA metabolism, posttranscriptional regulation, and germ line protection (Gallo, Wang, Motegi, & Seydoux, 2010; Voronina, 2013).

Numerous examples of preloaded specification exist within vertebrates as well. *Xenopus* displays a prominent germ plasm, called the Balbiani body, which translocates to the vegetal cortex during oogenesis. The *Xenopus* germ plasm contains numerous germ line-associated RNAs and proteins, in addition to a collection of mitochondria, which are commonly associated with germ granule structures in different species. Nanos1 RNA is transcribed during oogenesis and incorporates into the germ plasm and, following fertilization, is inherited by germ line-fated vegetal blastomeres (Forristall, Pondel, Chen, & King, 1995; MacArthur, Bubunenko, Houston, & King, 1999). The zebrafish also segregates its PGCs by preloading, displaying remarkable similarities with the frog including a vegetally localized Balbiani body during oogenesis. Nanos and Vasa RNAs, as well as that of Dazl, another translational regulator, assemble into the zebrafish Balbiani body. Following fertilization, the germ plasm material translocates to the animal pole of the zygote and becomes enriched at the distal ends of cleavage furrows. The four blastomeres that inherit this material are directed to germ line fate (Kosaka, Kawakami, Sakamoto, & Inoue, 2007). A gene called *Bucky ball* is required for germ plasm assembly and transport and, while it does not bear evolutionary homology to Oskar, its protein product performs an analogous function (Bontems et al., 2009).

11. EVOLUTIONARY TRANSITION OF PGC SEGREGATION IN THE ECHINODERMS

Echinoderms provide an intriguing test case for the evolution of preloaded germ line specification, supported by a rich fossil record (Fig. 2). Expression pattern and embryological analyses indicate that the ancestral mode of germ line segregation in the echinoderms was by induction (Wessel et al., 2014). In the sea star, embryological investigations suggest that a posterior coelomic pouch (*posterior enterocoel, PE*) in the larvae is required for fertility in the adult (Fig. 3B); when this structure is removed by microsurgery, juveniles contain fewer putative PGCs by cytological criteria (Inoue, Kiyomoto, & Shirai, 1992). Patterns of germ line gene expression also support an inductive mode for the sea star. Transcripts for Vasa and Piwi are ubiquitously distributed in early embryos, with no indications of a localized germ plasm. However, these transcripts become progressively restricted to vegetal cells that form the PE as gastrulation progresses. In contrast, Nanos expression is activated embryonically within the PE cells, reminiscent of Nanos activation in the mouse PGCs after stimulation by Bmp and Wnt signaling (Fresques, Zazueta-Novoa, Reich, & Wessel, 2014). Both Wnt3 and Bmp2/4 are expressed in territories close to the PE progenitors, suggesting that these pathways could have a conserved role for germ line induction, though this premise remains to be tested functionally.

In contrast, the echinoid echinoderms display some features of preloaded PGC segregation. Echinoids comprise a relatively recently diverging clade that includes the euechinoids, or sea urchins and sand dollars, and the cidaroids, or pencil urchins (Fig. 3B). A derived trait of this clade is the micromere lineage, which forms by an asymmetric division at the fourth cleavage. This division produces a quartet of small cells at the vegetal pole called the *micromeres*. The micromeres then divide asymmetrically again to produce the more animal-oriented large micromeres and the vegetal *small micromeres* (hereafter abbreviated *sMics*; Fig. 3B). This separation of fates between the large and small micromeres is particularly dramatic, because the large micromeres are unipotent and will only construct the larval skeleton, while the sMics are the likely PGCs (Wessel et al., 2014).

Numerous lines of molecular evidence support the concept that the sMics are the bona fide PGCs formed by preloaded specification. Upon formation, the sMics are highly enriched for Vasa protein, which occurs via posttranslational mechanisms including ubiquitination and, perhaps, spindle association (Gustafson, Yajima, Juliano, & Wessel, 2011; Yajima & Wessel, 2011a). Shortly after their formation, the sMics express Nanos2, which along with the forkhead transcription factor FoxY, is one of only two known genes embryonically and selectively expressed in the sMics prior to gastrulation (Juliano, Yajima, & Wessel, 2010; Materna, Swartz, & Smith, 2013; Fig. 6). Nanos is uniquely detectable in the sMics by *in situ* hybridization as early as the 32/64-cell stage. As in the sea star, Piwi mRNA is maternally supplied and ubiquitous in early embryos, but becomes restricted posttranscriptionally to the sMics by gastrula stages (Swartz et al., 2014; Yajima, Gustafson, Song, & Wessel, 2014). Intriguingly, the vegetal egg cortex and subsequently the sMics of the sea urchin *Hemicentrotus pulcherrimus* are enriched for mitochondrial rRNAs outside of the mitochondria themselves (Ogawa et al., 1999). Similar observations have been made for the germ plasms of *Drosophila* and *Xenopus*, but the significance of extramitochondrial rRNA is unknown (Kobayashi, Amikura, & Mukai, 1998; Kobayashi, Amikura, & Okada, 1993). The sMics show signs of autonomy associated with preloaded PGC segregation: when the

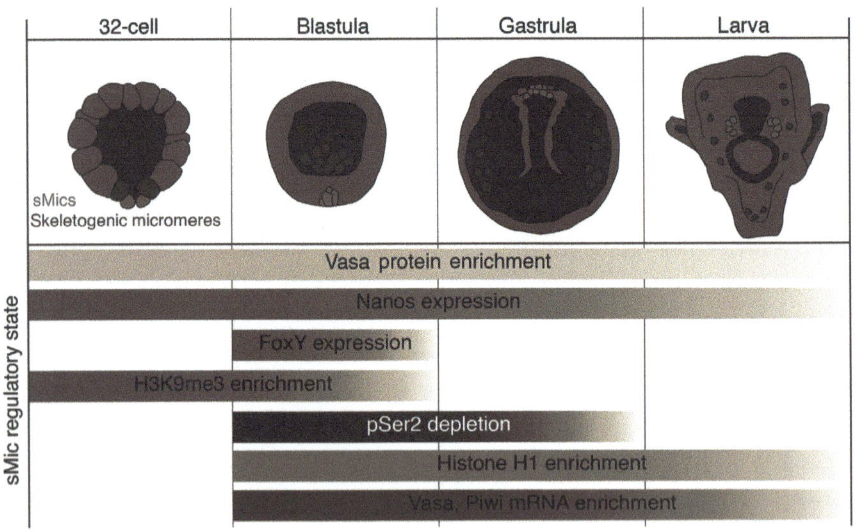

Figure 6 Temporal dynamics of sMic regulatory state. Location of the sMic lineage is indicated in green, from 32-cell stage into the larva. Colored bars indicate the presence or absence of gene expression, or enrichment or depletion for specific markers. *Adapted from Wessel et al. (2014).* (See the color plate.)

micromeres, representing the parent lineage, are surgically removed and cultured in isolation, they divide again asymmetrically to form the sMics and upregulate both Vasa protein and Nanos RNA on a timescale consistent with the intact embryo (Yajima & Wessel, 2012). After their creation, the sMics divide only once to yield eight descendants for all of embryogenesis and early larval stages. Thus, the sMics display all characteristics associated with preloaded PGCs, specified much more precociously than in the sea star. Yet, no morphologically apparent germ plasm has been identified, neither in eggs nor embryos. Lack of evidence is not proof, but these embryos appear to specify germ cells by preloading mechanisms, with a distinct from those seen in other model organisms.

Given the preloaded characteristics of the sMics, the embryonic expression of Nanos2 is unique. Three Nanos paralogs are present in the sea urchin (termed Nanos1–3), which are expressed at different times in the life cycle. Nanos1 is ovary-specific, while Nanos2 is detectable from shortly after the sMics are created and into the early coelomic pouches. Nanos3 is transiently expressed at the tip of the developing gut late in gastrulation (Juliano, Yajima, et al., 2010). The regulation of Nanos1 and 3 is unexplored, though some traction has been gained with the early sMic paralog, Nanos2. A late requirement of the *Nanos2* gene appears to be FoxY, whose knockdown results in a twofold reduction in Nanos2 levels at the onset of gastrulation (Song & Wessel, 2012). It is not clear, however, whether Nanos2 is a direct FoxY target. FoxY is directly activated by the Delta/Notch (D/N) signaling pathway, which is an upstream inducer of mesoderm in the sea urchin (Materna & Davidson, 2012; Materna et al., 2013). Neither FoxY nor D/N perturbations affect Nanos2 expression in the sMics before gastrulation; therefore, the earliest inputs into this gene remain a critical open question. An important caveat is that there exists significant maternally supplied FoxY protein in the early embryo, which may be refractory to knockdown. FoxY may serve a supporting role for the sMics though its possible role as a direct regulator of Nanos has not been ruled out. After its initial sMic expression, FoxY expression shifts into adjacent mesodermal precursors and its effect on nanos expression in the sMics then may be indirect. Knockdown of FoxY or D/N perturbation both completely prevent coelomic pouch formation, which contain the final niche for the sMics. D/N signaling through FoxY may therefore be important for establishing the somatic gonad required for GSC maintenance.

Given the early activation of Nanos2 in the sMics, it seems likely that a maternally supplied factor should be responsible. A strong candidate is the

Wnt/β-catenin pathway (discussed above). Nuclear beta-catenin, the transcriptional effector of the Wnt pathway, is indeed highly enriched in vegetal blastomeres of the sea urchin embryo (Logan et al., 1999). Additionally, the micromere lineage expresses three Wnt ligands prior to gastrulation that may reinforce the maternally activated Wnt pathway (Cui et al., 2014). Lithium chloride treatment, which upregulates nuclear beta-catenin throughout the embryo, results in an increase in overall Vasa protein levels throughout the embryo (Voronina et al., 2008). Future investigation should further test whether the Wnt pathway regulates Nanos expression directly, particularly in light of conserved roles for Wnt and BMP signaling in germ lines of other species.

Consistent with germ line segregation in other preloaded systems, sea urchin Nanos2 is heavily regulated posttranscriptionally. The 3′UTR contains a stability and translational control element, termed the GNARLE (global Nanos-associated RNA lability element). The GNARLE element is sufficient to confer sMic localization of injected mRNA reporters, independently of the Nanos open reading frame (ORF) sequence (Oulhen et al., 2013). Intriguingly, 3′UTR-mediated control of Nanos localization has also been found in zebrafish Nanos1 and mouse Nanos3 (Köprunner, Thisse, Thisse, & Raz, 2001; Suzuki, Saba, Sada, & Saga, 2010). This observation implies that transcriptional regulation alone is insufficient for selective Nanos2 expression in the sMics, and that other posttranscriptional systems are required for separation between PGC and mesodermal-associated gene expression at the sea urchin vegetal plate. Instead, Nanos2 may be downstream of a more general mesodermal transcription factor (such as β-catenin), but degradation processes in the cytoplasm further refine its localization. While these processes have not yet been elucidated, they may involve small RNAs or RNA-binding proteins as observed in zebrafish and *Drosophila.*

The cytological features of the sMics, their gene expression profile, and the fact that they do not contribute to the embryo or larva, but instead are set aside for the adult rudiment, make them candidates for PGCs in the sea urchin. However, their precise function has been a point of controversy for some time, perhaps because the definitive resolution of their fate requires a challenging lineage trace. While transgenesis is possible in the sea urchin, the long generation times of most species (up to 2 years for the popular *S. purpuratus*) make the maintenance of stable lines impractical (Arnone et al., 1997). Classic embryological approaches, however, have been used to test whether the micromere lineage is required for fertility of the adult sea urchin. Surprisingly, development can proceed following removal of the micromeres before the fifth cleavage. Such embryos gastrulate, and the resultant larvae even form skeletons, by way of compensatory mesodermal cells (Ettensohn, Kitazawa, Cheers, Leonard, & Sharma, 2007). When raised to adulthood, these manipulated embryos do indeed yield fertile animals (Ransick, Cameron, & Davidson, 1996). It was concluded from this experiment that the micromere lineage contains no obligate germ cell determinants. However, a subsequent study found that when, instead, the sMics are removed at the subsequent cleavage, the resultant adults are infertile (Yajima & Wessel, 2011b). These seemingly disparate results are actually quite compatible when one considers the organizing capability of the micromeres. When transplanted ectopically to the animal pole of a recipient embryo, a second axis is induced with a complete gut (Ransick & Davidson, 1993). Conversely, experiments in which sMics were transplanted to the animal cap indicated only very weak organizing activity (Kurihara & Amemiya, 2005). Taken together, these results suggest that the micromeres repress the germ line program in what will normally become somatic blastomeres, but upon their removal, another cell lineage can compensate for their loss. Since the sMics lack organizing capability, their removal does not induce compensation. In support of this premise, ectopic Vasa protein strongly accumulates throughout the entire embryo when micromeres are removed but not when sMics are removed (Voronina et al., 2008; Yajima & Wessel, 2011b). Furthermore, following micromere removal, Nanos2 mRNA accumulates in a mesodermal territory that normally neighbors the sMics (Fujii et al., 2009).

12. PUTTING YOUR GERM CELLS IN THE FREEZER: POSSIBLE ADVANTAGES FOR PRELOADED SPECIFICATION

While there are apparent similarities in the germ plasms of different systems, mapping these characters to the phylogenetic tree strongly implies that germ line preloading evolved independently multiple times. Furthermore, examination of the genes involved reveals a lack of homology (e.g., the insect-specific Oskar, the vertebrate-specific Bucky ball), pointing instead to convergent evolution. Multiple realizations of preloading imply a strong selective pressure for the strategy. A consequence of early segregation, particularly before the MET, is that the PGCs must sit and wait for the rest of the embryo to catch up. That is, the PGCs require a somatic support structure—the gonad—to migrate toward and colonize. Until colonization occurs, the PGCs remain locked down and insulated from differentiation cues. Why not specify the germ cells after the MET, in the same location where the mature GSCs will ultimately reside?

Some hypotheses have been proposed. The work of Weismann and others leading to the Modern Synthesis suggested that early segregation of PGCs followed by cell cycle quiescence could confer a protective advantage against the accumulation of DNA replication errors (Buss, 1987). In rapidly developing organisms such as *Drosophila*, these hypotheses make intuitive sense; complete genome replication and cleavage in the syncytial blastoderm can occur in as little as 4 min (Farrell & O'Farrell, 2014). *D. melanogaster* can develop normally without a spindle assembly checkpoint machinery, implying that its evolution has perhaps favored rapid development over absolute fidelity (Buffin, Emre, & Karess, 2007). Over many cell cycle generations, one can envision the accumulation of deleterious mutations and, by setting aside the germ line early in embryogenesis, somatic mutations are unable to gain access to the germ line and thus will not be transmitted to subsequent generations. This premise seems like a reasonable selection mechanism but must be balanced by the fact that in many animals, mammals included, the stem cells that give rise to eggs and sperm will divide enormous numbers of times to give millions of oocytes and sperm. Mitotic quiescence in early PGCs only delays the onset of the potentially damaging rapid cell divisions, unless the early cell divisions lack a quality control mechanism that instead is present later.

Johnson and colleagues have proposed a complementary hypothesis that early germ line segregation enhances the evolvability of the species. They suggest that by establishing the germ line in isolation instead of as part of the mesoderm, constraints upon the embryo are relaxed. Consequently, somatic tissues can diversify with less risk of compromising the species' reproductive capability (Johnson, Richardson, Bachvarova, & Crother, 2011). These authors suggest that clades with preloaded germ cell specification should speciate (and/or acquire greater morphological diversity) more rapidly than their sister clades. However, we note here that such a speciation would not occur without the imposition of appropriate selective pressures. Nature has performed this experiment for us several times, and indeed, there may be a correlation between germ line preloading and enhanced speciation. Several examples include the teleost fishes, such as zebrafish, versus inductive, more basally branching ray-finned fishes, such as the sturgeon; and anuran amphibians such as *Xenopus*, which uses preloading, versus the inductive urodeles like the axolotl. In these examples, the preloaded clade does indeed appear to have greater species diversity than the inductive sister (Johnson et al., 2011). A phylogenetic analysis also suggests that genes in preloading animals tend to evolve more rapidly than their inductive sisters. Evans et al. found that gene trees between anuran and urodele amphibians and mammals often do not recapitulate the proper species relationships, and when they are incongruent, urodele (inductive) sequences tend to cluster with vertebrate sequences, while the anuran (preloaded) sequences are more divergent (Evans, Wade, Chapman, Johnson, & Loose, 2014). This creative analysis may support the premise that germ line preloading correlates with speciation; however, incongruent trees were heavily affected by the choice of outgroup to which they were rooted. Future investigation might also seek to directly measure rates of gene evolution within the preloading-based echinoids, versus other echinoderms, for which transcriptomes in many species are becoming available. Because the sea urchin GRN is well characterized, one could test whether particular subnetworks, such as that specifying mesoderm, display different evolutionary rates than others.

We cannot conclude from these correlations that germ line preloading is a universally optimal strategy. Instead, it is more productive to ask what unique attributes of these clades' environments influenced the convergence. One possibility is that preloading facilitates rapid development; consider that all of the examples provided here, including *Drosophila*, *C. elegans*, *Xenopus*, the sea urchin, and zebrafish each develop through a larval form. By getting germ line specification "out of the way" early, the embryo can then direct its investments toward reaching a motile, feeding form as quickly as possible. Precocial development may be particularly beneficial when the animal is subjected to high predation, or rapid changes in the environment. Put another way, while acquisition of a preloaded germ line may correlate with speciation, it may not be the causal agent. Instead, we suggest the alternative possibility that a preloaded germ line, enhanced speciation, and larval development are parallel consequences of some other environmental pressure toward rapidity.

13. ARE METs DIFFERENT BETWEEN THE GERM LINE AND SOMA? TRANSCRIPTIONAL REPRESSION IN PGCs

Careful study of GRNs in the sea urchin has shown us that transcriptional asymmetries initiate very early in embryogenesis. The embryo is poised for these asymmetries largely thanks to the maternal β-catenin pathway, which is activated in early vegetal blastomeres (as reviewed above). The GRNs put in place by these early asymmetries prime different parts of the embryonic genome for deployment in different embryonic territories. In comparison to the rest of the embryo, however, the sMics are poorly understood in terms of their transcriptional regulation. With a strong collective effort of the community to identify all spatially restricted developmental transcription factors, it is surprising that FoxY is the only one known to be expressed in the sMics before gastrulation (Juliano et al., 2006; Materna et al., 2013; Song & Wessel, 2012; Swartz et al., 2014). Furthermore, its knockdown leads to no obvious defects in the specification of the sMics. It is therefore plausible that the dearth of known transcription factors in the sMics reflects something of biological significance, such as transcriptional repression.

This premise is consistent with the PGCs of other species that undergo periods of broad transcriptional repression, including *Drosophila*, *C. elegans*, *X. laevis*, and the ascidian *Ciona intestinalis* (Nakamura & Seydoux, 2008; Shirae-Kurabayashi, Matsuda, & Nakamura, 2011). These studies have all benefited from the generation of antibodies that specifically recognize phosphorylations of the heptapeptide repeat of the C-terminal domain (Seydoux & Dunn, 1997). Antibodies to phosphorylated Serine 2 (pSer2), which is associated with transcriptional elongation, have revealed that early PGCs transcribe at very low levels compared to their somatic neighbors. Surprisingly, each of these species achieves transcriptional repression by distinct mechanisms. *Drosophila* and *C. elegans* have both found solutions by interfering with the P-TEFb complex. This complex contains CyclinT and Cdk9, and acts by phosphorylating the C-terminal heptapeptide repeats of RNA polymerase II (RNAPII) at Serine 2. In *Drosophila*, *Pole granule component* (pgc) is localized within the germ plasm, and prevents P-TEFb recruitment by directly binding CyclinT (Hanyu-Nakamura, Sonobe-Nojima, Tanigawa, Lasko, & Nakamura, 2008). In *C. elegans*, a protein called PIE-1 interferes with Cdk9 activity and is thought to function by mimicking the RNAPII C-terminal tail. Both Pgc and PIE-1 are species-specific genes, an example of evolutionary convergence that is intimately associated with having a preloaded germ line. In *Xenopus*, Nanos is required for pSer2 depletion, though this is almost certainly an indirect effect, given that Nanos is a cytoplasmic regulator of mRNA (Lai, Singh, & King, 2012).

Each of these species uses a preloaded mode of PGC segregation, and has also acquired a mechanism for general transcriptional repression. This correlation may imply that transcriptional repression is a prerequisite for having a preloaded germ line. A possible reason for this requirement is insulation from differentiation. To be kept developmentally naïve, PGCs must not respond to the differentiation cues intended for somatic cells. When transcriptional repression is perturbed in *Drosophila*, *Ciona*, and *Xenopus*, PGCs misexpress somatic genes and can adopt a somatic fate (Hayashi, Hayashi, & Kobayashi, 2004; Lai et al., 2012; Shirae-Kurabayashi et al., 2011). In addition to these preloaded PGC animals, the mouse (which uses induction) also shows signs of broad transcriptional repression, although with some differences. Its PGCs are transcriptionally active at their initial segregation, but lose pSer2 intensity as they migrate toward the somatic gonad. The mechanism of this is unknown, but must involve different regulation than observed in the preloading-based systems. Mouse PGCs are arrested in G2 of the cell cycle during migration, and their transcriptional repression thus may be cell cycle dependent (Seki et al., 2007).

The sea urchin sMics also display signs of broad transcriptional repression. Immediately after their creation, the sMic nuclei are enriched for the heterochromatin mark H3K9 trimethylation, and this enrichment persists into blastula stages. Subsequently, the sMic nuclei become depleted for elongating RNAPII based on immunofluorescence with phosphospecific antibodies (Swartz et al., 2014; Fig. 6). The mechanism of this depletion is unknown, but one possibility is that a methyltransferase is either localized to, or more active within, the sMics than in the somatic blastomeres. The subsequent phase of RNA polymerase repression could be dependent upon the earlier phase of H3K9 trimethylation. There also could be cell cycle dependency; BrdU incorporation studies have found that sMics display a prolonged S-phase from their birth until migration (Tanaka & Dan, 1990). They begin to divide at the early gastrula stage during translocation, and this mitotic activity could be a cause for their reduced RNAPII pSer2.

The sMics are also enriched for a maternally supplied form of linker histone H1 based on immunofluorescence data (Fig. 6). While this work was carried out before the sequencing of the genome, and the precise identity of this H1 ortholog has not yet been identified, the sMics retain this variant form from their creation through migration and into the coelomic pouches (Pehrson & Cohen, 1986). The retention of this linker histone may simply reflect mitotic quiescence of the sMics—slow cell cycling would prevent dilution of cleavage stage histones, while somatic cells dilute out their maternally supplied histones among the many progeny cells. However, this H1 variant could also actively confer transcriptional repression. Such a role would be consistent with the function of linker histones, which compact chromatin into higher order structure. Furthermore, a histone H1 variant called dBigH1 was recently reported to regulate EGA in *Drosophila*. dBigH1 is exchanged for a shorter variant upon cellularization of the blastoderm and activation of embryonic transcription, but is retained in the PGCs. dBigH1 loss-of-function mutants precociously activate transcription in both the somatic and germ line cells (Pérez-Montero, Carbonell, Morán, Vaquero, & Azorín, 2013). Thus, histone variants may play a conserved role in regulating the timing of EGA in PGCs and somatic cells.

In addition to these clues in the nuclei, some cell surface changes suggest that the sMics may not respond to differentiation signals. Scanning electron microscopy suggested that the sMic plasma membranes lack the characteristic microvilli of blastomeres and are "smoother" than somatic neighbors (Dale, Yazaki, & Tosti, 1997). This observation could reflect a broad membrane rearrangement in the micromere lineage. Perhaps in support of this premise, the micromeres, and subsequently the sMics, have reduced ABC/multidrug transporter activity based on efflux of fluorescent reporter molecules (Campanale & Hamdoun, 2012). A reduction of multidrug activity in the sMics is surprising, since protection of the germ line from toxicants would seemingly confer an advantage. However, reduction of efflux activity may sensitize the sMics to migration cues, as inhibition of MDRs results in improper coelomic pouch homing. Loss of efflux activity may be a consequence of a greater plasma membrane modification hinted at by the earlier electron microscopy observations; perhaps, the sMics broadly internalize transmembrane proteins. Furthermore, the sMics are enriched for the mRNA of a Sprouty family member. Sprouty family proteins are negative regulators of receptor tyrosine kinase signaling and of proliferation (Kim & Bar-Sagi, 2004). Taken together, these observations depict the sMics as a cell type that "covers its ears" and ignores differentiation signals. What turns them "back on" after gastrulation may be the loss of Nanos, transiently expressed in this cell, but the mechanism of this premise is unclear also.

14. DIFFERENTIAL STABILITY OF mRNA IN THE GERM LINE AND SOMA

The sMics have a unique propensity for retaining inherited RNA. Transcriptomic analyses, and specific examination of Vasa and Seawi RNAs, suggest that the sMics generally inherit maternally supplied RNA, rather than transcribing them *de novo*. These RNAs are typically ubiquitously distributed throughout the early embryo, and retained in the sMics, but turned over in somatic cells later. This pattern of acquisition is consistent with the observation that the sMics are transcriptionally repressed (Swartz et al., 2014; Voronina et al., 2008; Yajima et al., 2014). Furthermore, exogenous mRNA reporters injected into zygotes, while eventually degraded in somatic cells, are retained several days longer in the sMics (Gustafson & Wessel, 2010; Oulhen & Wessel, 2013). This RNA retention phenomenon occurs independently of sequence information, because several reporters, containing mCherry, GFP, different UTRs, and other cross-species-derived sequences, all are similarly retained in the sMics. Instead, RNA retention is probably reflective of generally increased stability for mRNA in the sMics relative to somatic cells. The sea urchin embryo achieves this differential stability at least in part by repressing accumulation of the major deadenylase CNOT6 in the sMics, while expressing it in all other blastomeres (Fig. 7). Without its full nuclease activity, the CNOT complex would be less capable of shortening poly-A tails, reducing their likelihood as substrates for various degradation processes. CNOT6 is directly depleted in the sMics by Nanos2 repression of its mRNA. The CNOT6 3′UTR contains two consensus PRE motifs, which are each required for CNOT6 mRNA to be degraded in the sMics. Importantly, Nanos knockdown prevents CNOT6 degradation while forcing expression of CNOT6 in the sMics disrupts their ability to retain the germ line determinants Vasa and Piwi. Conversely, broad knockdown of CNOT6 expands the domain of retention of both Piwi RNA and exogenous RNA. These results highlight an important difference in RNA stability between germ line and soma in the sea urchin embryo that is important for inheriting germ line molecules.

Differential RNA degradation and protection has been noted in several different species, particularly in those that use preloaded strategies. *Hsp83* and *Nanos* RNAs provide examples in the *Drosophila* embryo. Both of these RNAs are broadly distributed throughout the oocyte, egg, and syncytial embryo; in the case of *Nanos*, only about 5% of its total transcripts are enriched at the early posterior pole (Bergsten & Gavis, 1999). Thus, some other process must fine-tune their localization to the PGCs. Analysis of the 3′UTRs of *Nanos* and *Hsp83* has turned up *cis* elements that confer lability in somatic cells (Bashirullah et al., 1999; Dahanukar & Wharton, 1996). The RNA-binding protein Smaug has been identified as the maternally supplied factor required to degrade *Hsp83* RNA in *Drosophila*. Smaug targets *Hsp83* for degradation by the CNOT complex and represses translation of unlocalized *Nanos* (Jeske, Moritz, Anders, & Wahle, 2011; Semotok et al., 2005). The primary recognition element for Smaug association was

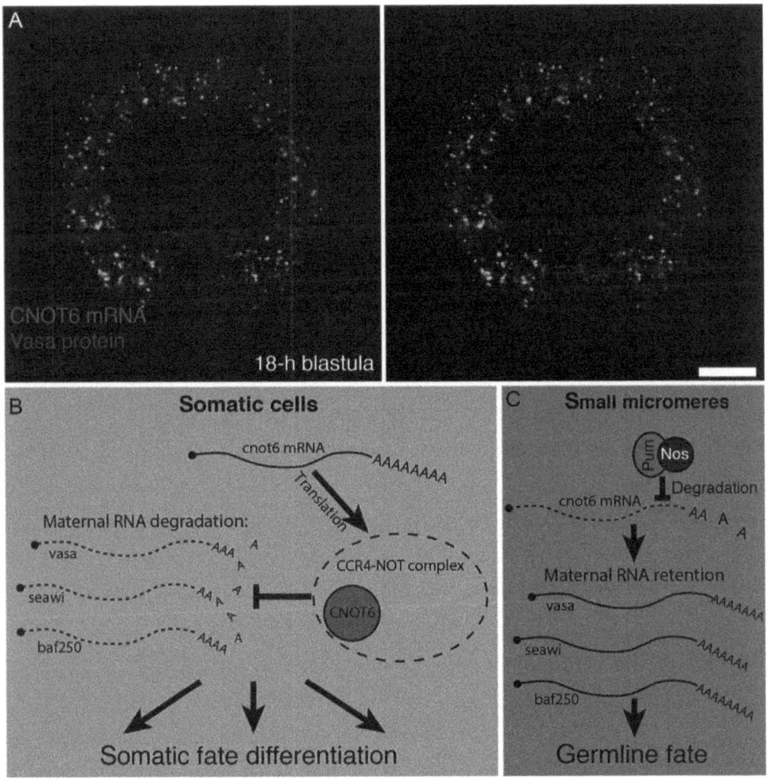

Figure 7 Time Capsule model for (A) CNOT6 mRNA (green) is present in all somatic cells but depleted in the sMics (red, Vasa protein) in the 18 h.p.f. blastula. (B and C) Model for differential RNA stability between the somatic cells and sMics. (B) CNOT6 deadenylase is broadly present, and destabilizes maternal transcripts such as Vasa, Seawi, and Baf250 by shortening their poly-A tails. (B) These RNAs are protected in the sMics, because Nanos (Nos) and its partner Pumilio (Pum) repress the accumulation of CNOT6. Scale bar = 20 μm. *Reproduced from Wessel et al. (2014).* (See the color plate.)

mapped to a region within the *Hsp83* ORF termed the *Hsp83* mRNA instability element (Semotok et al., 2008). The piRNA pathway has been also been implicated in the turnover of maternal *Nanos* mRNA in *Drosophila* somatic cells by interacting with the CNOT complex (Rouget et al., 2010). In the zebrafish, Nanos1 RNA is also differentially regulated between PGCs and soma by a microRNA: miR-430 (Mishima et al., 2006).

Furthermore, specific 3′UTR elements confer protection to certain RNAs in the PGCs. Such elements have been identified in *Drosophila Hsp83*; when swapped with the *Hsp70* 3′UTR, *Hsp83* is degraded in both somatic cells and PGCs simultaneously. Thus, the RNA degradation pathway is active in the PGCs, but certain RNAs are masked from its action. The *trans*-acting factors that mediate protection in the fly have not yet been definitively identified, but protective factors have been discovered in vertebrates. One such factor is Dnd1, an RNA-binding protein that is essential for germ line development in zebrafish and mice (Weidinger et al., 2003; Youngren et al., 2005). In zebrafish, Dnd1 binds the 3′UTR of Nanos in direct proximity to miR-430 sites and occludes access by the miRNA machinery (Kedde et al., 2007). Another conserved germ line factor, Dazl, also opposes miR-430 by enhancing polyadenylation of Nanos mRNA (Takeda, Mishima, Fujiwara, Sakamoto, & Inoue, 2009). In each of these cases, specific mRNAs are protected by the addition of *cis*- and *trans*-acting factors. Thus, the sea urchin is the only known example which creates a general bottleneck in RNA turnover by the depletion of CNOT6. As both RNA-binding proteins and miRNAs use the CNOT complex to effect target degradation, in the future it will be important to investigate whether deadenylase depletion is a conserved phenomenon in germ cells. Expression analysis in the sea star, which uses an inductive PGC system, suggests that mesodermal progenitors have reduced CNOT6 (Fresques et al., 2014).

15. METs ARE DELAYED IN PGCs

The emerging picture is that animals that specify their germ line by preloading display a delay in both EGA and maternal transcript degradation in their PGCs. Some genomic approaches using fluorescence-activated cell sorting (FACS)-isolated PGCs have directly tested the timing and extent of delays for EGA and maternal transcriptome degradation. *Drosophila* PGCs do not upregulate embryonically transcribed transcripts until 3–5 h postfertilization, compared to the major somatic EGA, which occurs at 2 h (Siddiqui et al., 2012). A 1–3 h delay in *Drosophila* time is substantial, as this period spans when the PGCs were created, into gastrulation, until just before the PGCs begin migrating to the somatic gonad. Furthermore, the *Drosophila* PGCs are delayed for degradation of maternal RNA by several hours. At least some mechanistic aspects of EGA and transcript degradation are similar between the germ line and soma, such as the shared requirement for the RNA-binding protein, Smaug (Siddiqui et al., 2012). Similarly, FACS of sea urchin sMics indicated that transcription is inhibited in the sMics from their creation, and they retain inherited mRNA substantially longer than somatic cells (Swartz et al., 2014). Future studies should examine how that repression is removed. One possibility is that, once the sMics have finished their migration into the coelomic pouches, transcriptional activity will resume, perhaps based on degradation of Nanos protein.

16. A CONTINUUM OF MATERNAL AND EMBRYONIC CONTRIBUTIONS TO DEVELOPMENT

We have highlighted a range of differences in maternal versus embryonic control of development among diverse animal species. In rapidly developing organisms such as the fruit fly *D. melanogaster*, the frog *X. laevis*, and the zebrafish *D. rerio*, the early embryo subsists entirely on maternally provided factors for a protracted period of many cell cycles. It is only later in development that the embryonic genome activates. In contrast, organisms such as the sea urchin and the mouse initiate transcription immediately after fertilization. A more conserved feature of animal development instead appears to be a broad degradation of the maternal transcriptome, which is critical for cell fate specification. In animals with a classically defined MBT, like the fruit fly, frog, and zebrafish, maternal transcriptome degradation and EGA are coupled, and occur with remodeling of the cell cycle. However, in the sea urchin, maternal transcriptome degradation does not appear to occur until many hours after the embryonic genome has activated, suggesting that these two processes can be decoupled.

Given the pervasiveness of global RNA turnover events throughout the animal tree, we have proposed that this MET may have been essential for the transition from a unicellular ancestor to a multicellular organism with specialized cell fates (Swartz et al., 2014). We have highlighted the importance of miRNA and RNA-binding protein-mediated turnover of maternal mRNA in the sea urchin and other organisms; blocking these processes broadly perturbs cell fate specification in development. The miRNA pathway also has a well-appreciated requirement for the differentiation of many adult stem cells in myogenesis, skin homeostasis, and hematopoiesis (Gangaraju & Lin, 2009). Global RNA turnover "clears the slate" for somatic fate specification. In contrast to the somatic cells in the sea urchin embryo which clear the slate by use of at least CNOT6, the PGCs (i.e., the sMics) protect and retain inherited mRNA and remain transcriptionally inert—in other words, they are insulated from the MET which occurs in the soma. This mechanism of germ line—soma separation, which we call the Time Capsule mode in the sea urchin, bears striking similarity to differential RNA stability between germ line and soma observed in *Drosophila* and zebrafish, suggesting that it may be a deeply conserved strategy (Bashirullah et al., 1999; Mishima et al., 2006; Fig. 7). With the transition to multicellular animals, the true novelty was the soma—the specialized parts devoid of reproductive potential—since the unicellular ancestor played the role of both germ line and soma. The diverse METs of many species documented here may be functional remnants of how that primordial separation may have occurred: a broad degradation of RNA to allow division of labor and insulation of a select subset of cells to remain reproductively potent.

REFERENCES

Aramaki, S., Hayashi, K., Kurimoto, K., Ohta, H., Yabuta, Y., Iwanari, H., et al. (2013). A mesodermal factor, T, specifies mouse germ cell fate by directly activating germline determinants. *Developmental Cell*, *27*(5), 516–529.

Arnone, M. I., Bogarad, L. D., Collazo, A., Kirchhamer, C. V., Cameron, R. A., Rast, J. P., et al. (1997). Green Fluorescent Protein in the sea urchin: New experimental approaches to transcriptional regulatory analysis in embryos and larvae. *Development (Cambridge, England)*, *124*(22), 4649–4659.

Bashirullah, A., Halsell, S. R., Cooperstock, R. L., Kloc, M., Karaiskakis, A., Fisher, W. W., et al. (1999). Joint action of two RNA degradation pathways controls the timing of maternal transcript elimination at the midblastula transition in Drosophila melanogaster. *The EMBO Journal*, *18*(9), 2610–2620.

Bergsten, S. E., & Gavis, E. R. (1999). Role for mRNA localization in translational activation but not spatial restriction of nanos RNA. *Development (Cambridge, England)*, *126*(4), 659–669. The Company of Biologists Limited.

Bhandari, D., Raisch, T., Weichenrieder, O., Jonas, S., & Izaurralde, E. (2014). Structural basis for the Nanos-mediated recruitment of the CCR4-NOT complex and translational repression. *Genes & Development*, *28*(8), 888–901.

Bontems, F., Stein, A., Marlow, F., Lyautey, J., Gupta, T., Mullins, M. C., et al. (2009). Bucky ball organizes germ plasm assembly in zebrafish. *Current Biology*, *19*(5), 414–422.

Boveri, T. (1893). An organism produced sexually without characteristics of the mother (T.H. Morgan, Trans.) *American Naturalist*, *27*, 222–232.

Buffin, E., Emre, D., & Karess, R. E. (2007). Flies without a spindle checkpoint. *Nature Cell Biology*, *9*(5), 565–572.

Bushati, N., Stark, A., Brennecke, J., & Cohen, S. M. (2008). Temporal reciprocity of miRNAs and their targets during the maternal-to-zygotic transition in Drosophila. *Current Biology*, *18*(7), 501–506.

Buss, L. W. (1987). *The evolution of individuality*. Princeton: Princeton University Press.

Campanale, J. P., & Hamdoun, A. (2012). Programmed reduction of ABC transporter activity in sea urchin germline progenitors. *Development (Cambridge, England)*, *139*(4), 783–792.

Chatfield, J., O'Reilly, M.-A., Bachvarova, R. F., Ferjentsik, Z., Redwood, C., Walmsley, M., et al. (2014). Stochastic specification of primordial germ cells from mesoderm precursors in axolotl embryos. *Development (Cambridge, England)*, *141*(12), 2429–2440.

Chen, L., Dumelie, J. G., Li, X., Cheng, M. H., Yang, Z., Laver, J. D., et al. (2014). Global regulation of mRNA translation and stability in the early Drosophila embryo by the Smaug RNA-binding protein. *Genome Biology*, *15*(1), R4. BioMed Central Ltd.

Chen, D., Zheng, W., Lin, A., Uyhazi, K., Zhao, H., & Lin, H. (2012). Pumilio 1 suppresses multiple activators of p53 to safeguard spermatogenesis. *Current Biology*, *22*(5), 420–425.

Collart, M. A., & Panasenko, O. O. (2012). The Ccr4–not complex. *Gene*, *492*(1), 42–53.

Conklin, E. (1905). Organ-forming substances in the eggs of ascidians. *Biological Bulletin*, *8*(4), 205–230.

Core, L. J., Waterfall, J. J., & Lis, J. T. (2008). Nascent RNA sequencing reveals widespread pausing and divergent initiation at human promoters. *Science (New York)*, *322*(5909), 1845–1848.

Crampton, H. E., & Wilson, E. B. (1896). Experimental studies on gasteropod development. *Development Genes and Evolution*, *3*(1), 1–19. Springer.

Cui, M., Siriwon, N., Li, E., Davidson, E. H., & Peter, I. S. (2014). Specific functions of the Wnt signaling system in gene regulatory networks throughout the early sea urchin embryo. *Proceedings of the National Academy of Sciences of the United States of America*, *111*(47), E5029–E5038.

Dahanukar, A., & Wharton, R. P. (1996). The Nanos gradient in Drosophila embryos is generated by translational regulation. *Genes & Development*, *10*(20), 2610–2620.

Dale, B., Yazaki, I., & Tosti, E. (1997). Polarized distribution of L-type calcium channels in early sea urchin embryos. *The American Journal of Physiology*, *273*(3 Pt. 1), C822–C825.

Dan, K., Tanaka, S., Yamazaki, K., & Kato, Y. (1980). Cell cycle study up to the time of hatching in the embryos of the sea urchin, Hemicentrotus pulcherrimus. *Development, Growth & Differentiation*, *22*(3), 589–598.

Davidson, E. H. (1986). *Gene activity in early development.* Orlando, FL: Academic Press. Eric H. Davidson—Google Books.

Donoughe, S., Nakamura, T., Ewen-Campen, B., Green, D. A., Henderson, L., & Extavour, C. G. (2014). BMP signaling is required for the generation of primordial germ cells in an insect. *Proceedings of the National Academy of Sciences of the United States of America*, *111*(11), 4133–4138.

Edgar, B. A., & O'Farrell, P. H. (1989). Genetic control of cell division patterns in the Drosophila embryo. *Cell*, *57*(1), 177–187.

Epel, D. (1967). Protein synthesis in sea urchin eggs: A "late" response to fertilization. *Proceedings of the National Academy of Sciences of the United States of America*, *57*(4), 899. National Academy of Sciences.

Ephrussi, A., & Lehmann, R. (1992). Induction of germ cell formation by oskar. *Nature*, *358*(6385), 387–392.

Ettensohn, C. A., Kitazawa, C., Cheers, M. S., Leonard, J. D., & Sharma, T. (2007). Gene regulatory networks and developmental plasticity in the early sea urchin embryo: Alternative deployment of the skeletogenic gene regulatory network. *Development (Cambridge, England)*, *134*(17), 3077–3087.

Evans, T., Wade, C. M., Chapman, F. A., Johnson, A. D., & Loose, M. (2014). Acquisition of germ plasm accelerates vertebrate evolution. *Science (New York)*, *344*(6180), 200–203.

Ewen-Campen, B., Donoughe, S., Clarke, D. N., & Extavour, C. G. (2013). Germ cell specification requires zygotic mechanisms rather than germ plasm in a basally branching insect. *Current Biology*, *23*(10), 835–842.

Ewen-Campen, B., Jones, T. E. M., & Extavour, C. G. (2013). Evidence against a germ plasm in the milkweed bug Oncopeltus fasciatus, a hemimetabolous insect. *Biology Open*, *2*(6), 556–568.

Ewen-Campen, B., Schwager, E. E., & Extavour, C. G. M. (2010). The molecular machinery of germ line specification. *Molecular Reproduction and Development*, 77(1), 3–18.

Ewen-Campen, B., Srouji, J. R., Schwager, E. E., & Extavour, C. G. (2012). Oskar predates the evolution of germ plasm in insects. *Current biology*, *22*(23), 2278–2283.

Extavour, C. G. M. (2007). Evolution of the bilaterian germ line: Lineage origin and modulation of specification mechanisms. *Integrative and Comparative Biology*, *47*(5), 770–785.

Extavour, C. G., & Akam, M. (2003). Mechanisms of germ cell specification across the metazoans: Epigenesis and preformation. *Development (Cambridge, England)*, *130*(24), 5869–5884.

Extavour, C. G., Pang, K., Matus, D. Q., & Martindale, M. Q. (2005). vasa and nanos expression patterns in a sea anemone and the evolution of bilaterian germ cell specification mechanisms. *Evolution & Development*, 7(3), 201–215.

Farrell, J. A., & O'Farrell, P. H. (2014). From egg to gastrula: How the cell cycle is remodeled during the Drosophila mid-blastula transition. *Annual Review of Genetics*, *48*, 269–294.

Forristall, C., Pondel, M., Chen, L., & King, M. L. (1995). Patterns of localization and cytoskeletal association of two vegetally localized RNAs, Vg1 and Xcat-2. *Development (Cambridge, England)*, *121*(1), 201–208.

Fresques, T., Zazueta-Novoa, V., Reich, A., & Wessel, G. M. (2014). Selective accumulation of germ-line associated gene products in early development of the sea star and distinct differences from germ-line development in the sea urchin. *Developmental Dynamics*, *243*(4), 568–587.

Fujii, T., Sakamoto, N., Ochiai, H., Fujita, K., Okamitsu, Y., Sumiyoshi, N., et al. (2009). Role of the nanos homolog during sea urchin development. *Developmental Dynamics*, *238*(10), 2511–2521.

Gallo, C. M., Wang, J. T., Motegi, F., & Seydoux, G. (2010). Cytoplasmic partitioning of P granule components is not required to specify the germline in C. elegans. *Science (New York)*, *330*(6011), 1685–1689.

Gangaraju, V. K., & Lin, H. (2009). MicroRNAs: Key regulators of stem cells. *Nature Reviews Molecular Cell Biology*, *10*(2), 116–125.

Giraldez, A. J. (2005). MicroRNAs regulate brain morphogenesis in zebrafish. *Science (New York)*, *308*(5723), 833–838.

Giraldez, A. J., Mishima, Y., Rihel, J., Grocock, R. J., Van Dongen, S., Inoue, K., et al. (2006). Zebrafish MiR-430 promotes deadenylation and clearance of maternal mRNAs. *Science (New York)*, *312*(5770), 75–79.

Gross, P. R., & Cousineau, G. H. (1963). Effects of actinomycin D on macromolecule synthesis and early development in sea urchin eggs. *Biochemical and Biophysical Research Communications*, *10*(4), 321–326.

Gross, P. R., Malkin, L. I., & Moyer, W. A. (1964). Templates for the first proteins of embryonic development. *Proceedings of the National Academy of Sciences of the United States of America*, *51*, 407–414.

Gruidl, M. E., Smith, P. A., Kuznicki, K. A., McCrone, J. S., Kirchner, J., Roussell, D. L., et al. (1996). Multiple potential germ-line helicases are components of the germ-line-specific P granules of Caenorhabditis elegans. *Proceedings of the National Academy of Sciences of the United States of America*, *93*(24), 13837–13842.

Gustafson, E. A., & Wessel, G. M. (2010). Exogenous RNA is selectively retained in the small micromeres during sea urchin embryogenesis. *Molecular Reproduction and Development*, 77(10), 836.

Gustafson, E. A., Yajima, M., Juliano, C. E., & Wessel, G. M. (2011). Post-translational regulation by gustavus contributes to selective Vasa protein accumulation in multipotent cells during embryogenesis. *Developmental Biology*, *349*(2), 440–450.

Hamatani, T., Carter, M. G., Sharov, A. A., & Ko, M. S. H. (2004). Dynamics of global gene expression changes during mouse preimplantation development. *Developmental Cell*, *6*(1), 117–131.

Hanyu-Nakamura, K., Sonobe-Nojima, H., Tanigawa, A., Lasko, P., & Nakamura, A. (2008). Drosophila Pgc protein inhibits P-TEFb recruitment to chromatin in primordial germ cells. *Nature*, *451*(7179), 730–733.

Harvey, E. B. (1940). A comparison of the development of nucleate and non-nucleate eggs of Arbacia punctulata. *Biological Bulletin*, *79*(1), 166–187.

Hayashi, Y., Hayashi, M., & Kobayashi, S. (2004). Nanos suppresses somatic cell fate in Drosophila germ line. *Proceedings of the National Academy of Sciences of the United States of America, 101*(28), 10338–10342.

Illmensee, K., & Mahowald, A. P. (1974). Transplantation of posterior polar plasm in Drosophila. Induction of germ cells at the anterior pole of the egg. *Proceedings of the National Academy of Sciences of the United States of America, 71*(4), 1016–1020.

Inoue, C., Kiyomoto, M., & Shirai, H. (1992). Germ cell differentiation in starfish: The posterior enterocoel as the origin of germ cells in Asterina pectinifera. *Development, Growth & Differentiation, 34*(4), 413–418. Wiley Online Library.

Irie, N., Weinberger, L., Tang, W. W. C., Kobayashi, T., Viukov, S., Manor, Y. S., et al. (2015). SOX17 is a critical specifier of human primordial germ cell fate. *Cell, 160*(1–2), 253–268.

Jeske, M., Moritz, B., Anders, A., & Wahle, E. (2011). Smaug assembles an ATP-dependent stable complex repressing nanos mRNA translation at multiple levels. *The EMBO Journal, 30*(1), 90–103.

Johnson, A. D., Richardson, E., Bachvarova, R. F., & Crother, B. I. (2011). Evolution of the germ line-soma relationship in vertebrate embryos. *Reproduction (Cambridge, England), 141*(3), 291–300.

Juliano, C. E., Reich, A., Liu, N., Götzfried, J., Zhong, M., Uman, S., et al. (2014). PIWI proteins and PIWI-interacting RNAs function in Hydra somatic stem cells. *Proceedings of the National Academy of Sciences of the United States of America, 111*(1), 337–342.

Juliano, C. E., Swartz, S. Z., & Wessel, G. M. (2010). A conserved germline multipotency program. *Development (Cambridge, England), 137*(24), 4113–4126.

Juliano, C. E., Voronina, E., Stack, C., Aldrich, M., Cameron, A. R., & Wessel, G. M. (2006). Germ line determinants are not localized early in sea urchin development, but do accumulate in the small micromere lineage. *Developmental Biology, 300*(1), 406–415.

Juliano, C. E., Yajima, M., & Wessel, G. M. (2010). Nanos functions to maintain the fate of the small micromere lineage in the sea urchin embryo. *Developmental Biology, 337*(2), 220–232.

Kedde, M., Strasser, M. J., Boldajipour, B., Oude Vrielink, J. A. F., Slanchev, K., le Sage, C., et al. (2007). RNA-binding protein Dnd1 inhibits microRNA access to target mRNA. *Cell, 131*(7), 1273–1286.

Kim, H. J., & Bar-Sagi, D. (2004). Modulation of signalling by Sprouty: A developing story. *Nature Reviews Molecular Cell Biology, 5*(6), 441–450.

Kobayashi, S., Amikura, R., & Mukai, M. (1998). Localization of mitochondrial large ribosomal RNA in germ plasm of Xenopus embryos. *Current Biology, 8*(20), 1117–1120.

Kobayashi, S., Amikura, R., & Okada, M. (1993). Presence of mitochondrial large ribosomal RNA outside mitochondria in germ plasm of Drosophila melanogaster. *Science (New York), 260*(5113), 1521–1524.

Köprunner, M., Thisse, C., Thisse, B., & Raz, E. (2001). A zebrafish nanos-related gene is essential for the development of primordial germ cells. *Genes & Development, 15*(21), 2877–2885.

Kosaka, K., Kawakami, K., Sakamoto, H., & Inoue, K. (2007). Spatiotemporal localization of germ plasm RNAs during zebrafish oogenesis. *Mechanisms of Development, 124*(4), 279–289.

Kurihara, H., & Amemiya, S. (2005). Developmental potential of small micromeres in sea urchin embryos. *Zoological Science, 22*(8), 845–852. Zoological Society of Japan.

Lai, F., Singh, A., & King, M. L. (2012). Xenopus Nanos1 is required to prevent endoderm gene expression and apoptosis in primordial germ cells. *Development (Cambridge, England), 139*(8), 1476–1486.

Laubichler, M. D., & Davidson, E. H. (2008). Boveri's long experiment: Sea urchin merogones and the establishment of the role of nuclear chromosomes in development. *Developmental Biology, 314*, 1–11.

Lawson, K. A., Dunn, N. R., Roelen, B. A., Zeinstra, L. M., Davis, A. M., Wright, C. V., et al. (1999). Bmp4 is required for the generation of primordial germ cells in the mouse embryo. *Genes & Development*, *13*(4), 424–436.

Lee, M. T., Bonneau, A. R., Takacs, C. M., Bazzini, A. A., DiVito, K. R., Fleming, E. S., et al. (2013). Nanog, Pou5f1 and SoxB1 activate zygotic gene expression during the maternal-to-zygotic transition. *Nature*, *503*(7476), 360–364.

Liu, N., Han, H., & Lasko, P. (2009). Vasa promotes Drosophila germline stem cell differentiation by activating mei-P26 translation by directly interacting with a (U)-rich motif in its 3'UTR. *Genes & Development*, *23*, 2742–2752.

Logan, C. Y., Miller, J. R., Ferkowicz, M. J., & McClay, D. R. (1999). Nuclear beta-catenin is required to specify vegetal cell fates in the sea urchin embryo. *Development (Cambridge, England)*, *126*(2), 345–357.

Lund, E., Liu, M., Hartley, R. S., Sheets, M. D., & Dahlberg, J. E. (2009). Deadenylation of maternal mRNAs mediated by miR-427 in Xenopus laevis embryos. *RNA*, *15*(12), 2351–2363.

MacArthur, H., Bubunenko, M., Houston, D. W., & King, M. L. (1999). Xcat RNA is a translationally sequestered germ plasm component in Xenopus. *Mechanisms of Development*, *84*(1–2), 75–88.

Magnúsdóttir, E., Dietmann, S., Murakami, K., Günesdogan, U., Tang, F., Bao, S., et al. (2013). A tripartite transcription factor network regulates primordial germ cell specification in mice. *Nature Cell Biology*, *15*(8), 905–915.

Mani, S. R., & Juliano, C. E. (2013). Untangling the web: The diverse functions of the PIWI/piRNA pathway. *Molecular Reproduction and Development*, *80*(8), 632–664.

Masuda, M., & Sato, H. (1984). Asynchronization of cell division is concurrently related with ciliogenesis in sea urchin blastulae. *Development, Growth & Differentiation*, *26*(3), 281–294. Wiley Online Library.

Materna, S. C., & Davidson, E. H. (2012). A comprehensive analysis of Delta signaling in pre-gastrular sea urchin embryos. *Developmental Biology*, *364*(1), 77–87.

Materna, S. C., Swartz, S. Z., & Smith, J. (2013). Notch and Nodal control forkhead factor expression in the specification of multipotent progenitors in sea urchin. *Development (Cambridge, England)*, *140*(8), 1796–1806.

Mishima, Y., Giraldez, A. J., Takeda, Y., Fujiwara, T., Sakamoto, H., Schier, A. F., et al. (2006). Differential regulation of germline mRNAs in soma and germ cells by zebrafish miR-430. *Current Biology*, *16*(21), 2135–2142.

Müller, W. A., Teo, R., & Frank, U. (2004). Totipotent migratory stem cells in a hydroid. *Developmental Biology*, *275*(1), 215–224.

Nakamura, A., & Seydoux, G. (2008). Less is more: Specification of the germline by transcriptional repression. *Development (Cambridge, England)*, *135*(23), 3817–3827.

Nelson, M. R., Leidal, A. M., & Smibert, C. A. (2004). Drosophila Cup is an eIF4E-binding protein that functions in Smaug-mediated translational repression. *The EMBO Journal*, *23*(1), 150–159.

Ogawa, M., Amikura, R., Akasaka, K., Kinoshita, T., Kobayashi, S., & Shimada, H. (1999). Asymmetrical distribution of mitochondrial rRNA into small micromeres of sea urchin embryos. *Zoological Science*, *16*(3), 445–451. BioOne.

Ohinata, Y., Ohta, H., Shigeta, M., Yamanaka, K., Wakayama, T., & Saitou, M. (2009). A signaling principle for the specification of the germ cell lineage in mice. *Cell*, *137*(3), 571–584.

Oliveri, P., Tu, Q., & Davidson, E. H. (2008). Global regulatory logic for specification of an embryonic cell lineage. *Proceedings of the National Academy of Sciences of the United States of America*, *105*(16), 5955–5962.

Oulhen, N., & Wessel, G. M. (2013). Retention of exogenous mRNAs selectively in the germ cells of the sea urchin requires only a 5"-cap and a 3"-UTR. *Molecular Reproduction and Development*, *80*(7), 561–569.

Oulhen, N., Yoshida, T., Yajima, M., Song, J. L., Sakuma, T., Sakamoto, N., et al. (2013). The 3'UTR of nanos2 directs enrichment in the germ cell lineage of the sea urchin. *Developmental Biology*, *377*(1), 275–283.

Pehrson, J. R., & Cohen, L. H. (1986). The fate of the small micromeres in sea urchin development. *Developmental Biology*, *113*(2), 522–526.

Peng, C. J., & Wikramanayake, A. H. (2013). Differential regulation of disheveled in a novel vegetal cortical domain in sea urchin eggs and embryos: Implications for the localized activation of canonical Wnt signaling. *PLoS One*, *8*(11), e80693.

Pérez-Montero, S., Carbonell, A., Morán, T., Vaquero, A., & Azorín, F. (2013). The embryonic linker histone H1 variant of drosophila, dBigH1, regulates zygotic genome activation. *Developmental Cell*, *26*(6), 578–590. Elsevier.

Peter, I., & Davidson, E. H. (2015). *Genomic control process: Development and evolution*. Waltham, Massachusetts: Academic Press.

Rabinowitz, J. S., Chan, X. Y., Kingsley, E. P., Duan, Y., & Lambert, J. D. (2008). Nanos is required in somatic blast cell lineages in the posterior of a mollusk embryo. *Current Biology*, *18*(5), 331–336.

Ransick, A., Cameron, R. A., & Davidson, E. H. (1996). Postembryonic segregation of the germ line in sea urchins in relation to indirect development. *Proceedings of the National Academy of Sciences of the United States of America*, *93*(13), 6759–6763.

Ransick, A., & Davidson, E. (1993). A complete second gut induced by transplanted micromeres in the sea urchin embryo. *Science (New York)*, *259*(5098), 1134–1138.

Reich, A., Dunn, C., Akasaka, K., & Wessel, G. (2015). Phylogenomic analyses of Echinodermata support the sister groups of Asterozoa and Echinozoa. *PLoS One*, *10*(3), e0119627.

Rouget, C., Papin, C., Boureux, A., Meunier, A.-C., Franco, B., Robine, N., et al. (2010). Maternal mRNA deadenylation and decay by the piRNA pathway in the early Drosophila embryo. *Nature*, *467*(7319), 1128–1132.

Seki, Y., Yamaji, M., Yabuta, Y., Sano, M., Shigeta, M., Matsui, Y., et al. (2007). Cellular dynamics associated with the genome-wide epigenetic reprogramming in migrating primordial germ cells in mice. *Development (Cambridge, England)*, *134*(14), 2627–2638.

Semotok, J. L., Cooperstock, R. L., Pinder, B. D., Vari, H. K., Lipshitz, H. D., & Smibert, C. A. (2005). Smaug recruits the CCR4/POP2/NOT deadenylase complex to trigger maternal transcript localization in the early Drosophila embryo. *Current Biology*, *15*(4), 284–294.

Semotok, J. L., Luo, H., Cooperstock, R. L., Karaiskakis, A., Vari, H. K., Smibert, C. A., et al. (2008). Drosophila maternal Hsp83 mRNA destabilization is directed by multiple SMAUG recognition elements in the open reading frame. *Molecular and Cellular Biology*, *28*(22), 6757–6772.

Seydoux, G., & Braun, R. E. (2006). Pathway to totipotency: Lessons from germ cells. *Cell*, *127*(5), 891–904.

Seydoux, G., & Dunn, M. A. (1997). Transcriptionally repressed germ cells lack a subpopulation of phosphorylated RNA polymerase II in early embryos of Caenorhabditis elegans and Drosophila melanogaster. *Development (Cambridge, England)*, *124*(11), 2191–2201.

Shermoen, A. W., & O'Farrell, P. H. (1991). Progression of the cell cycle through mitosis leads to abortion of nascent transcripts. *Cell*, *67*(2), 303–310.

Shirae-Kurabayashi, M., Matsuda, K., & Nakamura, A. (2011). Ci-Pem-1 localizes to the nucleus and represses somatic gene transcription in the germline of Ciona intestinalis embryos. *Development (Cambridge, England)*, *138*(14), 2871–2881. Retrieved from: http://eutils.ncbi.nlm.nih.gov/entrez/eutils/elink.fcgi?dbfrom=pubmed&id=21693510&retmode=ref&cmd=prlinks.

Siddiqui, N. U., Li, X., Luo, H., Karaiskakis, A., Hou, H., Kislinger, T., et al. (2012). Genome-wide analysis of the maternal-to-zygotic transition in Drosophila primordial germ cells. *Genome Biology*, *13*(2), R11.

Smibert, C. A., Wilson, J. E., Kerr, K., & Macdonald, P. M. (1996). smaug protein represses translation of unlocalized nanos mRNA in the Drosophila embryo. *Genes & Development*, *10*(20), 2600–2609.

Smith, J. L., Wilson, J. E., & Macdonald, P. M. (1992). Overexpression of oskar directs ectopic activation of nanos and presumptive pole cell formation in Drosophila embryos. *Cell*, *70*(5), 849–859.

Song, J. L., Stoeckius, M., Maaskola, J., Friedländer, M., Stepicheva, N., Juliano, C., et al. (2012). Select microRNAs are essential for early development in the sea urchin. *Developmental Biology*, *362*(1), 104–113.

Song, J. L., & Wessel, G. M. (2012). The forkhead transcription factor FoxY regulates Nanos. *Molecular Reproduction and Development*, *79*(10), 680–688.

Suzuki, H., Saba, R., Sada, A., & Saga, Y. (2010). The Nanos3-3'UTR is required for germ cell specific NANOS3 expression in mouse embryos. *PLoS One*, *5*(2), e9300.

Suzuki, H., Tsuda, M., Kiso, M., & Saga, Y. (2008). Nanos3 maintains the germ cell lineage in the mouse by suppressing both Bax-dependent and -independent apoptotic pathways. *Developmental Biology*, *318*(1), 133–142.

Swartz, S. Z., Chan, X. Y., & Lambert, J. D. (2008). Localization of Vasa mRNA during early cleavage of the snail Ilyanassa. *Development Genes and Evolution*, *218*(2), 107–113.

Swartz, S. Z., Reich, A. M., Oulhen, N., Raz, T., Milos, P. M., Campanale, J. P., et al. (2014). Deadenylase depletion protects inherited mRNAs in primordial germ cells. *Development (Cambridge, England)*, *141*(16), 3134–3142.

Tadros, W., Goldman, A. L., Babak, T., Menzies, F., Vardy, L., Orr-Weaver, T., et al. (2007). SMAUG is a major regulator of maternal mRNA destabilization in Drosophila and its translation is activated by the PAN GU kinase. *Developmental Cell*, *12*(1), 143–155.

Tadros, W., & Lipshitz, H. D. (2009). The maternal-to-zygotic transition: A play in two acts. *Development (Cambridge, England)*, *136*(18), 3033–3042.

Takeda, Y., Mishima, Y., Fujiwara, T., Sakamoto, H., & Inoue, K. (2009). DAZL relieves miRNA-mediated repression of germline mRNAs by controlling poly(A) tail length in zebrafish. *PLoS One*, *4*(10), e7513.

Tam, P. P., & Zhou, S. X. (1996). The allocation of epiblast cells to ectodermal and germ-line lineages is influenced by the position of the cells in the gastrulating mouse embryo. *Developmental Biology*, *178*(1), 124–132.

Tanaka, S., & Dan, K. (1990). Study of the lineage and cell cycle of small micromeres in embryos of the sea urchin, Hemicentrotus pulcherrimus. (small micromeres/cell cycle/cell lineage/unequal cleavage/sea urchin). *Development, Growth & Differentiation*, *32*(2), 145–156.

Tritschler, F., Huntzinger, E., & Izaurralde, E. (2010). Role of GW182 proteins and PABPC1 in the miRNA pathway: A sense of déjà vu. *Nature Reviews Molecular Cell Biology*, *11*(5), 379–384.

Tsuda, M., Sasaoka, Y., Kiso, M., Abe, K., Haraguchi, S., Kobayashi, S., et al. (2003). Conserved role of nanos proteins in germ cell development. *Science (New York)*, *301*(5637), 1239–1241.

Tu, Q., Cameron, R. A., & Davidson, E. H. (2014). Quantitative developmental transcriptomes of the sea urchin Strongylocentrotus purpuratus. *Developmental Biology*, *385*(2), 160–167.

Updike, D., & Strome, S. (2010). P granule assembly and function in Caenorhabditis elegans germ cells. *Journal of Andrology*, *31*(1), 53–60.

Voronina, E. (2013). The diverse functions of germline P-granules in Caenorhabditis elegans. *Molecular Reproduction and Development*, *80*(8), 624–631.
Voronina, E., Lopez, M., Juliano, C. E., Gustafson, E., Song, J. L., Extavour, C., et al. (2008). Vasa protein expression is restricted to the small micromeres of the sea urchin, but is inducible in other lineages early in development. *Developmental Biology*, *314*(2), 276–286.
Wang, Y., Zayas, R. M., Guo, T., & Newmark, P. A. (2007). Nanos function is essential for development and regeneration of planarian germ cells. *Proceedings of the National Academy of Sciences of the United States of America*, *104*(14), 5901–5906.
Wei, Z., Angerer, R. C., & Angerer, L. M. (2006). A database of mRNA expression patterns for the sea urchin embryo. *Developmental Biology*, *300*(1), 476–484.
Weidinger, G., Stebler, J., Slanchev, K., Dumstrei, K., Wise, C., Lovell-Badge, R., et al. (2003). dead end, a novel vertebrate germ plasm component, is required for zebrafish primordial germ cell migration and survival. *Current Biology*, *13*(16), 1429–1434.
Wessel, G. M., Brayboy, L., Fresques, T., Gustafson, E. A., Oulhen, N., Ramos, I., et al. (2014). The biology of the germ line in echinoderms. *Molecular Reproduction and Development*, *81*(8), 679–711.
Wikramanayake, A. H., Huang, L., & Klein, W. H. (1998). beta-Catenin is essential for patterning the maternally specified animal-vegetal axis in the sea urchin embryo. *Proceedings of the National Academy of Sciences of the United States of America*, *95*(16), 9343–9348.
Wikramanayake, A. H., Peterson, R., Chen, J., Huang, L., Bince, J. M., McClay, D. R., et al. (2004). Nuclear beta-catenin-dependent Wnt8 signaling in vegetal cells of the early sea urchin embryo regulates gastrulation and differentiation of endoderm and mesodermal cell lineages. *Genesis*, *39*(3), 194–205.
Xiol, J., Spinelli, P., Laussmann, M. A., Homolka, D., Yang, Z., Cora, E., et al. (2014). RNA clamping by Vasa assembles a piRNA amplifier complex on transposon transcripts. *Cell*, *157*(7), 1698–1711.
Yajima, M., Gustafson, E. A., Song, J. L., & Wessel, G. M. (2014). Piwi regulates Vasa accumulation during embryogenesis in the sea urchin. *Developmental Dynamics*, *243*(3), 451–458.
Yajima, M., & Wessel, G. M. (2011a). The DEAD-box RNA helicase Vasa functions in embryonic mitotic progression in the sea urchin. *Development (Cambridge, England)*, *138*(11), 2217–2222.
Yajima, M., & Wessel, G. M. (2011b). Small micromeres contribute to the germline in the sea urchin. *Development (Cambridge, England)*, *138*(2), 237–243.
Yajima, M., & Wessel, G. M. (2012). Autonomy in specification of primordial germ cells and their passive translocation in the sea urchin. *Development (Cambridge, England)*, *139*(20), 3786–3794.
Yajima, M., & Wessel, G. M. (2015). Essential elements for translation: The germline factor Vasa functions broadly in somatic cells. *Development (Cambridge, England)*, *142*, 1960–1970.
Youngren, K. K., Coveney, D., Peng, X., Bhattacharya, C., Schmidt, L. S., Nickerson, M. L., et al. (2005). The Ter mutation in the dead end gene causes germ cell loss and testicular germ cell tumours. *Nature*, *435*(7040), 360–364.

CHAPTER SIX

The Maternal-to-Zygotic Transition During Vertebrate Development: A Model for Reprogramming

Valeria Yartseva*[,1], Antonio J. Giraldez*[,†,1]
*Department of Genetics, Yale University School of Medicine, New Haven, Connecticut, USA
†Yale Stem Cell Center, Yale University School of Medicine, New Haven, Connecticut, USA
[1]Corresponding authors: e-mail address: valeria.yartseva@yale.edu; antonio.giraldez@yale.edu

Contents

Current Topics in Developmental Biology, Volume 113
ISSN 0070-2153
http://dx.doi.org/10.1016/bs.ctdb.2015.07.020

Abstract

Cellular transitions occur at all stages of organismal life from conception to adult regeneration. Changing cellular state involves three main features: activating gene expression necessary to install the new cellular state, modifying the chromatin status to stabilize the new gene expression program, and removing existing gene products to clear out the previous cellular program. The maternal-to-zygotic transition (MZT) is one of the most profound changes in the life of an organism. It involves gene expression remodeling at all levels, including the active clearance of the maternal oocyte program to adopt the embryonic totipotency. In this chapter, we provide an overview of molecular mechanisms driving maternal mRNA clearance during the MZT, describe the developmental consequences of losing components of this gene regulation, and illustrate how remodeling of gene expression during the MZT is common to other cellular transitions with parallels to cellular reprogramming.

1. INTRODUCTION

The debate regarding the origin of life complexity dates back to Greek philosophical writings. Through observation of chick embryos, Aristotle postulated that development starts with a uniform egg that gradually develops complexity. However, the seventeenth century was dominated by the preformationist theory of heredity, which favored the idea that the sperm and/or the egg contained a small, but fully formed, individual called the "homunculus" that grew over time (reviewed in Maienschein, 2012). In fact, early microscopists reported observing homunculi, supporting the preformationist theory of heredity (reviewed in Magner, 2002).

These contrasting ideas prompted nineteenth Century embryologists to experimentally test each model. In his seminal experiment, Wilhelm Roux, credited as one of the founders of embryology, destroyed two of the four cells in an early frog blastula and found that only half of the embryo eventually formed, supporting the idea that early embryonic cell fate is predetermined (reviewed in Maienschein, 2012). However, when Hans Driesch separated two-cell sea urchin embryos, two normal, but smaller urchins formed, demonstrating instead that the earliest embryonic cell fate is undetermined (reviewed in Maienschein, 2012). This work led to the current paradigm that life starts with a naïve state that gradually develops complexity through sequentially transitioning to more differentiated cell types, giving rise to tissues and organs in the embryo.

The paradigm that life starts from a naïve state is seemingly paradoxical given that embryos derive from a union of two rather specialized differentiated cells, an egg and a spermatozoon. Thus the first step of development requires the reprogramming of differentiated gametes to a transiently totipotent zygote. This reprogramming event during embryonic development occurs when fertilization triggers the maternal-to-zygotic transition (MZT).

The foundation toward understanding how such reprogramming occurs was facilitated with the discoveries that both nuclear and cytoplasmic activities control embryonic development. To investigate whether nuclear or cytoplasmic information drives development, Theodor Boveri fertilized enucleate urchin oocytes with sperm of a different species and showed that resulting larvae possesses features of both parents (Laubichler & Davidson, 2008). However, it was unknown whether DNA information in gametes is specialized and was lost as cells differentiated during development. To address this question, Sir John Gurdon transplanted the nucleus of a differentiated intestinal epithelial cell into an enucleate oocyte and showed that a mature frog develops (Gurdon, 1962). This experiment demonstrated that DNA in all cells within an individual remains the same as differentiation proceeds and additionally demonstrated that oocyte cytoplasm is endowed with factors capable of reprogramming a somatic nucleus back to its naïve state. Together, the concerted efforts between nuclear information and cytoplasmic material establish the totipotent cellular state required to make new life.

In retrospect, Boveri's observations also explain the two hallmarks of MZT during early embryonic development. Initially, the transcriptionally silent zygote utilizes mRNAs and proteins inherited in the egg cytoplasm to carry out cellular functions. Subsequently, the zygotic genome begins transcription, maternal mRNAs are actively cleared, and developmental control is transferred to the nucleus. Together, cytoplasmic and nuclear activities enable oocyte reprogramming during the maternal-to-zygotic transition (Lee, Bonneau, & Giraldez, 2014).

Oocyte reprogramming during embryogenesis is analogous to somatic cell reprogramming to pluripotency *in vitro*; both involve transitioning from differentiated to pluripotent identity (Giraldez, 2010), summarized in Fig. 1. During the MZT, the previously silenced zygotic genome starts transcription to activate the new genetic program (Lee, Bonneau, et al., 2014),

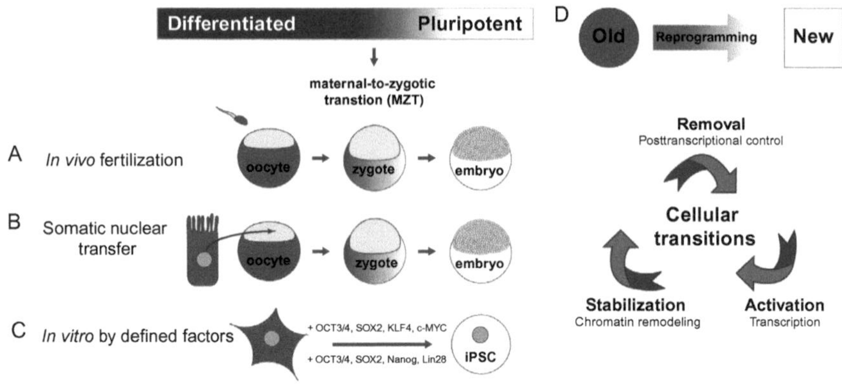

Figure 1 Features of cellular reprogramming. (A–C) Types of cellular reprogramming to pluripotency. (A) *In vivo* fusion of oocyte and spermatazoon initiates the MZT during which the zygote is reprogrammed to a transiently totipotent embryo. (B) Nucleus from a differentiated cell is reprogrammed to a totipotent embryo when transplanted into enucleate fertilized oocyte (Gurdon, 1962). (C) *In vitro*, forced expression of four transcription factors OCT3/4, SOX2, KLF4, and c-MYC (Takahashi & Yamanaka, 2006) or OCT3/4, SOX2, Nanog, and Lin28 (Yu et al., 2007) in differentiated cells induces a fraction of cells to reprogram to a pluripotent-like state called induced pluripotent stem cell (iPSC). (D) Model of cellular reprogramming: reprogramming between two cellular states involves (1) activation of the new program through gene transcription, (2) stabilization of that program through chromatin remodeling, and (3) removal of the previous state by posttranscriptional mechanisms.

the chromatin is remodeled to stabilize the pluripotent state (Zhou & Dean, 2015), and maternal instructions in the form of mRNAs and proteins are actively cleared to remove the previous cellular identify (Giraldez, 2010; Tadros & Lipshitz, 2009; Walser & Lipshitz, 2011). Likewise, reprogramming to pluripotency *in vitro* requires activation of the pluripotency program (Takahashi & Yamanaka, 2006; Yu et al., 2007), chromatin remodeling (Apostolou & Hochedlinger, 2013), and less well-understood posttranscriptional mechanisms to erase the differentiated gene expression program.

Recent reviews on maternal mRNA clearance during MZT within this book and elsewhere (Barckmann & Simonelig, 2013; Colegrove-Otero, Minshall, & Standart, 2005; Langley, Smith, Stemple, & Harvey, 2014; Walser & Lipshitz, 2011) highlight known factors involved in maternal mRNA clearance. Here, we focus on recent advances in the field, common themes in the mechanisms of maternal mRNA clearance across animals, and how this process closely parallels other cellular reprogramming events. We end by describing developmental contexts where maternal clearance is compromised. We propose that maternal mRNA clearance is a requirement to enable the acquisition of the pluripotent state and may even be a common feature of many cellular transitions.

2. MECHANISMS OF MATERNAL mRNA CLEARANCE DURING THE MZT

2.1 Scope of Maternal mRNA Destabilization During the MZT

The maternal-to-zygotic transition occurs in all animals (Tadros & Lipshitz, 2009) and in plants (Baroux, Autran, Gillmor, Grimanelli, & Grossniklaus, 2008; Xin, Zhao, & Sun, 2012), indicating that this transition may be a universal feature of multicellular life. Beginning with a mostly transcriptionally silent embryo, the MZT involves the activation of the zygotic genome and the clearance of maternal mRNAs. Mechanisms regulating the activation of the zygotic genome were recently reviewed (Lee, Bonneau, et al., 2014) and highlight the interplay between zygotic transcription and maternal mRNA clearance. Maternal mRNA clearance during the MZT is a dramatic remodeling of the transcriptional landscape with 30–40% maternal mRNAs eliminated in different species (Baugh, Hill, Slonim, Brown, & Hunter, 2003; De Renzis, Elemento, Tavazoie, & Wieschaus, 2007; Hamatani, Carter, Sharov, & Ko, 2004) and up to 60% of maternal mRNA levels are considerably reduced (Thomsen, Anders, Janga, Huber, & Alonso, 2010). In order to understand how maternal mRNAs are regulated during MZT, it is useful to first review which mRNA features impact its stability.

2.2 Steps in Eukaryotic mRNA Regulation

Following transcription, gene expression in the cytoplasm depends on protein synthesis rate and on the stability of the cognate mRNA. Protein synthesis rate and mRNA stability are influenced by a combination of three main mRNA features: the mRNA sequence, the 7-methylguanylate (m^7G) cap at the 5′ end, and the length of the 3′ poly(A) tail.

Sequences and chemical modifications within the mRNA encode recognition sites for factors that positively and negatively regulate mRNA stability, translation, and localization to permit cell-specific gene expression, recently reviewed in Fu, Dominissini, Rechavi, and He (2014), Gebauer, Preiss, and Hentze (2012), and Medioni, Mowry, and Besse (2012). Mechanistically, binding factors either lead to endonucleolytic cleavage, followed by XRN1 and Exosome complex-mediated hydrolysis from both unprotected mRNA ends, or recruit PARN or CCR4–NOT1 complex to stimulate deadenylation (Beelman & Parker, 1995; Decker & Parker, 1994; Schoenberg & Maquat, 2012), which leads to decapping for some mRNAs (Decker & Parker, 1994) and serves as the rate-limiting step for many mRNA degradation pathways (Wahle & Winkler, 2013).

The poly(A) tail, situated at the 3′ extremity of mRNAs, is bound by poly(A)-binding proteins (PABPs) to stabilize the 3′ end (Bernstein, Peltz, & Ross, 1989) and interacts with translation initiation factor eIF4G bound to the 5′ cap to stimulate translation (Weill, Belloc, Bava, & Méndez, 2012). Proteins bound to 3′UTR elements regulate poly(A) tail length (Charlesworth, Meijer, & De Moor, 2013) and they are dynamically regulated during embryonic development (Richter, 1996, 1999; Richter & Lasko, 2011).

Finally, capped mRNAs are protected from 5′-to-3′ XRN1-mediated exonucleolytic decay (Murthy, Park, & Manley, 1991). Cap hydrolysis via DCP2 leads to mRNA destabilization and can be regulated globally or for a subset of mRNAs (Cowling, 2010; Franks & Lykke-Andersen, 2009; Liu & Kiledjian, 2006). In some cases, 5′-to-3′ mRNA degradation occurs cotranslationally (Hu, Sweet, Chamnongpol, Baker, & Coller, 2009; Pelechano, Wei, & Steinmetz, 2015). Additionally, efficient translation requires m^7G cap interaction with translation initiation factor eIF4E, which is dynamically regulated in development and disease (Richter & Sonenberg, 2005).

Together these mRNA features are mechanistically linked to coordinate posttranscriptional gene regulation for individual mRNAs as well as for coregulated groups of transcripts. This network of mRNA regulation dominates gene control during oogenesis and early embryogenesis, occurring in the absence of transcription. Deadenylation of target mRNAs is a common convergence point of many maternal mRNA clearance mechanisms and will be discussed first.

2.3 Maternal Clearance Mechanisms Involving Poly(A) Tail: Smaug

In *Drosophila*, Smaug is a multifunctional, highly conserved protein that is translationally activated by the Pan gu (PNG) Ser/Thr kinase following egg activation (Tadros et al., 2007). Smaug binds RNA via a sterile alpha motif (SAM) domain, with specificity that is shared with yeast Vts1 and is likely conserved from yeast to humans (Aviv et al., 2003). Smaug recognizes its targets via binding stem loop structures called Smaug recognition elements (SREs) (Dahanukar, Walker, & Wharton, 1999; Smibert, Lie, Shillinglaw, Henzel, & Macdonald, 1999) and recruits the CCR4/POP2/NOT-deadenylase complex to initiate poly(A) tail shortening and consequent mRNA elimination (Semotok et al., 2005). Interestingly, SREs in the *nanos* 3′UTR result in translational repression (Nelson, Leidal, & Smibert, 2004), while for *Hsp83*, which has several SREs in the coding region, the result is deadenylation via CCR4/POP2/NOT, followed by mRNA elimination (Semotok et al., 2005, 2008), suggesting that SRE location on the transcript may influence target mRNA fate. However, transcriptome-wide mapping of Smaug binding in *Drosophila* embryos revealed that the overwhelming majority of SREs are in coding regions, regardless of whether the mRNA is translationally repressed but not destabilized by Smaug (like *nanos*) or destabilized but not repressed (like *Hsp83*) (Chen et al., 2014). Interestingly, *nanos* SREs associate with an ATP-dependent complex (Jeske, Moritz, Anders, & Wahle, 2011) that requires Smaug-mediated microRNA-independent recruitment of Ago1 for repression (Pinder & Smibert, 2013), indicating that additional features distinguish these two modes of posttranscriptional regulation.

2.4 Maternal Clearance Mechanisms Involving Poly(A) Tail: Pumilio

The RNA-binding protein Pumilio has been implicated in deadenylation and clearance of maternal mRNAs. In developmental contexts, Puf-domain family members remain most appreciated for their diverse roles in translational repression (Quenault, Lithgow, & Traven, 2011; Vardy & Orr-Weaver, 2007). In *Drosophila*, the Pumilio-binding element (PBE) sequence is enriched in destabilized maternal mRNAs (De Renzis et al., 2007; Thomsen et al., 2010), implicating Pumilio as a regulator of maternal mRNA clearance in this species. In fact, when Pumilio's RNA-binding domain is expressed in the female germline, it is found to interact with over 900 mRNAs, yet only 130 of these increased and 243 decreased in abundance in a *pumilio* mutants (Gerber, Luschnig, Krasnow, Brown, & Herschlag, 2006), a globally modest effect on mRNA stability. However, a recent analysis of endogenous Pumilio identified over 600 bound mRNAs and found that these are highly enriched for transcripts that are translationally repressed and degraded during the MZT (Laver et al., 2015). Mechanistically, Pumilio binds POP2 of the CCR4–POP2–NOT–deadenylase complex, accelerates reporter mRNA deadenylation, and antagonizes poly(A)-binding protein (PABP) activity (Weidmann, Raynard, Blewett,

Van Etten, & Goldstrohm, 2014). Interestingly, full length Pumilio represses and destabilizes reporter mRNA, while the N-terminal portion predominantly causes repression (Weidmann & Goldstrohm, 2012), suggesting that its role in mRNA destabilization relates to the C-terminal domain's binding partners or activity. Despite the extensive work on Pumilio function in *Drosphila* as well as its stem cell function (discussed below), the extent of Pumilio-mediated maternal mRNA clearance across species remains to be determined.

2.5 Maternal Clearance Mechanisms Involving Poly(A) Tail: EDEN-BP

In *Xenopus*, fertilization triggers sequence-specific and regulated deadenylation of target maternal mRNAs. Embryonic deadenylation element (EDEN) is a U(A/G) dinucleotide repeat and serves as the recognition site for EDEN-binding protein (EDEN-BP) (Paillard et al., 1998) to deadenylate select target mRNAs. EDEN deadenylation capacity is enhanced by an $(AUU)_3$ sequence located in close proximity to the poly(A) tail (Audic, Omilli, & Osborne, 1998) and by the AUUUA sequence (Ueno & Sagata, 2002), suggesting the existence of a combinatorial sequence code to allow for target-specific deadenylation dynamics. Additionally, while EDEN-BP levels remain constant from fertilization to the tadpole stage in *Xenopus* (Gautier-Courteille et al., 2004), it is dephosphorylated following fertilization, which corresponds to target deadenylation (Detivaud, Pascreau, Karaiskou, Osborne, & Kubiak, 2003), indicating that fertilization-dependent posttranslational control is required for EDEN-BP activity. EDEN-BP oligomerization via a 27-amino acid region is also required for target mRNA binding and deadenylation (Cosson et al., 2006), implicating posttranslational control as an additional layer of regulating maternal mRNA deadenylation.

EDEN-BP is homologous to human CUG-BP (Timchenko et al., 1996), which recruits PARN to deadenylate and destabilize specific target mRNAs (Moraes, Wilusz, & Wilusz, 2006). Additionally, the action of embryonic poly(A)-binding protein (ePAB), the predominant poly(A)-binding protein in *Xenopus* oocytes and early embryos, is critical for deadenylation rate in *Xenopus* embryos. ePAB immunodepletion results in increased deadenylation, while overexpression inhibits it (Voeltz, Ongkasuwan, Standart, & Steitz, 2001), suggesting that ePAB stabilizes poly(A) tails and that deadenylation requires antagonizing ePAB binding to its targets or its activity.

The role of EDEN-BP for deadenylation varies across species. EDEN-BP binds 158 maternal mRNAs in *Xenopus* egg extracts that are enriched for genes involved in cell cycle and oocyte maturation (Graindorge et al., 2008), suggesting that EDEN-BP functions to repress the oogenesis program. Interestingly, in aged *Xenopus* oocytes mRNAs that are normally deadenylated after fertilization, undergo precocious deadenylation during oocyte maturation (Kosubek, Klein-Hitpass, Rademacher, Horsthemke, & Ryffel, 2010). This suggests that decreased fertility with age involves dysregulation of molecular mechanisms mediating deadenylation during early embryogenesis. EDEN-dependent translational repression activity is conserved between *Xenopus* and *Drosophila*. However, EDEN reporters are not deadenylated in *Drosophila* (Ezzeddine et al., 2002), indicating species-specific usage of the same sequences for different modes of regulation. In fact, *Drosophila* Bruno, which resembles human CUG-BP (Webster, Liang, Berg, Lasko, & Macdonald, 1997), recruits Cup to translationally repress, but not deadenylate *oskar* mRNA (Nakamura, Sato, & Hanyu-Nakamura, 2004; Wilhelm, Hilton, Amos, & Henzel, 2003), suggesting that the role of EDEN-BP in deadenylation has diverged in vertebrates.

2.6 Maternal Clearance Mechanisms Involving Poly(A) Tail: microRNAs

The interaction of microRNAs with target mRNAs causes translational repression, deadenylation, and mRNA destabilization (Bartel, 2009). Investigating microRNA action in a developmental context revealed detailed features of this mRNA destabilization mechanism. In zebrafish, *miR-430* is transcribed zygotically (Giraldez et al., 2005; Lee et al., 2013) and induces deadenylation and mRNA destabilization of several hundred transcripts (Giraldez et al., 2006). Identification of the temporal sequence of events for microRNA-mediated regulation was made possible by ribosome profiling (Ingolia, Ghaemmaghami, Newman, & Weissman, 2009) to measure the precise location and density of ribosome occupancy genome-wide during the MZT (Bazzini, Lee, & Giraldez, 2012). These experiments showed that *miR-430* reduces ribosome occupancy on target mRNAs before inducing complete deadenylation and mRNA destabilization. Indeed, *miR-430* reduces ribosome density uniformly, consistent with inhibition of translation initiation rather than ribosome drop-off (Bazzini et al., 2012). While *miR-430* ultimately triggers deadenylation of its targets, the nonsteady state context of the developing embryo allows the uncoupling of translational repression from the decay of these mRNAs. However, it appears that many of these mRNAs are being deadenylated soon after the microRNA is expressed (Subtelny, Eichhorn, Chen, Sive, & Bartel, 2014). microRNAs have also been found to play a role during the MZT in *Xenopus*, *Drosophila* and, possibly, mammals (discussed below).

2.7 Methods of Measuring Poly(A) Tail Length

The importance of polyadenylation for mRNA regulation mechanisms during early embryogenesis motivated the development of methods to accurately measure poly(A) tail length for large numbers of mRNAs. Poly(A) tail length measurement for individual mRNAs was initially achieved by cDNA synthesis with oligo(dT) or cleavage of upstream sequence to release the poly(A) tail (Murray & Schoenberg, 2008). The high-resolution poly(A) tail (Hire-PAT) assay enables single nucleotide resolution and quantification of polyA tail length for individual mRNAs and, using this technique (Bazzini et al., 2012), it was shown that microRNA-mediated translational repression takes place before complete mRNA deadenylation. Estimation of poly(A) tail length in a population of mRNAs is possible through affinity chromatography of RNA on poly(U) beads, its differential elution at increasing temperatures or salt concentrations, followed by microarray or sequencing (Beilharz & Preiss, 2007; Du & Richter, 2005; Meijer et al., 2007).

More recently, high-throughput methods have enabled poly(A) tail measurements transcriptome wide. First, poly(A)-tail length profiling by sequencing (PAL-seq) measures fluorescence signal after reverse transcription of the poly(A) tail as a proxy for its length (Subtelny et al., 2014). This method identified a positive correlation between poly(A) tail length and translation efficiency during early embryonic development in several species, which diminished at gastrulation, suggesting a developmental switch in translational control during the MZT. Second, TAIL-seq directly sequences the 3′ mRNA ends to determine the position of the poly(A) tail start, allowing highly accurate and high-throughput measurement of mRNA poly(A) tail lengths (Chang, Lim, Ha, & Kim, 2014). This method identified widespread uridylation and guanylation of poly(A) tails in cells, but the function of these modifications during development awaits further analysis. Polyadenylation is a highly dynamic process during embryogenesis that is directly linked to translational regulation and mRNA stability. These new methods make it possible to investigate the mechanisms that dictate poly(A) tail length during embryonic development.

2.8 Role of Decapping in Maternal mRNA Clearance

In somatic cells, mRNA decapping and decay are tightly coupled to deadenylation (Parker & Song, 2004); however, this is not always the case during early development. In *Xenopus*, maternal mRNAs with AU-rich elements (ARE) in their 3′UTRs are deadenylated following egg activation, but are eliminated only after the mid-blastula transition (MBT) (Audic, Omilli, & Osborne, 1997; Voeltz & Steitz, 1998). This indicates that deadenylation and decay are uncoupled during early embryonic development. What developmental cue prevents mRNAs primed for destabilization to remain stable until the MBT? Interestingly decapping activity in *Xenopus* is first detected at the MBT (Gillian-Daniel, Gray, Aström, Barkoff, & Wickens, 1998; Zhang, Williams, Wormington, Stevens, & Peltz, 1999), suggesting that for maternal mRNAs that are deadenylated maternally via AU-rich elements, zygotic activation of decapping activity could be the trigger for mRNA destabilization.

Decapping enzymes are regulated at the level of their catalytic activity and localization. The catalytic activity of Dcp2 is stimulated by several enhancers of decapping (Arribas-Layton, Wu, Lykke-Andersen, & Song, 2013; Jonas & Izaurralde, 2013; Ling, Qamra, & Song, 2011), suggesting the potential for developmental regulation. Metazoan-specific EDC4 functions as a scaffold to facilitate Dcp2 interaction with its cofactor, Dcp1, to activate decapping (Chang, Bercovich, Loh, Jonas, & Izaurralde, 2014). Zygotic synthesis of decapping enhancers could be a feasible developmental strategy to trigger destabilization of maternal mRNAs with AREs at the MBT, but maintain their stability during oogenesis. Additionally, in *C. elegans*, Dcp2 localizes to cytoplasmic foci, potential sites of mRNA decay, while DcpS is distributed throughout the cytoplasm (Lall, Piano, & Davis, 2005). Thus, localization of mRNAs and of decapping proteins may be regulated to initiate mRNA degradation.

It remains to be discovered what mechanisms activate decapping and what their contribution is during maternal mRNA clearance. However, a large fraction of destabilized maternal mRNAs first require a zygotic deadenylation trigger such as *miR-430* (Bazzini et al., 2012; Giraldez et al., 2006), indicating that decapping is not the limiting factor in global degradation of maternal mRNAs.

2.9 Maternal and Zygotic Modes of Maternal mRNA Clearance

A major segregating factor in the mechanisms of maternal mRNA clearance is their dependence on zygotic transcription (Fig. 2). This observation was

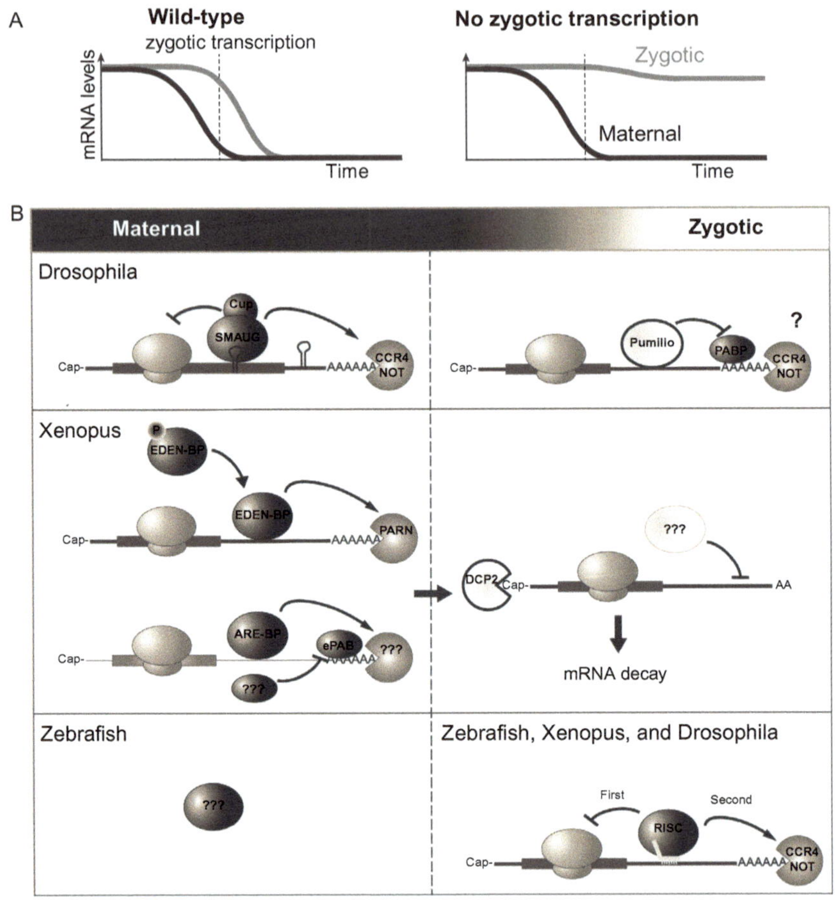

Figure 2 Maternal and zygotic mechanisms of maternal mRNA clearance. (A) Maternal mRNAs under the regulation of the "maternal mode" mechanisms (red) will be destabilized independently of zygotic transcription, while mRNAs under the "zygotic mode" mechanisms (blue) will be stable in the absence of zygotic transcription. (B) Examples of characterized pathways of maternal and zygotic mode mechanisms across species. Maternal mode factors (red), zygotic mode factors (blue), and maternal factors activated after zygotic transcription (red with blue outline). (See the color plate.)

first possible in *Drosophila*, where egg activation and fertilization are uncoupled processes (Bashirullah et al., 1999; Tadros et al., 2003). For example, *nanos* mRNAs is degraded upon egg activation, indicating that this decay mechanism uses only maternal instructions (maternal mode). Conversely, *bicoid* mRNA is degraded only after zygotic transcription (zygotic mode), while *Hsp83* mRNA utilizes both clearance mechanisms (Bashirullah et al., 1999). In other species, maternal and zygotic modes of regulation can be distinguished either temporally or by blocking zygotic transcription (Ferg et al., 2007; Hamatani et al., 2004). The maternal mode of mRNA decay occurs before or independently of zygotic transcription. While the zygotic mode occurs after zygotic transcription and is blocked when zygotic transcription is inhibited, summarized in Fig. 2A. This indicates that some pathways of maternal mRNA clearance are inherited in the oocyte cytoplasm, while others are synthesized *de novo*, further highlighting the contributions of both the maternal cytoplasm and zygotic nucleus to the MZT.

Zygotic transcription can be detected using multiple methods. Application of RNA Polymerase II inhibitors such as alpha-amanitin (Lindell, Weinberg, Morris, Roeder, & Rutter, 1970; reviewed in Bensaude, 2011), cause embryonic arrest soon after the onset of zygotic transcription. Using this method, it was determined that zygotic transcription is required beyond the two-cell stage in mouse (Flach, Hjohnson, Braude, Taylor, & Bolton, 1982; Golbus, Calarco, & Epstein, 1973; Warner & Versteegh, 1974), 4–8 cell stage in human (Braude, Bolton, & Moore, 1988), 5–8 cell stage in cat (Hoffert, Anderson, Wildt, & Roth, 1997), 8–16 cell stage in rabbit (Manes, 1973), 100-cell stage in *C. elegans* (Edgar, Wolf, & Wood, 1994), following nuclear cycle 13 in *Drosophila* (Edgar, Kiehle, & Schubiger, 1986) and at the mid-blastula transition (MBT) in Zebrafish and *Xenopus* (Kane et al., 1996; Newport & Kirschner, 1982). More sensitive methods to identify the onset and the identity of early zygotic transcripts includes measuring (1) RNA accumulation containing paternal single-nucleotide polymorphisms (SNPs) (Harvey et al., 2013; Sawicki, Magnuson, & Epstein, 1981), (2) intronic sequences in unprocessed zygotic mRNAs (Lee et al., 2013), and (3) new transcripts after labeling with 4-thiouridine (4SU) (Heyn et al., 2014). Defining the timing of zygotic transcription across species facilitated distinguishing mRNAs under maternal or zygotic modes of clearance across species.

2.10 Proportion of Maternal and Zygotic Modes Across Species

While most animals experience both maternal and zygotic modes of maternal mRNA clearance, the proportion of each mode of clearance utilized varies across animals (Tadros & Lipshitz, 2009; Walser & Lipshitz, 2011). High-throughput mRNA profiling at different developmental stages or coupled with chemical inhibitors of *de novo* zygotic transcription enabled the identification of mRNAs under these distinct modes of clearance across species. In *Drosophila*, over 1000 maternal mRNAs are cleared following egg activation, two-thirds of which depend on Smaug for elimination (Tadros et al., 2007). An additional 563 stabilized mRNAs were discovered in the background of specific chromosomal arm deletions (De Renzis et al., 2007), suggesting that these mRNAs are also subject to the zygotic mode of clearance and require genes encoded within these deletions for degradation.

In mouse, the majority of mRNAs are destabilized prior to zygotic transcription. Oocyte maturation between the germinal vesicle (GV) breakdown and Meiosis II (MII) stages triggers destabilization of almost 3000 mRNAs (Su et al., 2007). These transcripts are enriched for genes involved in ATP production such as oxidative phosphorylation and ubiquinone biosynthesis (Su et al., 2007; Zeng, Baldwin, & Schultz, 2004), likely reflecting the changing metabolic needs of the maturing oocytes. Degradation of maternal mRNA during oocyte maturation is carefully regulated and evidence for this is that over 9200 mRNAs remain remarkably stable during this time (Su et al., 2007), potentially implicating coexistence of stabilization mechanisms to protect mRNAs utilized after fertilization. Interestingly mammalian oocytes require high levels of cAMP to remain arrested in the GV stage (Horner et al., 2003). Selective degradation of the ATP production machinery in order to generate cAMP could be a mechanism to enable oocytes to progress to MII stage (Su et al., 2007). An additional almost 2300 maternal mRNAs are eliminated immediately following fertilization, consistent with a maternal mode of clearance, while almost 500 mRNAs are cleared at the two-cell stage, consistent with a zygotic mode of regulation (Hamatani et al., 2004). It would be interesting to investigate whether alpha-amanitin treatment selectively stabilizes these mRNAs to determine whether this cluster requires zygotic transcription for clearance.

Comprehensive analysis of the *C. elegans* transcriptome during the oocyte-to-embryo transition (OET) showed that roughly 25% of the maternal mRNA pool (~1900 mRNAs) is eliminated between the mature oocyte and one-cell stage (Stoeckius et al., 2014), indicating dramatic mRNA turnover during the MZT, and prior to the MBT, similar to that in *Drosophila* (Tadros et al., 2007). Destabilized mRNAs are enriched for a poly(C) motif, which is also sufficient to destabilize reporter mRNAs, and binds poly(C)-binding protein (PCBP). Additionally, endo-siRNAs but not microRNAs have been implicated in this regulation (Stoeckius et al., 2014). An additional ~30% of the remaining maternal mRNAs is destabilized around the four-cell stage (Baugh et al., 2003). MicroRNAs are implicated in maternal and zygotic mode deadenylation, but not the decay of maternal mRNAs in this species (Alvarez-Saavedra & Horvitz, 2010; Wu et al., 2010), which parallels EDEN-BP in *Xenopus*.

In Zebrafish, the zygotic modes of maternal mRNA clearance likely dominate. The zygotic expression of *miR-430* triggers repression, deadenylation, and clearance of several hundred maternal mRNAs (Bazzini et al., 2012; Giraldez et al., 2006) that have *miR-430* target sites; up to 40% of maternal mRNAs are potentially regulated by this mechanism (Giraldez et al., 2006). Interestingly, several studies have reported early sizable destabilization of maternal mRNAs in Zebrafish prior to zygotic transcription (Aanes et al., 2011; Mathavan et al., 2005; Rabani et al., 2014), suggesting the action of early-acting maternal-mode decay mechanisms. However, these studies rely on poly(A) selected RNA-seq (Aanes et al., 2011; Mathavan et al., 2005; Rabani et al., 2014) to draw these conclusions. Polyadenylated mRNAs are more efficiently captured during poly(A) selection protocols (Harvey et al., 2013). Because it is likely that the poly(A) tail length of different maternal mRNAs varies (see above), conclusions about early decay or transcription can be misleading. Leveraging ribosome depletion to sequence total RNA (Lee et al., 2013) and taking advantage of exogenous spike-ins (Lovén et al., 2012) will offer accurate measurement of the timing and the dynamics of mRNA clearance.

2.11 Shared Features of Maternal mRNA Clearance Mechanisms Across Animals

We have seen that maternal mRNA clearance is universal yet diverse across animals. This diversity likely stems from different developmental requirements such as timing and unique physiological environments. While the regulated targets may vary across species, the underlying mechanisms are likely conserved. For example, in the parasitic worm *Ascaris suum* maternal mRNA clearance is temporally coordinated despite it lacking a transcriptionally quiescent maternal stage. Unlike other known metazoans, favorable environmental cues rather than fertilization triggers pronuclear fusion and embryonic progression in this species (Wang, Garrey, & Davis, 2014). However, maternal mRNA clearance still occurs in distinct waves in *Ascaris suum*, with over 1100 genes degraded immediately after fertilization and an additional 1662 mRNAs degraded by the 10-cell stage (Wang, Garrey, et al., 2014). The protracted development in this species likely accounts for its unique MZT features. However, the temporal coordination of maternal mRNA clearance in this species suggests that the same mechanisms are likely in place as in other animals.

Additionally, zygotic expression of microRNAs to clear maternal mRNAs is a common strategy used by most animals and highlights the conserved nature of maternal clearance mechanisms. The first microRNA identified to function in the clearance of maternal mRNAs was zebrafish *miR-430* (Giraldez et al., 2006). Interestingly, *Xenopus* exhibits zygotic expression of *miR-427*, which shares its seed sequence with zebrafish *miR-430*, and also destabilizes maternal mRNAs (Lund, Liu, Hartley, Sheets, & Dahlberg, 2009; Rosa, Spagnoli, & Brivanlou, 2009). In *Drosophila*, a different set of miRNAs (the *miR-309* family) plays an analogous role in maternal mRNA clearance (Bushati, Stark, Brennecke, & Cohen, 2008), suggesting convergent evolution of the same mechanism.

In mammals, the scope and the diversity of microRNA function for maternal mRNA clearance is still poorly understood. In mouse, *miR-290* shares its seed sequence with *miR-430/427/302* family microRNAs and is expressed at high levels in early embryos and in embryonic stem cells (Tang et al., 2007; Zeng & Schultz, 2005), suggesting a role in maternal mRNA clearance similar to other vertebrates. Genetic inactivation of the *miR-290/295* cluster in mouse results in partially penetrant embryonic lethality, primordial germ cell migration defects, and female sterility for surviving homozygous mutants (Medeiros et al., 2011), pointing to a predominant role for *miR-290/295* cluster in germ line development. Interestingly, while genetic inactivation of *miR-302*, which shares its seed with *miR-290*, results in failure in neural tube closure, *miR-302/miR-209* double mutant is early embryonic lethal (Parchem et al., 2015). This suggests that the *miR-290/302* family microRNAs have a redundant role for early embryonic development in mouse and specific roles during later embryogenesis. In bovine embryos, early zygotic expression of *miR-212* negatively regulates *FIGLA* mRNA levels (Tripurani et al., 2013), suggesting that microRNAs have a role in maternal mRNA clearance in other mammals. The extent of *miR-212* and other mammalian microRNA involvement in maternal mRNA clearance awaits further investigation.

2.12 Regulation of microRNAs During the MZT

We have seen that embryogenesis depends heavily on coordinated control of maternally provided mRNAs, but the importance of microRNA regulation during the MZT has only recently come to light. For example, *miR-430* in zebrafish undergoes rapid synthesis soon after zygotic genome activation initiates, but precursor levels stop accumulating after gastrulation (Giraldez et al., 2005), indicating that *miR-430* locus transcription is suppressed at this stage. Nanog binds the *miR-430* promoter and is required for *miR-430* expression (Lee et al., 2013), but it remains to be discovered how Nanog function is blocked after the MZT.

Interestingly, mature *miR-430* persists up to 2 days of development (Giraldez et al., 2005), and its activity is regulated in a tissue specific manner. *miR-430* balances the nodal signaling pathway through regulation of both the pathway agonist, *squint*, and the antagonist, *lefty* (Choi, Giraldez, & Schier, 2007). *miR-430* also regulates Sdf1 chemokine signaling to ensure accurate primordial germ cell (PGC) migration (Staton, Knaut, & Giraldez, 2011). In the germ line, the RNA-binding proteins, Dead end 1 (Dnd1) (Kedde et al., 2007) and Deleted in azoospermia like (Dazl) (Takeda, Mishima, Fujiwara, Sakamoto, & Inoue, 2009), inhibit *miR-430* action to set up differential gene expression between the soma and germ line (Mishima et al., 2006). Additionally, the RNA-binding protein, TDP-43, disrupts microRNA incorporation into RISC to limit its activity (King et al., 2014). However, it remains to be elucidated exactly how microRNA activity is regulated during development.

At the posttranscriptional level, maternal microRNAs undergo regulated clearance. Maternal microRNAs are heavily adenylated at the 3′ end from invertebrates to mammals (Lee, Choi, et al., 2014), suggesting a deeply conserved function of this mechanism for embryonic development. In *Drosophila* oocytes and embryos, microRNAs are polyadenylated by a noncanonical poly(A) polymerase, Wispy, which directs their destabilization after fertilization (Lee, Choi, et al., 2014). Maternal *wispy* mutants arrest prior to pronuclear fusion in *Drosophila* (Brent, MacQueen, & Hazelrigg, 2000), and Wispy's role in active clearance of microRNAs via deadenylation may be required for appropriate gene expression during embryogenesis (Lee, Choi, et al., 2014). Mechanistically, Wispy interacts with Ago1, which may allow selective adenylation of microRNAs (Lee, Choi, et al., 2014). Interestingly, zygotic microRNAs such as *miR-430* in zebrafish seem to have a long half-life (Giraldez et al., 2005) and serve important functions during embryonic development (as discussed earlier), indicating that zygotic microRNAs are either protected from Wispy-mediated polyadenylation or that this pathway is no longer active after zygotic transcription. Together these studies illustrate how microRNA activity and stability is tightly regulated during development.

2.13 Endonucleolytic Cleavage During the MZT

Endonucleolytic activity exists in *Xenopus* and *Drosophila* embryos; however, its role during embryogenesis has not been determined. It is clear that the

Xlhbox2/HoxB7 mRNA undergoes endonucleolytic cleavage in *Xenopus* oocytes because cleavage intermediates have been detected using Northern and RNAseH assays (Brown & Harland, 1990). The poly(A) tail length of the mRNA remains constant in these degradation intermediates (Brown & Harland, 1990) and decay persists in the presence of cycloheximide (Brown, Zipkin, & Harland, 1993), indicating that the mechanism neither involves deadenylation nor requires translation, respectively. A 90-nucleotide region in the *Xlhbox2/HoxB7* 3′UTR is sufficient for cleavage (Brown & Harland, 1990), indicating that this mechanism is sequence specific. This endonucleolytic activity is also present in *Drosophila* embryo lysates (Brown et al., 1993), suggesting conservation between distantly related species. Interestingly, injecting RNA corresponding to the endonuclease recognition site and nearby flanking region in excess accelerates cleavage in endogenous and reporter mRNA (Brown et al., 1993), indicative of endonuclease inhibitor presence. It is possible that active silencing of endonuclease activity plays a role during embryogenesis. Widespread action of endonucleolytic cleavage in transcriptionally silent oocytes and early embryos could result in precocious mRNA decay, eliminate the potential to reuse mRNAs, and be harmful for embryogenesis. Specific endonuclease inhibitor(s) that are active during embryonic development have yet to be identified.

Endonucleolytic cleavage mechanisms have also been proposed to regulate maternal mRNAs. In mouse, genetic inactivation of *Dicer*, an enzyme required for biogenesis of microRNAs and processing of dsRNAs, results in the failure to complete Meiosis I, stabilization of 1300 mRNAs, and upregulation of retrotransposons (Murchison et al., 2007). However, DGCR8, required for processing primary (pri-) microRNAs, is dispensable for mouse oocyte (Ma et al., 2010; Suh et al., 2010) and preimplantation development (Suh et al., 2010). The phenotype of Dicer mutants is independent of microRNAs and likely results from misregulation of transposons (Murchison et al., 2007). Additionally, in *C. elegans* embryos, endo-siRNAs regulate ~10% of destabilized maternal mRNAs while microRNAs do not appear to contribute to maternal mRNA destabilization (Stoeckius et al., 2014).

2.14 Role of Coding Sequence in mRNA Decay

Recent evidence suggests that the coding sequence can have an impact on mRNA stability. Transcriptome-wide analysis in yeast found that mRNAs that utilize optimal codons have increased half-life and, strikingly, that optimizing codon usage in an unstable mRNA, *LSM8*, increases its half-life nearly eightfold, from 2.5 to 18.7 min (Presnyak et al., 2015). This finding is intriguing because it offers an explanation of how groups of mRNAs could be coregulated independently of harboring common *cis*-regulatory elements. Additionally, genes in common pathways share codon usage characteristics. For example, stable mRNAs encoding glycolytic enzymes have 86% optimized codons, a disproportionately large fraction, while unstable mRNAs, such as those encoding factors involved in the pheromone response have 43% optimal codons (Presnyak et al., 2015), prompting the authors to speculate that species could evolve specific codon content in related gene groups to enable their coregulation. In fact, common *cis*-regulatory elements are not enriched in unstable mRNAs in yeast (Geisberg, Moqtaderi, Fan, Ozsolak, & Struhl, 2014); usage of rare codons could, therefore, explain the instability of mRNAs with seemingly unrelated sequence signatures.

However, that study does not exclude the possibility that primary sequence determinants drive mRNA destabilization in this context. Coding sequence is known to harbor destabilization elements. For example, highly unstable *c-fos* mRNA harbors several destabilization elements within its coding region since specific deletions prolong mRNA half-life (Schiavi et al., 1994). These coding-region destabilization elements require translation and involve deadenylation since inhibiting translation with Anisomycin increases mRNA half-life and increases the poly(A) tail length (Schiavi et al., 1994). Interestingly, the sequence of the *c-fos* coding region, not its amino acid composition, encodes the destabilization element because an out-of-frame *c-fos* variant is equally unstable (Wellington, Greenberg, & Belasco, 1993). While the authors of the yeast codon-usage study showed that the correlation between mRNA stability and optimal codon usage disappears when the same mRNAs are translated out-of-frame *in silico* (Presnyak et al., 2015), half-life measurements for out-of-frame translated mRNAs were not reported. It is, thus, possible that the alteration to the primary sequence necessary to generate codon-optimized mRNAs in this study resulted in increased mRNA stability. Comparing half-lives of codon optimized, nonoptimized, and out-of-frame translated optimized mRNAs will help to uncover the contribution of this mechanism to maternal mRNA regulation.

Evidence of a global correlation of rare codons with mRNA instability motivates investigation of whether unstable maternal mRNAs correlate with suboptimal codon usage across species. It is possible that maternal mRNAs destined for clearance utilize rare codons to favor mRNA destabilization during the MZT.

2.15 Cooperativity and Redundancy in Maternal mRNA Clearance Mechanisms

Destabilized maternal mRNAs are regulated by multiple cooperating mechanisms. In *Drosophila*, *nanos* mRNA harbors Smaug recognition elements (SREs) (Dahanukar et al., 1999; Smibert et al., 1999) and predicted piRNA-binding sites in different regions of its 3′UTR (Rouget et al., 2010), suggesting that piRNAs and Smaug could cooperate in maternal mRNA clearance. Indeed, mutants for Smaug or the piRNA effector protein, Aubergine, exhibit stabilization of *nanos* mRNA (Rouget et al., 2010; Tadros et al., 2007). Additionally, in zebrafish, *miR-430* is zygotically encoded and targets several hundred mRNAs for clearance (Bazzini et al., 2012; Giraldez et al., 2006), but evidence suggests that it does not always function in isolation. Knockdown of TATA-binding factor (TBP), a transcription preinitiation complex component, using a translation-blocking morpholino has no effect on *miR-430* transcription or processing, but results in stabilization of a subset of *miR-430*-target mRNAs (Ferg et al., 2007). This observation demonstrates that, to destabilize a subset of its target mRNAs, *miR-430* cooperates with an additional, as-of-yet unidentified, factor(s) that require zygotic activation. In *Xenopus*, EDEN-dependent deadenylation is necessary, but not sufficient for *c-mos* mRNA deadenylation, and ARE-like sequences enhance deadenylation in a position-dependent manner relative to EDEN (Audic et al., 1998; Ueno & Sagata, 2002). Cooperative mechanisms of mRNA clearance may enable specificity or precise timing for target mRNA destabilization.

Additionally, redundancy is built into maternal mRNA clearance during the MZT. Pumilio directly interacts with hundreds of mRNAs, yet *pumilio* mutants have been reported not to exhibit global defects in mRNA destabilization (Gerber et al., 2006), implicating the action of a redundant factor for Pumilio target mRNAs. Similarly, in *C. elegans*, maternal mRNAs destabilized during the oocyte-to-embryo transition are enriched for a poly(C) motif (Stoeckius et al., 2014). This motif is necessary and sufficient for destabilization in reporter assays, but knockdown of all poly(C)-binding protein (PCBP) paralogs does not result in stabilization of mRNAs harboring the poly(C) motif (Stoeckius et al., 2014). Likewise, the RNA-binding protein, Brain Tumor (Brat), associates with nearly 1100 mRNAs in *Drosophila* embryos and its binding motif is sufficient to trigger mRNA destabilization in Brat-dependent manner using reporters, but the *brat* mutant exhibits stabilization of only a subset of its target mRNAs (Laver et al., 2015). The absence of widespread stabilization of targets mRNAs in mutants or knockdown of the corresponding *trans*-factors, suggests redundancy in maternal mRNA clearance mechanisms.

2.16 Combinatorial Code in Maternal mRNA Clearance

Cooperativity and redundancy in maternal mRNA clearance mechanisms suggests the existence of a combinatorial code that governs mRNA fate during development (Fig. 3). While some elements of this regulatory code have been identified during the MZT reviewed in (Walser & Lipshitz, 2011) the complexity of this regulatory mechanisms suggest that there are additional elements of this code that likely regulate mRNA stability. Large-scale approaches have allowed probing of mRNA structure (Ding et al., 2014;

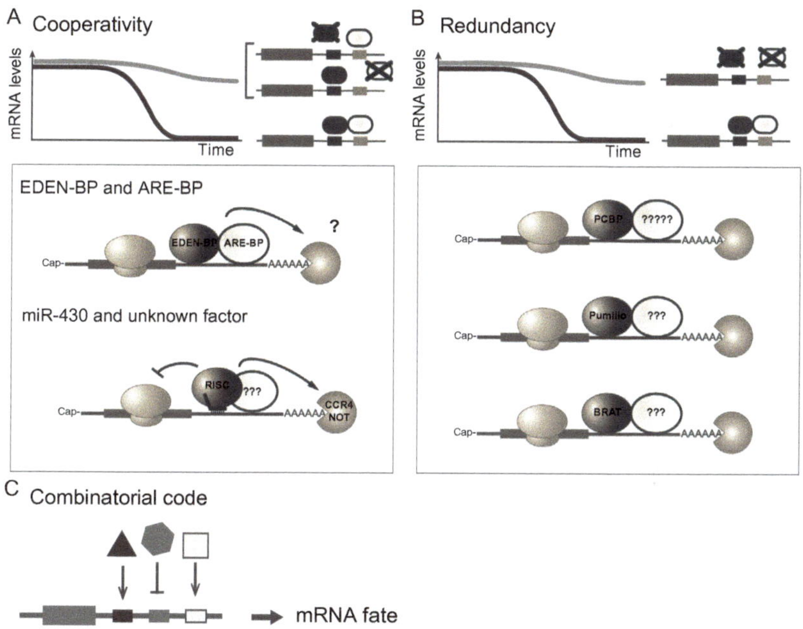

Figure 3 Combinatorial code in maternal mRNA clearance. (A) Cooperative mechanisms require both factors to destabilize mRNA, such that depletion of either one results in stabilization of the target mRNAs. Examples include EDEN-BP together with ARE-BP and miR-430 together with as-of-yet unidentified factor(s) (Ferg et al., 2007; Ueno & Sagata, 2002). (B) Redundant mechanisms require either factor, such that depletion of one of the factors does not affect the stability of target mRNAs. Examples include PCBP in *C. elegans* (Stoeckius et al., 2014) and Pumilio (Gerber et al., 2006) and BRAT (Laver et al., 2015) in Drosophila. Depleting both factors is required to block mRNA destabilization. (C) Model of combinatorial code for maternal mRNA clearance. Individual transcripts harbor multiple regulatory elements that affect mRNA stability. The combination of all signals acting on the mRNA determines mRNA fate.

Rouskin, Zubradt, Washietl, Kellis, & Weissman, 2014; Spitale et al., 2015; Wan et al., 2014), RNA modifications (Batista et al., 2014; Geula et al., 2015), and identification of sequence elements that cause mRNA decay in cell culture (Goodarzi et al., 2012; Oikonomou, Goodarzi, & Tavazoie, 2014) and in yeast (Geisberg et al., 2014). Technological advances that enable the application of these methods *in vivo* within the embryo will define the function of individual elements and the regulatory code (sequence, structure and RNA modifications) to understand the posttranscriptional regulatory networks driving the embryonic transition to pluripotency.

3. CONSEQUENCES OF FAILURE OF MATERNAL mRNA CLEARANCE

There is a growing consensus that degradation of maternal mRNAs is instructive for development and essential to successfully undergoing the MZT (DeRenzo & Seydoux, 2004; Giraldez, 2010; Stitzel & Seydoux, 2007; Tadros & Lipshitz, 2009). Model organisms have offered tractable means to investigate the functional relevance of maternal mRNA clearance during embryonic development (Table 1). Additionally, high-throughput gene expression profiling experiments support the relevance of maternal mRNA clearance for early development.

3.1 Loss of Maternal mRNA Clearance in Model Organisms

Components of the maternal mRNA clearance machinery are required for normal embryogenesis. In *Drosophila*, *smaug* mutants exhibit stabilization of ~1000 mRNAs, which is two-thirds of the transcripts that undergo the maternal mode maternal mRNA clearance in this species (Tadros et al., 2007). *smaug* mutants fail to undergo cellularization at the MBT and do not activate high levels of zygotic transcription (Benoit et al., 2009), suggesting that maternal mRNA clearance is critical for these developmental processes. Additionally, mutants for the piRNA effector protein, Aubergine, exhibit stabilization of *nanos* mRNA and defects in head development (Rouget et al., 2010), suggesting a function of piRNAs in the regulation of maternal mRNAs. However, given the complexity of piRNA populations in the embryo, it is unclear whether the effects of Aubergine loss could be secondary to disrupting other processes, such as DNA damage or the activation of the zygotic program. In mouse, mutations in CCCH tandem zinc finger protein, Zfp36l2, results in embryonic arrest at the

Table 1 Loss of Function Phenotypes of Maternal mRNA Clearance Machinery

Species	Factor	Phenotype	Reference
C. elegans	PCBP	No phenotype	Stoeckius et al. (2014)
	miR-35–42, miR-51–56, miR-58/80–82	Embryonic lethality; locomotion, body size, and reproductive defects	Alvarez-Saavedra & Horvitz (2010) and Wu et al. (2010)
Drosophila	Smaug	Stabilization of ~1000 maternal mRNAs, failure to undergo cellularization, failure to activate zygotic genes	Tadros et al. (2007) and Benoit et al. (2009)
	Aubergine	Stabilization of nanos, head development defects	Rouget et al. (2010)
	Brain tumor (BRAT)	Stabilization of ~20% target mRNAs	Laver et al. (2015)
	Pumilio	Targets mRNAs ~10% stabilized; failure to maintain female germ line	Forbes and Lehmann (1998), Lin and Spradling (1997), and Gerber et al. (2006)
	miR-309	20% embryonic lethality	Bushati et al. (2008)
Zebrafish	miR-430	Stabilization of several hundred mRNAs, abnormal cell movements during morphogenesis (in MZDicer mutant)	Giraldez et al. (2006) and Bazzini et al. (2012)
Xenopus	EDEN-BP	Somitogenesis defects (in morpholino knockdown)	Paillard et al. (1998) and Gautier-Courteille et al. (2004)
	miR-427	Axis formation defects (in morpholino knockdown)	Rosa et al. (2009) and Lund et al. (2009)
	ARE-BP	Unknown	Voeltz and Steitz (1998)
	ePABP	Increased deadenylation of target mRNAs (immunodepletion)	Voeltz et al. (2001)
Mouse	ZFP36L2	Embryonic lethal at two-cell stage	Ramos et al. (2004)
	miR-290/295	Partially penetrant embryonic lethality, female sterilty in surviving mutants	Medeiros et al. (2011)
	miR-302a-d; miR-302/miR-209	failure in neural tube closure; early embryonic arrest	Parchem et al. (2015)
Cow	miR-212	Unknown	Tripurani et al. (2013)

two-cell stage (Ramos et al., 2004), indicating a failure to complete the MZT. Zfp36l2 is an RNA-binding protein that recognizes AREs in the 3′UTRs of its target mRNAs to initiate mRNA degradation (Lai, Carballo, Thorn, Kennington, & Blackshear, 2000), thus it is conceivable that this protein functions to target maternal mRNAs for clearance during MZT in mouse. The targets of Zfp36l2 during mouse preimplantation development remain to be identified, as well as the sequence motifs it recognizes, and whether the embryonic lethal mutant phenotype results directly from the loss of its maternal mRNA clearance function.

MicroRNAs are directly involved in the clearance of their target mRNAs and the loss of microRNA-mediated maternal mRNA clearance results in developmental defects across different species. The first microRNA involved in maternal mRNA clearance was discovered using a maternal-zygotic mutant for Dicer (MZ*dicer*), an enzyme required for canonical microRNA biogenesis (Giraldez et al., 2006). Absence of *miR-430*-mediated mRNA clearance in the MZ*dicer* results in cell movement defects during gastrulation (Giraldez et al., 2006), implicating maternal mRNA clearance to be important for early embryonic processes. In *Xenopus*, *miR-427* has the same seed sequence as zebrafish *miR-430* (Rosa et al., 2009), is highly transcribed at the MBT, and causes deadenylation and decay of *cyclin A1* and *cyclin B2* mRNAs (Lund et al., 2009). Inhibition of *miR-427* function with anti-*miR-427* morpholinos causes dramatic defects in axis formation resulting, at least in part, from misregulation of the nodal pathway (Rosa et al., 2009), but it is possible that additional target mRNAs contribute to this phenotype. In *Drosophila*, a cluster of *miR-309* microRNAs regulates maternal mRNA clearance and genetic deletion of this microRNA cluster results in 20% embryonic lethality that is rescued with a transgene encoding this genomic locus (Bushati et al., 2008). MicroRNAs are directly involved in destabilizing target mRNAs, thus developmental defects resulting from loss of microRNA-mediated maternal mRNA clearance demonstrate the functional importance of maternal mRNA clearance for animal development.

3.2 Maternal mRNA Clearance During Human Preimplantation Development

Maternal mRNA clearance likely plays an instructive role in human preimplantation development. This step involves several waves of maternal mRNA clearance with ~1700 mRNAs eliminated by day 2 (four-cell stage), ~700 mRNAs between day 2 and 3 (four- to eight-cell stage), and additional ~2700 mRNAs by day 5 (blastocyst stage) (Zhang et al.,

2009). Additionally, single-cell RNA sequencing of preimplantation human embryos identified that ~10% of the expressed maternal mRNAs (1941 out of 22,687) and 2% of the lncRNAs (185 out of 8701) are eliminated between the four- and eight-cell stage (Yan et al., 2013). These studies demonstrate that early human development is characterized by dramatic turnover of mRNAs and ncRNAs and that this occurs in distinct waves, suggesting that this is a regulated process. Indeed, maternal mRNAs that are eliminated early, by the two-cell stage, are enriched for different gene categories (cell cycle, transcription regulation) than mRNAs eliminated at later stages (protein phosphorylation, cell morphogenesis) (Vassena et al., 2011; Yan et al., 2013), arguing for specificity of maternal clearance during each successive wave. Furthermore, clearance of maternal mRNAs is likely a requirement for preimplantation human development because IVF-derived human embryos that fail show evidence of zygotic gene expression but failure to downregulate maternal transcripts (Dobson et al., 2004; Wong et al., 2010). These studies demonstrate that successful reprogramming corresponds with clearance of maternal mRNAs and suggest that this process plays an instructive role during reprogramming.

3.3 The MZT in Interspecies Somatic Nuclear Transfer Embryos

Somatic nuclear transfer (Fig. 1A) within a species (Gurdon, 1962; Wilmut, Schnieke, McWhir, Kind, & Campbell, 1997) allows successful reprogramming of the differentiated nucleus and the completion of the MZT. However, the efforts to clone animals combining an oocyte and a nucleus from different species (interspecies somatic nuclear transfer, or iSNT) have shown limited success. iSNT was reported as early as 1886 to not be possible between toad and frog in either direction (reviewed in Laubichler & Davidson, 2008). The only successful example of iSNT was the cloning of the endangered gaur bull (*Bos gaurus*) using enucleated oocytes of domestic cow (*Bos taurus*) (Lanza et al., 2000). What limits the reprogramming potential when a nucleus is in a foreign oocyte?

Transcriptome analyses show that iSNT embryos fail prior to completing the MZT and that incomplete maternal mRNA clearance may be the culprit. Development could only be recapitulated until the 8–16 cell stage when a rhesus fibroblast nucleus was fused with bovine enucleated oocytes (Wang et al., 2011). Zygotic genome activation occurs during the 6–8 cell stage in rhesus (Schramm & Bavister, 1999) and 8–16 cell stage in cow (Camous, Kopecny, & Flechon, 1986) indicating that developmental arrest in rhesus-bovine iSNT embryos occurs prior to MZT completion. All of these defects could be due to a complete failure to activate the zygotic genome. However, comparison of gene expression between cow embryos produced by *in vitro* fertilization (IVF) and failed rhesus iSNT embryos using microarrays showed that zygotic genes are activated, but over 1500 maternal mRNAs are not cleared (Wang et al., 2011). A similar approach was used in efforts to clone the endangered Przewalski gazelle, but no viable embryos developed (Zuo et al., 2014). Transcriptome analysis showed that successful IVF-derived cow embryos cleared 1515 mRNAs, while failed gazelle iSCNT embryos cleared only 343 mRNAs (Zuo et al., 2014), demonstrating a dramatic defect in maternal mRNA clearance. While these defects could result from the failed transcription of key zygotic genes, failed recognition and clearance of the maternal mRNAs by the heterologous maternal and zygotic programs could also contribute to embryonic reprogramming during the MZT.

4. MZT CONNECTION TO OTHER TRANSITIONS AND REPROGRAMMING

4.1 Unicellular to Multicellular Transition

Active clearance of the previous mRNA landscape may be a general feature of cellular transitions in multicellular organisms. For example, differentiation in a simple model of multicellularity, the slime mold *Dictyostelium discoideum*, involves dramatic mRNA turnover. The life cycle of this organism involves a unicellular growth state and differentiation to multicellular, aggregated stage consisting of two cell types (Kessin, 2001). Starvation triggers the unicellular to multicellular transition in *D. discoideum* and corresponds to changes in cell fate decisions and morphology (Clarke & Gomer, 1995). Time-course transcriptional profiling using microarrays following starvation in *D. discoideum* showed that the unicellular to multicellular transition involves dramatic changes in gene expression within 6–8 h poststarvation and corresponds to the multicellular transition (Van Driessche et al., 2002). Several hundred mRNAs are dramatically downregulated during this transition and the authors speculate that these genes may function to repress the differentiated, multicellular state (Van Driessche et al., 2002).

Active mRNA clearance mechanisms likely control the unicellular to multicellular transition in *D. discoideum*. mRNA half-lives range from 50 min to 10 h in this species (Casey, Palnik, & Jacobson, 1983). Given that several hundred mRNAs are eliminated within 6–8 h after starvation, a subset of these mRNAs likely undergoes regulated mRNA destabilization during the unicellular to multicellular transition. Distinguishing between active and passive mechanisms of mRNA clearance requires comparing mRNA half-lives in the unicellular state versus the starvation-induced differentiated state in the presence of transcription inhibitors. Identification of molecular triggers of mRNA destabilization in such a simple system will reveal mechanisms regulating multicellular transitions and advance our understanding of how similar mechanisms may be involved in cellular transitions in metazoans.

4.2 Maternal mRNA Clearance Is Analogous to Reprogramming *in Vitro*

The MZT is in many ways analogous to cellular reprogramming during the induced pluripotency transition *in vitro* (Giraldez, 2010; Lee, Bonneau, et al., 2014) and summarized in Fig. 4. During the MZT, endogenous factors activate zygotic genes and maternal instructions are cleared to facilitate the transition of differentiated gametes to a totipotent state. *In vitro*, forced expression of reprogramming factors, induces the transition of differentiated somatic cells to adopt a pluripotent identity (Takahashi & Yamanaka, 2006;

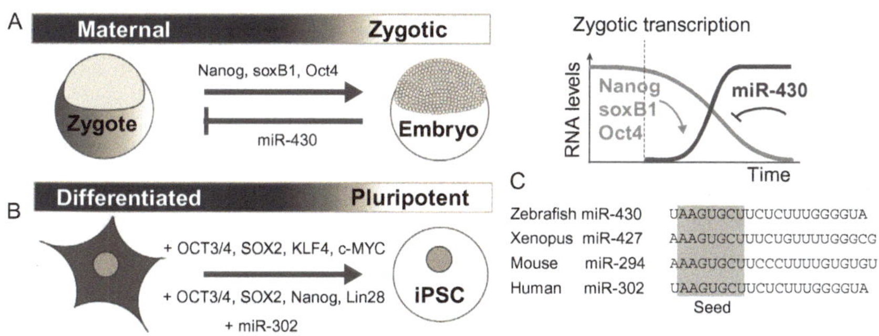

Figure 4 The MZT is analogous to *in vitro* pluripotency reprogramming. (A) In zebrafish, the pluripotency factors Nanog, SoxB1, and Oct4 activate zygotic gene transcription, including miR-430, which clears maternal mRNAs (Giraldez et al., 2006; Lee et al., 2013). Together, the activation of the zygotic genome and the clearance of maternal mRNAs facilitate oocyte reprogramming to the zygotic state. (B) Forced expression of pluripotency factors reprograms somatic cells to induced pluripotent cells (Takahashi & Yamanaka, 2006; Yu et al., 2007) and miR-302 (orthologous to miR-430) is sufficient for reprogramming (Anokye-Danso et al., 2011; Miyoshi et al., 2011). (C) miR-430/302/294 family microRNAs are highly conserved, share seed sequence, and are expressed in stem cells and early embryos (Houbaviy, Murray, & Sharp, 2003; Suh et al., 2004).

Yu et al., 2007). Below, we highlight recent studies that describe shared features of posttranscriptional regulation for pluripotent cells and embryos during the MZT.

4.3 microRNA Function in Reprogramming

Reprogramming during development and *in vitro* both exploit the highly conserved *miR430/290/302* family of microRNAs to erase the previous transcriptional landscape. Orthologs of these microRNAs are abundantly expressed in early embryogenesis in zebrafish (Giraldez et al., 2005), *Xenopus* (Lund et al., 2009), and in mammalian stem cells and embryos (Houbaviy et al., 2003; Suh et al., 2004), suggesting that they influence early developmental events. In the context of iPSC reprogramming, the addition of *miR-302/294* together with Oct4, Sox2, and Klf4 increases fibroblast reprogramming efficiency by 10-fold in mouse (Judson, Babiarz, Venere, & Blelloch, 2009) and in human fibroblasts (Subramanyam et al., 2011), implicating this microRNA as a core component of the pluripotency network. In fact, expression of the *miR302/367* cluster alone appears sufficient to reprogram mouse and human fibroblasts to iPSCs two orders of magnitude more efficiently than the OSKM cocktail (Anokye-Danso et al., 2011; Lin et al., 2011; Miyoshi et al., 2011).

What makes this microRNA family such a potent reprogramming factor? *miR-430/302/294* family members rescue the Dgcr8 mouse knockout ES cell proliferation defect through downregulation of several G1/S transition regulators (Wang et al., 2008), implicating a role in cell proliferation for reprogramming regulation. However, in addition to proliferation these reprogramming microRNAs also regulate apoptosis, chromatin remodelers, and the mesenchymal to epithelial transition (MET) (Anokye-Danso, Snitow, & Morrisey, 2012). Additionally, *miR-181* family microRNAs enhances OSK-mediated fibroblast reprogramming efficiency by three fold in mouse (Judson, Greve, Parchem, & Blelloch, 2013). Interestingly, no synergistic increase in reprogramming was observed for the combination of *miR-294* and *miR-181*, suggesting that these microRNA converge on common pathways downstream of their direct targets (Judson et al., 2013). Likewise *miR-430* targets several hundred different mRNAs for clearance during development (Giraldez et al., 2006). The diversity and the scope of regulation exerted by these microRNA families suggest that its function may be in erasing the preexisting transcriptional landscape as an instructive strategy to facilitate the installation of the pluripotency program.

4.4 Pumilio Function in Stem Cell Maintenance

Across metazoans species, Pumilio has a role in repressing differentiation. In *Drosophila pumilio* mutants, germline stem cells fail to undergo asymmetric divisions and consequently differentiate into egg chambers, indicating a failure to maintain the self-renewal potential in the germline stem cells (Forbes & Lehmann, 1998; Lin & Spradling, 1997). Likewise, genetic inactivation of both Pumilio homologs, *fbf-1* and *fbf-2*, in *C. elegans* leads to failure in germline maintenance, manifested in the adult germline consisting exclusively of sperm (Crittenden et al., 2002). Human Pumilio2 is expressed in embryonic stem cells, in ovary and testis, while Pumilio1 is expressed ubiquitously (Moore et al., 2003), suggesting a conserved function for Pumilio2 in stem cell maintenance. Drawing on these examples, and on function of Pumilio in posttranscriptional regulation of maternal mRNA in *Drosophila* (discussed above), this multipurpose protein could play a role in clearing the oocyte's transcriptional history to enable a transient, pluripotent state.

4.5 RNA Modifications as Markers of Decay in Stem Cells

RNA modification has recently been implicated in mediating mRNA turnover in embryonic stem cells and for maintaining the pluripotent identify. N^6-methyl-adenosine (m^6A) modification in RNA is a substrate for YTHDF proteins (Dominissini et al., 2012; Wang, Lu, et al., 2014), which direct mRNAs to processing bodies (P-bodies) (Wang, Lu, et al., 2014) and directly interact with Pop2 in the Pop2–Ccr4–Not1-deadenylase complex to direct mRNA deadenylation and destabilization (Kang et al., 2014). Global mapping of m^6A modification in mouse and human ES cells showed that thousands of mRNAs, including components of the pluripotency network, and ncRNAs are modified with m^6A and that this correlates with mRNA instability (Batista et al., 2014; Geula et al., 2015). Genetic inactivation of the methylation "writer," Mettl3, resulted in a global decrease in m^6A, stabilization of pluripotency factors such as Nanog, and prevented exit from self-renewal (Batista et al., 2014). Likewise, Mettl3 mouse knockout ES cells have prolonged expression of pluripotency factors and fail to undergo proper lineage priming, which consequently leads to embryonic lethality (Geula et al., 2015). RNA modification provides a potentially effective way to mark maternal mRNA to facilitate selective destabilization of these transcripts at MZT. In some cases, m^6A modification even alters the binding affinity of RNA-binding proteins (Liu et al., 2015), providing a dynamic strategy to regulate large numbers of transcripts.

4.6 Poly(C) Destabilization Motif in Stem Cells and Embryos

Embryos and pluripotent stem cells share unique features of posttranscriptional regulatory mechanisms that are distinct from somatic cells. Global analysis of mRNA decay rates reveals that posttranscriptional regulation is different between differentiated mouse embryonic fibroblasts (MEFs) and induced pluripotent stem cells (iPSCs) derived from these MEFs. Transcripts bearing 3′UTR C-rich sequence elements, many of which encode transcription factors, are significantly less stable in iPS cells than in MEFs. Intriguingly, two poly(C)-binding proteins that recognize this type of element are reciprocally expressed in iPS and HFF cells (Neff, Lee, Wilusz, Tian, & Wilusz, 2012), suggesting that poly(C)-binding proteins direct mRNA instability in iPSC. Interestingly, the poly(C) motif is also enriched in unstable maternal mRNAs during the oocyte-to-embryo transition (OET) in *C. elegans* (Stoeckius et al., 2014). However, depletion of all poly(C)-binding proteins in this species does not globally affect the stability of mRNAs harboring the poly(C) motif (Stoeckius et al., 2014), indicating that this mRNA destabilization mechanism requires additional component(s). Identification of these additional components will likely also elucidate the unique features of posttranscriptional mRNA regulation in pluripotent cells that permit selective destabilization of mRNAs in pluripotent stem cells, but not in differentiated somatic cells. This further suggests that posttranscriptional mRNA regulatory machinery could be leveraged to improve reprogramming efficiency *in vitro* by pushing somatic cells toward pluripotency.

5. CONCLUDING REMARKS

In this chapter, we have described the current understanding of the molecular mechanisms regulating maternal mRNA clearance across species. Through some of these examples we describe the parallels that exist between cellular and developmental reprograming, including microRNA regulation (*miR-430/302/295*), RNA-binding proteins (Pumilio and PCBP), and even transcription factors such as Nanog, Oct4, Sox2 (Fig. 4). Based on the key parallels between cellular and developmental reprograming, we postulate that many of these mechanisms might be shared during the maternal-to-zygotic transition to clear the previous maternal program and to help establish the new zygotic program of development.

The advent of new methods to measure RNA and protein modifications genome-wide will allow us to investigate their role in the regulation of maternal mRNAs and the modulation of RNA-binding affinity. Likewise, methods to determine RNA structure, dynamics of poly(A) tail length, translational regulation, and RNA-binding protein activity using high-throughput methods is opening exciting avenues to identify the code that mediates the posttranscriptional regulation of maternal instructions during the MZT. Finally, translational activation following fertilization could modify mRNA half-life based on codon usage (Presnyak et al., 2015) to regulate mRNA stability during the MZT. Future studies in these areas are likely to provide not only the key mechanisms regulating gene expression during cellular transitions but also during reprogramming and cell differentiation.

While all animals undergo the MZT, there is a surprising degree of variation in the targets that undergo mRNA clearance during development. Variation in targets across different species likely reflects different reproductive strategies and developmental requirements. Species employ a diverse, yet conserved set of RNA-binding proteins and noncoding RNAs to regulate maternal mRNAs during the MZT. Several questions remain to be answered. What mechanisms activate the different pathways of mRNA clearance and translational regulation? What instructive roles does maternal clearance play in development? To what extent does translational and posttranslational control contribute to regulating each step?

We also lack an understanding of how different decay pathways interact to facilitate mRNA clearance in a coordinated and specific manner. We have seen examples of mRNAs under the regulation of cooperative and redundant mechanisms (Fig. 3). Therefore, mRNA fate can be calculated according to the combined action of several stabilizing and destabilizing factors acting on the same transcript. Understanding how this regulatory code is organized and interpreted to allow for the coordinated and specific elimination of hundreds of mRNAs, remain fundamental challenges for the field.

Change is a key constant in development. Multicellular life continually uses cellular transitions to generate tissue complexity during development and regeneration in adulthood. The MZT is a dramatic example of a cellular transition of a differentiated oocyte toward a transient totipotent embryo. Across animals, hundreds of mRNAs are regulated in a coordinated manner during the MZT, making this an ideal context to identify posttranscriptional mechanisms that shape gene expression.

ACKNOWLEDGMENTS

We thank Dr. Miler T. Lee, Dr. Carter M. Takacs, and Dr. Howard D. Lipshitz for critical reading of the manuscript and all members of the Giraldez Laboratory for discussion. V.Y. has been supported by the NIH Training Grants T32 GM007223 and T32 HD007149, the Edward L. Tatum Fellowship (Yale University), and the Yale Medical Research Scholars Program. The Giraldez Lab is supported by National Institutes of Health grants R01-GM081602-07, R01-GM101108-4, R01-GM102251-4, R01-GM103789-3, R01-HD074078-4, and R21-HD073768-2.

REFERENCES

Aanes, H., Winata, C. L., Lin, C. H., Chen, J. P., Srinivasan, K. G., Lee, S. G. P., et al. (2011). Zebrafish mRNA sequencing deciphers novelties in transcriptome dynamics during maternal to zygotic transition. *Genome Research, 21*(8), 1328–1338.

Alvarez-Saavedra, E., & Horvitz, H. R. (2010). MicroRNAs are not essential for development or viability. *Current Biology, 20*(4), 367–373.

Anokye-Danso, F., Snitow, M., & Morrisey, E. E. (2012). How microRNAs facilitate reprogramming to pluripotency. *Journal of Cell Science, 125*, 1–9.

Anokye-Danso, F., Trivedi, C. M., Juhr, D., Gupta, M., Cui, Z., Tian, Y., et al. (2011). Highly efficient miRNA-mediated reprogramming of mouse and human somatic cells to pluripotency. *Cell Stem Cell, 8*(4), 376–388.

Apostolou, E., & Hochedlinger, K. (2013). Chromatin dynamics during cellular reprogramming. *Nature, 502*(7472), 462–471.

Arribas-Layton, M., Wu, D., Lykke-Andersen, J., & Song, H. (2013). Structural and functional control of the eukaryotic mRNA decapping machinery. *Biochimica et Biophysica Acta, 1829*(6–7), 580–589.

Audic, Y., Omilli, F., & Osborne, H. B. (1997). Postfertilization deadenylation of mRNAs in Xenopus laevis embryos is sufficient to cause their degradation at the blastula stage. *Molecular and Cellular Biology, 17*(1), 209–218.

Audic, Y., Omilli, F., & Osborne, H. B. (1998). Embryo deadenylation element-dependent deadenylation is enhanced by a cis element containing AUU repeats. *Molecular and Cellular Biology, 18*(12), 6879–6884.

Aviv, T., Lin, Z., Lau, S., Rendl, L. M., Sicheri, F., & Smibert, C. A. (2003). The RNA-binding SAM domain of Smaug defines a new family of post-transcriptional regulators. *Nature Structural Biology, 10*(8), 614–621.

Barckmann, B., & Simonelig, M. (2013). Control of maternal mRNA stability in germ cells and early embryos. *Biochimica et Biophysica Acta, 1829*(6–7), 714–724.

Baroux, C., Autran, D., Gillmor, C. S., Grimanelli, D., & Grossniklaus, U. (2008). The maternal to zygotic transition in animals and plants. *Cold Spring Harbor Symposia on Quantitative Biology, 73*, 89–100.

Bartel, D. P. (2009). MicroRNAs: Target recognition and regulatory functions. *Cell, 136*(2), 215–233.

Bashirullah, A., Halsell, S. R., Cooperstock, R. L., Kloc, M., Karaiskakis, A., Fisher, W. W., et al. (1999). Joint action of two RNA degradation pathways controls the timing of maternal transcript elimination at the midblastula transition in Drosophila melanogaster. *EMBO Journal, 18*(9), 2610–2620.

Batista, P. J., Molinie, B., Wang, J., Qu, K., Zhang, J., Li, L., et al. (2014). m^6A RNA modification controls cell fate transition in mammalian embryonic stem cells. *Cell Stem Cell, 15*(6), 707–719.

Baugh, L. R., Hill, A. A., Slonim, D. K., Brown, E. L., & Hunter, C. P. (2003). Composition and dynamics of the Caenorhabditis elegans early embryonic transcriptome. *Development, 130*(5), 889–900.

Bazzini, A. A., Lee, M. T., & Giraldez, A. J. (2012). Ribosome profiling shows that miR-430 reduces translation before causing mRNA decay in zebrafish. *Science, 336*(6078), 233–237.

Beelman, C. A., & Parker, R. (1995). Degradation of mRNA in eukaryotes. *Cell, 81*, 179–183.

Beilharz, T. H., & Preiss, T. (2007). Widespread use of poly (A) tail length control to accentuate expression of the yeast transcriptome. *RNA, 13*(7), 982–997.

Benoit, B., He, C. H., Zhang, F., Votruba, S. M., Tadros, W., Westwood, J. T., et al. (2009). An essential role for the RNA-binding protein Smaug during the Drosophila maternal-to-zygotic transition. *Development, 136*(6), 923–932.

Bensaude, O. (2011). Inhibiting eukaryotic transcription. *Transcription, 2*(3), 103–108.

Bernstein, P., Peltz, S. W., & Ross, J. (1989). The poly(A)-poly(A)-binding protein complex is a major determinant of mRNA stability in vitro. *Molecular and Cellular Biology, 9*(2), 659–670.

Braude, P., Bolton, V., & Moore, S. (1988). Human gene expression first occurs between the four- and eight-cell stages of preimplantation development. *Nature, 332*(6163), 459–461.

Brent, A. E., MacQueen, A., & Hazelrigg, T. (2000). The Drosophila wispy gene is required for RNA localization and other microtubule-based events of meiosis and early embryogenesis. *Genetics, 154*(4), 1649–1662.

Brown, B. D., & Harland, R. M. (1990). Endonucleolytic cleavage of a maternal homeo box mRNA in Xenopus oocytes. *Genes and Development, 4*(11), 1925–1935.

Brown, B. D., Zipkin, I. D., & Harland, R. M. (1993). Sequence-specific endonucleolytic cleavage and protection of mRNA in Xenopus and Drosophila. *Genes and Development, 7*(8), 1620–1631.

Bushati, N., Stark, A., Brennecke, J., & Cohen, S. M. (2008). Temporal reciprocity of miRNAs and their targets during the maternal-to-zygotic transition in drosophila. *Current Biology, 18*(7), 501–506.

Camous, S., Kopecny, V., & Flechon, J. E. (1986). Autoradiographic detection of the earliest stage of [3H]-uridine incorporation into the cow embryo. *Biology of the Cell, 58*, 195–200.

Casey, L., Palnik, C. M., & Jacobson, A. (1983). RNA half-life in *Dictyostelium discoideum*. *Developmental Biology, 95*, 239–243.

Chang, C. T., Bercovich, N., Loh, B., Jonas, S., & Izaurralde, E. (2014). The activation of the decapping enzyme DCP2 by DCP1 occurs on the EDC4 scaffold and involves a conserved loop in DCP1. *Nucleic Acids Research, 42*(8), 5217–5233.

Chang, H., Lim, J., Ha, M., & Kim, V. N. (2014). TAIL-seq: Genome-wide determination of poly(A) tail length and 3′ end modifications. *Molecular Cell, 53*(6), 1044–1052.

Charlesworth, A., Meijer, H. A., & De Moor, C. H. (2013). Specificity factors in cytoplasmic polyadenylation. *Wiley Interdisciplinary Reviews: RNA, 4*(4), 437–461.

Chen, L., Dumelie, J. G., Li, X., Cheng, M. H., Yang, Z., Laver, J. D., et al. (2014). Global regulation of mRNA translation and stability in the early Drosophila embryo by the Smaug RNA-binding protein. *Genome Biology, 15*(1), R4.

Choi, W.-Y., Giraldez, A. J., & Schier, A. F. (2007). Target protectors reveal dampening and balancing of nodal agonist and antagonist by miR-430. *Science, 318*(5848), 271–274.

Clarke, M., & Gomer, R. H. (1995). PSF and CMF, autocrine factors that regulate gene expression during growth and early development of Dictyostelium. *Experientia, 51*(12), 1124–1134.

Colegrove-Otero, L. J., Minshall, N., & Standart, N. (2005). RNA-binding proteins in early development. *Critical Reviews in Biochemistry and Molecular Biology, 40*, 21–73.

Cosson, B., Gautier-Courteille, C., Maniey, D., Aït-Ahmed, O., Lesimple, M., Osborne, H. B., et al. (2006). Oligomerization of EDEN-BP is required for specific mRNA deadenylation and binding. *Biology of the Cell, 98*(11), 653–665.

Cowling, V. H. (2010). Regulation of mRNA cap methylation. *The Biochemical Journal, 425*(2), 295–302.

Crittenden, S. L., Bernstein, D. S., Bachorik, J. L., Thompson, B. E., Gallegos, M., Petcherski, A. G., et al. (2002). A conserved RNA-binding protein controls germline stem cells in Caenorhabditis elegans. *Nature, 417*(6889), 660–663.

Dahanukar, A., Walker, J. A., & Wharton, R. P. (1999). Smaug, a novel RNA-binding protein that operates a translational switch in Drosophila. *Molecular Cell, 4*(2), 209–218.

Decker, C. J., & Parker, R. (1994). Mechanisms of mRNA degradation in eukaryotes. *Trends in Biochemical Sciences, 19*(8), 336–340.

De Renzis, S., Elemento, O., Tavazoie, S., & Wieschaus, E. F. (2007). Unmasking activation of the zygotic genome using chromosomal deletions in the Drosophila embryo. *PLoS Biology, 5*(5), 1036–1051.

DeRenzo, C., & Seydoux, G. (2004). A clean start: Degradation of maternal proteins at the oocyte-to-embryo transition. *Trends in Cell Biology, 14*(8), 420–426.

Detivaud, L., Pascreau, G., Karaiskou, A., Osborne, H. B., & Kubiak, J. Z. (2003). Regulation of EDEN-dependent deadenylation of Aurora A/Eg2-derived mRNA via phosphorylation and dephosphorylation in Xenopus laevis egg extracts. *Journal of Cell Science, 116*(Pt. 13), 2697–2705.

Ding, Y., Tang, Y., Kwok, C. K., Zhang, Y., Bevilacqua, P. C., & Assmann, S. M. (2014). In vivo genome-wide profiling of RNA secondary structure reveals novel regulatory features. *Nature, 505*(7485), 696–700.

Dobson, A. T., Raja, R., Abeyta, M. J., Taylor, T., Shen, S., Haqq, C., et al. (2004). The unique transcriptome through day 3 of human preimplantation development. *Human Molecular Genetics, 13*(14), 1461–1470.

Dominissini, D., Moshitch-Moshkovitz, S., Schwartz, S., Salmon-Divon, M., Ungar, L., Osenberg, S., et al. (2012). Topology of the human and mouse m^6A RNA methylomes revealed by m^6A-seq. *Nature, 485*(7397), 201–206.

Du, L., & Richter, J. D. (2005). Activity-dependent polyadenylation in neurons. *RNA, 11*(9), 1340–1347.

Edgar, B., Kiehle, C., & Schubiger, G. (1986). Cell cycle control by the nucleo-cytoplasmic ratio in early Drosophila development. *Cell, 44*(2), 365–372.

Edgar, L. G., Wolf, N., & Wood, W. B. (1994). Early transcription in Caenorhabditis elegans embryos. *Development, 120*(2), 443–451.

Ezzeddine, N., Paillard, L., Capri, M., Maniey, D., Bassez, T., Ait-Ahmed, O., et al. (2002). EDEN-dependent translational repression of maternal mRNAs is conserved between Xenopus and Drosophila. *Proceedings of the National Academy of Sciences of the United States of America, 99*(1), 257–262.

Ferg, M., Sanges, R., Gehrig, J., Kiss, J., Bauer, M., Lovas, A., et al. (2007). The TATA-binding protein regulates maternal mRNA degradation and differential zygotic transcription in zebrafish. *The EMBO Journal, 26*(17), 3945–3956.

Flach, G., Hjohnson, M., Braude, P. R., Taylor, R. A. S., & Bolton, V. N. (1982). The transition from maternal to embryonic control in the 2-cell mouse embryo. *EMBO Journal, 1*(6), 681–686.

Forbes, A., & Lehmann, R. (1998). Nanos and Pumilio have critical roles in the development and function of Drosophila germline stem cells. *Development, 125*(4), 679–690.

Franks, T. M., & Lykke-Andersen, J. (2009). The control of mRNA decapping and P-body formation. *Molecular Cell, 32*(5), 605–615.

Fu, Y., Dominissini, D., Rechavi, G., & He, C. (2014). Gene expression regulation mediated through reversible m^6A RNA methylation. *Nature Reviews. Genetics, 15*(5), 293–306.

Gautier-Courteille, C., Le Clainche, C., Barreau, C., Audic, Y., Graindorge, A., Maniey, D., et al. (2004). EDEN-BP-dependent post-transcriptional regulation of gene expression in Xenopus somitic segmentation. *Development*, *131*(24), 6107–6117.

Gebauer, F., Preiss, T., & Hentze, M. W. (2012). From cis-regulatory elements to complex RNPs and back. *Cold Spring Harbor Perspectives in Biology*, *4*(7), 1–14.

Geisberg, J. V., Moqtaderi, Z., Fan, X., Ozsolak, F., & Struhl, K. (2014). Global analysis of mRNA isoform half-lives reveals stabilizing and destabilizing elements in yeast. *Cell*, *156*(4), 812–824.

Gerber, A. P., Luschnig, S., Krasnow, M. A., Brown, P. O., & Herschlag, D. (2006). Genome-wide identification of mRNAs associated with the translational regulator PUMILIO in Drosophila melanogaster. *Proceedings of the National Academy of Sciences of the United States of America*, *103*(12), 4487–4492.

Geula, S., Moshitch-Moshkovitz, S., Dominissini, D., Mansour, A. A., Kol, N., Salmon-Divon, M., et al. (2015). m^6A mRNA methylation facilitates resolution of naïve pluripotency toward differentiation. *Science*, *347*(6225), 1002–1006.

Gillian-Daniel, D. L., Gray, N. K., Aström, J., Barkoff, A., & Wickens, M. (1998). Modifications of the 5′ cap of mRNAs during Xenopus oocyte maturation: Independence from changes in poly(A) length and impact on translation. *Molecular and Cellular Biology*, *18*(10), 6152–6163.

Giraldez, A. J. (2010). MicroRNAs, the cell's Nepenthe: Clearing the past during the maternal-to-zygotic transition and cellular reprogramming. *Current Opinion in Genetics and Development*, *20*(4), 369–375.

Giraldez, A. J., Cinalli, R. M., Glasner, M. E., Enright, A. J., Thomson, J. M., Baskerville, S., et al. (2005). MicroRNAs regulate brain morphogenesis in zebrafish. *Science*, *308*(5723), 833–838.

Giraldez, A. J., Mishima, Y., Rihel, J., Grocock, R. J., Van Dongen, S., Inoue, K., et al. (2006). Zebrafish MiR-430 promotes deadenylation and clearance of maternal mRNAs. *Science*, *312*(5770), 75–79.

Golbus, M. S., Calarco, P. G., & Epstein, C. J. (1973). The effects of inhibitors of RNA synthesis (alpha-amanitin and actinomycin D) on preimplantation mouse embryogenesis. *The Journal of Experimental Zoology*, *186*(2), 207–216.

Goodarzi, H., Najafabadi, H. S., Oikonomou, P., Greco, T. M., Fish, L., Salavati, R., et al. (2012). Systematic discovery of structural elements governing stability of mammalian messenger RNAs. *Nature*, *485*(7397), 264–268.

Graindorge, A., Le Tonquèze, O., Thuret, R., Pollet, N., Osborne, H. B., & Audic, Y. (2008). Identification of CUG-BP1/EDEN-BP target mRNAs in Xenopus tropicalis. *Nucleic Acids Research*, *36*(6), 1861–1870.

Gurdon, J. (1962). The developmental capacity of nuclei taken from intestinal epithelium cells of feeding tadpoles. *Journal of Embryology and Experimental Morphology*, *10*, 622–640.

Hamatani, T., Carter, M. G., Sharov, A. A., & Ko, M. S. H. (2004). Dynamics of global gene expression changes during mouse preimplantation development. *Developmental Cell*, *6*(1), 117–131.

Harvey, S. A., Sealy, I., Kettleborough, R., Fenyes, F., White, R., Stemple, D., et al. (2013). Identification of the zebrafish maternal and paternal transcriptomes. *Development*, *140*(13), 2703–2710.

Heyn, P., Kircher, M., Dahl, A., Kelso, J., Tomancak, P., Kalinka, A. T., et al. (2014). The earliest transcribed zygotic genes are short, newly evolved, and different across species. *Cell Reports*, *6*(2), 285–292.

Hoffert, K. A., Anderson, G. B., Wildt, D. E., & Roth, T. L. (1997). Transition from maternal to embryonic control of development in IVM/IVF domestic cat embryos. *Molecular Reproduction and Development*, *48*(2), 208–215.

Horner, K., Livera, G., Hinckley, M., Trinh, K., Storm, D., & Conti, M. (2003). Rodent oocytes express an active adenylyl cyclase required for meiotic arrest. *Developmental Biology, 258*(2), 385–396.

Houbaviy, H. B., Murray, M. F., & Sharp, P. A. (2003). Embryonic stem cell-specific MicroRNAs. *Developmental Biology, 5*(2), 351–358.

Hu, W., Sweet, T. J., Chamnongpol, S., Baker, K. E., & Coller, J. (2009). Co-translational mRNA decay in Saccharomyces cerevisiae. *Nature, 461*(7261), 225–229.

Ingolia, N. T., Ghaemmaghami, S., Newman, J. R. S., & Weissman, J. S. (2009). Genome-wide analysis in vivo of translation with nucleotide resolution using ribosome profiling. *Science, 324*(5924), 218–223.

Jeske, M., Moritz, B., Anders, A., & Wahle, E. (2011). Smaug assembles an ATP-dependent stable complex repressing nanos mRNA translation at multiple levels. *The EMBO Journal, 30*(1), 90–103.

Jonas, S., & Izaurralde, E. (2013). The role of disordered protein regions in the assembly of decapping complexes and RNP granules. *Genes and Development, 27*(24), 2628–2641.

Judson, R. L., Babiarz, J. E., Venere, M., & Blelloch, R. (2009). Embryonic stem cell-specific microRNAs promote induced pluripotency. *Nature Biotechnology, 27*(5), 459–461.

Judson, R. L., Greve, T. S., Parchem, R. J., & Blelloch, R. (2013). MicroRNA-based discovery of barriers to dedifferentiation of fibroblasts to pluripotent stem cells. *Nature Structural & Molecular Biology, 20*(10), 1227–1235.

Kane, D. A., Hammerschmidt, M., Mullins, M. C., Maischein, H. M., Brand, M., van Eeden, F. J., et al. (1996). The zebrafish epiboly mutants. *Development, 123*, 47–55.

Kang, H.-J., Jeong, S.-J., Kim, K.-N., Baek, I.-J., Chang, M., Kang, C.-M., et al. (2014). A novel protein, Pho92, has a conserved YTH domain and regulates phosphate metabolism by decreasing the mRNA stability of PHO4 in Saccharomyces cerevisiae. *The Biochemical Journal, 457*(3), 391–400.

Kedde, M., Strasser, M. J., Boldajipour, B., Oude Vrielink, J. A. F., Slanchev, K., le Sage, C., et al. (2007). RNA-binding protein Dnd1 inhibits microRNA access to target mRNA. *Cell, 131*(7), 1273–1286.

Kessin, R. H. (2001). *Dictyostelium: Evolution, cell biology, and the development of multicellularity.* Cambridge, UK: Cambridge University Press.

King, I. N., Yartseva, V., Salas, D., Kumar, A., Heidersbach, A., Ando, D. M., et al. (2014). The RNA-binding protein TDP-43 selectively disrupts microRNA-1/206 incorporation into the RNA-induced silencing complex. *Journal of Biological Chemistry, 289*(20), 14263–14271.

Kosubek, A., Klein-Hitpass, L., Rademacher, K., Horsthemke, B., & Ryffel, G. U. (2010). Aging of Xenopus tropicalis eggs leads to deadenylation of a specific set of maternal mRNAs and loss of developmental potential. *PLoS One, 5*(10), e13532.

Lai, W. S., Carballo, E., Thorn, J. M., Kennington, E. A., & Blackshear, P. J. (2000). Interactions of CCCH zinc finger proteins with mRNA. Binding of tristetraprolin-related zinc finger proteins to AU-rich elements and destabilization of mRNA. *Journal of Biological Chemistry, 275*(23), 17827–17837.

Lall, S., Piano, F., & Davis, R. E. (2005). Caenorhabditis elegans decapping proteins: Localization and functional analysis of Dcp1, Dcp2, and DcpS during embryogenesis. *Molecular Biology of the Cell, 16*, 1018–1032.

Langley, A. R., Smith, J. C., Stemple, D. L., & Harvey, S. A. (2014). New insights into the maternal to zygotic transition. *Development, 141*(20), 3834–3841.

Lanza, R. P., Cibelli, J. B., Diaz, F., Moraes, C. T., Farin, P. W., Farin, C. E., et al. (2000). Cloning of an endangered species (Bos gaurus) using interspecies nuclear transfer. *Cloning, 2*(2), 79–90.

Laubichler, M. D., & Davidson, E. H. (2008). Establishment of the role of nuclear chromosomes in development. *Developmental Biology, 314*(1), 1–11.

Laver, J. D., Li, X., Ray, D., Cook, K. B., Hahn, N. A., Nabeel-Shah, S., et al. (2015). Brain tumor is a sequence-specific RNA-binding protein that directs maternal mRNA clearance during the Drosophila maternal-to-zygotic transition. *Genome Biology*, *16*, 94.

Lee, M. T., Bonneau, A. R., & Giraldez, A. J. (2014). Zygotic genome activation during the maternal-to-zygotic transition. *Annual Review of Cell and Developmental Biology*, *30*(1), 581–613.

Lee, M. T., Bonneau, A. R., Takacs, C. M., Bazzini, A. A., DiVito, K. R., Fleming, E. S., et al. (2013). Nanog, Pou5f1 and SoxB1 activate zygotic gene expression during the maternal-to-zygotic transition. *Nature*, *503*(7476), 360–364.

Lee, M., Choi, Y., Kim, K., Jin, H., Lim, J., Nguyen, T. A., et al. (2014). Adenylation of maternally inherited MicroRNAs by Wispy. *Molecular Cell*, *56*(5), 696–707.

Lin, H., & Spradling, A. C. (1997). A novel group of pumilio mutations affects the asymmetric division of germline stem cells in the Drosophila ovary. *Development*, *124*(12), 2463–2476.

Lin, S. L., Chang, D. C., Lin, C. H., Ying, S. Y., Leu, D., & Wu, D. T. S. (2011). Regulation of somatic cell reprogramming through inducible mir-302 expression. *Nucleic Acids Research*, *39*(3), 1054–1065.

Lindell, T. J., Weinberg, F., Morris, P. W., Roeder, R. G., & Rutter, W. J. (1970). Specific inhibition of nuclear RNApolymerase II by alpha-amanitin. *Science*, *170*, 447–449.

Ling, S. H. M., Qamra, R., & Song, H. (2011). Structural and functional insights into eukaryotic mRNA decapping. *Wiley Interdisciplinary Reviews: RNA*, *2*(2), 193–208.

Liu, N., Dai, Q., Zheng, G., He, C., Parisien, M., & Pan, T. (2015). N6-methyladenosine-dependent RNA structural switches regulate RNA–protein interactions. *Nature*, *518*(7540), 560–564.

Liu, H., & Kiledjian, M. (2006). Decapping the message: A beginning or an end. *Biochemical Society Transactions*, *34*, 35–38.

Lovén, J., Orlando, D. A., Sigova, A. A., Lin, C. Y., Rahl, P. B., Burge, C. B., et al. (2012). Revisiting global gene expression analysis. *Cell*, *151*(3), 476–482.

Lund, E., Liu, M., Hartley, R. S., Sheets, M. D., & Dahlberg, J. E. (2009). Deadenylation of maternal mRNAs mediated by miR-427 in Xenopus laevis embryos. *RNA*, *15*(12), 2351–2363.

Ma, J., Flemr, M., Stein, P., Berninger, P., Malik, R., Zavolan, M., et al. (2010). microRNA activity is suppressed in mouse oocytes. *Current Biology*, *20*(3), 265–270.

Magner, L. N. (2002). *A history of the life sciences* (3rd ed.). New York, NY: Marcel Dekker.

Maienschein, J. (2012). Epigenesis and preformationism. In E. N. Zalta (Ed.), *The Stanford encyclopedia of philosophy* (Spring 2012 edition). Retrieved from http://plato.stanford.edu/archives/spr2012/entries/epigenesis/

Manes, C. (1973). The participation of the embryonic genome during early cleavage in the rabbit. *Developmental Biology*, *32*(2), 453–459.

Mathavan, S., Lee, S. G. P., Mak, A., Miller, L. D., Murthy, K. R. K., Govindarajan, K. R., et al. (2005). Transcriptome analysis of zebrafish embryogenesis using microarrays. *PLoS Genetics*, *1*(2), 0260–0276.

Medeiros, L. A., Dennis, L. M., Gill, M. E., Houbaviy, H., Markoulaki, S., Fu, D., et al. (2011). Mir-290-295 deficiency in mice results in partially penetrant embryonic lethality and germ cell defects. *Proceedings of the National Academy of Sciences of the United States of America*, *108*(34), 14163–14168.

Medioni, C., Mowry, K., & Besse, F. (2012). Principles and roles of mRNA localization in animal development. *Development*, *139*(18), 3263–3276.

Meijer, H. A., Bushell, M., Hill, K., Gant, T. W., Willis, A. E., Jones, P., et al. (2007). A novel method for poly(A) fractionation reveals a large population of mRNAs with a short poly(A) tail in mammalian cells. *Nucleic Acids Research*, *35*, e132.

Mishima, Y., Giraldez, A. J., Takeda, Y., Fujiwara, T., Sakamoto, H., Schier, A. F., et al. (2006). Differential regulation of germline mRNAs in soma and germ cells by zebrafish miR-430. *Current Biology, 16*(22), 2135–2142.

Miyoshi, N., Ishii, H., Nagano, H., Haraguchi, N., Dewi, D. L., Kano, Y., et al. (2011). Reprogramming of mouse and human cells to pluripotency using mature microRNAs. *Cell Stem Cell, 8*(6), 633–638.

Moore, F. L., Jaruzelska, J., Fox, M. S., Urano, J., Firpo, M. T., Turek, P. J., et al. (2003). Human Pumilio-2 is expressed in embryonic stem cells and germ cells and interacts with DAZ (deleted in AZoospermia) and DAZ-like proteins. *Proceedings of the National Academy of Sciences of the United States of America, 100*(2), 538–543.

Moraes, K. C. M., Wilusz, C. J., & Wilusz, J. (2006). CUG-BP binds to RNA substrates and recruits PARN deadenylase. *RNA, 12*(6), 1084–1091.

Murchison, E. P., Stein, P., Xuan, Z., Pan, H., Zhang, M. Q., Schultz, R. M., et al. (2007). Critical roles for Dicer in the female germline. *Genes and Development, 21*(6), 682–693.

Murray, E. L., & Schoenberg, D. R. (2008). Assays for determining poly(a) tail length and the polarity of mRNA decay in mammalian cells. *Methods in enzymology, 488*, 483–504.

Murthy, K. G., Park, P., & Manley, J. L. (1991). A nuclear micrococcal-sensitive, ATP-dependent exoribonuclease degrades uncapped but not capped RNA substrates. *Nucleic Acids Research, 19*(10), 2685–2692.

Nakamura, A., Sato, K., & Hanyu-Nakamura, K. (2004). Drosophila cup is an eIF4E binding protein that associates with Bruno and regulates oskar mRNA translation in oogenesis. *Developmental Cell, 6*(1), 69–78.

Neff, A. T., Lee, J. Y., Wilusz, J., Tian, B., & Wilusz, C. J. (2012). Global analysis reveals multiple pathways for unique regulation of mRNA decay in induced pluripotent stem cells. *Genome Research, 22*(8), 1457–1467.

Nelson, M. R., Leidal, A. M., & Smibert, C. A. (2004). Drosophila cup is an eIF4E-binding protein that functions in Smaug-mediated translational repression. *The EMBO Journal, 23*(1), 150–159.

Newport, J., & Kirschner, M. (1982). A major developmental transition in early Xenopus embryos: I. Characterization and timing of cellular changes at the midblastula stage. *Cell, 30*(3), 675–686.

Oikonomou, P., Goodarzi, H., & Tavazoie, S. (2014). Systematic identification of regulatory elements in conserved 3′ UTRs of human transcripts. *Cell Reports*, 7(1), 281–292.

Paillard, L., Omilli, F., Legagneux, V., Bassez, T., Maniey, D., & Osborne, H. B. (1998). EDEN and EDEN-BP, a cis element and an associated factor that mediate sequence-specific mRNA deadenylation in Xenopus embryos. *EMBO Journal, 17*(1), 278–287.

Parchem, R. J., Moore, N., Fish, J. L., Parchem, J. G., Braga, T. T., Shenoy, A., et al. (2015). miR-302 Is required for timing of neural differentiation, neural tube closure, and embryonic viability. *Cell Reports*. Advance online publication. http://dx.doi.org/10.1016/j.celrep.2015.06.074.

Parker, R., & Song, H. (2004). The enzymes and control of eukaryotic mRNA turnover. *Nature Structural & Molecular Biology, 11*(2), 121–127.

Pelechano, V., Wei, W., & Steinmetz, L. M. (2015). Widespread co-translational RNA decay reveals ribosome dynamics. *Cell, 161*(6), 1400–1412.

Pinder, B. D., & Smibert, C. A. (2013). microRNA-independent recruitment of Argonaute 1 to nanos mRNA through the Smaug RNA-binding protein. *EMBO Reports, 14*(1), 80–86.

Presnyak, V., Alhusaini, N., Chen, Y., Martin, S., Morris, N., Kline, N., et al. (2015). Codon optimality is a major determinant of mRNA stability. *Cell, 160*(6), 1111–1124.

Quenault, T., Lithgow, T., & Traven, A. (2011). PUF proteins: Repression, activation and mRNA localization. *Trends in Cell Biology, 21*(2), 104–112.

Rabani, M., Raychowdhury, R., Jovanovic, M., Rooney, M., Stumpo, D. J., Pauli, A., et al. (2014). Resource high-resolution sequencing and modeling identifies distinct dynamic RNA regulatory strategies. *Cell*, *159*, 1–13.
Ramos, S. B. V., Stumpo, D. J., Kennington, E. A., Phillips, R. S., Bock, C. B., Ribeiro-Neto, F., et al. (2004). The CCCH tandem zinc-finger protein Zfp36l2 is crucial for female fertility and early embryonic development. *Development*, *131*(19), 4883–4893.
Richter, J. (1996). Dynamics of poly(A) addition and removal during development. *Cold Spring Harbor Monograph Archive*, *30*(1973), 481–503.
Richter, J. D. (1999). Cytoplasmic polyadenylation in development and beyond. *Microbiology and Molecular Biology Reviews*, *63*(2), 446–456.
Richter, J. D., & Lasko, P. (2011). Translational control in oocyte development. *Cold Spring Harbor Perspectives in Biology*, *3*(9), 1–14.
Richter, J. D., & Sonenberg, N. (2005). Regulation of cap-dependent translation by eIF4E inhibitory proteins. *Nature*, *433*(7025), 477–480.
Rosa, A., Spagnoli, F. M., & Brivanlou, A. H. (2009). The miR-430/427/302 family controls mesendodermal fate specification via species-specific target selection. *Developmental Cell*, *16*(4), 517–527.
Rouget, C., Papin, C., Boureux, A., Meunier, A.-C., Franco, B., Robine, N., et al. (2010). Maternal mRNA deadenylation and decay by the piRNA pathway in the early Drosophila embryo. *Nature*, *467*(7319), 1128–1132.
Rouskin, S., Zubradt, M., Washietl, S., Kellis, M., & Weissman, J. S. (2014). Genome-wide probing of RNA structure reveals active unfolding of mRNA structures in vivo. *Nature*, *505*(7485), 701–705.
Sawicki, J. A., Magnuson, T., & Epstein, C. J. (1981). Evidence for expression of the paternal genome in the two-cell mouse embryo. *Nature*, *294*(5840), 450–451.
Schiavi, S. C., Wellington, C. L., Shyu, A. Bin, Chen, C. Y. A., Greenberg, M. E., & Belasco, J. G. (1994). Multiple elements in the c-fos protein-coding region facilitate mRNA deadenylation and decay by a mechanism coupled to translation. *Journal of Biological Chemistry*, *269*(5), 3441–3448.
Schoenberg, D. R., & Maquat, L. E. (2012). Regulation of cytoplasmic mRNA decay. *Nature Reviews. Genetics*, *13*(6), 246–259.
Schramm, R. D., & Bavister, B. D. (1999). Onset of nucleolar and extranucleolar transcription and expression of fibrillarin in macaque embryos developing in vitro. *Biology of Reproduction*, *60*(3), 721–728.
Semotok, J. L., Cooperstock, R. L., Pinder, B. D., Vari, H. K., Lipshitz, H. D., & Smibert, C. A. (2005). Smaug recruits the CCR4/POP2/NOT deadenylase complex to trigger maternal transcript localization in the early drosophila embryo. *Current Biology*, *15*, 284–294.
Semotok, J. L., Luo, H., Cooperstock, R. L., Karaiskakis, A., Vari, H. K., Smibert, C. A., et al. (2008). Drosophila maternal Hsp83 mRNA destabilization is directed by multiple SMAUG recognition elements in the open reading frame. *Molecular and Cellular Biology*, *28*(22), 6757–6772.
Smibert, C. A., Lie, Y. S., Shillinglaw, W., Henzel, W. J., & Macdonald, P. M. (1999). Smaug, a novel and conserved protein, contributes to repression of nanos mRNA translation in vitro. *RNA*, *5*(12), 1535–1547.
Spitale, R. C., Flynn, R. A., Zhang, Q. C., Crisalli, P., Lee, B., Jung, J.-W., et al. (2015). Structural imprints in vivo decode RNA regulatory mechanisms. *Nature*, *519*(7544), 486–490.
Staton, A. A., Knaut, H., & Giraldez, A. J. (2011). miRNA regulation of Sdf1 chemokine signaling provides genetic robustness to germ cell migration. *Nature Genetics*, *43*(3), 204–211.
Stitzel, M. L., & Seydoux, G. (2007). Regulation of the oocyte-to-zygote transition. *Science*, *316*(5823), 407–408.

Stoeckius, M., Grün, D., Kirchner, M., Ayoub, S., Torti, F., Piano, F., et al. (2014). Global characterization of the oocyte-to-embryo transition in Caenorhabditis elegans uncovers a novel mRNA clearance mechanism. *The EMBO Journal*, *33*(16), 1751–1766.

Su, Y. Q., Sugiura, K., Woo, Y., Wigglesworth, K., Kamdar, S., Affourtit, J., et al. (2007). Selective degradation of transcripts during meiotic maturation of mouse oocytes. *Developmental Biology*, *302*(1), 104–117.

Subramanyam, D., Lamouille, S., Judson, R. L., Liu, J. Y., Bucay, N., Derynck, R., et al. (2011). Multiple targets of miR-302 and miR-372 promote reprogramming of human fibroblasts to induced pluripotent stem cells. *Nature Biotechnology*, *29*(5), 443–448.

Subtelny, A. O., Eichhorn, S. W., Chen, G. R., Sive, H., & Bartel, D. P. (2014). Poly(A)-tail profiling reveals an embryonic switch in translational control. *Nature*, *508*(7494), 66–71.

Suh, N., Baehner, L., Moltzahn, F., Melton, C., Chen, J., & Blelloch, R. (2010). MicroRNA function is globally suppressed in mouse oocytes and early embryos. *Current Biology*, *20*(3), 271–277.

Suh, M.-R., Lee, Y., Kim, J. Y., Kim, S.-K., Moon, S.-H., Lee, J. Y., et al. (2004). Human embryonic stem cells express a unique set of microRNAs. *Developmental Biology*, *270*(2), 488–498.

Tadros, W., Goldman, A. L., Babak, T., Menzies, F., Vardy, L., Orr-Weaver, T., et al. (2007). SMAUG is a major regulator of maternal mRNA destabilization in Drosophila and its translation is activated by the PAN GU kinase. *Developmental Cell*, *12*(1), 143–155. http://dx.doi.org/10.1016/j.devcel.2006.10.005.

Tadros, W., Houston, S. A., Bashirullah, A., Cooperstock, R. L., Semotok, J. L., Reed, B. H., et al. (2003). Regulation of maternal transcript destabilization during egg activation in drosophila. *Genetics*, *164*(3), 989–1001.

Tadros, W., & Lipshitz, H. D. (2009). The maternal-to-zygotic transition: A play in two acts. *Development*, *136*(18), 3033–3042.

Takahashi, K., & Yamanaka, S. (2006). Induction of pluripotent stem cells from mouse embryonic and adult fibroblast cultures by defined factors. *Cell*, *126*(4), 663–676.

Takeda, Y., Mishima, Y., Fujiwara, T., Sakamoto, H., & Inoue, K. (2009). DAZL relieves miRNA-mediated repression of germline mRNAs by controlling poly(A) tail length in zebrafish. *PLoS One*, *4*(10), e7513.

Tang, F., Kaneda, M., O'Carroll, D., Hajkova, P., Barton, S. C., Sun, Y. A., et al. (2007). Maternal microRNAs are essential for mouse zygotic development. *Genes and Development*, *21*(6), 644–648.

Thomsen, S., Anders, S., Janga, S. C., Huber, W., & Alonso, C. R. (2010). Genome-wide analysis of mRNA decay patterns during early Drosophila development. *Genome Biology*, *11*(9), R93.

Timchenko, L. T., Miller, J. W., Timchenko, N. A., Devore, D. R., Datar, K. V., Lin, L., et al. (1996). Identification of a (CUG)(n) triplet repeat RNA-binding protein and its expression in myotonic dystrophy. *Nucleic Acids Research*, *24*(22), 4407–4414.

Tripurani, S. K., Wee, G., Lee, K. B., Smith, G. W., Wang, L., & Yao, J. (2013). MicroRNA-212 post-transcriptionally regulates oocyte-specific basic-helix-loop-helix transcription factor, factor in the germline alpha (FIGLA), during bovine early embryogenesis. *PLoS One*, *8*(9), e76114.

Ueno, S., & Sagata, N. (2002). Requirement for both EDEN and AUUUA motifs in translational arrest of Mos mRNA upon fertilization of Xenopus eggs. *Developmental Biology*, *250*(1), 156–167.

Van Driessche, N., Shaw, C., Katoh, M., Morio, T., Sucgang, R., Ibarra, M., et al. (2002). A transcriptional profile of multicellular development in Dictyostelium discoideum. *Development*, *129*(7), 1543–1552.

Vardy, L., & Orr-Weaver, T. L. (2007). Regulating translation of maternal messages: Multiple repression mechanisms. *Trends in Cell Biology*, *17*(11), 547–554.

Vassena, R., Boué, S., González-Roca, E., Aran, B., Auer, H., Veiga, A., et al. (2011). Waves of early transcriptional activation and pluripotency program initiation during human preimplantation development. *Development*, *138*(17), 3699–3709.

Voeltz, G. K., Ongkasuwan, J., Standart, N., & Steitz, J. A. (2001). A novel embryonic poly(A) binding protein, ePAB, regulates mRNA deadenylation in Xenopus egg extracts. *Genes and Development*, *15*(6), 774–788.

Voeltz, G. K., & Steitz, J. A. (1998). AUUUA sequences direct mRNA deadenylation uncoupled from decay during Xenopus early development. *Molecular and Cellular Biology*, *18*(12), 7537–7545.

Wahle, E., & Winkler, G. S. (2013). RNA decay machines: Deadenylation by the Ccr4-Not and Pan2-Pan3 complexes. *Biochimica et Biophysica Acta*, *1829*(6–7), 561–570.

Walser, C. B., & Lipshitz, H. D. (2011). Transcript clearance during the maternal-to-zygotic transition. *Current Opinion in Genetics and Development*, *21*(4), 431–443.

Wan, Y., Qu, K., Zhang, Q. C., Flynn, R. A., Manor, O., Ouyang, Z., et al. (2014). Landscape and variation of RNA secondary structure across the human transcriptome. *Nature*, *505*(7485), 706–709.

Wang, Y., Baskerville, S., Shenoy, A., Babiarz, J. E., Baehner, L., & Blelloch, R. (2008). Embryonic stem cell specific MicroRNAs regulate the G1/S transition and promote rapid proliferation. *Nature Genetics*, *40*(12), 1478–1483.

Wang, J., Garrey, J., & Davis, R. E. (2014). Transcription in pronuclei and one- to four-cell embryos drives early development in a nematode. *Current Biology*, *24*(2), 124–133.

Wang, X., Lu, Z., Gomez, A., Hon, G. C., Yue, Y., Han, D., et al. (2014). N6-methyladenosine-dependent regulation of messenger RNA stability. *Nature*, *505*(7481), 117–120.

Wang, K., Otu, H. H., Chen, Y., Lee, Y., Latham, K., & Cibelli, J. B. (2011). Reprogrammed transcriptome in rhesus-bovine interspecies somatic cell nuclear transfer embryos. *PLoS One*, *6*(7), e22197.

Warner, C. M., & Versteegh, L. R. (1974). In vivo and in vitro effect of alpha-amanitin on preimplantation mouse embryo RNA polymerase. *Nature*, *248*(450), 678–680.

Webster, P. J., Liang, L., Berg, C. A., Lasko, P., & Macdonald, P. M. (1997). Translational repressor bruno plays multiple roles in development and is widely conserved. *Genes and Development*, *11*(19), 2510–2521.

Weidmann, C. A., & Goldstrohm, A. C. (2012). Drosophila Pumilio protein contains multiple autonomous repression domains that regulate mRNAs independently of Nanos and brain tumor. *Molecular and Cellular Biology*, *32*(2), 527–540.

Weidmann, C. A., Raynard, N. A., Blewett, N. H., Van Etten, J., & Goldstrohm, A. C. (2014). The RNA binding domain of Pumilio antagonizes poly-adenosine binding protein and accelerates deadenylation. *RNA*, *20*(8), 1298–1319.

Weill, L., Belloc, E., Bava, F.-A., & Méndez, R. (2012). Translational control by changes in poly(A) tail length: Recycling mRNAs. *Nature Structural & Molecular Biology*, *19*(6), 577–585.

Wellington, C. L., Greenberg, M. E., & Belasco, J. G. (1993). The destabilizing elements in the coding region of c-fos mRNA are recognized as RNA. *Molecular and Cellular Biology*, *13*(8), 5034–5042.

Wilhelm, J. E., Hilton, M., Amos, Q., & Henzel, W. J. (2003). Cup is an eIF4E binding protein required for both the translational repression of oskar and the recruitment of Barentsz. *Journal of Cell Biology*, *163*(6), 1197–1204.

Wilmut, I., Schnieke, A. E., McWhir, J., Kind, A. J., & Campbell, K. H. S. (1997). Viable offspring derived from fetal and adult mammalian cells. *Nature*, *385*, 810–813.

Wong, C. C., Loewke, K. E., Bossert, N. L., Behr, B., De Jonge, C. J., Baer, T. M., et al. (2010). Non-invasive imaging of human embryos before embryonic genome activation predicts development to the blastocyst stage. *Nature Biotechnology*, *28*(10), 1115–1121.

Wu, E., Thivierge, C., Flamand, M., Mathonnet, G., Vashisht, A. A., Wohlschlegel, J., et al. (2010). Pervasive and cooperative deadenylation of 3′UTRs by embryonic microRNA families. *Molecular Cell*, *40*(4), 558–570.

Xin, H. P., Zhao, J., & Sun, M. X. (2012). The maternal-to-zygotic transition in higher plants. *Journal of Integrative Plant Biology*, *54*(9), 610–615.

Yan, L., Yang, M., Guo, H., Yang, L., Wu, J., Li, R., et al. (2013). Single-cell RNA-Seq profiling of human preimplantation embryos and embryonic stem cells. *Nature Structural & Molecular Biology*, *20*(9), 1131–1139.

Yu, J., Vodyanik, M. A., Smuga-Otto, K., Antosiewicz-Bourget, J., Frane, J. L., Tian, S., et al. (2007). Induced pluripotent stem cell lines derived from human somatic cells. *Science*, *318*(5858), 1917–1920.

Zeng, F., Baldwin, D. A., & Schultz, R. M. (2004). Transcript profiling during preimplantation mouse development. *Developmental Biology*, *272*(2), 483–496.

Zeng, F., & Schultz, R. M. (2005). RNA transcript profiling during zygotic gene activation in the preimplantation mouse embryo. *Developmental Biology*, *283*(1), 40–57.

Zhang, S., Williams, C. J., Wormington, M., Stevens, A., & Peltz, S. W. (1999). Monitoring mRNA decapping activity. *Methods: A Companion to Methods in Enzymology*, *17*, 46–51.

Zhang, P., Zucchelli, M., Bruce, S., Hambiliki, F., Stavreus-Evers, A., Levkov, L., et al. (2009). Transcriptome profiling of human pre-implantation development. *PLoS One*, *4*(11), e7844.

Zhou, L., & Dean, J. (2015). Reprogramming the genome to totipotency in mouse embryos. *Trends in Cell Biology*, *25*(2), 82–91.

Zuo, Y., Gao, Y., Su, G., Bai, C., Liu, K., Li, Q., et al. (2014). Irregular transcriptome reprogramming probably caused embryo developmental failure of interspecies nuclear transfer between the Przewalski's gazelle and the bovine. *BMC Genomics*, *15*(1), 1–14.

CHAPTER SEVEN

Building the Future: Post-transcriptional Regulation of Cell Fate Decisions Prior to the *Xenopus* Midblastula Transition

Michael D. Sheets[1]

Department of Biomolecular Chemistry, School of Medicine and Public Health, University of Wisconsin, Madison, Wisconsin, USA

[1]Corresponding author: e-mail address: mdsheets@wisc.edu

Contents

Current Topics in Developmental Biology, Volume 113
ISSN 0070-2153
http://dx.doi.org/10.1016/bs.ctdb.2015.06.008

Abstract

In all animals, a critical period in early development is when embryonic cells switch from relying solely upon maternally deposited RNAs and proteins to relying upon molecules encoded by the zygotic genome. *Xenopus* embryos have served as a model for examining this switch, as well as the maternally controlled stages that prepare for it. In *Xenopus*, the robust activation of zygotic transcription occurs at the 12th cleavage division and is referred to as the midblastula transition (MBT). Prior to MBT, gene expression is regulated by post-transcriptional events including mRNA and protein localization, protein post-translational modification, and mRNA translation. After the MBT, appropriate transcriptional regulation of the zygotic genome becomes critical and predominates. However, it is important to realize that the first key cell fate decisions that have profound impacts on development occur prior to the MBT and these are governed by regulating the expression of maternally deposited regulatory mRNAs and proteins. In this chapter, I will discuss post-transcriptional mechanisms that function during the maternal stages of *Xenopus* development with an emphasis on mechanisms known to directly modulate cell fate decisions. Emerging approaches and technologies that will help better understand this phase of development will also be discussed.

1. INTRODUCTION

1.1 The Midblastula Transition

In *Xenopus* and many animal species, the initial cell divisions that follow fertilization are rapid and synchronous (Langley, Smith, Stemple, & Harvey, 2014). The rapid pace results from a simplified cell cycle that consists of alternating periods of DNA replication (S-phase) and genome segregation (M-phase), but lacks the elongated G1 and G2 phases of typical cell cycles. During this period there is no transcription and no growth of the embryo. In addition, maternal stores of mRNAs, proteins, and metabolites provide

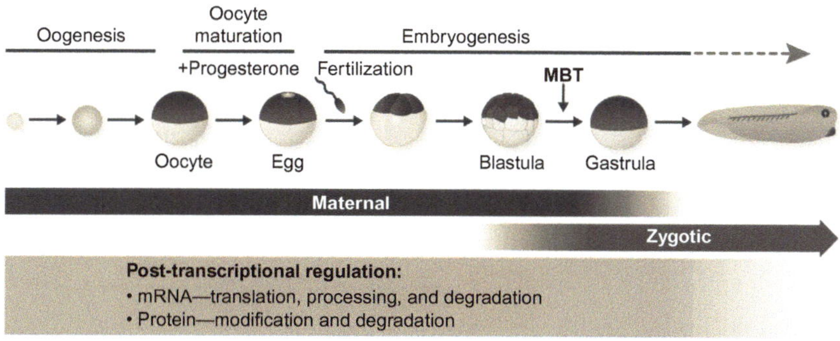

Figure 1 The maternal stages of *Xenopus* development depend upon post-transcriptional regulatory mechanisms. Summary diagram of oogenesis, oocyte maturation, fertilization, and cleavage stages of embryogenesis. The early stages of *Xenopus* development are totally dependent upon maternal mRNAs and proteins that are produced during oogenesis. It is the regulated use of these maternal components via post-transcriptional mechanisms that mediate the earliest cell fate decisions. The transcription of a relatively small number of genes is activated at the early blastula stage. When embryos reach the midblastula transition (MBT), they undergo many changes including the robust activation of zygotic transcription. Once transcription begins, the maternal contributions to development gradually decrease as zygotic processes become more prevalent.

all the components necessary for these stages of development (Fig. 1). After 12 divisions, *Xenopus* embryonic cells undergo major changes in their behavior. The period of development when these changes are initiated is referred to as the midblastula transition (MBT). At the MBT the embryonic cell cycle slows by adding defined G1 and G2 phases, the cells begin to divide asynchronously and they become motile for the first time. The MBT is also when robust transcription of the zygotic genome is first activated during *Xenopus* embryogenesis. This transcription signals the beginning of an important period of development when the embryo switches from its dependence upon maternal products to their replacement by and dependence on newly synthesized zygotic mRNAs and proteins.

Early embryologists first recognized the phenomenon of the MBT as an important developmental event and modern interest can be traced to the seminal papers of Newport and Kirschner (1982a, 1982b). This work was critical for providing important new information about the MBT and its regulation, but it also spurred researchers in other organisms to begin analyzing the maternal–zygotic transition. Over the last few years, interest in the MBT and specifically how it is regulated has been rekindled (Langley et al., 2014).

1.2 Regulation of MBT

The slowing of the cell cycle and the rate of DNA replication that accompanies MBT suggests a functional connection between these processes. Collart, Allen, Bradshaw, Smith, and Zegerman (2013) investigated this connection and identified *Xenopus* replication factors (cut5, drf1, treslin, and recq4) whose levels changed during the blastula and gastrula stages and as such represent candidate regulators. When these factors are overexpressed, embryonic cells fail to exhibit a slowing of the cell cycle normally observed at MBT and instead the cells continue dividing rapidly and synchronously. Analysis of the treated embryos demonstrated that developmental defects were accompanied by the increased rates of replication initiation. In addition to the cell cycle defects, overexpression of the four replication factors delayed the expression of many transcripts that normally begin to accumulate at the MBT. Thus, these results connect the rate of DNA replication to both cell cycle changes and activation of the zygotic genome that coincides with the MBT.

Just after fertilization, the embryo prior to the first cleavage division consists of a single nucleus contained in a vast cytoplasm. During subsequent cleavage stages, the nuclei remain relatively similar in size while the volume of cytoplasm in each cell is significantly reduced at each cell division. A recent study carefully measured the volumes of the nucleus and the cytoplasm of embryonic cells during development (Jevtic & Levy, 2015). It was suggested that a specific nucleus to cytoplasm (N/C) ratio is an important determinant of when the MBT commences. To test this idea, different factors such as nuclear lamins and reticulons that are known to drive changes in nuclear volume were overexpressed in developing *Xenopus* embryos. Increasing the N/C ratio in pre-MBT embryos leads to premature activation of zygotic gene transcription and early onset of longer cell cycles, while decreasing the N/C ratio had the opposite affect.

Murphy and Michael used *Xenopus* egg extracts to investigate how changing the N/C ratio affects replication (Murphy & Michael, 2013). Increasing the amount of nuclear components lengthened the time required for DNA replication of sperm nuclei. Further analysis indicated that the protein phosphatase PP2A and its regulatory subunit, B55α are limiting factors for replication. Together these results demonstrate that nuclear size also contributes to the regulation of MBT timing.

A recent study investigated the idea that the timing of the *Xenopus* MBT is controlled by a maternal inhibitor that is gradually neutralized by the increasing amounts of DNA that arise due to replication (Amodeo, Jukam, Straight, & Skotheim, 2015). *Xenopus* egg extract was used to identify histones H3 and H4 as concentration-dependent inhibitors of transcription. *In vivo* experiments where the concentration of histone H3 was reduced in cleavage-stage embryos induced precocious transcriptional activation and cell cycle lengthening. In contrast, increasing the concentration of H3 and H4 in embryos caused the inordinate shortening of post-MBT cell cycles. These results suggest that the ratio of free histones to DNA is a critical factor that regulates the timing of MBT and the associated events, either through effects on DNA replication or transcription or both (Amodeo et al., 2015).

1.3 Maternal Control of Cell Fates

Fertilization activates a complex developmental program that guides the formation of specific differentiated tissues and ultimately the formation of functional adult structures. In *Xenopus*, the initial part of this program that occurs prior to the MBT relies entirely upon maternal RNAs and proteins that are generated during oogenesis (Heasman, 2006a, 2006b; Houston, 2012). Cleavage-stage embryos use these maternal components to form the regionalized differences between the first embryonic cells. These differences are responsible for the formation of distinct groups of progenitor cells, such as the germ layers that undergo further specializations as development progresses. It must be emphasized that while the newly synthesized zygotic gene products are critically important, it is the maternal components and their regulation that provide the foundation upon which the first cell fate decisions are made and upon which further development depends (Fig. 1).

1.4 Building the Foundation: Oogenesis and the Formation of Asymmetries

In *Xenopus*, the foundation for embryonic development is created during oogenesis (Gerhart, 1980; Keller, 1991). Oocytes develop over a period of months and when fully grown have a distinctive darkly pigmented animal hemisphere and a pale vegetal hemisphere. These visual differences mark the animal–vegetal axis of the oocyte that is carried over to the egg during oocyte maturation and then passed on to the embryo after fertilization. In addition, growing oocytes are transcriptionally active and they synthesize maternal mRNAs. Some of these mRNAs are used immediately and translated into proteins, while others are translationally repressed and stored for later use. One class of stored mRNAs encodes cell cycle regulators, such as the cyclin B1 mRNA. These stored mRNAs are translationally activated during oocyte maturation to synthesize proteins that facilitate the maturation process (Cragle & MacNicol, 2014; Minshull et al., 1989; Sheets, Fox, Hunt, Vande Woude, & Wickens, 1994). Other maternal mRNAs are stored until after fertilization and then activated at the appropriate time in development when the proteins they encode are needed (Wickens, 1990). Some of these stored mRNAs are localized to specific regions of the oocyte. In particular, one such group of mRNAs is localized to the vegetal cortex of the oocyte, while other stored mRNAs are transported and concentrated in the animal hemisphere (Houston, 2013; King, Messitt, & Mowry, 2005; Medioni, Mowry, & Besse, 2012; Rebagliati, Weeks, Harvey, & Melton, 1985; Weeks & Melton, 1987; Weeks, Rebagliati, Harvey, & Melton, 1985). These localized mRNAs contribute to the important molecular distinctions that define the animal–vegetal axis of oocytes, eggs, and embryos. More importantly, the proteins encoded by many of these localized mRNAs create regional differences between cells that influence cell fate decisions in the embryo.

1.5 Localized mRNAs, Maternal Determinants, and Germ Layer Formation

The establishment of the primary germ layers during the initial stages of *Xenopus* development is one of the critical processes controlled by localized mRNAs (Houston, 2013; King et al., 2005; Medioni et al., 2012). The partitioning of different mRNAs to distinct regions of the oocyte and egg establishes molecular differences between cells during embryogenesis (Heasman, 2006a; Houston, 2012; White & Heasman, 2008). After fertilization, vegetally localized mRNAs become concentrated in embryonic vegetal cells that give rise to endoderm, while animally localized mRNAs become concentrated in embryonic animal cells that give rise to ectoderm.

The importance of localized maternal mRNAs for the formation of specific germ layers is supported by loss-of-function experiments that deplete specific maternal mRNAs from developing embryos (Heasman, Holwill, & Wylie, 1991; Olson, Hulstrand, & Houston, 2012; Schneider, Hulstrand, & Houston, 2010). For example, the vegetally localized VegT mRNA encodes a T-box transcription factor (Lustig, Kroll, Sun, & Kirschner, 1996; Zhang & King, 1996). *Xenopus* embryos depleted of maternal VegT mRNA lack endoderm (Zhang et al., 1998). In contrast, the Foxi2 maternal mRNA is concentrated in animal cells and it encodes a transcription factor (Cha, McAdams, Kormish, Wylie, & Kofron, 2012). The animal cells are precursors to ectoderm and neurectoderm. Depletion of maternal Foxi2 mRNA from embryos causes defects in ectoderm formation by disrupting the activation of ectoderm-specific genes (Cha et al., 2012). Thus, specific localized mRNAs provide important links between the molecular asymmetries formed in the oocyte and the germ layers formed by distinct cells at the beginnings of embryogenesis.

1.6 Creating Embryonic Asymmetries: Cortical Rotation and the Organizer-Inducing Center

Localized mRNAs also provide the molecular basis for establishing the asymmetries that form during the maternal stages of embryogenesis and drive the vertebrate body plan (Houston, 2012, 2013). During the first cell cycle following fertilization, the *Xenopus* egg undergoes a cytoplasmic rearrangement where the outer cortex rotates 30° with respect to the heavy inner yolk mass (Gerhart, 1980; Gerhart et al., 1989; Gerhart, Ubbels, Black, Hara, & Kirschner, 1981). This process, referred to as cortical rotation, creates new molecular asymmetries in the embryo in the horizontal dimension, the so-called dorsal/ventral axis of the embryo that is perpendicular to the animal/vegetal axis. Cells along the path of the upward displacement of the cortex form the organizer and anterior structures of the embryo, while cells on the opposite side of the embryo form posterior structures. While it is not clear what molecules are transported and/or activated by cortical rotation, it is clear that the Wnt signaling pathway is activated in cells along the pathway of cortical rotation (Brannon, Gomperts, Sumoy, Moon, & Kimelman, 1997; Heasman et al., 1994; Larabell et al., 1997; McCrea, Brieher, & Gumbiner, 1993; Rowning et al., 1997; Schohl & Fagotto, 2002; Wylie et al., 1996; Yost et al., 1996). Wnt signaling results in the stabilization of β-catenin protein, which is followed by β-catenin's translocation to the nucleus where it activates the transcription of genes that establish the organizer, a central inducing center that forms and functions after zygotic transcription begins.

The organizer cells perform important functions that drive embryonic development; they emit signals that drive the cell movements of gastrulation and signals that pattern the adjacent cells of the germ layers (Cho, Blumberg, Steinbeisser, & De Robertis, 1991; De Robertis, Blum, Niehrs, & Steinbeisser, 1992; Gerhart et al., 1989; Gerhart, Doniach, & Stewart, 1991; Harland & Gerhart, 1997; Lemaire, Garrett, & Gurdon, 1995; Lemaire & Kodjabachian, 1996; Smith & Harland, 1992). During the blastula and gastrula stages, the embryo in the animal–vegetal dimension is partitioned into groups of cells that will form the three germ layers: ectoderm, mesoderm, and endoderm. Each layer represents a group of progenitor cells whose fate is restricted to derivatives characteristic of that layer. For example, cells of the ectoderm germ layer will give rise to ectoderm and neuroectoderm cell types. The particular germ layer derivatives that form are a function of cell proximity to the organizer cells and their signals. Organizer signals promote cells to differentiate as anterior derivatives of each germ layer. For example, ectodermal cells exposed to organizer signals form anterior neural structures—the anterior parts of the nervous system that include the brain and the anterior spinal cord. In contrast, cells of the ectodermal germ layer that do not receive organizer signals form primitive ectoderm and posterior neural derivatives such as posterior spinal cord. Organizer signaling directs similar asymmetry in the form of anterior/posterior patterning of each germ layer. Thus, the organizer has its roots in the maternal mRNA localization events that are established during oogenesis and then elaborated upon by the embryonic events of cortical rotation (Gerhart et al., 1989; Heasman, 2006a, 2006b).

In this brief overview of *Xenopus* development, I have tried to highlight the importance of maternal mRNAs and proteins to creating asymmetries that contribute to the initial cell fate determination events that occur in the embryo. These events begin during oogenesis and continue throughout the first hours of development that follow fertilization. During this critical initial phase of development prior to the MBT, the genome is largely transcriptionally silent; regulatory events are limited to post-transcriptional control mechanisms. The mechanisms that control the translation and stability of specific mRNAs and the mechanisms that control the modification and proteolysis of specific maternal proteins are the subject of this chapter.

2. REGULATION OF MATERNAL mRNAs IN THE *XENOPUS* EMBRYO

The processes that regulate the steady-state levels of specific mRNAs and their use as templates for translation are important control mechanisms in all cells (Godwin, Kojima, Green, & Wilusz, 2013; Goldstrohm & Wickens, 2008; Valencia-Sanchez, Liu, Hannon, & Parker, 2006; Wilusz, Wang, & Peltz, 2001). These mechanisms function in coordination with transcriptional events to ensure the proper amount of each gene product is available to cells under the appropriate conditions. In the pre-MBT

Xenopus embryo where transcription is absent or at most very limited, mRNA regulatory mechanisms are central to controlling the abundance of specific proteins. In the following sections, I will discuss these mechanisms with an emphasis on those known to influence cell fates.

2.1 Embryo-Specific Polyadenylation

The addition of poly(A) to stored maternal mRNAs is a common mechanism to activate their translation (Ivshina, Lasko, & Richter, 2014; Weill, Belloc, Bava, & Mendez, 2012; Wickens, 1990). During *Xenopus* development, there are broad categories of different poly(A) addition mechanisms based on when poly(A) addition occurs. The most widely studied events were discovered over 30 years ago and occur during oocyte maturation (Dworkin & Dawid, 1980; Dworkin & Dworkin-Rastl, 1985; Dworkin & Hershey, 1981). These maturation-specific polyadenylation reactions have been extensively studied and as a result we have a significant knowledge of the mRNA substrates, the sequence elements, the regulatory proteins, and the signaling processes that temporally control these reactions (Cragle & MacNicol, 2014; Fox, Sheets, & Wickens, 1989; Ivshina et al., 2014; McGrew, Dworkin-Rastl, Dworkin, & Richter, 1989; Weill et al., 2012; Wickens, 1990).

The other category is represented by mRNAs that are polyadenylated during postfertilization stages of maternal development. Much less is known about these mechanisms in comparison to the polyadenylation that occurs during oocyte maturation. For example, translation of the maternal Cl1, Cl2, and activin receptor mRNAs in *Xenopus* embryos is activated following fertilization, coincident with their polyadenylation (Paris, Osborne, Couturier, Le Guellec, & Philippe, 1988; Paris & Philippe, 1990; Simon & Richter, 1994; Simon, Tassan, & Richter, 1992; Simon, Wu, & Richter, 1996). This embryonic polyadenylation requires 3′UTR sequence elements, termed *e*mbryonic *c*ytoplasmic *p*olyadenylation *e*lements (eCPEs). These elements are distinct from the sequences that direct poly(A) addition during maturation, indicating that different translational regulatory mechanisms operate during oocyte maturation and embryonic development. It is unclear what proteins bind to eCPEs and mediate embryonic polyadenylation, but some evidence suggests that eCPEs are bound by the ElrA protein (Wu, Good, & Richter, 1997), a member of the ELAV family of RNA-binding proteins. ElrA proteins have broad functions in RNA regulation across vertebrate and invertebrate species (Antic & Keene, 1997).

In addition to eCPE-mediated polyadenylation, a C-rich sequence element referred to as a C-CPE mediates the polyadenylation of certain mRNAs during embryogenesis (Paillard, Maniey, Lachaume, Legagneux, & Osborne, 2000; Vishnu, Sumaroka, Klein, & Liebhaber, 2011). C-CPEs are bound by the *p*oly(r*C*)-*b*inding *p*rotein *2* (PCBP2, also called αCP2, hnRNP-E2), which is part of a complex that includes known polyadenylation factors, such as poly(A) polymerase. While it is clear that this activity is present in *Xenopus* embryos, it is unclear how many maternal mRNAs are substrates for C-CPE-mediated polyadenylation.

Other maternal mRNAs, such as the mRNA encoding the laminin B protein, are polyadenylated during embryogenesis but lack easily recognizable eCPE or C-CPE sequences. This observation suggests that embryonic polyadenylation and translational activation prior to the MBT may occur by multiple parallel pathways (Ralle, Gremmels, & Stick, 1999).

More recently, genomic methods have been used to analyze polyadenylation in *Xenopus* embryos and have corroborated many of the findings of the original investigations as well as providing new insights (Collart et al., 2014; Graindorge, Thuret, Pollet, Osborne, & Audic, 2006; Paranjpe, Jacobi, van Heeringen, & Veenstra, 2013; Subtelny, Eichhorn, Chen, Sive, & Bartel, 2014). One of the important contributions of the genome level studies is identifying the sheer scope of regulated polyadenylation since hundreds of mRNAs receive poly(A) at or after fertilization. These studies also identified cohorts of maternal mRNAs that exhibit the same pattern of regulation with regard to when during development they are polyadenylated. These results provide critical information for mechanistic studies to identify the RNA sequences and proteins that mediate regulated polyadenylation as well as providing information to investigate the functional relevance of the encoded proteins for development.

2.2 Regulation of mRNA Degradation and the MBT

In somatic cells once an mRNA has lost its poly(A) tail, it is rapidly degraded by processing factors such as those of the CCR4–NOT complex (Godwin et al., 2013; Goldstrohm & Wickens, 2008; van Hoof & Parker, 2002). However, during the maternal stages of *Xenopus* development, mRNA deadenylation and degradation are uncoupled (Audic, Omilli, & Osborne, 1997, 1998; Duval et al., 1990; Paillard et al., 1998). There are numerous examples of mRNAs that are deadenylated at fertilization or during the cleavage stages but remain stable for hours, both from detailed analysis of specific mRNAs and genomic studies that analyze mRNA populations (Graindorge et al., 2008; Paris et al., 1988; Sagata, Oskarsson, Copeland, Brumbaugh, & Vande Woude, 1988; Sheets et al., 1994; Subtelny et al., 2014; Tchang, Vriz, & Mechali, 1991). In this maternal context, the main function of deadenylation is to promote translational inactivation (see Section 2.3).

mRNA degradation processes become activated once development proceeds through the MBT and deadenylated mRNAs are degraded by mechanisms that appear similar to those found in most somatic cells. Treatment of embryos with cycloheximide blocks the activation of mRNA degradation functions at the MBT, while transcriptional inhibitors have no effect on the activation (Duval et al., 1990). These results suggest that the mRNA degradation functions at the MBT are controlled by temporally regulating either the synthesis or activation of a limiting component encoded by a maternal mRNA. Thus, at the MBT, a significant change in mRNA metabolism occurs in which the embryo switches from a maternal regulatory mode that preserves mRNAs regardless of their poly(A) state to a zygotic mode of regulation more typical of somatic cells in which the majority of deadenylated mRNAs are degraded. The major issue to be addressed is defining how the degradation activities are regulated.

2.3 Embryo-Specific mRNA Deadenylation

The *Xenopus* Eg1, Eg2, Eg5, and c-mos mRNAs are members of a class of maternal mRNAs that are deadenylated and translationally repressed shortly after fertilization (Paillard et al., 1998; Paris et al., 1988; Paris & Philippe, 1990; Sheets et al., 1994). This deadenylation depends upon *e*mbryo *d*eadenylation *e*leme*n*t (EDEN) sequence motifs that are composed of GU-rich sequences that are present in the 3′UTRs of deadenylated mRNAs. The EDEN sequences serve as binding sites for the RNA-binding protein EDEN-*b*inding *p*rotein (EDEN-BP, also called CUG-BP1 and CELF1) (Audic et al., 1997, 1998; Bouvet et al., 1994; Paillard et al., 1998). The temporal regulation of this deadenylation depends upon specific signaling events triggered by fertilization that phosphorylate and activate EDEN-BP (Detivaud, Pascreau, Karaiskou, Osborne, & Kubiak, 2003). Genomic experiments identified hundreds of EDEN-BP target mRNAs (Graindorge et al., 2008). Many of these target mRNAs encode proteins that are involved in oocyte maturation and cell cycle progression, suggesting that EDEN-BP-dependent deadenylation is important for post-transcriptionally regulating the remodeling of the early cell cycle that occurs at fertilization.

3. miRNA REGULATION OF mRNAs DURING EARLY *XENOPUS* DEVELOPMENT

microRNAs (miRNAs) are small noncoding RNAs that post-translationally regulate gene expression by serving as the mRNA recognition component of the RNA-induced silencing complexes (Pasquinelli, 2012). In animal development miRNAs serve many different roles, such as influencing cell fate determination and body axis formation (Inui, Montagner, & Piccolo, 2012; Pauli, Rinn, & Schier, 2011). The following examples illustrate the emerging importance of miRNAs as post-transcriptional regulators of cell fates during the earliest stages of *Xenopus* development.

3.1 miR-15/16 Regulation of Nodal Signaling

Formation and maintenance of the organizer-inducing center in *Xenopus* embryos relies upon restricting signaling by the Nodal pathway to the dorso-anterior cells of the embryo and the organizer progenitor cells (Harland & Gerhart, 1997; Heasman, 2006b).

There are many different mechanisms that contribute to restricting Nodal signaling and these include the post-transcriptional regulation of mRNAs encoding Nodal pathway components. Expression of miR-15/16 RNAs is confined to cells on the ventral side of *Xenopus* embryos (Martello et al., 2007). These miRNAs specifically inhibit mRNA of the Nodal receptor mRNA (acvr2a), reduce expression of the receptor protein and as a consequence reduce Nodal signaling in ventral cells. This regionalized inhibition contributes to preferential activation of Nodal signaling on the dorso-anterior side of the embryo and facilitates the asymmetric positioning of the organizer.

Another mechanism that restricts Nodal signaling involves regulated processing of the miR-15/16 precursors (Martello et al., 2007). While ventral cells preferentially express miR-15/16 precursors, the dorso-anterior blastomeres that give rise to the organizer also express these precursors. However, Wnt signaling is active in these dorso-anterior cells and some evidence suggests that this signaling inhibits processing of the precursors to mature miR-15/16 miRNAs in this part of the embryo. In the absence of miR-15/16 Nodal signaling is unimpeded in the cells that give rise to the organizer.

3.2 miR-427 Regulation of Nodal Signaling

miR-427 is expressed beginning with the onset of zygotic transcription and has the opposite effect on Nodal signaling and the organizer compared to the miR-15/16 (Rosa, Spagnoli, & Brivanlou, 2009). miR-427 targets the mRNAs encoding the Nodal antagonist Lefty. Overexpression of miR-427 prolongs Nodal signaling and increases the size of the organizer while blocking miR-427 has the opposite effect and decreases the size of the organizer. miR-427 also potentially targets the mRNAs encoding the Nodal ligands, *Xnr5* and *Xnr6*. Presumably, inhibiting the expression of these ligands would cause a reduction in Nodal signaling, suggesting that either they are not genuine targets of miR-427 or that the inhibition of Lefty mRNA predominates.

The regulation of Nodal signaling by miR-15/16 and miR-427 indicates that these miRNAs perform key regulatory functions and influence cell fates during embryonic development, but many important questions remain. For example, it is unclear whether these miRNAs are exerting their affects during the maternal or the early zygotic periods of *Xenopus* development. In addition, miRNAs typically affect their mRNA targets by repressing translation and/or promoting mRNA degradation, but is it unclear how miR-15/16 and miR-427 affect their mRNA targets.

4. TRANSLATIONAL CONTROL OF mRNAs ENCODING CELL FATE DETERMINANTS

In *Xenopus*, many important events that control axis formation and cell fate decisions occur during the maternally controlled stages of development that follow fertilization (Fig. 1). With the exception of the activin receptor mRNA, the mRNAs and mechanisms discussed above were chosen for analysis solely because of the embryonic nature of their regulation. Subsequent studies have emphasized maternal mRNAs that encode important cell fate regulatory proteins whose translation is regulated during early embryogenesis. The underlying motivation is that the mechanisms that control the translation of such mRNAs represent important control points in early development. These mRNAs and their regulation will be discussed in Section 4.1.

4.1 Wnt11 mRNA

Maternal loss-of-function studies indicate a critical role for Wnt11 in *Xenopus* axis formation (Tao et al., 2005). Embryos depleted of Wnt11 mRNA lack most dorso-anterior structures, while overexpression of Wnt11 has the opposite effect: embryos develop with an excess of anterior cell types. These observations suggest that the maternal Wnt11 mRNA has an important role in the events required for the formation of the organizer. Consistent with this idea, analysis of the Wnt11 mRNA suggests that its translation is connected to cortical rotation (Schroeder, Condic, Eisenberg, & Yost, 1999). In cleavage-stage embryos, the Wnt11 mRNA is uniformly distributed between dorso-anterior and ventral-posterior blastomeres. However, Wnt11 mRNA in dorso-anterior cells is polyadenylated more extensively compared to the mRNA in ventral cells. The presence of a poly(A) tail on maternal mRNAs in oocytes, eggs, and embryos is known to stimulate translation. Therefore, the observed differential polyadenylation suggests that translation of Wnt11 is activated in dorso-anterior cells. In support of this idea, the Wnt11 mRNA in dorso-anterior cells is preferentially associated with polyribosomes compared to other cells. Significantly, both Wnt11 polyadenylation and translation are sensitive to treatments known to disrupt cortical rotation, such as exposure to low doses of UV light. These results suggest that cortical rotation activates the polyadenylation and translation of the Wnt11 mRNA in dorsal blastomeres, the cells that will give rise to the organizer (Schroeder et al., 1999). However, other results suggest that the Wnt11 mRNA itself is transported to the dorsal cells during cortical rotation (Tao et al., 2005). The subsequent translational activation would still be sufficient to activate Wnt signaling in the cells that will give rise to the organizer. Resolution of this issue will require additional experiments.

4.2 VegT mRNA

The VegT mRNA encodes a T-box transcription factor and the mRNA is localized to the vegetal cortex of fully grown stage 6 oocytes (Lustig et al., 1996; Stennard, Carnac, & Gurdon, 1996; Zhang & King, 1996). Embryos depleted of VegT maternal mRNA do not form endoderm and they exhibit defects in the production of signals needed to induce mesoderm (Zhang et al., 1998). Localized mRNAs are translationally repressed while they are transported during oogenesis (King et al., 2005; Medioni et al., 2012). At subsequent stages of development the translation of localized mRNAs is activated, but these activation processes are poorly understood. An exception is the XSeb4R protein that acts as a positive regulator of VegT mRNA translation and stability in embryos (Souopgui et al., 2008). XSeb4R is an

RRM domain-containing protein that binds to 3′UTR sequences in the VegT mRNA. The mechanistic basis of how XSeb4R enhances translation and stability is unknown.

4.3 Temporal Control of Translation: The Bone Morphogenetic Protein Pathway

Signaling by the bone morphogenetic protein (BMP) pathway is important for multiple aspects of animal development (Moustakas & Heldin, 2009; Ramel & Hill, 2012). In *Xenopus*, BMP signaling functions in select regions of embryos to guide important cell fate decisions. This signaling is first activated at the MBT via a BMP pathway assembled from maternal signaling proteins (Faure, Lee, Keller, ten Dijke, & Whitman, 2000). Because the translation of maternal mRNAs is highly regulated, this suggests that controlling the synthesis of BMP pathway components via regulated translation could be used to control the timing of signaling.

To investigate this idea, the mRNAs encoding proteins of the *Xenopus* BMP pathway were analyzed using a polyribosome association assay to monitor translational activity (Fritz & Sheets, 2001). A diverse array of translational behaviors is observed. All of the mRNAs analyzed are inefficiently associated with polyribosomes in oocytes and then each mRNA is recruited to polyribosomes at a different developmental stage (Fig. 2). For example, the mRNAs encoding the Smad1 transcription factor and the ALK2 receptor (a type I BMP receptor) are recruited to polyribosomes during oocyte maturation. In contrast, the mRNA encoding the BMP7 ligand is recruited during the blastula stage of embryogenesis. The BMP-receptor1A mRNA (BMPR1A, also called ALK3) is only recruited to polyribosomes after the onset of zygotic transcription during the MBT. Thus, the maternal mRNAs of the BMP pathway are temporally regulated at the post-transcriptional level. The mRNAs are initially repressed and then, as development proceeds, specific mRNAs become activated at characteristic times. The BMPR1A mRNA is not translationally activated until the MBT, suggesting that synthesis of the BMPR1A receptor could be rate limiting for formation of a functional pathway and BMP signaling.

As discussed above, the addition of poly(A) is a prevalent mechanism used to control the translation of specific maternal mRNAs (See Section 2.1). Notably, the poly(A) tail of each mRNA is lengthened coincident with it polyribosome recruitment (Fig. 3). For example, the poly(A) tail of Smad1 mRNA is lengthened during oocyte maturation,

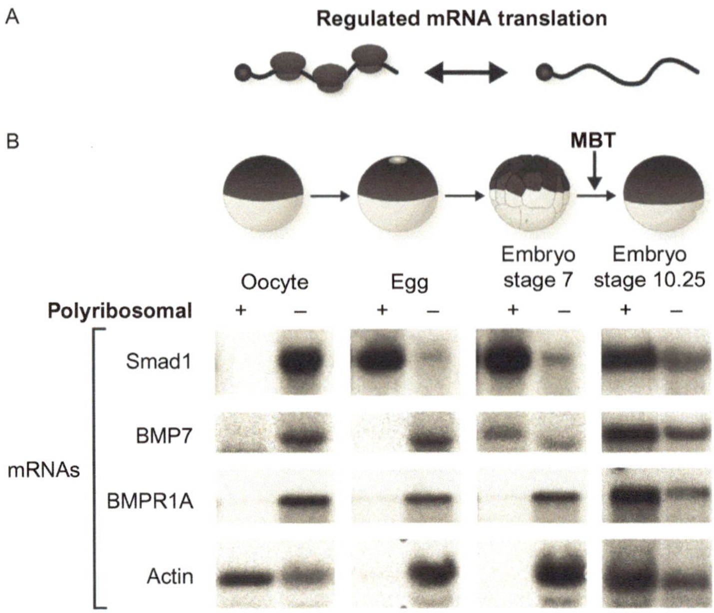

Figure 2 Translational control of maternal mRNAs. (A) During the early stages of *Xenopus* development, mRNA translation is highly regulated. This diagram depicts these general events in which mRNAs switch between being ribosome associated and translationally active versus translationally repressed. (B) mRNAs encoding proteins of the BMP pathway are translationally regulated. Polyribosome association assays were used to monitor the translational activity of specific mRNAs. RNAs from the polyribosome and nonpolyribosome fractions of oocytes, eggs, stage 7 embryos, and stage 10.25 embryos were analyzed for the presence of specific mRNAs using RNA blot hybridization. The examples shown illustrate the diverse patterns of regulation observed. For example, the Smad1 mRNA is repressed in oocytes and then translationally activated during oocyte maturation. In addition, the BMP7 mRNA is repressed in oocytes and eggs and becomes translationally activated during cleavage stages. The results shown also demonstrate that while the translation of these mRNAs is regulated during the maternally controlled stages of embryogenesis, once the MBT occurs the regulation is significantly diminished and the mRNAs are translated at similar levels. *Adapted from Fritz and Sheets (2001).*

while the BMP7 mRNA becomes associated with polyribosomes and polyadenylated during embryogenesis. These results suggest that the temporal control of polyadenylation regulates the translation of each mRNA (Fritz & Sheets, 2001).

The timing of each mRNA's polyadenylation suggests specific regulatory mechanisms. For example, the Smad1 mRNA is polyadenylated during

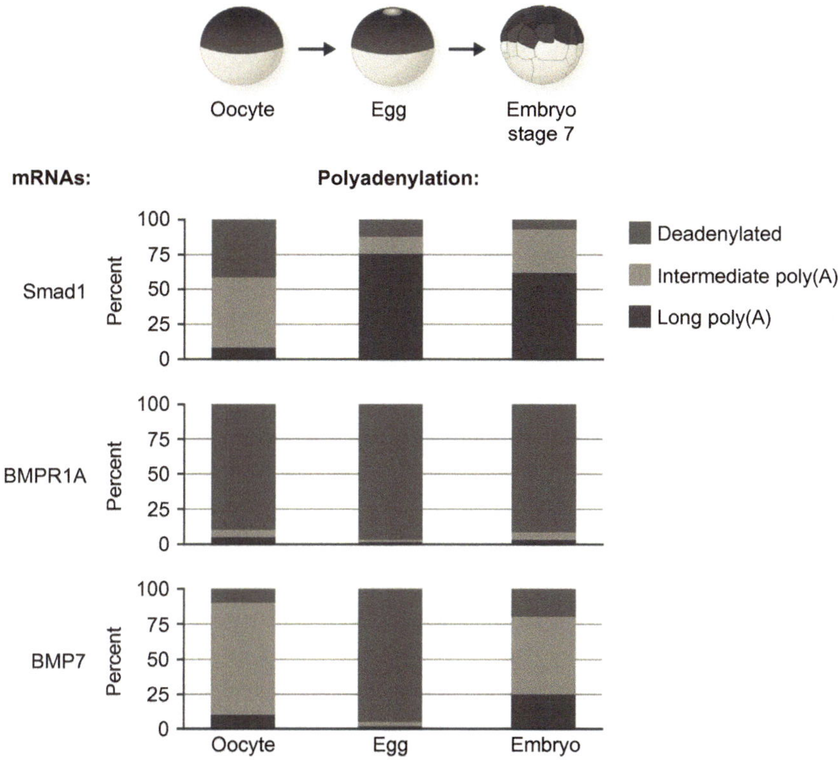

Figure 3 Regulation of maternal mRNA polyadenylation. One prevalent mechanism used to regulate the translation of maternal mRNAs in *Xenopus* and other organisms is the cytoplasmic addition and removal of 3′ poly(A); mRNAs that have elongated tails are translated more efficiently than mRNAs that have no poly(A) or shorts tails (Cragle & MacNicol, 2014). The dynamic nature of poly(A) metabolism and its relationship to translational behavior have been studied by in-depth analysis of individual mRNAs and by genomic approaches. The graph shown plots the length of poly(A) tails present on three *Xenopus* maternal mRNAs at different stages of development. The translational activation of each mRNA parallels the presence of a long poly(A) tail. Translation of the Smad1 mRNA is activated during oocyte maturation when its poly(A) tail is elongated. In addition, the BMP7 mRNA is translationally activated during the blastula stages, coincident with the elongation of its poly(A) tail. Finally, the BMPR1A mRNA is not translated at any of the stages analyzed and it is deadenylated or has only a short poly(A) tail. These and other results illustrate that the timing of poly(A) elongation during maternally controlled stages of *Xenopus* embryonic development is a predominant mechanism for temporally controlling the translation of specific mRNAs. *Adapted from Fritz and Sheets (2001).*

oocyte maturation, suggesting that CPE/CPEB-dependent mechanisms function on this mRNA during maturation. In contrast, the *Xenopus* BMP7 mRNA is polyadenylated following fertilization. Other mRNAs polyadenylated during this period of development contain sequence elements called eCPEs in their 3′UTRs (Simon & Richter, 1994; Simon et al., 1992). However, BMP7 mRNA lacks such sequences suggesting that this embryo-specific polyadenylation process may be controlled by novel mechanisms.

4.4 Spatial Control of Translation: The Cripto-1 mRNA

Signaling by the Nodal pathway is critical for vertebrate development and Cripto proteins are secreted coreceptors important for this signaling (Klauzinska et al., 2014; Nagaoka et al., 2012; Schier, 2009). In *Xenopus*, depleting the maternal Cripto-1 mRNA from embryos changes cell fates and disrupts axis formation (Schier, 2009; Tao et al., 2005). The importance of maternal Cripto-1 mRNA for *Xenopus* development prompted investigations into Cripto-1 mRNA translation, which is regulated both temporally and spatially (Zhang et al., 2009). Temporally, the Cripto-1 mRNA is initially repressed in oocytes and eggs, but after fertilization a fraction of the message becomes translationally active. Spatially, while the Cripto-1 mRNA is uniformly distributed throughout the *Xenopus* embryo, the Cripto-1 protein only accumulates in cells of the marginal zone and animal hemisphere (Dorey & Hill, 2006). Polyribosome analysis reveals that the differences in Cripto-1 protein accumulation are due to translational regulation. Cripto-1 mRNA is specifically repressed in vegetal cells and activated in animal cells. Thus, Cripto-1 mRNA translation is spatially regulated in the *Xenopus* embryo (Zhang et al., 2009; Fig. 4) and this regulation could be functionally important for pathways that depend upon Cripto-1 for signaling.

To investigate the regulation of the Cripto-1 mRNA, luciferase reporter mRNAs containing the Cripto-1 3′UTR were assayed after injection into either animal cells or vegetal cells of blastula stage embryos (Zhang et al., 2009). These studies reveal that vegetal cells efficiently repress reporter mRNA translation. This repression is specific, as reporter mRNAs with other 3′UTRs are translated equally by both cell types. Additional experiments identified a subregion of the Cripto-1 3′UTR referred to as the translational control element (TCE) that is sufficient for repression.

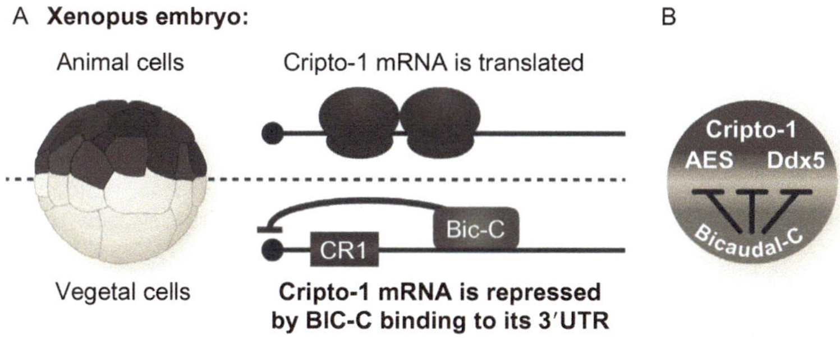

Figure 4 Spatial accumulation of the Cripto-1 protein in *Xenopus* embryos is controlled by regulated translation of the Cripto-1 mRNA. In *Xenopus*, the Cripto-1 mRNA is uniformly distributed throughout the egg and all embryonic cells. The Cripto-1 protein is only produced in animal cells of embryos after fertilization, but not in vegetal cells (Dorey & Hill, 2006). Differential accumulation of the Cripto-1 protein in *Xenopus* embryos is due to the spatially regulated translation of the Cripto-1 mRNA. Translation is activated in animal cells and repressed in vegetal cells (Zhang et al., 2009). The Bicaudal-C (Bic-C) protein is responsible for repression functions in the vegetal cells (Zhang et al., 2013; Zhang, Park, Blaser, & Sheets, 2014). The Bic-C mRNA is localized to the vegetal cortex of developing oocytes and, as a consequence, after fertilization Bic-C is restricted to vegetal cells (Wessely & De Robertis, 2000).

4.5 Bicaudal-C: A Lineage-Specific Translational Repressor

Restricting Cripto-1 protein accumulation to the correct embryonic cells requires both mechanisms that activate Cripto-1 mRNA translation in animal cells and mechanisms that repress Cripto-1 mRNA translation in vegetal cells. Bicaudal-C (Bic-C) was identified as the key RNA-binding protein responsible for vegetal cell-specific translational repression of Cripto-1 mRNA (Zhang et al., 2013; Fig. 4). Two pieces of experimental evidence led to this conclusion. First, the *Xenopus* Bic-C mRNA is localized to the vegetal cortex of oocytes and as such is restricted to vegetal cells of developing embryos. This suggests that the Cripto-1 mRNA is not repressed in animal cells because these cells lack the Bic-C repressor. Importantly, ectopic expression of Bic-C leads to Cripto-1 mRNA repression in these cells. Specifically, reporter mRNAs containing the Cripto-1 3′UTR are repressed efficiently in animal cells forced to express Bic-C. This repression is specific as it depends upon the presence of the TCE from the 3′UTR of the Cripto-1 mRNA. The second prediction is that Bic-C directly binds the Cripto-1 mRNA. Immunoprecipitation of Bic-C followed by analysis of the associated RNA demonstrates that the endogenous Cripto-1 mRNA as well as the Cripto-1 reporter mRNA are both efficiently bound by Bic-C. Additionally, biochemical experiments identified a single 32-nucleotide Bic-C-binding site within the TCE of the Cripto-1 mRNA 3′UTR (Zhang, Park, et al., 2014). This binding site is predicted to form a stem-loop structure. Significantly, this minimal Bic-C-binding site is *sufficient* for both Bic-C binding and translational repression *in vivo*. These results demonstrate that Bic-C is vegetal cell-specific repressor responsible for spatially regulating Cripto-1 mRNA translation.

4.6 Defining the Bic-C Maternal Regulatory Network

The identification of Cripto-1 mRNA as a Bic-C target provided new molecular insights into Bic-C functions in the early embryo. RNA binding proteins like Bic-C typically regulate multiple mRNAs, however few targets of Bic-C were known. To address this deficit, Bic-C was immunoprecipitated from *Xenopus* embryos and the associated RNAs were analyzed with RNA-Seq to identify 62 new Bic-C targets (Zhang et al., 2013). Many of the putative Bic-C target mRNAs encode proteins that function in developmentally important pathways. These include the AES/GRG5 transcription factor (Costa et al., 2013) and the Ddx5 RNA helicase (Guturi, Sarkar, Bhowmik, Das, & Ghosh, 2014) are known regulatory proteins implicated in Wnt signaling. Other targets encode post-transcriptional regulatory proteins such as neuroguidin (Jung, Lorenz, & Richter, 2006; Udagawa et al., 2012) and tristetraproline (Brooks & Blackshear, 2013) each of which have their own RNA targets. Together, these results suggest that Bic-C is at the center of a post-transcriptional regulatory network in the *Xenopus* embryo. The function of this network is predicted to establish the proper spatial distribution of specific maternal proteins in the embryo that is essential for normal development.

5. REGULATION OF MATERNAL PROTEINS IN THE *XENOPUS* EMBRYO

5.1 Maternal Signaling Proteins and Their Activation by Phosphorylation

When and where signaling pathways are first activated during *Xenopus* embryonic development has been addressed with immunostaining experiments using antibodies specific for the phosphorylated and activated pathway components, including activated forms of Smad1 (the BMP pathway), Smad2 (the Nodal pathway), MAPK (the FGF pathway), and β-catenin (the Wnt pathway) signaling proteins. Results indicate that robust signaling by the BMP, Nodal, and FGF pathways coincides with the MBT (Faure et al., 2000; Heasman, 2006b; LaBonne & Whitman, 1997; Lee, Heasman, & Whitman, 2001; Schohl & Fagotto, 2002), and only low levels of signaling can be detected in the preceding maternally controlled stages. In contrast, signaling by the Wnt pathway is activated prior to the MBT.

5.2 Temporal Control of TGFβ Signaling and the MBT

Members of the transforming growth factor-β (TGFβ) family of ligands, including Nodal, activin, and the BMPs, regulate a variety of key processes during the early stages of *Xenopus* development, such as guiding specific cell differentiation events and orchestrating the cell movements of gastrulation (Moustakas & Heldin, 2009; Ramel & Hill, 2012). TGFβ ligands transduce signals through heteromeric receptor complexes containing Type I and Type II serine/threonine kinase receptors and intracellular Smad proteins. In response to ligand binding, the Type II receptor phosphorylates the Type I receptor, which in turn phosphorylates the receptor-regulated Smads (R-Smads) at a C-terminal SXS motif. Smad1 and Smad5 are substrates for the BMP receptors, whereas Smad2 and Smad3 are substrates for the Nodal and activin receptors. Activated R-Smads translocate to the nucleus after associating with Smad4. In the nucleus, the activated Smad complexes recruit additional factors, such as the FoxH1 transcriptional activator, as well as general coactivators (e.g., p300 and CBP) or corepressors (e.g., SnoN), to regulate the transcription of specific target genes.

Analysis of the BMP and Nodal/Activin pathways and their activation at MBT suggest different modes of regulation. For example, transcriptional inhibitors do not block Smad1 phosphorylation indicating that activation of BMP signaling relies upon assembly of the BMP pathway from maternal components (Faure et al., 2000). Most components of the BMP pathway are encoded by maternal mRNAs and their translation is highly regulated (see Section 4.3). This raises the possibility that the regulated translation of an mRNA encoding a limiting component contributes to control of the timing of BMP signaling. In contrast, activation of Smad2 phosphorylation requires active transcription suggesting that an intact and functional Nodal/Activin signaling pathway is not present maternally and that an unknown rate-limiting component must be synthesized from a zygotic gene (Faure et al., 2000). However, it is not that simple. Providing pre-MBT cells with Activin can precociously activate signaling as assayed by the formation of phospho-Smad2, but precocious activation is not observed by adding Nodal-related ligands (Lee et al., 2001).

Other results suggest that nuclear import may restrict signaling. In most cell types, phosphorylated Smads translocate to the nucleus to mediate changes in gene expression. In *Xenopus* embryos, imaging of fluorescently tagged Smad proteins reveals that Smad proteins remain in the cytoplasm of embryonic cells and translocate to the nucleus only at the MBT (Saka, Hagemann, Piepenburg, & Smith, 2007). This suggests that regulated nuclear import of phospho-Smads controls the timing of signaling, at least for Smad2 activated pathways. It is unknown whether a similar nuclear import restriction applies to the Smads activated during BMP signaling. Regulated nuclear import of *Xenopus* proteins in pre-MBT embryos has been observed, but the underlying mechanisms are unknown (Dreyer, 1989). In summary, while it is clear that embryonic cells have multiple strategies to temporally restrict TGFβ signaling, precisely how these restrictions are established in the early, maternally controlled stage of embryogenesis and then released at MBT remains to be determined.

5.3 Temporal Control of Wnt Signaling and the MBT

As discussed earlier, maternal loss-of-function experiments demonstrate that Wnt signaling is critical for embryonic development (Houston, 2012). It is clear from experiments monitoring phosphorylated β-catenin when and where signaling occurs in the *Xenopus* embryo (Schohl & Fagotto, 2002). Wnt signaling is activated by cortical rotation shortly after fertilization in cleavage-stage embryos and signaling is localized predominantly to the vegetal cells on the dorso-anterior side of the embryo. Thus Wnt signaling is activated early in development, well before MBT, and restricted to a specific group of cells, those that will induce and give rise to the organizer-inducing center.

The MBT marks the time during *Xenopus* development when transcription is first activated robustly for a large number of genes. However, during the cleavage stages prior to MBT there are a limited number of genes that are transcriptionally activated, such as the *Xenopus* *n*odal-*r*elated 5 and 6 (Xnr-5 and 6) genes (Skirkanich, Luxardi, Yang, Kodjabachian, & Klein, 2011; Yang, Tan, Darken, Wilson, & Klein, 2002). The activation of these genes requires Wnt signaling and specifically β-catenin. In most contexts, activated β-catenin translocates to the nucleus where it functions along with other transcription factors to activate specific gene targets. In the case of cleavage-stage *Xenopus* embryos and the Xnr-5 and 6 genes, β-catenin functions to modify the chromatin template of these genes to facilitate their transcription (Blythe, Cha, Tadjuidje, Heasman, & Klein, 2010). Thus, the β-catenin activation of the Xnr-5 and 6 genes and potentially other early Wnt gene targets represent some of the earliest zygotic events of *Xenopus* development.

6. REGULATED PROTEOLYSIS AND THE CONTROL OF MATERNAL SIGNALING PATHWAYS

6.1 TGF-β Signaling

Regulated proteolysis influences TGFβ signaling in multiple ways by targeting different pathway components for degradation (Datto & Wang, 2005; Xu, Liu, & Derynck, 2012). For example, after activation, specific signaling proteins such as the Smads are degraded to reset the pathway to its initial state and allow reactivation to occur. In other cases, proteolysis is required for pathway activation, such as the degradation of specific repressor proteins is necessary for the induction of TGFβ-responsive genes.

6.2 SMURF Proteins

*Sm*ad *u*biquitin *r*egulatory *f*actor 1 and 2 (Smurf1 and 2) are E3 ubiquitin ligases that regulate signaling by TGFβ ligands (Inoue & Imamura, 2008; Zhu, Kavsak, Abdollah, Wrana, & Thomsen, 1999). Smurfs contain WW domains that mediate protein–protein interactions and a C-terminal HECT domain for ubiquitin ligation. Smurf1 was originally identified as an E3 ubiquitin ligase that interacts with Smad1 and Smad5 via the Smurf1 WW domain and the Smad PY motif in its linker region to induce degradation. Smurf1 binding inhibits Smad1 from interacting with the nuclear translocation factors and promotes its cytoplasmic retention. Smurf2 interacts with both Smad1 and Smad2 to induce their ubiquitin-mediated degradation. In addition, Smurfs target other proteins for degradation such as certain TGFβ receptors and negative regulators of the pathway, such as the SnoN transcriptional repressor (Bonni et al., 2001; Tan, He, Lin, Kiss, & Liu, 2008). The degradation of distinct Smurf targets is thought to influence the initiation, strength, and/or duration of TGFβ signaling.

The Smurfs were discovered and in *Xenopus* embryos and this context provided the first functional insights into these proteins (Zhu et al., 1999). Loss-of-function experiments in *Xenopus* combined with the use of dominant-negative mutants have been used to assess the function of

Smurf1 during early frog embryogenesis (Alexandrova & Thomsen, 2006). Smurf1 regulates neural patterning but not mesendoderm development. A subsequent study that focused on the roles of Smurf2 in early *Xenopus* development found that, while Smurf1 and Smurf2 exhibit distinct activities, they also share some functional overlap and work in concert to properly regulate development (Das & Chang, 2012). Both Smurfs are encoded by maternal mRNAs. The Smurf1 mRNA is localized to the animal pole of eggs and blastula stage embryos, and the majority of the mRNA is degraded at the MBT. This regulation raises the possibility that the Smurf protein may be present at high levels in pre-MBT embryos and contribute to the restriction of TGFβ signaling during this period.

6.3 Ectodermin and Smad4 Monoubiquitination

Ectodermin was identified in a screen performed in *Xenopus* embryos designed to identify molecules opposing the differentiation of ectodermal cells (Dupont et al., 2005). The Ectodermin protein is encoded by a maternal mRNA that is concentrated in cells of the animal pole. Some of these cells are fated to form the ectoderm, and Ectodermin is required to protect these cells from Nodal/TGFβ signals that could induce them to form endoderm and mesoderm. Within the presumptive ectodermal cells, Ectodermin restricts BMP signaling and balances their differentiation into epidermal and neural lineages.

Ectodermin (also called Tif1g/TRIM33) was initially proposed to be a SMAD4 polyubiquitination ligase, but later studies demonstrated that it is a ligase for the monoubiquitination of SMAD4 (Dupont et al., 2009; Morsut et al., 2010). This modification of SMAD4 does not lead to degradation but it disrupts the ability of SMAD4 to form complexes with SMAD2 and their transcriptional functions in the nucleus. FAM/Usp9x was identified as an enzyme that specifically deubiquitinates Smad4 and it is critical for both TGFβ and BMP responsiveness in *Xenopus* embryos and human cells (Dupont et al., 2009). Mechanistically, the deubiquitination of Smad4 by FAM/Usp9x enables the formation of productive SMAD2–SMAD4 complexes that increase gene expression.

6.4 Wnt Signaling—Regulated Proteolysis of β-Catenin

In the canonical Wnt pathway, signaling is initiated when Wnt ligands bind to the Frizzled and LRP5/6 receptor proteins that trigger a series of events that culminating in the accumulation and translocation to the nucleus of β-catenin (Nusse, 2012; van Amerongen & Nusse, 2009). In the nucleus β-catenin functions with TCF transcription factors to regulate the expression of specific genes, most notably the genes of Spemann's organizer. In the Wnt pathway, proteosomal degradation of specific components is integral to the signaling process.

β-Catenin is restricted to cells receiving Wnt signals by a group of proteins that mediate its proteolysis and are referred to as the destruction complex (Hikasa & Sokol, 2013; MacDonald, Tamai, & He, 2009). Components of this complex include proteins such as axin, adenomatous polyposis coli, and glycogen synthase kinase 3β. In the absence of a Wnt ligand, the destruction complex ubiquitinates β-catenin and targets it for proteolytic degradation. The activation of Wnt signaling disrupts the formation of the destruction complex, and β-catenin accumulates and translocates to the nucleus to activate transcription.

The *Xenopus* β-transducin repeat containing protein (β-TRCP) is an E3 ubiquitin ligase encoded by a maternal mRNA (Lagna, Carnevali, Marchioni, & Hemmati-Brivanlou, 1999; Liu et al., 1999; Marikawa & Elinson, 1998). This maternally encoded E3 ligase functions as a negative regulator of Wnt signaling as a component of the destruction complex. Its role is to recognize β-catenin and target it for ubiquitination and proteolysis.

6.5 Notch Regulation of β-Catenin

Recent results suggest that signaling by the Notch pathway also regulates β-catenin proteolysis (Acosta, Lopez, Revinski, & Carrasco, 2011). Disrupting Notch signaling during the maternally controlled stages of development before MBT gives rise to embryos with exaggerated neural derivatives, effects that are consistent with the expansion of β-catenin signaling. Additional experiments provide evidence that Notch signaling directs the proteolysis of β-catenin and contributes to confinement of β-catenin to a subset of embryonic cells.

6.6 Proteolytic Degradation of Disheveled

The discovery of β-catenin and its regulated proteolysis was seminal for understanding signaling by the Wnt pathway (Hikasa & Sokol, 2013; MacDonald et al., 2009). Subsequent studies revealed that β-catenin is not unique and that the signaling functions of other pathway components are also regulated by proteolysis.

Disheveled proteins (Dvl) are conserved and key components of the Wnt signaling pathway. The NEDD4L protein (*n*eural precursor cell *e*xpressed, *d*evelopmentally *d*own-regulated *4*-like, also called NEDD4-2) is a ubiquitin ligase that regulates proteolysis of Dvl proteins. NEDD4L proteins contain four WW domains, and a C-terminal HECT ubiquitin-ligase domain. The WW domains mediate NEDD4L's recognition and binding to substrates containing PPXY (PY) motifs. The third WW domain of NEDD4L directly interacts with the PY motif of Dvl2 and this binding promotes the polyubiquitination and degradation of Dvl2 in cultured cells and *Xenopus* embryos. In *Xenopus* embryos, the NEDD4L protein is encoded by a maternal mRNA and both the mRNA and protein are enriched in cells of the animal hemisphere (Ding, Zhang, Xu, Tao, & Chen, 2013; Zhang, Ding, Chen, & Tao, 2014). Loss-of-function experiments in *Xenopus* embryos, employing morpholinos, demonstrate that NEDD4L depletion causes defects in gastrulation, neural tube closure, and specification of the mesoderm. These defects can be attributed in large part to the misregulation of Dvl protein expression. Thus, the regulation of Dvl protein levels by NEDD4L directed proteolysis is essential for several important processes in *Xenopus* embryos (Ding et al., 2013; Zhang, Ding, et al., 2014).

In *Xenopus* embryos and cultured cells, the KLHL12 protein (kelch-like family member 12) negatively regulates Dvl functions by recruiting Dvl to the Cullin-3 ubiquitin-ligase complex, promoting its ubiquitination and degradation (Angers et al., 2006). Overexpression of KLHL12 in *Xenopus* embryos blocks Dvl-regulated events such as convergent extension and Dvl-mediated transcriptional activation of specific genes.

7. THE *XENOPUS* MATERNAL PROTEOME

There is a long history of using proteomic approaches to analyze *Xenopus* oocytes, eggs, and embryos and to identify developmentally relevant proteins. In the early era two-dimensional gel electrophoresis was used to identify proteins that changed in abundance in different samples (Ballantine, Woodland, & Sturgess, 1979; Bravo & Knowland, 1979). The underlying premise was that such changes potentially reflected regulation important for development. While proteins with such properties were observed, most could not be definitively identified and this drawback severely limited the value of this approach.

More recently, advances in mass spectrometry have provided new power to proteomics analysis, providing a global view of the proteins present in

Xenopus eggs and embryos and how these change during development. For example, one study identified over 11,000 maternal proteins present in eggs and obtained estimates of each protein's relative concentration (Wuhr et al., 2014). In another study, the proteome of embryos at several stages was analyzed and the identified proteins classified according to abundance changes seen during development (Sun et al., 2014). One regulatory class included the RecQ4 and Cut5 proteins whose reductions may be important for changes in the cell cycle that occur at the MBT. A consistent finding from these studies is that there are significant differences when comparing the concentrations of individual mRNAs and their corresponding proteins (Smits et al., 2014). This discordance supports the view that many maternal mRNAs are translationally repressed in the egg and stored for activation at later stages of development. While these changes were attributed to translational regulatory processes, it is also plausible that the differences are due to differential proteolysis of the proteins. Resolving this issue will require direct analysis of the translational activity of individual mRNAs.

From the studies outlined above, it is clear that the comparative proteomics of *Xenopus* eggs and embryos has begun to provide new insights into the biological processes occurring as the embryo transitions from a strictly maternal to a zygotic program of development during the MBT. Equally important, this approach has led to the identification of proteins that are specifically regulated on a translational or post-translational level and have essential functions during early development.

8. THE MATERNAL PHOSPHOPROTEOME

The central importance of reversible phosphorylation as a regulatory mechanism has led to the development of methods to globally identify the phosphorylation events present in a particular cell type. The application of these methods provides a comprehensive view of protein phosphorylation in specific cells and can also identify specific modifications that change under different conditions to reveal potential regulatory events. In *Xenopus* oocytes, eggs, and embryos, there are well-documented examples of phosphorylation events, but outside of these examples, there is relatively limited information about the phosphorylation of proteins involved in developmental processes.

To begin to address this deficit, electron capture dissociation mass spectrometry was used to analyze the *Xenopus* maternal phosphoproteome and identified over 1400 different phosphorylation sites on 654 different proteins from oocytes, eggs, and embryos (McGivern, Swaney, Coon, & Sheets, 2009). The identified phosphoproteins are known to function in a diverse array of cellular processes. For example, the eIF4E-binding protein (4E-BP2) regulates translation by inhibiting the activity of eukaryotic translation initiation factor 4E (eIF4E) (Kong & Lasko, 2012). The phosphorylation of 4E-BP2 reverses its inhibition of eIF4E, resulting in a general stimulation of translation. Phosphoproteomic analysis indicated that *Xenopus* 4E-BP2 is phosphorylated on multiple residues and these modifications change during development (Fig. 5; McGivern et al., 2009). These results suggest that *Xenopus* oocytes, eggs, and embryos may possess different translational environments due to differences in the phosphorylation of the 4E-BP2 protein.

This study provides the foundation for addition analyses that seek to comprehensively identify all phosphorylation events and how they change during the maternal stages of *Xenopus* development. These results also provide an informational foundation for functional studies of specific phosphorylation events. *Xenopus* offers many advantages for functional studies in a developmental context, either by the *in vivo* expression of protein variants coupled to phenotypic assays or by analyzing the activity of modified proteins exploiting *in vitro* extracts (Amaya, Musci, & Kirschner, 1991; Murray & Kirschner, 1989). These approaches provide relatively straightforward strategies to functionally evaluate specific phosphorylation events that are not possible in other embryos or cell types. Experiments of this type

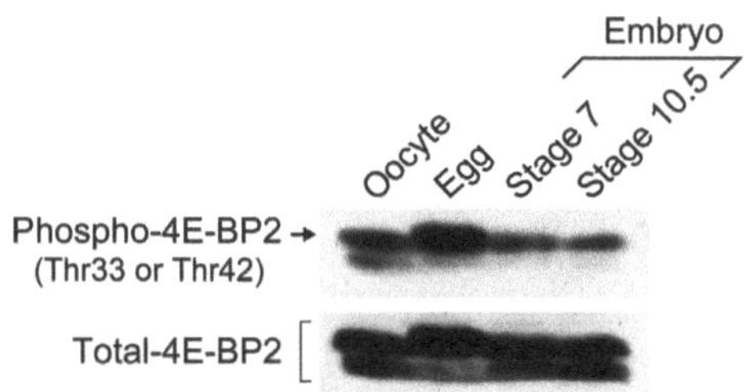

Figure 5 Reversible phosphorylation of maternal proteins. Reversible phosphorylation is a fundamental post-transcriptional regulatory mechanism used to control the amounts and activities of specific proteins such as the *Xenopus* eIF4E-binding protein two (4E-BP2). Proteins from *Xenopus* oocytes, eggs, and embryos were analyzed by immunoblotting and antibodies that recognize 4E-BP2 regardless of its modification (bottom), and antibodies that recognize only phosphorylated forms of 4E-BP2 (top). This revealed that while the total amount of 4E-BP2 is relatively constant (bottom), 4E-BP2 phosphorylation undergoes dynamic changes at the different stages of *Xenopus* development. *Adapted from McGivern et al. (2009).*

could provide new insights into the regulatory events that mediate cell fate decisions and as well as the regulatory events that facilitate the major changes that occur during important developmental events such as the maternal to zygotic transition.

9. METABOLISM AND METABOLOMICS

The MBT represents a time of dramatic changes for the embryo and embryonic cells. While it is most often associated with the activation of transcription and the lengthening of the cell cycle, there are other events, such as changes in metabolism, which occur during MBT that have received less attention. The analysis of small molecules and metabolites in *Xenopus* oocytes, eggs, and embryos is such a relatively new area of investigation that it is challenging to relate the results from recent analyses to more traditional studies of development. However, these types of experiments have begun to provide new insights into how metabolism is intimately connected to developmental events such as the MBT.

For example, changes in the glycolytic cycle that occur as the embryo progresses through the MBT were observed over 25 years ago (Dworkin & Dworkin-Rastl, 1987, 1992). This led to the idea that such metabolic changes are needed to meet the energy demands of embryonic cells as they activate transcription, remodel the cell cycle, and initiate new cell behaviors. More recently, the powerful tools of mass spectrometry have been applied to *Xenopus* embryos to detect and monitor metabolites during development. Vastag et al. analyzed several different metabolites before, during and after the MBT and observed significant changes in their abundance (Vastag et al., 2011). One of their most intriguing observations is that the maternally supplied dATP pool decreases during the MBT. This suggests that the slowing and lengthening of the cell cycle that occur at the MBT is due to the decrease in the dNTPs available to support replication. Another study monitored acetyl CoA levels in *Xenopus* embryos and observed that this metabolite increases significantly at the same time as proteins such as the core histones are acetylated (Tsuchiya, Pham, Hu, Ohnuma, & Gout, 2014). This suggests that increases in acetyl CoA could facilitate acetylation events that affect development. For example, changes in acetyl CoA could facilitate the formation of modified histones and the formation of modified chromatin templates for gene expression. Shrestha et al. used laser ablation electrospray ionization (Shrestha et al., 2014) to analyze *Xenopus* eggs and embryos and monitor regional differences in metabolites.

They found clear differences in the metabolites present in the animal and vegetal regions of embryos. This indicates that small molecules are likely to make an important contribution to establishing the differences in the animal–vegetal axis of the embryo.

10. CONCLUSIONS

The MBT represents an important hallmark of *Xenopus* development that has intrigued biologists for decades. While it has been primarily associated with the major activation of transcription, there are many other processes that also change at this time. As described above, there are changes in mRNA processing and mRNA stability mechanisms as well as changes in important signaling events, all intimately connected to the activation of zygotic transcription. To fully understand the events of the MBT and their influence on embryonic patterning requires defining the component networks (e.g., mRNAs and proteins) and activity networks (mRNA translation and protein modification) present in the *Xenopus* embryo and how they change. While the results thus far have provided important clues and glimpses of these systems, additional studies are required to provide a comprehensive view of each network and how they influence development. The ultimate goal is to determine how the networks function together as an integrated system to direct the fates of specific subsets of embryonic cells and guide the regionalized patterning of the *Xenopus* embryo.

ACKNOWLEDGMENTS

I thank Laura Vanderploeg for preparing the figures. I thank Catherine Fox for suggestions and editorial insights. Work in the Sheets lab is supported by NSF grant 1050395 and NIH grants (R21HD069345 and R21HD076828).

REFERENCES

Acosta, H., Lopez, S. L., Revinski, D. R., & Carrasco, A. E. (2011). Notch destabilises maternal beta-catenin and restricts dorsal-anterior development in Xenopus. *Development (Cambridge, England)*, *138*, 2567–2579.

Alexandrova, E. M., & Thomsen, G. H. (2006). Smurf1 regulates neural patterning and folding in Xenopus embryos by antagonizing the BMP/Smad1 pathway. *Developmental Biology*, *299*, 398–410.

Amaya, E., Musci, T. J., & Kirschner, M. W. (1991). Expression of a dominant negative mutant of the FGF receptor disrupts mesoderm formation in Xenopus embryos. *Cell*, *66*, 257–270.

Amodeo, A. A., Jukam, D., Straight, A. F., & Skotheim, J. M. (2015). Histone titration against the genome sets the DNA-to-cytoplasm threshold for the Xenopus midblastula

transition. *Proceedings of the National Academy of Sciences of the United States of America, 112*, E1086–E1095.
Angers, S., Thorpe, C. J., Biechele, T. L., Goldenberg, S. J., Zheng, N., MacCoss, M. J., et al. (2006). The KLHL12-cullin-3 ubiquitin ligase negatively regulates the Wnt-beta-catenin pathway by targeting dishevelled for degradation. *Nature Cell Biology, 8*, 348–357.
Antic, D., & Keene, J. D. (1997). Embryonic lethal abnormal visual RNA-binding proteins involved in growth, differentiation, and post-transcriptional gene expression. *American Journal of Human Genetics, 61*, 273–278.
Audic, Y., Omilli, F., & Osborne, H. B. (1997). Postfertilization deadenylation of mRNAs in Xenopus laevis embryos is sufficient to cause their degradation at the blastula stage. *Molecular and Cellular Biology, 17*, 209–218.
Audic, Y., Omilli, F., & Osborne, H. B. (1998). Embryo deadenylation element-dependent deadenylation is enhanced by a cis element containing AUU repeats. *Molecular and Cellular Biology, 18*, 6879–6884.
Ballantine, J. E., Woodland, H. R., & Sturgess, E. A. (1979). Changes in protein synthesis during the development of Xenopus laevis. *Journal of Embryology and Experimental Morphology, 51*, 137–153.
Blythe, S. A., Cha, S. W., Tadjuidje, E., Heasman, J., & Klein, P. S. (2010). beta-Catenin primes organizer gene expression by recruiting a histone H3 arginine 8 methyltransferase, Prmt2. *Developmental Cell, 19*, 220–231.
Bonni, S., Wang, H. R., Causing, C. G., Kavsak, P., Stroschein, S. L., Luo, K., et al. (2001). TGF-beta induces assembly of a Smad2-Smurf2 ubiquitin ligase complex that targets SnoN for degradation. *Nature Cell Biology, 3*, 587–595.
Bouvet, P., Omilli, F., Arlot-Bonnemains, Y., Legagneux, V., Roghi, C., Bassez, T., et al. (1994). The deadenylation conferred by the 3′ untranslated region of a developmentally controlled mRNA in Xenopus embryos is switched to polyadenylation by deletion of a short sequence element. *Molecular and Cellular Biology, 14*, 1893–1900.
Brannon, M., Gomperts, M., Sumoy, L., Moon, R. T., & Kimelman, D. (1997). A beta-catenin/XTcf-3 complex binds to the siamois promoter to regulate dorsal axis specification in Xenopus. *Genes & Development, 11*, 2359–2370.
Bravo, R., & Knowland, J. (1979). Classes of proteins synthesized in oocytes, eggs, embryos, and differentiated tissues of Xenopus laevis. *Differentiation, 13*, 101–108.
Brooks, S. A., & Blackshear, P. J. (2013). Tristetraprolin (TTP): Interactions with mRNA and proteins, and current thoughts on mechanisms of action. *Biochimica et Biophysica Acta, 1829*, 666–679.
Cha, S. W., McAdams, M., Kormish, J., Wylie, C., & Kofron, M. (2012). Foxi2 is an animally localized maternal mRNA in Xenopus, and an activator of the zygotic ectoderm activator Foxi1e. *PLoS One, 7*, e41782.
Cho, K. W., Blumberg, B., Steinbeisser, H., & De Robertis, E. M. (1991). Molecular nature of Spemann's organizer: The role of the Xenopus homeobox gene goosecoid. *Cell, 67*, 1111–1120.
Collart, C., Allen, G. E., Bradshaw, C. R., Smith, J. C., & Zegerman, P. (2013). Titration of four replication factors is essential for the Xenopus laevis midblastula transition. *Science (New York), 341*, 893–896.
Collart, C., Owens, N. D., Bhaw-Rosun, L., Cooper, B., De Domenico, E., Patrushev, I., et al. (2014). High-resolution analysis of gene activity during the Xenopus mid-blastula transition. *Development (Cambridge, England), 141*, 1927–1939.
Costa, A. M., Pereira-Castro, I., Ricardo, E., Spencer, F., Fisher, S., & da Costa, L. T. (2013). GRG5/AES interacts with T-cell factor 4 (TCF4) and downregulates Wnt signaling in human cells and zebrafish embryos. *PLoS One, 8*, e67694.

Cragle, C. E., & MacNicol, A. M. (2014). From oocyte to fertilizable egg –Regulated mRNA translation and the control of maternal gene expression. In M. Kloc & J. Z. Kubiak, (Eds.), *Xenopus development*. Hoboken, NJ, USA: John Wiley & Sons. http://dx.doi.org/10.1002/9781118492833.

Das, S., & Chang, C. (2012). Regulation of early Xenopus embryogenesis by Smad ubiquitination regulatory factor 2. *Developmental Dynamics*, *241*, 1260–1273.

Datto, M., & Wang, X. F. (2005). Ubiquitin-mediated degradation a mechanism for fine-tuning TGF-beta signaling. *Cell*, *121*, 2–4.

De Robertis, E. M., Blum, M., Niehrs, C., & Steinbeisser, H. (1992). Goosecoid and the organizer. *Development. Supplement*, 167–171.

Detivaud, L., Pascreau, G., Karaiskou, A., Osborne, H. B., & Kubiak, J. Z. (2003). Regulation of EDEN-dependent deadenylation of aurora A/Eg2-derived mRNA via phosphorylation and dephosphorylation in Xenopus laevis egg extracts. *Journal of Cell Science*, *116*, 2697–2705.

Ding, Y., Zhang, Y., Xu, C., Tao, Q. H., & Chen, Y. G. (2013). HECT domain-containing E3 ubiquitin ligase NEDD4L negatively regulates Wnt signaling by targeting dishevelled for proteasomal degradation. *The Journal of Biological Chemistry*, *288*, 8289–8298.

Dorey, K., & Hill, C. S. (2006). A novel cripto-related protein reveals an essential role for EGF-CFCs in nodal signalling in Xenopus embryos. *Developmental Biology*, *292*, 303–316.

Dreyer, C. (1989). Fate and nuclear localization of germinal vesicle proteins during embryogenesis. *Developmental Biology*, *6*, 31–57.

Dupont, S., Mamidi, A., Cordenonsi, M., Montagner, M., Zacchigna, L., Adorno, M., et al. (2009). FAM/USP9x, a deubiquitinating enzyme essential for TGFbeta signaling, controls Smad4 monoubiquitination. *Cell*, *136*, 123–135.

Dupont, S., Zacchigna, L., Cordenonsi, M., Soligo, S., Adorno, M., Rugge, M., et al. (2005). Germ-layer specification and control of cell growth by Ectodermin, a Smad4 ubiquitin ligase. *Cell*, *121*, 87–99.

Duval, C., Bouvet, P., Omilli, F., Roghi, C., Dorel, C., LeGuellec, R., et al. (1990). Stability of maternal mRNA in Xenopus embryos: Role of transcription and translation. *Molecular and Cellular Biology*, *10*, 4123–4129.

Dworkin, M. B., & Dawid, I. B. (1980). Use of a cloned library for the study of abundant poly(A)+RNA during Xenopus laevis development. *Developmental Biology*, *76*, 449–464.

Dworkin, M. B., & Dworkin-Rastl, E. (1985). Changes in RNA titers and polyadenylation during oogenesis and oocyte maturation in Xenopus laevis. *Developmental Biology*, *112*, 451–457.

Dworkin, M. B., & Dworkin-Rastl, E. (1987). Metabolic regulation during early frog development. Identification of proteins labeled by 32P-glycolytic intermediates. *The Journal of Biological Chemistry*, *262*, 17038–17045.

Dworkin, M. B., & Dworkin-Rastl, E. (1992). Glycogen breakdown in cleaving Xenopus embryos is limited by ADP. *Molecular Reproduction and Development*, *32*, 354–362.

Dworkin, M. B., & Hershey, J. W. (1981). Cellular titers and subcellular distributions of abundant polyadenylate-containing ribonucleic acid species during early development in the frog Xenopus laevis. *Molecular and Cellular Biology*, *1*, 983–993.

Faure, S., Lee, M. A., Keller, T., ten Dijke, P., & Whitman, M. (2000). Endogenous patterns of TGF beta superfamily signaling during early Xenopus development. *Development (Cambridge, England)*, *127*, 2917–2931.

Fox, C. A., Sheets, M. D., & Wickens, M. P. (1989). Poly(A) addition during maturation of frog oocytes: Distinct nuclear and cytoplasmic activities and regulation by the sequence UUUUUAU. *Genes & Development*, *3*, 2151–2162.

Fritz, B. R., & Sheets, M. D. (2001). Regulation of the mRNAs encoding proteins of the BMP signaling pathway during the maternal stages of Xenopus development. *Developmental Biology, 236*, 230–243.
Gerhart, J. C. (1980). Mechanisms regulating pattern formation in the amphibian egg and early embryo. In R. F. Goldberger (Ed.), *Biological Regulation and Development* (pp. 133–316). New York: Plenum Press.
Gerhart, J., Danilchik, M., Doniach, T., Roberts, S., Rowning, B., & Stewart, R. (1989). Cortical rotation of the Xenopus egg: Consequences for the anteroposterior pattern of embryonic dorsal development. *Development (Cambridge, England), 107*(Suppl.), 37–51.
Gerhart, J., Doniach, T., & Stewart, R. (1991). Organizing the Xenopus organizer. In R. E. Keller (Ed.), *Gastrulation* (pp. 57–77). New York: Plenum Press.
Gerhart, J., Ubbels, G., Black, S., Hara, K., & Kirschner, M. (1981). A reinvestigation of the role of the grey crescent in axis formation in Xenopus laevis. *Nature, 292*, 511–516.
Godwin, A. R., Kojima, S., Green, C. B., & Wilusz, J. (2013). Kiss your tail goodbye: The role of PARN, nocturnin, and angel deadenylases in mRNA biology. *Biochimica et Biophysica Acta, 1829*, 571–579.
Goldstrohm, A. C., & Wickens, M. (2008). Multifunctional deadenylase complexes diversify mRNA control. *Nature Reviews Molecular Cell Biology, 9*, 337–344.
Graindorge, A., Le Tonqueze, O., Thuret, R., Pollet, N., Osborne, H. B., & Audic, Y. (2008). Identification of CUG-BP1/EDEN-BP target mRNAs in Xenopus tropicalis. *Nucleic Acids Research, 36*, 1861–1870.
Graindorge, A., Thuret, R., Pollet, N., Osborne, H. B., & Audic, Y. (2006). Identification of post-transcriptionally regulated Xenopus tropicalis maternal mRNAs by microarray. *Nucleic Acids Research, 34*, 986–995.
Guturi, K., Sarkar, M., Bhowmik, A., Das, N., & Ghosh, M. (2014). DEAD-box protein p68 is regulated by ss-catenin/transcription factor 4 to maintain a positive feedback loop in control of breast cancer progression. *Breast Cancer Research, 16*, 496.
Harland, R., & Gerhart, J. (1997). Formation and function of Spemann's organizer. *Annual Review of Cell and Developmental Biology, 13*, 611–667.
Heasman, J. (2006a). Maternal determinants of embryonic cell fate. *Seminars in Cell & Developmental Biology, 17*, 93–98.
Heasman, J. (2006b). Patterning the early Xenopus embryo. *Development (Cambridge, England), 133*, 1205–1217.
Heasman, J., Crawford, A., Goldstone, K., Garner-Hamrick, P., Gumbiner, B., McCrea, P., et al. (1994). Overexpression of cadherins and underexpression of beta-catenin inhibit dorsal mesoderm induction in early Xenopus embryos. *Cell, 79*, 791–803.
Heasman, J., Holwill, S., & Wylie, C. C. (1991). Fertilization of cultured Xenopus oocytes and use in studies of maternally inherited molecules. *Methods in Cell Biology, 36*, 213–230.
Hikasa, H., & Sokol, S. Y. (2013). Wnt signaling in vertebrate axis specification. *Cold Spring Harbor Perspectives in Biology, 5*, a007955.
Houston, D. W. (2012). Cortical rotation and messenger RNA localization in Xenopus axis formation. *Wiley Interdisciplinary Reviews. Developmental Biology, 1*, 371–388.
Houston, D. W. (2013). Regulation of cell polarity and RNA localization in vertebrate oocytes. *International Review of Cell and Molecular Biology, 306*, 127–185.
Inoue, Y., & Imamura, T. (2008). Regulation of TGF-beta family signaling by E3 ubiquitin ligases. *Cancer Science, 99*, 2107–2112.
Inui, M., Montagner, M., & Piccolo, S. (2012). miRNAs and morphogen gradients. *Current Opinion in Cell Biology, 24*, 194–201.
Ivshina, M., Lasko, P., & Richter, J. D. (2014). Cytoplasmic polyadenylation element binding proteins in development, health, and disease. *Annual Review of Cell and Developmental Biology, 30*, 393–415.

Jevtic, P., & Levy, D. L. (2015). Nuclear size scaling during Xenopus early development contributes to midblastula transition timing. *Current Biology*, *25*, 45–52.

Jung, M. Y., Lorenz, L., & Richter, J. D. (2006). Translational control by neuroguidin, a eukaryotic initiation factor 4E and CPEB binding protein. *Molecular and Cellular Biology*, *26*, 4277–4287.

Keller, R. (1991). Early embryonic development of Xenopus laevis. *Methods in Cell Biology*, *36*, 61–113.

King, M. L., Messitt, T. J., & Mowry, K. L. (2005). Putting RNAs in the right place at the right time: RNA localization in the frog oocyte. *Biology of the Cell*, *97*, 19–33.

Klauzinska, M., Castro, N. P., Rangel, M. C., Spike, B. T., Gray, P. C., Bertolette, D., et al. (2014). *Seminars in Cancer Biology*, *29*, 51–58. http://dx.doi.org/10.1016/j.semcancer.08.003.

Kong, J., & Lasko, P. (2012). Translational control in cellular and developmental processes. *Nature Reviews Genetics*, *13*, 383–394.

LaBonne, C., & Whitman, M. (1997). Localization of MAP kinase activity in early Xenopus embryos: Implications for endogenous FGF signaling. *Developmental Biology*, *183*, 9–20.

Lagna, G., Carnevali, F., Marchioni, M., & Hemmati-Brivanlou, A. (1999). Negative regulation of axis formation and Wnt signaling in Xenopus embryos by the F-box/WD40 protein beta TrCP. *Mechanisms of Development*, *80*, 101–106.

Langley, A. R., Smith, J. C., Stemple, D. L., & Harvey, S. A. (2014). New insights into the maternal to zygotic transition. *Development (Cambridge, England)*, *141*, 3834–3841.

Larabell, C. A., Torres, M., Rowning, B. A., Yost, C., Miller, J. R., Wu, M., et al. (1997). Establishment of the dorso-ventral axis in Xenopus embryos is presaged by early asymmetries in beta-catenin that are modulated by the Wnt signaling pathway. *The Journal of Cell Biology*, *136*, 1123–1136.

Lee, M. A., Heasman, J., & Whitman, M. (2001). Timing of endogenous activin-like signals and regional specification of the Xenopus embryo. *Development (Cambridge, England)*, *128*, 2939–2952.

Lemaire, P., Garrett, N., & Gurdon, J. B. (1995). Expression cloning of Siamois, a Xenopus homeobox gene expressed in dorsal-vegetal cells of blastulae and able to induce a complete secondary axis. *Cell*, *81*, 85–94.

Lemaire, P., & Kodjabachian, L. (1996). The vertebrate organizer—Structure and molecules (Review). *Trends in Genetics*, *12*, 525–531.

Liu, C., Kato, Y., Zhang, Z., Do, V. M., Yankner, B. A., & He, X. (1999). beta-Trcp couples beta-catenin phosphorylation-degradation and regulates Xenopus axis formation. *Proceedings of the National Academy of Sciences of the United States of America*, *96*, 6273–6278.

Lustig, K. D., Kroll, K. L., Sun, E. E., & Kirschner, M. W. (1996). Expression cloning of a Xenopus T-related gene (Xombi) involved in mesodermal patterning and blastopore lip formation. *Development (Cambridge, England)*, *122*, 4001–4012.

MacDonald, B. T., Tamai, K., & He, X. (2009). Wnt/beta-catenin signaling: Components, mechanisms, and diseases. *Developmental Cell*, *17*, 9–26.

Marikawa, Y., & Elinson, R. P. (1998). beta-TrCP is a negative regulator of Wnt/beta-catenin signaling pathway and dorsal axis formation in Xenopus embryos. *Mechanisms of Development*, *77*, 75–80.

Martello, G., Zacchigna, L., Inui, M., Montagner, M., Adorno, M., Mamidi, A., et al. (2007). MicroRNA control of Nodal signalling. *Nature*, *449*, 183–188.

McCrea, P. D., Brieher, W. M., & Gumbiner, B. M. (1993). Induction of a secondary body axis in Xenopus by antibodies to beta-catenin. *Journal of Cell Biology*, *123*, 477–484.

McGivern, J. V., Swaney, D. L., Coon, J. J., & Sheets, M. D. (2009). Toward defining the phosphoproteome of Xenopus laevis embryos. *Developmental Dynamics*, *238*, 1433–1443.

McGrew, L. L., Dworkin-Rastl, E., Dworkin, M. B., & Richter, J. D. (1989). Poly(A) elongation during Xenopus oocyte maturation is required for translational recruitment and is mediated by a short sequence element. *Genes & Development*, *3*, 803–815.

Medioni, C., Mowry, K., & Besse, F. (2012). Principles and roles of mRNA localization in animal development. *Development (Cambridge, England)*, *139*, 3263–3276.

Minshull, J., Pines, J., Golsteyn, R., Standart, N., Mackie, S., Colman, A., et al. (1989). The role of cyclin synthesis, modification and destruction in the control of cell division. *Journal of Cell Science. Supplement*, *12*, 77–97.

Morsut, L., Yan, K. P., Enzo, E., Aragona, M., Soligo, S. M., Wendling, O., et al. (2010). Negative control of Smad activity by ectodermin/Tif1gamma patterns the mammalian embryo. *Development (Cambridge, England)*, *137*, 2571–2578.

Moustakas, A., & Heldin, C. H. (2009). The regulation of TGFbeta signal transduction. *Development (Cambridge, England)*, *136*, 3699–3714.

Murphy, C. M., & Michael, W. M. (2013). Control of DNA replication by the nucleus/cytoplasm ratio in Xenopus. *The Journal of Biological Chemistry*, *288*, 29382–29393.

Murray, A. W., & Kirschner, M. W. (1989). Cyclin synthesis drives the early embryonic cell cycle. *Nature*, *339*, 275–280.

Nagaoka, T., Karasawa, H., Castro, N. P., Rangel, M. C., Salomon, D. S., & Bianco, C. (2012). An evolving web of signaling networks regulated by Cripto-1. *Growth Factors*, *30*, 13–21.

Newport, J., & Kirschner, M. (1982a). A major developmental transition in early Xenopus embryos: I. Characterization and timing of cellular changes at the midblastula stage. *Cell*, *30*, 675–686.

Newport, J., & Kirschner, M. (1982b). A major developmental transition in early Xenopus embryos: II. Control of the onset of transcription. *Cell*, *30*, 687–696.

Nusse, R. (2012). *Cold Spring Harbor Perspectives in Biology*, *4*(5), pii: a011163. http://dx.doi.org/10.1101/cshperspect.a011163.

Olson, D. J., Hulstrand, A. M., & Houston, D. W. (2012). Maternal mRNA knock-down studies: Antisense experiments using the host-transfer technique in Xenopus laevis and Xenopus tropicalis. *Methods in Molecular Biology*, *917*, 167–182.

Paillard, L., Maniey, D., Lachaume, P., Legagneux, V., & Osborne, H. B. (2000). Identification of a C-rich element as a novel cytoplasmic polyadenylation element in Xenopus embryos. *Mechanisms of Development*, *93*, 117–125.

Paillard, L., Omilli, F., Legagneux, V., Bassez, T., Maniey, D., & Osborne, H. B. (1998). EDEN and EDEN-BP, a cis element and an associated factor that mediate sequence-specific mRNA deadenylation in Xenopus embryos. *The EMBO Journal*, *17*, 278–287.

Paranjpe, S. S., Jacobi, U. G., van Heeringen, S. J., & Veenstra, G. J. (2013). A genome-wide survey of maternal and embryonic transcripts during Xenopus tropicalis development. *BMC Genomics*, *14*, 762.

Paris, J., Osborne, H. B., Couturier, A., Le Guellec, R., & Philippe, M. (1988). Changes in the polyadenylation of specific stable RNA during the early development of Xenopus laevis. *Gene*, *72*, 169–176.

Paris, J., & Philippe, M. (1990). Poly(A) metabolism and polysomal recruitment of maternal mRNAs during early Xenopus development. *Developmental Biology*, *140*, 221–224.

Pasquinelli, A. E. (2012). MicroRNAs and their targets: Recognition, regulation and an emerging reciprocal relationship. *Nature Reviews Genetics*, *13*, 271–282.

Pauli, A., Rinn, J. L., & Schier, A. F. (2011). Non-coding RNAs as regulators of embryogenesis. *Nature Reviews Genetics*, *12*, 136–149.

Ralle, T., Gremmels, D., & Stick, R. (1999). Translational control of nuclear lamin B1 mRNA during oogenesis and early development of Xenopus. *Mechanisms of Development*, *84*, 89–101.

Ramel, M. C., & Hill, C. S. (2012). Spatial regulation of BMP activity. *FEBS Letters*, *586*, 1929–1941.

Rebagliati, M. R., Weeks, D. L., Harvey, R. P., & Melton, D. A. (1985). Identification and cloning of localized maternal RNAs from Xenopus eggs. *Cell*, *42*, 769–777.

Rosa, A., Spagnoli, F. M., & Brivanlou, A. H. (2009). The miR-430/427/302 family controls mesendodermal fate specification via species-specific target selection. *Developmental Cell, 16*, 517–527.

Rowning, B. A., Wells, J., Wu, M., Gerhart, J. C., Moon, R. T., & Larabell, C. A. (1997). Microtubule-mediated transport of organelles and localization of beta-catenin to the future dorsal side of Xenopus eggs. *Proceedings of the National Academy of Sciences of the United States of America, 94*, 1224–1229.

Sagata, N., Oskarsson, M., Copeland, T., Brumbaugh, J., & Vande Woude, G. F. (1988). Function of c-mos proto-oncogene product in meiotic maturation in Xenopus oocytes. *Nature, 335*, 519–525.

Saka, Y., Hagemann, A. I., Piepenburg, O., & Smith, J. C. (2007). Nuclear accumulation of Smad complexes occurs only after the midblastula transition in Xenopus. *Development (Cambridge, England), 134*, 4209–4218.

Schier, A. F. (2009). Nodal morphogens. *Cold Spring Harbor Perspectives in Biology, 1*, a003459.

Schneider, P. N., Hulstrand, A. M., & Houston, D. W. (2010). Fertilization of Xenopus oocytes using the host transfer method. *Journal of Visualized Experiments, 45*, pii: 1864. http://dx.doi.org/10.3791/1864.

Schohl, A., & Fagotto, F. (2002). beta-catenin, MAPK and Smad signaling during early Xenopus development. *Development (Cambridge, England), 129*, 37–52.

Schroeder, K. E., Condic, N. L., Eisenberg, L. M., & Yost, H. J. (1999). Spatially regulated translation in embryos: Asymmetric expression of maternal Wnt-11 along the dorsal-ventral axis in Xenopus. *Developmental Biology, 214*, 288–297.

Sheets, M. D., Fox, C. A., Hunt, T., Vande Woude, G., & Wickens, M. (1994). The 3′-untranslated regions of c-mos and cyclin mRNAs stimulate translation by regulating cytoplasmic polyadenylation. *Genes & Development, 8*, 926–938.

Shrestha, B., Sripadi, P., Reschke, B. R., Henderson, H. D., Powell, M. J., Moody, S. A., et al. (2014). Subcellular metabolite and lipid analysis of Xenopus laevis eggs by LAESI mass spectrometry. *PLoS One, 9*, e115173.

Simon, R., & Richter, J. D. (1994). Further analysis of cytoplasmic polyadenylation in Xenopus embryos and identification of embryonic cytoplasmic polyadenylation element-binding proteins. *Molecular and Cellular Biology, 14*, 7867–7875.

Simon, R., Tassan, J. P., & Richter, J. D. (1992). Translational control by poly(A) elongation during Xenopus development: Differential repression and enhancement by a novel cytoplasmic polyadenylation element. *Genes & Development, 6*, 2580–2591.

Simon, R., Wu, L., & Richter, J. D. (1996). Cytoplasmic polyadenylation of activin receptor mRNA and the control of pattern formation in Xenopus development. *Developmental Biology, 179*, 239–250.

Skirkanich, J., Luxardi, G., Yang, J., Kodjabachian, L., & Klein, P. S. (2011). An essential role for transcription before the MBT in Xenopus laevis. *Developmental Biology, 357*, 478–491.

Smith, W. C., & Harland, R. M. (1992). Expression cloning of noggin, a new dorsalizing factor localized to the Spemann organizer in Xenopus embryos. *Cell, 70*, 829–840.

Smits, A. H., Lindeboom, R. G., Perino, M., van Heeringen, S. J., Veenstra, G. J., & Vermeulen, M. (2014). Global absolute quantification reveals tight regulation of protein expression in single Xenopus eggs. *Nucleic Acids Research, 42*, 9880–9891.

Souopgui, J., Rust, B., Vanhomwegen, J., Heasman, J., Henningfeld, K. A., Bellefroid, E., et al. (2008). The RNA-binding protein XSeb4R: A positive regulator of VegT mRNA stability and translation that is required for germ layer formation in Xenopus. *Genes & Development, 22*, 2347–2352.

Stennard, F., Carnac, G., & Gurdon, J. B. (1996). The Xenopus T-box gene, Antipodean, encodes a vegetally localised maternal mRNA and can trigger mesoderm formation. *Development (Cambridge, England), 122*, 4179–4188.

Subtelny, A. O., Eichhorn, S. W., Chen, G. R., Sive, H., & Bartel, D. P. (2014). Poly(A)-tail profiling reveals an embryonic switch in translational control. *Nature*, *508*, 66–71.

Sun, L., Bertke, M. M., Champion, M. M., Zhu, G., Huber, P. W., & Dovichi, N. J. (2014). Quantitative proteomics of Xenopus laevis embryos: Expression kinetics of nearly 4000 proteins during early development. *Scientific Reports*, *4*, 4365.

Tan, R., He, W., Lin, X., Kiss, L. P., & Liu, Y. (2008). Smad ubiquitination regulatory factor-2 in the fibrotic kidney: Regulation, target specificity, and functional implication. *American Journal of Physiology. Renal Physiology*, *294*, F1076–F1083.

Tao, Q., Yokota, C., Puck, H., Kofron, M., Birsoy, B., Yan, D., et al. (2005). Maternal wnt11 activates the canonical wnt signaling pathway required for axis formation in Xenopus embryos. *Cell*, *120*, 857–871.

Tchang, F., Vriz, S., & Mechali, M. (1991). Post-transcriptional regulation of c-myc RNA during early development of Xenopus laevis. *FEBS Letters*, *291*, 177–180.

Tsuchiya, Y., Pham, U., Hu, W., Ohnuma, S., & Gout, I. (2014). Changes in acetyl CoA levels during the early embryonic development of Xenopus laevis. *PLoS One*, *9*, e97693.

Udagawa, T., Swanger, S. A., Takeuchi, K., Kim, J. H., Nalavadi, V., Shin, J., et al. (2012). Bidirectional control of mRNA translation and synaptic plasticity by the cytoplasmic polyadenylation complex. *Molecular Cell*, *47*, 253–266.

Valencia-Sanchez, M. A., Liu, J., Hannon, G. J., & Parker, R. (2006). Control of translation and mRNA degradation by miRNAs and siRNAs. *Genes & Development*, *20*, 515–524.

van Amerongen, R., & Nusse, R. (2009). Towards an integrated view of Wnt signaling in development. *Development (Cambridge, England)*, *136*, 3205–3214.

van Hoof, A., & Parker, R. (2002). Messenger RNA degradation: Beginning at the end. *Current Biology*, *12*, R285–R287.

Vastag, L., Jorgensen, P., Peshkin, L., Wei, R., Rabinowitz, J. D., & Kirschner, M. W. (2011). Remodeling of the metabolome during early frog development. *PLoS One*, *6*, e16881.

Vishnu, M. R., Sumaroka, M., Klein, P. S., & Liebhaber, S. A. (2011). The poly(rC)-binding protein alphaCP2 is a noncanonical factor in X. laevis cytoplasmic polyadenylation. *RNA (New York)*, *17*, 944–956.

Weeks, D. L., & Melton, D. A. (1987). A maternal mRNA localized to the animal pole of Xenopus eggs encodes a subunit of mitochondrial ATPase. *Proceedings of the National Academy of Sciences of the United States of America*, *84*, 2798–2802.

Weeks, D. L., Rebagliati, M. R., Harvey, R. P., & Melton, D. A. (1985). Localized maternal mRNAs in Xenopus laevis eggs. *Cold Spring Harbor Symposia on Quantitative Biology*, *50*, 21–30.

Weill, L., Belloc, E., Bava, F. A., & Mendez, R. (2012). Translational control by changes in poly(A) tail length: Recycling mRNAs. *Nature Structural & Molecular Biology*, *19*, 577–585.

Wessely, O., & De Robertis, E. M. (2000). The Xenopus homologue of Bicaudal-C is a localized maternal mRNA that can induce endoderm formation. *Development (Cambridge, England)*, *127*, 2053–2062.

White, J. A., & Heasman, J. (2008). Maternal control of pattern formation in Xenopus laevis. *Journal of Experimental Zoology*, *310*, 73–84.

Wickens, M. (1990). In the beginning is the end: Regulation of poly(A) addition and removal during early development. *Trends in Biochemical Sciences*, *15*, 320–324.

Wilusz, C. J., Wang, W., & Peltz, S. W. (2001). Curbing the nonsense: The activation and regulation of mRNA surveillance. *Genes & Development*, *15*, 2781–2785.

Wu, L., Good, P. J., & Richter, J. D. (1997). The 36-kilodalton embryonic-type cytoplasmic polyadenylation element-binding protein in xenopus laevis is ElrA, a member of the ELAV family of RNA-binding proteins. *Molecular and Cellular Biology*, *17*, 6402–6409.

Wuhr, M., Freeman, R. M., Jr., Presler, M., Horb, M. E., Peshkin, L., Gygi, S. P., et al. (2014). Deep proteomics of the Xenopus laevis egg using an mRNA-derived reference database. *Current Biology, 24*, 1467–1475.

Wylie, C., Kofron, M., Payne, C., Anderson, R., Hosobuchi, M., Joseph, E., et al. (1996). Maternal beta-catenin establishes a "dorsal signal" in early Xenopus embryos. *Development (Cambridge, England), 122*, 2987–2996.

Xu, P., Liu, J., & Derynck, R. (2012). Post-translational regulation of TGF-beta receptor and Smad signaling. *FEBS Letters, 586*, 1871–1884.

Yang, J., Tan, C., Darken, R. S., Wilson, P. A., & Klein, P. S. (2002). Beta-catenin/Tcf-regulated transcription prior to the midblastula transition. *Development (Cambridge, England), 129*, 5743–5752.

Yost, C., Torres, M., Miller, J. R., Huang, E., Kimelman, D., & Moon, R. T. (1996). The axis-inducing activity, stability, and subcellular distribution of beta-catenin is regulated in Xenopus embryos by glycogen synthase kinase 3. *Genes & Development, 10*, 1443–1454.

Zhang, Y., Cooke, A., Park, S., Dewey, C. N., Wickens, M., & Sheets, M. D. (2013) Bicaudal-C spatially controls translation of vertebrate maternal mRNAs. *RNA (New York), 19*(11), 1575–1582. http://dx.doi.org/10.1261/rna.041665.113.

Zhang, Y., Ding, Y., Chen, Y. G., & Tao, Q. (2014). NEDD4L regulates convergent extension movements in Xenopus embryos via Disheveled-mediated non-canonical Wnt signaling. *Developmental Biology, 392*, 15–25.

Zhang, Y., Forinash, K. D., McGivern, J., Fritz, B., Dorey, K., & Sheets, M. D. (2009). Spatially restricted translation of the xCR1 mRNA in Xenopus embryos. *Molecular and Cellular Biology, 29*, 3791–3802.

Zhang, J., Houston, D. W., King, M. L., Payne, C., Wylie, C., & Heasman, J. (1998). The role of maternal VegT in establishing the primary germ layers in Xenopus embryos. *Cell, 94*, 515–524.

Zhang, J., & King, M. L. (1996). Xenopus VegT RNA is localized to the vegetal cortex during oogenesis and encodes a novel T-box transcription factor involved in mesodermal patterning. *Development (Cambridge, England), 122*, 4119–4129.

Zhang, Y., Park, S., Blaser, S., & Sheets, M. D. (2014). Determinants of RNA binding and translational repression by the Bicaudal-C regulatory protein. *The Journal of Biological Chemistry, 289*, 7497–7504.

Zhu, H., Kavsak, P., Abdollah, S., Wrana, J. L., & Thomsen, G. H. (1999). A SMAD ubiquitin ligase targets the BMP pathway and affects embryonic pattern formation. *Nature, 400*, 687–693.

CHAPTER EIGHT

The *Xenopus* Maternal-to-Zygotic Transition from the Perspective of the Germline

Jing Yang*, Tristan Aguero†, Mary Lou King†,[1]

*Department of Comparative Biosciences, College of Veterinary Medicine, University of Illinois at Urbana-Champaign, Urbana, Illinois, USA

†Department of Cell Biology, University of Miami, Miller School of Medicine, Miami, Florida, USA

[1]Corresponding author: e-mail address: mking@med.miami.edu

Contents

Abstract

In *Xenopus*, the germline is specified by the inheritance of germ-plasm components synthesized at the beginning of oogenesis. Only the cells in the early embryo that receive germ plasm, the primordial germ cells (PGCs), are competent to give rise to the gametes. Thus, germ-plasm components continue the totipotent potential exhibited by the oocyte into the developing embryo at a time when most cells are preprogrammed for somatic differentiation as dictated by localized maternal determinants. When zygotic transcription begins at the mid-blastula transition, the maternally set program for somatic differentiation is realized. At this time, genetic control is ceded to the zygotic genome, and developmental potential gradually becomes more restricted within the primary germ layers. PGCs are a notable exception to this paradigm and remain transcriptionally silent until the late gastrula. How the germ-cell lineage retains full potential while somatic cells become fate restricted is a tale of translational repression, selective degradation of somatic maternal determinants, and delayed activation of zygotic transcription.

Current Topics in Developmental Biology, Volume 113
ISSN 0070-2153
http://dx.doi.org/10.1016/bs.ctdb.2015.07.021

1. INTRODUCTION

Primordial germ cells (PGCs), sole precursors of the gametes in the adult animal, are able to differentiate into all cell lineages, including themselves. Thus, PGCs are considered the "stem cells" of the species. PGC specification is an early event in embryonic development and relies on a unique set of germ-cell determinants. In mammals, germ-cell determinants are induced within a small population of cells prior to gastrulation by BMP signaling (reviewed in Saitou & Yamaji, 2012). In species including worm, fly, fish, and frog, these germ-cell determinants are synthesized during early oogenesis and aggregate into germ plasm (reviewed in King, 2014; Seydoux & Braun, 2006). After fertilization, the germ plasm distributes to just a few blastomeres and those cells acquire the PGC fate. Although initiation of germline development can occur through these two distinct mechanisms, many mammalian PGC-specific genes are components of the maternally synthesized germ plasm in other species. Interestingly, a number of these genes encode RNA-binding proteins that function in regulating zygotic genome activity, totipotency, proliferation, differentiation, and migration of PGCs. They do so through controlling translation and mRNA stability as discussed in this chapter.

In classic studies, Nieuwkoop and Faber (1956) described "cytoplasmic inclusions" or germ plasm at the vegetal pole of eggs and early embryos of *Xenopus laevis* that segregated exclusively into the germline. Both loss- and gain-of-function experiments supported the conclusion that germ plasm is necessary for PGC formation (Nieuwkoop & Sutasurya, 1976; Smith, 1966; Wakahara, 1977, 1978). Recent compelling experiments have shown that germ plasm is both required and sufficient to specify the germline in *Xenopus*. Germ plasm introduced into animal pole blastomeres caused them to enter the germ-cell lineage and give rise to functioning gametes, albeit, in small numbers (Tada, Mochii, Orii, & Watanabe, 2012). The only requirement for proper migration was that these animal pole blastomeres be placed back into the endodermal region.

Most of oogenesis is spent in G2 of the first meiotic prophase. At this time, high rates of transcription from all four copies of each chromosome achieve maximum transcript production. The products of these maternal transcripts are sufficient to meet the requirements of the embryo through early development in the absence of zygotic transcription. Fully-grown oocytes are arrested at this meiotic stage and can remain quiescent for a long period of time. In response to the hormone progesterone, oocytes are released from G2 prophase arrest, undergo maturation events including germinal vesicle breakdown (GVBD), and the completion of meiosis I. Matured oocytes now arrest at the second meiotic metaphase and, after passage through the oviducts, are capable of being fertilized. Sperm entry triggers completion of meiosis II. The female pronucleus then fuses with the male pronucleus, leading to the formation of the zygote and initiation of embryonic development (reviewed in Ferrell, 1999; Heasman, 2006).

Although the diploid genome is reconstituted at fertilization, in *Xenopus laevis*, the major activation of the zygotic genome occurs ~6 h later, midway through the blastula stage. This critical time is known as the mid-blastula transition (MBT) and it is distinguished by a shift in the cell cycle as well as *de novo* RNA synthesis. For the 12 divisions prior to the MBT, the cell cycle is synchronous, and occurs fairly rapidly, every 25 min. To achieve such rates, the gap phases are skipped and the cell cycle goes from mitosis to DNA synthesis to mitosis. At the MBT, the embryo is comprised of some 4000 cells, cell divisions slow and the gap phases enter the cycle. All major developmental decisions up until this point have been made from maternal transcripts. These decisions include the specification of the three major axes (animal/vegetal, dorsal-anterior/ventral, right/left), the three primary germ layers (ectoderm, mesoderm, and endoderm), and the germ-cell lineage (reviewed in Heasman, 2006; King, 2014; White & Heasman, 2008). However, the transition from maternal to zygotic genetic control of development begins before the MBT, at fertilization, when degradation of maternal messages begins and ends during gastrulation when zygotic transcripts are required for further embryonic development. Thus the maternal-to-zygotic transition (MZT) encompasses a longer developmental time period than the MBT (Langley, Smith, Stemple, & Harvey, 2014; Tadros & Lipshitz, 2009; Fig. 1).

The cleavage stages before the MBT accomplish the important task of generating enough cells to begin the process of regional differentiation according to germ layer identity. Not surprisingly, these early-stage blastomeres are maintained in a pluripotent state by the expression of the maternal factors, Oct 60 and Oct 25, orthologs of mammalian Oct 3/4 (Cao et al., 2004; Cao, Siegel, & Knochel, 2006; Cao, Siegel, Oswald, & Knochel, 2008; Hinkley, Martin, Leibham, & Perry, 1992; Whitfield, Heasman, & Wylie, 1993). Post-MBT, these factors gradually decline and are replaced by lineage-specific transcription factors such as Xsox17, Bix4, GATAs, and Xnrs, many of which are activated by the maternal transcription factor,

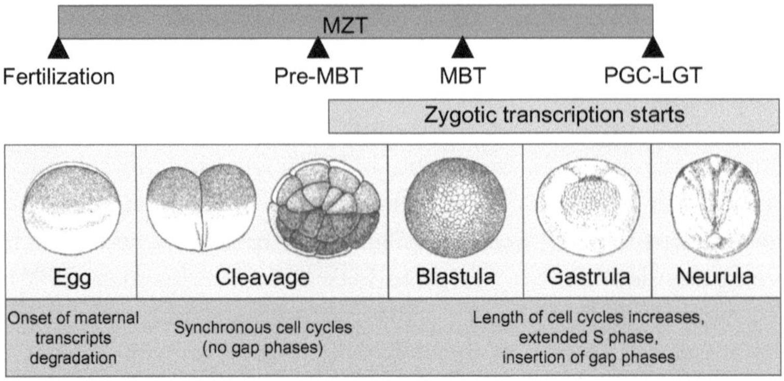

Figure 1 The maternal-to-zygotic transition (MZT). MZT is a period that occurs very early during embryonic development, when degradation of maternal transcripts is initiated and is complete with the first morphological change caused by zygotic transcription, gastrulation. After the 12 mitotic divisions, the synchronous cell cycles are followed by asynchronous cleavages, with the introduction of gap phases during a developmental event called the "mid-blastula transition" (MBT). Most zygotic genes are silent until large-scale zygotic gene activation (ZGA) that, in *Xenopus*, coincides with the MBT. Several genes required for embryo patterning are transcribed before the MBT (pre-MBT) starting at the 32-cell stage. PGC transcription begins later, during the transition between gastrula and neurula (*p*rimordial *g*erm *c*ell *l*ate-*g*astrula *t*ransition, PGC-LGT). Images of *Xenopus* embryos are shown to highlight the different developmental stages. (See the color plate.)

VegT, at the MBT (reviewed in Heasman, 2006; White & Heasman, 2008). Thus, as development proceeds and genetic control is ceded to the zygotic genome, developmental potential gradually becomes more restricted within the primary germ layers. The germline is a notable exception to this paradigm and raises the fundamental question as to how this lineage retains the potential for totipotency while the somatic cells surrounding them become fate restricted. There is strong evidence for overlapping "antidifferentiation" mechanisms involving repression at both transcriptional and translational levels that operate to preserve the germline through the MZT (Leatherman & Jongens, 2003; Venkatarama et al., 2010). In this chapter, we will examine the MZT within the context of germline specification.

2. GERM-PLASM RNAs AND CYTOSKELETAL DYNAMICS: STAGE VI OOCYTE

In Stage VI *Xenopus* oocytes, germ plasm is organized into numerous small islands. At the ultrastructural level, each of these germ-plasm islands contains mitochondria, endoplasmic reticulum, membraneless electron-dense materials and matrix. The electron-dense material or germinal granules, can be fibrillar or round-shaped and is a hallmark of germ plasm (Heasman, Quarmby, & Wylie, 1984). RNAs within the germ plasm are found with distinct localization patterns (Fig. 2). Germinal granules contain *nanos* RNA, while *Xpat* and *DeadSouth* are peripherally associated with granules. *Xdazl, Xlsirts, Xwnt11, DEADSouth,* and *fatvg* are found within the matrix as ribonucleoprotein particles (RNPs) (Kloc et al., 2002). The significance of this organization of germ-plasm mRNAs in relation to when they may be translated after fertilization is not known. In full-grown oocytes, the germ-plasm islands are finely dispersed in the subcortex at the vegetal pole, located between the cortical granules and yolk. This geographic location and the structural organization of the germ plasm are likely maintained by a highly organized cytoskeletal network, which is established during early stages of oogenesis (Gard, Cha, & King, 1997). At least, two types of intermediate filament are closely associated with germ plasm. These include a large amount of vimentin, which is colocalized with germ plasm (Godsave, Anderton, Heasman, & Wylie, 1984; Torpey, Heasman, & Wylie, 1990; Wylie, Brown, Godsave, Quarmby, & Heasman, 1985; Wylie, Heasman, Parke, Anderton, & Tang, 1986), and a well-developed network of cytokeratin filaments that can be easily detected in the vegetal cortex and subcortex (Gard, 1999; Kloc, Bilinski, & Dougherty, 2007;

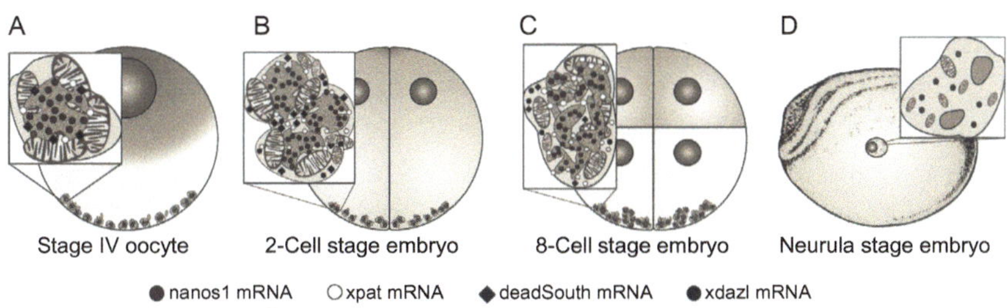

Figure 2 Schemes of germ-plasm structure during *Xenopus laevis* embryogenesis. At oocyte Stage VI, germ-plasm islands are associated with the vegetal cortex (A). Between fertilization and the 8-cell stage, germinal granules (GGs) undergo ultrastructural changes thought to represent a permissive state for translation. After fertilization, the germ plasm contains a large number of small (250 nm) spherical GGs (B). During early cleavage, GGs coalesce into progressively more complex aggregates at the apex of vegetal pole blastomeres (C). Germ plasm at the 8-cell stage contains a few irregularly shaped large GG (about 2000 nm) that, in cross section, show complex and variable morphology (C). Between the gastrula and neurula stages, large GGs disappear and primordial germ cells from the neurula stage contain only a few small (300 nm) GGs (D). (See the color plate.)

Kloc et al., 2005; Torpey, Heasman, & Wylie, 1992). Ultrastructural analysis reveals that some germ-plasm islands are surrounded and penetrated by the long cytokeratin filaments (Kloc et al., 2007). Although actin filaments (Gard, 1999) and a small number of microtubules (Gard, 1991, 1994; Pfeiffer & Gard, 1999) are present in the vegetal subcortex region, it is unclear whether they play a direct role in maintaining germ-plasm islands there. Recently, it has been reported that depletion of Vangl2 and aPKC, which results in a reduction in the acetylated microtubule cytoskeleton, disrupts vegetal localization of *VegT* and *Wnt11* mRNAs (Cha, Tadjuidje, Wylie, & Heasman, 2011). These results suggest that the microtubule cytoskeleton plays a role in maintaining at least some RNA components at the vegetal pole.

The cytokeratin network at the vegetal pole has been studied extensively. Evidence supports the cytokeratin network as a scaffold, allowing some vegetally localized maternal RNAs to associate with it. These include the noncoding RNAs *Xlsirts* (Kloc & Etkin, 1994) and *VegT* mRNA, which encodes the master regulator for *Xenopus* germ layer specification (Zhang et al., 1998). Both *Xlsirts* and *VegT* RNPs in Stage VI oocytes are found associated with the cytokeratin network (Kloc et al., 2005). Strikingly, both *Xlsirts* and *VegT* mRNAs play important structural roles in maintaining the architecture of the cytokeratin network in the vegetal cortex. Depletion of these RNAs disrupts the network in a transcript-specific manner and impairs the proper formation of the germinal granules and subsequent development of the germline (Kloc et al., 2007, 2005). In the case of *VegT*, Heasman and colleagues have provided convincing evidence that it is the *VegT* mRNA, not the VegT protein, which is required for maintaining the structure of the cytokeratin network. In *Xlsirts*- or *VegT*-depleted oocytes, vegetal localization of a number of maternal RNAs is disrupted, likely due to the defective cytokeratin network (Heasman, Wessely, Langland, Craig, & Kessler, 2001; Kloc & Etkin, 1994; Kloc et al., 2005).

3. GERM PLASM AND CYTOSKELETAL DYNAMICS: OOCYTE MATURATION/FERTILIZATION

During oocyte maturation and subsequent fertilization, the cytoskeleton in the vegetal pole undergoes a series of dramatic changes. While vimentin remains colocalized with germ plasm throughout these transitions (Godsave et al., 1984), other cytoskeletal architectures are largely erased during the process of oocyte maturation. For example, the cytokeratin network is disrupted nearly completely after GVBD, with only some cytokeratin punctuate foci remaining within the vegetal cortex (Kloc et al., 2005). In addition, actin filaments and microtubules become barely detectable. It is likely that disassembly of the existing cytoarchitecture in the fully-grown quiescent oocyte prepares the oocyte for fertilization. The fertilized egg quickly builds a new cytoskeletal system, which facilitates the rapid coalescence of germ-plasm islands found widely distributed throughout the vegetal subcortex before egg activation (Ressom & Dixon, 1988).

In addition to cytoskeletal dynamics, a few other changes associated with oocyte maturation have been observed. For example, endogenous aPKC protein, which is homogenously distributed in the oocyte, is translocated to the animal hemisphere of the egg and can no longer be detected at the vegetal pole (Cha et al., 2011; Nakaya et al., 2000). Hermes protein, a germ-plasm RNA-binding protein highly expressed in oocytes, is markedly downregulated after GVBD, and cannot be detected during early stages of embryonic development (Song et al., 2007; Zearfoss, Chan, Wu, Kloc, & Etkin, 2004). Interestingly, Hermes protein colocalizes with Xvelo, the homologue of zebrafish Bucky ball, which regulates assembly of the Balbiani body (Bb) in oocytes (Marlow & Mullins, 2008) and organizes germ plasm after fertilization (Bontems et al., 2009; Heim et al., 2014). Although the significance of these molecular events remains unclear, it is highly likely that some of these molecular changes trigger remodeling of the germ plasm that is required for the initiation of germline development after fertilization.

3.1 Postfertilization

To initiate PGC development, the germ-plasm islands (>500) that are widely dispersed in the oocyte's vegetal subcortex must coalesce into larger aggregates for segregation into a small number of blastomeres. This process begins at fertilization and with each division small aggregates are pushed into larger aggregates as the cleavage furrow advances. By the 32-cell stage, there are now on average 36 islands of germ plasm (Savage & Danilchik, 1993). During the first two *Xenopus* embryonic cell cycles, cleavage furrows are initiated in the animal hemisphere and move toward the vegetal pole. As a consequence of the first two embryonic cell divisions, germ-plasm aggregates become concentrated into the vegetal corner of each blastomere. In zebrafish embryos, germ-plasm RNPs first move toward the animal pole together with the cytoplasm during cytoplasmic streaming. Once streaming is completed, the blastoderm is created on top of the yolky mass. The large volume of yolk prevents the cleavage furrows from reaching the vegetal pole. For this reason, four large germ-plasm aggregates form at the periphery of the blastoderm as the zebrafish embryo divides. Interestingly, large germ-plasm aggregates can form in artificially activated eggs. Notably, it took 3–4 h for germ-plasm islands to coalesce in activated eggs, whereas the same process is largely completed by around 2 h postfertilization in normal embryos (Robb, Heasman, Raats, & Wylie, 1996). Considering that artificially activated eggs do not undergo cleavage, cleavage may not be absolutely required for coalescence of germ plasm.

A growing body of evidence suggests that formation of germ-plasm aggregates relies heavily on cytoskeletal dynamics after fertilization. Upon egg activation during fertilization, several cytoskeleton networks are established very rapidly in *Xenopus*. This includes a complex arrangement of cytokeratin network after egg activation (Kloc et al., 2005). Also, shortly after fertilization, a dense microtubule network is established in the entire fertilized egg. These microtubules, nucleated by the sperm centrosome and egg pronuclear-associated components, extend radially into the entire embryo. Treatments that block microtubule polymerization after egg activation (Ressom & Dixon, 1988; Savage & Danilchik, 1993), or that deplete the maternal kinesin-like protein, Xklp1, impair the aggregation process of the germ plasm. These studies have highlighted the importance of the microtubule network in germ-plasm aggregation.

It is worth mentioning that, shortly after fertilization, a unique microtubule network forms in the vegetal cortex. These microtubule arrays are organized into numerous parallel bundles and drive cortical rotation, a process that is essential for axis specification. This microtubule network exists only transiently and for a short period of time, and is disassembled prior to the beginning of the first embryonic division (Houliston & Elinson, 1991; Schroeder & Gard, 1992). In *Xenopus*, these microtubules are continuous with microtubules extending from the animal hemisphere. Polymerization of microtubules into the vegetal cortex is regulated by vegetally localized maternal factors such as Trim36 (Cuykendall & Houston, 2009; Olson, Oh, & Houston, 2015) and Dead-end (Dnd1) (Mei et al., 2013), both of which are actually components of the germ plasm. There is no evidence to suggest that these vegetal cortical microtubules are essential for germ-plasm aggregation. Recently, the zebrafish maternal-effect mutant, *Hecate*, has been characterized, which carries mutations in the glutamate receptor-interacting protein 2a (grip2a) gene. *Hecate* embryos show defects in the alignment and bundling of microtubules at the vegetal cortex. Such bundling abnormalities result in defective initiation of Wnt signaling during cleavage and subsequent axis specification. In contrast, the aggregation of germ plasm occurs normally in *hecate* mutants at the 4-cell stage. Even in severely ventralized *hecate* embryos, the average number of PGCs is unaffected at 24 h postfertilization (Ge et al., 2014). It is also worth pointing out that the defects in *hecate* embryos, which are maternal-effect mutant, does not necessarily rule out the possibility that zygotic Grip2 may be involved in germ-cell development. In *Xenopus*, injection of morpholino antisense oligos that block Grip2.1 translation, or overexpression of a dominant negative Grip2.1, reduces average PGC number and impairs the proper anterior-posterior positioning of PGCs in embryos (Tarbashevich, Koebernick, & Pieler, 2007). Nonetheless, the observation that germ-plasm coalesces as normal into four large masses in *hecate* embryos, which are defective in vegetal cortical microtubules, suggests that these aligned vegetal cortical microtubule arrays are unlikely to be required for germ-plasm aggregation.

Our current knowledge regarding how cleavage coordinates germ-plasm aggregation comes largely from studies in zebrafish. In zebrafish, germ-plasm RNPs exist as single particles before fertilization. After fertilization, these particles begin to form aggregates. This process is accelerated as the first embryonic cell cycle begins. As the astral microtubules extend from the sperm aster toward the periphery of the blastodisc, germ-plasm RNPs and associated actin move outwardly. As a result, germ-plasm RNPs are displaced to the outer edge of the blastodisc and the center of the blastodisc becomes free from germ-plasm RNPs. During cytokinesis, through the action of astral microtubules of the bipolar spindle, some germ-plasm RNPs are recruited to the cleavage furrow, forming two rod-like structures. These rod-like structures are pushed in opposite directions as the cleavage furrow advances outwardly from the center of the blastodisc, leading to distal compaction of germ-plasm RNP aggregates at both ends of the mature cleavage furrow. This is reminiscent of astral microtubule-dependent germ-plasm transportation during *Drosophila* germ-cell development (Lerit & Gavis, 2011). Subsequently, in zebrafish, similar processes occur repeatedly during the second and third cytokinesis. Ultimately, the majority of germ-plasm RNPs is packaged into four compact masses located at the periphery of the embryo (Eno & Pelegri, 2013; Nair et al., 2013; Theusch, Brown, & Pelegri, 2006). These large germ plasm aggregates are inherited by a few blastomeres, leading to the specification of the germline.

Several zebrafish maternal-effect mutants show defects in the aggregation of germ-plasm RNPs. These include *cellular island*, which carries mutation in Aurora B Kinase that is critical for cleavage furrow formation during cytokinesis (Yabe et al., 2009), and *motley*, a maternal-effect mutant deficient in the chromosomal passenger protein Birc5b/Survivin. Birc5b/Survivin regulates chromosomal passenger complex function, astral microtubule remodeling in the initiation of cytokinesis furrow ingression, and cortical microfilament dynamics (Nair et al., 2013). In addition, overexpression of *Bucky ball*, which encodes a RNA-binding protein important for Bb formation (Marlow & Mullins, 2008), induces abnormal germ-plasm aggregation, and generates ectopic PGCs in the zebrafish embryo (Bontems et al., 2009).

After coalescence of the germ plasm is completed, the location of zebrafish germ plasm remains unchanged in the embryo until morphogenesis begins. In *Xenopus*, until the 16-cell stage, germ plasm aggregates are located in the subcortex at the vegetal pole. As cleavage proceed, germ plasm changes its location from the vegetal subcortical regions to deep within the yolky endodermal mass (Ressom & Dixon, 1988; Savage & Danilchik, 1993). Interestingly, this movement can occur in artificially activated eggs and is independent of microtubules, microfilaments, or protein synthesis, although it takes longer in activated eggs as compared to fertilized embryos. This again suggests that cleavage may not be absolutely required for the germ-plasm dynamics in *Xenopus* embryos. In activated eggs, it seems that germ plasm is carried by cytoplasmic flow from the region of the vegetal pole toward the interior (Ressom & Dixon, 1988). By the blastula stage, four to five PGCs with concentrated germ plasm can be found in the endodermal mass. There, PGCs later divide a few times and eventually migrate out of the endoderm. At the tadpole stage, they migrate into the gonad where they proliferate and differentiate into gametes.

4. MATERNAL DETERMINANTS FOR TWO LINEAGES INHERITED BY THE SAME BLASTOMERES: A DILEMMA

While the overwhelming majority of mRNAs are distributed uniformly throughout the *Xenopus* oocyte, several hundred are found concentrated specifically at the vegetal pole. As alluded to earlier, many of these localized maternal RNAs are highly enriched in a cytokeratin-containing fraction, a characteristic that allowed their isolation (Mosquera, Forristall, Zhou, & King, 1993; Pondel & King, 1988). RNAs appear to localize using two different pathways during oogenesis while some use both pathways. In general, germ-plasm RNAs accumulate within the Bb along with mitochondria during the earliest oocyte stages. All RNAs identified thus far that localize into the Bb have functions in the germline (reviewed in King, Messitt, & Mowry, 2005). The Bb was shown to expand toward the vegetal pole, pushing germ plasm into the subcortex in Stage I to II oocytes (Wilk, Bilinski, Dougherty, & Kloc, 2005). RNAs encoding factors that function only in the soma such as Vg1 and VegT, localize within large particles transported along microtubules (reviewed in Pratt & Mowry, 2013). Localization of these RNAs occurs after germ-plasm RNAs have arrived at the vegetal cortex and for this reason, has been called the late pathway. However, not all RNAs that become restricted to PGCs during development are localized within the Bb germ plasm. A set of RNAs with dual functions in both the oocyte and PGCs (Dead-end) or in PGCs and the soma (Fatvg) is localized during the late pathway. In summary, by the end of oogenesis, maternal RNAs required to specify both germline and somatic lineages become localized to the oocyte vegetal cortex. The latter include VegT and Vg1 responsible for endoderm/mesoderm identity (Birsoy, Kofron, Schaible, Wylie, & Heasman, 2006; Zhang & King, 1996) and Xwnt11 and Trim36 for dorsal/ventral patterning (Cuykendall & Houston, 2009; Tao et al., 2005).

Such colocalization of determinants for different lineages presents a dilemma that must be resolved before the MBT. As the fertilized egg divides, germ plasm is asymmetrically segregated into only a few cells, while localized somatic RNAs are activated for translation and are passively transmitted into all vegetal cells, including PGCs (King et al., 2005; Lai, Singh, & King, 2012; Venkatarama et al., 2010). Thus, well before zygotic gene transcription begins 12 divisions later at the MBT (~4000 cells), determinants, including VegT, are available to initiate somatic gene-expression programs in PGCs (Fig. 3; Venkatarama et al., 2010). *VegT* RNA is actually translated earlier, after maturation, so its protein is available from the earliest stages of development. It is likely preloaded into nuclei and ready to initiate somatic gene programs at the MBT (Stennard, Zorn, Ryan, Garrett, & Gurdon, 1999). In spite of the presence of VegT, however, PGCs do not activate the endoderm gene-expression program, which includes the transcription factors, *Xsox17α* and *Bix4* (Casey et al., 1999; Hudson, Clements, Friday, Stott, & Woodland, 1997). Instead, they remain competent to form all cell types. How are PGCs protected from somatic fates at this time? What prevents VegT protein from activating the endoderm genetic program in PGCs at the MBT?

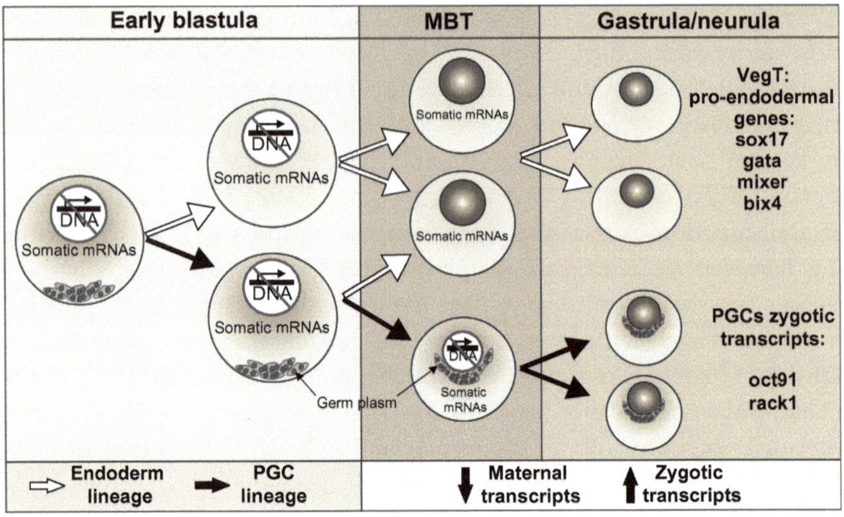

Figure 3 Scheme of primordial germ cell (PGC) segregation from soma. In *Xenopus*, the germline is specified through the inheritance of germ plasm formed during oogenesis and asymmetrically segregated into the future germ-cell lineage (black arrows). Endodermal somatic fate is established by the inheritance of maternal somatic transcripts, including VegT (white arrows). Germ cells repress somatic maternal mRNAs. PGCs delay starting mRNA transcription to ensure that somatic differentiation programs remain inactive at the time that zygotic transcription is initiated in the rest of the embryo. (See the color plate.)

5. MECHANISMS FOR PROTECTING THE GERMLINE DURING THE MZT

Two critical activities appear to be required in *Xenopus* PGCs to protect them from VegT activity and other somatic differentiation signals: (1) translational repression of VegT by Pumilio and Nanos, and (2) transient genome-wide suppression of mRNA transcription at the MBT. The latter activity ensures that somatic differentiation programs remain inactive when zygotic transcription is initiated in the rest of the embryo (Fig. 3). Nanos-family members are distinguished throughout the animal kingdom by a conserved CCHC Zinc-finger motif that is required for RNA binding (Hashimoto et al., 2010). Alone, Nanos has little RNA-sequence-specific binding activity, but in all species studied it acts as a translational repressor (reviewed in Lai & King, 2013; Lai, Zhou, Luo, Fox, & King, 2011). Like Nanos, Pumilio (PUM) is also a member of a conserved family of proteins found in all eukaryotes. The conserved C-terminal region contains eight tandem repeats that defines its RNA-binding domain and confers specificity in RNA recognition for binding. PUM specifically binds an 8-nt canonical sequence within the 3′UTRs (**UGUA**NAUA) of targeted RNAs for repression (Jaruzelska et al., 2003; Lai et al., 2012; Wang, Opperman, Wickens, & Hall, 2009; Wharton, Sonoda, Lee, Patterson, & Murata, 1998; Zamore, Williamson, & Lehmann, 1997; Zhang et al., 1997); hence, it is called the Pumilio Binding Element (PBE; formerly Nanos Response Element). The first four nucleotides (UGUA) are absolutely required and N could be A, U, or C. One change in the last three nucleotides (AUA) is also tolerated and those PBEs are referred to as noncanonical (ncPBE). The last three nucleotides most common in noncanonical PBEs are AUG and AAA (Table 1).

Although Nanos activity mediates repression and PUM selects the RNA target, there are reports that other regions of PUM can mediate repression alone (Padmanabhan & Richter, 2006; Weidmann & Goldstrohm, 2012). Two Pumilio family proteins have been characterized in *Xenopus* oocytes that differ significantly in their N-terminal regions, while their RNA-binding domains are virtually identical and are expected to bind the same set of RNAs (Ota, Kotani, & Yamashita, 2011). PUM has many targets, associates with proteins other than Nanos, including Dazl (Fox, Urano, & Reijo Pera, 2005; Padmanabhan & Richter, 2006), and is part of complexes in both somatic and germ cells. Thus, the challenge is to identify PGC RNAs that bind PUM, but which require either Nanos for repression or Dazl for activation. Recently, a set of RNAs involved in meiosis and repressed by Nanos2 has been identified in mouse male gonads by immunoprecipitation with Nanos antibody (Suzuki, Igarashi, Aisaki, Kanno, & Saga, 2010; Suzuki & Saga, 2008). While *Xenopus* has only one *nanos* gene, the mouse has three. Interestingly, the Nanos2-interacting proteins identified in this study include the CCR4-NOT deadenylation complex. These findings suggest that Nanos represses expression of these RNAs and leads to their degradation.

Compared to control *ornithine decarboxylases1* (*ODC*), both maternal *VegT* and *survivin* (*Xsurv*) RNA levels at Stage 8 are significantly lower in PGCs than in somatic cells and they are not detected by Stage 10 (Venkatarama et al., 2010). Maternally expressed Xsurv is a positive regulator of cell-cycle progression localized to the mitotic spindle (Murphy, Sabel, Sandler, & Dagle, 2002; Yan et al., 2005). Loss of Xsurv may reflect the difference in cell-cycle regulation in PGCs (Zust & Dixon, 1975). These differences in RNA levels cannot be accounted for by zygotic expression

Table 1 Candidate RNAs for Pumilio/Nanos Repression

PGC Transcripts	Canonical PBE: UGUANAUA	Noncanonical PBE: UGUANAUA[a]	Function
ddx25 (DeadSouth)	0	2	RNA helicase
Cdk9A	3	3	CTD-kinase, RNA synthesis
Cyclin B1[b]	1	0	Cell cycle regulator
Vg1	1	2	Growth factor, endoderm patterning, mesoderm induction
otx1	1	4	Transcription factor
pumilio1	1	7	Translation
rbpms2 (hermes)	0	2	Translation
sox7	2	5	Transcription factor
VegT[b]	1	2	Transcription factor, endoderm determinant
sybu	0	5	D/V axis
trim36	1	1	D/V axis
wnt11b[b]	1	5	D/V axis
xdazl	0	4	Translation (activator)
xpat	0	2	Germ-plasm development
xvelo	2	2	Cell polarity

[a]Noncanonical PBE: One change occurs in the last three nucleotides (AUA), most frequently AAA or AUG.
[b]Validated *in vivo* as repressed by Pumilio/Nanos or Pumilio.

of *VegT* RNA in the embryo sample. Specific antisense depletion of maternal VegT shows that most of VegT at Stage 10 is maternal RNA (Zhang et al., 1998). Therefore, it seems that *VegT* and *Xsurv* are regulated differently (less stable) in PGCs than in somatic cells. PGCs depleted of Nanos activity misexpress the downstream targets of VegT, Xsox17, and Bix4 (Lai et al., 2012). The phenotype can be rescued by injection of *Nanos* RNA. These results suggest that Nanos normally represses translation of the maternal *VegT* mRNA within the germline. Indeed, the 3′UTR of *VegT* contains one canonical PBE that binds PUM and is required for VegT repression (Table 1; Lai et al., 2012). Further, sometime between MBT and gastrulation, *VegT* is dramatically reduced in PGCs relative to the whole embryo, but not in *nanos*-depleted embryos (Lai et al., 2012). The elimination of somatic determinants from PGCs is part of the MZT for the germline. Thus, RNA degradation does indeed play a role in Nanos/PUM translational repression in *Xenopus* early embryos. Nanos functions to repress RNAs that normally specify endoderm and promote apoptosis, both of which are critical for preserving the germline.

What other germline RNAs may be regulated in this manner? Cyclin B1, a central regulator of the cell-cycle, is a prime example also targeted for repression by PUM/Nanos in other species (Lai et al., 2011). Pumilio also binds and represses *Xwnt11* but in this case does not require Nanos (Lai & King, 2013). Scores of other RNAs found within the germ plasm contain PBEs, but without direct testing for repressive activity, no conclusions can be drawn (see candidates in Table 1). Of interest is *Vg1* RNA, a somatic determinant involved in endoderm and mesoderm patterning (Birsoy et al., 2006). *Vg1* RNA contains 1 PBE and 2 ncPBEs and may be regulated as VegT is regulated. PUM mRNA itself has the highest number with 1 PBE and 6 ncPBEs, suggesting that a negative autoregulatory loop may operate in the germline. That is, PUM may be allowed to only reach a certain concentration within PGCs before its mRNA is repressed and degraded. Transcripts encoding Nanos, Dead-end, and all helicases except DeadSouth, do not have PBEs within their 3′UTRs.

6. PGCs ARE TRANSCRIPTIONALLY REPRESSED AT THE MBT

The second level of protection for PGCs during the MZT acts to prevent translated somatic determinants from causing inappropriate transcription. While somatic cells enter the MBT, PGCs are transcriptionally repressed and therefore unable to respond to maternally inherited signals like VegT. Similar to the *Drosophila* and *Caenorhabditis elegans* germline, the absence of new transcription at the MBT correlates with the unphosphorylated state of the carboxy-terminal domain in RNA polymerase II that must be phosphorylated for transcriptional elongation steps to occur. Zygotic transcription in PGCs does not occur until late gastrula stages, well after somatic endoderm cells have become fate restricted (Venkatarama et al., 2010; Wylie, Snape, Heasman, & Smith, 1987). What accounts for this delay? In *nanos* loss-of-function mutants in *Drosophila*, *C. elegans*, and *Xenopus*, the delay in transcription is lost and PGCs are transcriptionally active at the same time as somatic cells are at the MBT (Deshpande, Calhoun, Jinks, Polydorides, & Schedl, 2005; Lai et al., 2012; Schaner, Deshpande, Schedl, & Kelly, 2003). How does Nanos' function as a translational repressor in PGCs play a role in the transient genome-wide suppression of mRNA transcription at the MBT?

7. HOW DOES THE GERMLINE ESCAPE THE MBT?

It seems unlikely that PGCs have a different timing mechanism from somatic cells for initiating zygotic transcription. Global regulation of transcription must follow very conserved rules. It seems more likely that the mechanism governing the delay in transcription for both the soma and PGCs will be related. Studies on the regulation of MBT onset suggest that the first step is that genomic DNA is released from a repressed state. This event is correlated with the nuclear-to-cytoplasmic (N/C) ratio and depletion of a putative repressor (reviewed in Langley et al., 2014). Cell cycle regulation, therefore, is one component, driving changes in the N/C ratio although nuclear size also plays a role (Jevtic & Levy, 2015). As mentioned, a target of Nanos/PUM repression, apparently conserved across species, is cyclin B1. Cyclin B1 regulates the cell cycle in a concentration-dependent fashion (Kotani, Yasuda, Ota, & Yamashita, 2013; Mendez & Richter, 2001). By repressing the cyclin B1 mRNA, the levels of Cyclin B1 protein would be expected to decline in PGCs. Consistent with this observation, whereas endodermal cells can divide every 40–75 min, PGCs undergo cell divisions less frequently, dividing only during three discrete time periods by the time they exit the endoderm at tadpole stages. A second requirement for correct initiation of zygotic transcription depends on *de novo* translation and accumulation of key maternal transcription factors, which are in limited supply. In *Xenopus*, VegT was identified as being one of these critical factors (Skirkanich, Luxardi, Yang, Kodjabachian, & Klein, 2011). Translational repression of VegT, cyclinB1, and other maternal factors in PGCs by Nanos/PUM may contribute to the observed delay in transcription; sufficient levels of core factors are not achieved. In PGCs, VegT is translationally repressed early on and degraded by gastrulation (Lai et al., 2012). In zebrafish, the pluripotency factors, Oct 91 and Sox, were also found to be key maternal factors important in this regard (Harvey et al., 2013; Lee et al., 2013). Interestingly, maternal Sox7 is found in the germ plasm, like VegT. It, too, is a potential target for Nanos/PUM repression as the Sox7 3′UTR contains no less than 7 PBEs (Table 1).

PGCs have an open chromatin structure early that becomes condensed later. Global silencing of mRNA transcription could be explained at the level of chromatin structure. Dimethylated lysine 4 of histone H3 (H3K4) or hyperacetylated Histone H4 (Penta) are markers of transcriptionally active chromatin; methylation of lysine 9 of histone H3 (H3K9) marks inactive chromatin. When PGCs and somatic cells were isolated from embryos at MZT stages (Fig. 1) and compared for these histone modifications, no significant differences were found (Venkatarama et al., 2010). All cells stained positively for H4Penta and H3meK4, indicating an active chromatin structure. Interestingly, H1c linker protein, which restricts chromatin remodeling, is not expressed in PGCs while it is in the endoderm. PGCs may retain the more permissive oocyte linker variant, B4 (Venkatarama et al., 2010). For *Xenopus* PGCs, global transcriptional repression does not involve a dramatic alteration of chromatin architecture. Indeed, the mechanism involved prevents transcription in spite of "permissive" chromatin.

8. RNA DEGRADATION DURING THE MZT

Two important events occur during the MZT: turnover of maternal gene products and activation of the zygotic genome. In some species, for example mammals, these two events are accomplished in a very narrow time window shortly after fertilization. In mice, interfering with either degradation of maternal gene products or zygotic genome activation (ZGA) leads to developmental arrest around the 2-cell stage, suggesting that these two events are tightly coupled. Interestingly, degradation of maternal gene products and activation of the zygotic genome are not tightly coupled in many other species. In *Drosophila*, zebrafish, and *Xenopus*, degradation of maternal mRNAs lasts a long time. While some RNAs are degraded relatively rapidly, a large proportion of maternal mRNAs remains in the embryo even by the end of gastrulation (Bashirullah et al., 1999; Ferg et al., 2007; Mathavan et al., 2005). In these species, high-level ZGA occurs abruptly at the MBT, which corresponds to the 14th, 12th, and 10th embryonic cell division in *Drosophila*, *Xenopus*, and zebrafish, respectively. In *Drosophila*, turnover of maternal mRNAs is essential for ZGA. Mutation of Smaug, an RNA-binding protein essential for the destabilization of a large subset of the maternal mRNAs, causes a significant reduction in zygotic transcription at 2–3-h postfertilization (Benoit et al., 2009). In zebrafish, turnover of maternal RNAs does not seem to be necessary for ZGA. It has been reported that maternal-zygotic Dicer-mutant zebrafish embryos, which do not process precursor miRNAs into mature miRNAs, display mild defects during gastrulation, brain formation, somitogenesis, and heart development (Giraldez et al., 2005). Considering that inhibition of ZGA leads to the death of the embryo during gastrulation in both zebrafish and *Xenopus* (Kane et al., 1996; Sible, Anderson, Lewellyn, & Maller, 1997; Stack & Newport, 1997), the finding that maternal-zygotic Dicer mutants can survive for a number of days argues that interfering with maternal RNA degradation has no, or only a minimal, effect on ZGA in zebrafish. Thus, whether degradation of maternal gene products is a prerequisite for ZGA may be species specific.

Many maternal mRNAs are synthesized very early during oogenesis. In fact, in most species, more than 50% of protein-coding genes are expressed in the oocyte. In fully-grown oocytes, these maternal RNAs are very stable. After fertilization, a significant fraction of the maternal RNA pool is actively degraded. Existing evidence in the literature suggests that a significant amount of maternal mRNA turnover events relies on zygotic transcription, especially zygotically transcribed miRNAs (Bushati, Stark, Brennecke, & Cohen, 2008; Giraldez et al., 2006; Lund, Liu, Hartley, Sheets, & Dahlberg, 2009). The pioneering work of Newport and Kirschner elegantly demonstrated that, prior to the MBT, the zygotic genome is globally silenced. Large-scale zygotic transcription occurs at the MBT. Intriguingly, even though the major ZGA occurs at the MBT, the transcriptional machinery is present in pre-MBT stage embryos (Newport & Kirschner, 1982; Prioleau, Huet, Sentenac, & Mechali, 1994). In at least two independent studies in *Xenopus*, phosphorylated Ser2 in the CTD of RNA polymerase II was detected during early cleavage stages and indicated that Pol II-dependent transcription elongation could occur (Blythe, Cha, Tadjuidje, Heasman, & Klein, 2010; Collart, Ramis, Down, & Smith, 2009). These results suggested that at least a low level of RNA polymerase II-dependent transcription may occur pre-MBT. Indeed, it has been observed that a small number of zygotic genes are transcribed during pre-MBT stages when the zygotic genome is globally silenced in both *Drosophila* (Ali-Murthy, Lott, Eisen, & Kornberg, 2013; Edgar & Schubiger, 1986; Harrison, Botchan, & Cline, 2010; Harrison, Li, Kaplan, Botchan, & Eisen, 2011; Karr, Weir, Ali, & Kornberg, 1989; Liang et al., 2008; ten Bosch, Benavides, & Cline, 2006) and *Xenopus* (Blythe et al., 2010; Nakakura, Miura, Yamana, Ito, & Shiokawa, 1987; Shiokawa et al., 1989; Skirkanich et al., 2011; Suzuki, Ueno, & Hemmati-Brivanlou, 1997; Tan et al., 2013; Yang, Tan, Darken, Wilson, & Klein, 2002). The pre-MBT zygotic transcription is unlikely to be "transcriptional noise." Among 151 genes that are transcribed prior to the *Xenopus* MBT, many are associated with cellular events such as cell division, apoptosis, signal transduction, and phosphorylation. The pluripotency-associated gene, Oct 25, is also on the list of pre-MBT genes (Tan et al., 2013). Interestingly, some of the pre-MBT transcriptional events are precisely regulated by signaling pathways that are essential for early embryonic patterning. For example, at least a subset of pre-MBT transcription in *Xenopus* embryos relies on Wnt/β-catenin signaling (Blythe et al., 2010; Skirkanich et al., 2011; Yang et al., 2002). Inhibition of β-catenin-dependent pre-MBT transcription impairs axis specification, leading to ventralized embryos (Yang et al., 2002). Interestingly, we recently found that transcription of *Xenopus* miR-427, the homologue of miR-430, which plays key roles during the zebrafish MZT, can be detected as early as the 16-cell stage. Given that miR-427 is essential for deadenylation of maternal mRNAs during the *Xenopus* MZT (Lund et al., 2009), it seems likely that the low level of transcription occurring prior to the MBT may be important to initiate and coordinate the turnover of maternal RNAs. Later on the degradation machinery is accelerated by the major ZGA, leading to a rapid clearance of maternal products during gastrulation (Fig. 4).

9. STABILITY OF GERMLINE RNAs AND PROTEINS

Like all other maternal RNAs, germ-plasm RNAs are degraded during embryonic development. Unfortunately, in *Xenopus*, it is largely unclear when germ-plasm RNAs are degraded and how their turnover is regulated. So far, only a few studies have addressed these important questions. These studies suggest that germ-plasm RNAs are degraded through at least two different mechanisms. The first wave of germ-plasm RNA turnover occurs during gastrulation, and is most likely focused on the clearance of germ-plasm RNAs from the soma. Within PGCs, these maternal germ-plasm RNAs are protected and turnover is very slow. Little is known about the temporal control of germ-plasm RNAs in terms of translation and decay. However, based upon limited experimental evidence in the literature, we suggest a few general themes and highlight important questions that remain to be addressed.

The stability of germ-plasm RNAs must be regulated differently in the soma than in the germline as germ-plasm components can be detrimental to establishing somatic cell fates, but essential for PGC development. A good

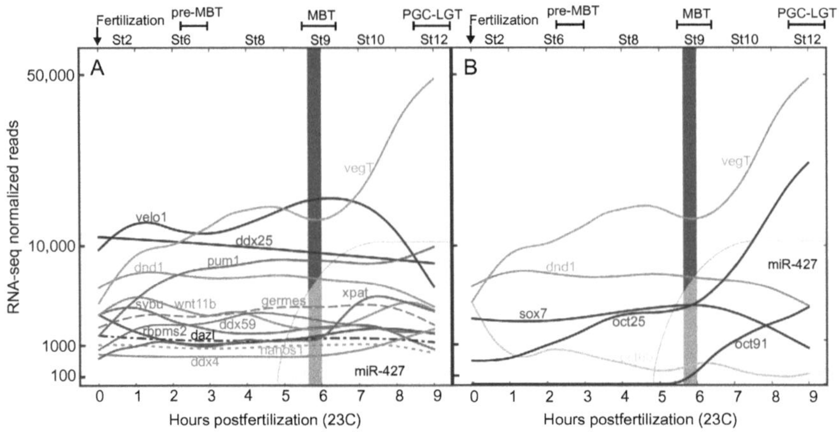

Figure 4 Transcriptional profiling in *Xenopus* early development. Representation of gene-expression patterns of (A) PGC-related genes and (B) pluripotency factors between fertilization and the onset of gastrulation. Before the MBT, positive/negative changes in reads are because of polyadenylation and deadenylation rather than degradation while, after the MBT, miR-427-dependent degradation occurs (Collart et al., 2014). Polyadenylation of maternally deposited transcripts occurs immediately after fertilization and is required for zygotic transcription. Two waves of zygotic genome activation occur: At the 32/64-cell stage (pre-MBT), ~150 new transcripts are synthesized while the bulk of transcription occurs hours later, at the MBT (Tan et al., 2013). The maternal determinant *VegT* is degraded in PGCs while it is polyadenlyated and translated in the soma and zygotically transcribed at the MBT (included in both graphics as a reference). miR-427 is essential for deadenylation of maternal mRNAs during the MZT (yellow curves in A and B). (A) Germline RNAs appear to be polyadenylated soon after fertilization and remain stable through the MBT. (B) Pluripotency factors generally decline after the MBT except for Oct 91, which is newly transcribed in the PGCs. *Modified from Collart et al. (2014).* (See the color plate.)

example is Nanos whose expression in the soma leads to abnormal development but its loss from germ plasm results in the failure of the germline while the soma is unaffected (Lai et al., 2012; Luo, Nerlick, An, & King, 2011). During early cleavage stages in zebrafish, nanos and other germ-plasm RNAs are also present in the soma and must be degraded there (Kedde et al., 2007). However, it is not clear whether that situation holds for *Xenopus*. The early history of germ-plasm RNAs is well established from whole mount *in situ* hybridization (WISH) analyses, but precise quantitation of these RNAs within germ plasm or within the ooplasm or soma has not been determined as this would require a separation of the two. For example, we know that in the oocyte, germ-plasm RNAs are either sequestered in the germinal granules or associated with the germ-plasm matrix. As described above, by the 8-cell stage, they coalesce into a few large aggregates that are inherited later by a few blastomeres, which consequently become PGCs. During these processes, germ-plasm RNAs are transported into the PGCs along with other components of the germ plasm. Once these processes are complete, highly concentrated germ-plasm RNAs, including *nanos1* (Forristall, Pondel, Chen, & King, 1995), *Xpat* (Hudson & Woodland, 1998), *Xdazl* (Houston, Zhang, Maines, Wasserman, & King, 1998), *DEADSouth* (MacArthur, Houston, Bubunenko, Mosquera, & King, 2000), and *dnd1* (Horvay, Claussen, Katzer, Landgrebe, & Pieler, 2006), can be detected in a few blastomeres by WISH. However, detection of highly concentrated germ-plasm RNAs in PGCs by WISH does not rule out the possibility that a certain amount of these RNAs fails to be transported into PGCs and is left in somatic cells. How efficient the process is at shuttling germline RNAs exclusively into PGCs is not known. Studies in *C. elegans* have revealed that, during segregation of the germline from soma, a certain amount of germ plasm is indeed left in the somatic cells and these RNAs are rapidly degraded (Zhang et al., 2009; Zhao, Tian, & Zhang, 2009). In *Xenopus* embryos, mechanisms capable of actively degrading germ-plasm RNAs in the soma have recently been identified (see below for details). Future work will have to establish in a more quantitative way the distribution and turnover of germ-plasm RNAs. Only then will it be possible to establish whether different mechanisms operate to control the stability of germ-plasm RNAs in the soma versus the germline.

Although only a few studies address how germ-plasm RNAs are degraded, the temporal and spatial expression patterns of many germ-plasm RNAs have been studied. RNA-seq analysis of RNAs isolated at different stages from *Xenopus tropicalis*, offers some insights (Collart et al., 2014). In this study, polyadenylated and nonpolyadenylated RNAs were quantitated. If we look at germ-plasm RNAs with *VegT* as a comparison, germline RNAs appear polyadenylated soon after fertilization and remain stable through the MBT (Fig. 4A). A general picture that emerges from these studies is that germ-plasm RNAs are degraded dramatically during gastrulation. For example, degradation of *Xpat* (Hudson & Woodland, 1998), *Xdazl* (Houston et al., 1998), *DEADSouth* (MacArthur et al., 2000), *dnd1* (Horvay et al., 2006), and *nanos1* (Lai et al., 2011), was detected easily by Northern Blot or RT-PCR analysis of whole gastrulae RNA. Interestingly, persistent expression of these germ-plasm RNAs can be detected in PGCs by WISH even as late as the tadpole stages (Kataoka et al., 2006). The significance of this degradation event during gastrulation is not fully understood.

In our opinion, it is critically important to clarify in which cell lineages this RNA turnover event occurs. If degradation of germ-plasm RNAs during gastrulation occurs mainly in PGCs, it would suggest that this turnover event is a critical step during the MZT within the germline. If it happens in somatic cells, it would suggest that this turnover event is more likely responsible for removal of germ-plasm components from the somatic tissues, which ensures proper segregation of the germline from soma. In principle, these two possibilities are not mutually exclusive.

Some recent studies have demonstrated that miR-427 and miR-18 play important roles in controlling the turnover of *DeadSouth* and *dnd1*, respectively. *Xenopus* DEADSouth, a homologue of mammalian DDX25, is a DEAD-box RNA helicase that is specifically localized in the germ plasm. Its function during PGC development can be substituted by VASA/DDX4 (Yamaguchi, Taguchi, Watanabe, & Orii, 2013). The 3′UTR of *DEADSouth* mRNA contains three miR-427 targeting sequences (Yamaguchi, Kataoka, Watanabe, & Orii, 2014). When the 3′UTR of *DEADSouth* is fused to a fluorescent protein, Venus, the reporter construct undergoes rapid turnover during gastrulation. By the tadpole stage, the reporter RNA is degraded completely in somatic tissues and Venus can be detected only in PGCs (Kataoka et al., 2006). Mutation in these miR-427 targeting sequences or knockdown of miR-427 interferes with clearance of the reporter RNA in somatic cells (Yamaguchi et al., 2014). Interestingly, the expression of miR-427 could not be detected in isolated PGCs (Yamaguchi et al., 2014). This raises the intriguing possibility that the miR-427-dependent turnover, which begins during gastrulation, does not occur in the PGCs. It is solely responsible for removal of *DEADSouth* mRNA from the somatic tissues. Therefore, it is unlikely that miR-427-dependent turnover plays a role during the MZT within the germline. In the case of *dnd1*, a regulatory element within the 3′UTR, which regulates vegetal localization of *dnd1* in the oocyte, plays an essential role in somatic *dnd1* mRNA clearance. This *dnd1* mRNA turnover event is mediated by miR-18, which binds to the GCACUU(U) sequence in this regulatory element of the *dnd1* 3′UTR. Interfering with the binding of miR-18 to *dnd1* significantly increases the level of endogenous *dnd1 RNA* in somatic tissues at the tadpole stage (Koebernick, Loeber, Arthur, Tarbashevich, & Pieler, 2010). As expected, when the *dnd1* 3′UTR is fused to GFP, it is competent to direct PGC-specific GFP expression. However, the PGC-specific GFP expression can be observed only when this miR-18-dependent somatic clearance event is complete (Dzementsei, Schneider, Janshoff, & Pieler,

2013; Koebernick et al., 2010). Taken together, these two lines of evidence support the idea that the dramatic decrease in the level of germ-plasm RNAs during gastrulation is a consequence of somatic RNA clearance.

The important question remains, how are maternal germ-plasm RNAs degraded within PGCs during germline development? Taking advantage of the *X. laevis* and *borealis* hybrid embryos, it was found that after the dramatic RNA degradation events during gastrulation, *nanos1* RNA, presumably located in PGCs, remained stable at least until the end of the tailbud stage. Zygotic transcription of *nanos1* does not occur in early-stage embryos (Lai et al., 2011). In the case of *DEADSouth*, maternal messages are removed nearly completely by the tailbud stage. Around the same time, embryos start to express zygotic *DEADSouth* (Yamaguchi et al., 2014). Other than these two RNAs, we do not know how other maternal germ-plasm RNAs are regulated during PGC development. Another issue is the half-life of the proteins encoded by germ-plasm RNAs. Nanos is present in germ plasm until PGCs leave the endoderm at Stage 39, well past when its RNA is detected. Xdazl protein has been detected even as late as tadpole Stage 52 and within germ cells of the developing gonads (Kataoka et al., 2006). Clearly, more analysis needs to be done to assess the gene-expression profiles within PGCs during germline development. Such information would lay the foundation to study how ZGA and removal of maternal germ-plasm RNAs are coordinated during germline development.

10. THE PGC LATE-GASTRULA TRANSITION

Germ plasm can be thought of as a mechanism to continue the potential for totipotency exhibited by the oocyte, forward into development. Germ plasm then, functions as an "escape hatch" for totipotency from the oocyte. Interestingly, some of the germ-plasm components that are passed on to PGCs as untranslated RNAs, are translated in the oocyte. These include the RNA-binding proteins Dead-end, DeadSouth, Xdazl, and RINGO/Spy. These proteins play a role in translational regulation or in regulating meiotic events. After maturation and fertilization, are these proteins degraded while being preserved as RNAs within the germ plasm? Is this one mechanism for carrying gametic identity forward through embryogenesis? What is the fate of oocyte proteins in the soma? We simply do not know the answers to these questions in *Xenopus*.

Segregation of the germline from the somatic endoderm occurs at gastrulation when the germ plasm moves to a perinuclear location; subsequent divisions result in both daughter cells receiving germ plasm. Before complete separation from the soma, PGCs are presumptive PGCs or pPGCs. Only after separation and during gastrulation do PGCs initiate their zygotic transcription program, or the PGC Late-Gastrula Transition (Fig. 3). Which maternal factor(s) activate the PGC program remains unknown. Importantly, *Xenopus* Oct 25, Oct 60, and Oct 91 are in the same POU subclass as the human pluripotency factor, Oct 3/4 (Hinkley et al., 1992), and Oct 3/4 acts as a functional homolog for Oct 25 or Oct 60 in rescue experiments (Cao et al., 2006). Maternal Oct 60 is expressed in isolated pPGCs prior to gastrulation and Oct 91 is expressed *de novo* in PGCs after they segregate from the endoderm (Hinkley et al., 1992; Venkatarama et al., 2010). The RNA-seq results from *X. tropicalis* for the Oct and Sox pluripotent factors show they generally decline after the MBT except for Oct 91, which is newly transcribed in PGCs (Fig. 4B; Collart et al., 2014; Tan et al., 2013).

Xenopus Oct 60 plays a key role in maintaining pluripotency by suppressing signaling in differentiation pathways including Activin/Nodal, BMP, and WNT (Cao et al., 2004, 2006, 2008; Whitfield et al., 1993). Oct 25 is not expressed in PGCs, while Oct 60 expression in pPGCs persists until gastrulation. Importantly, zygotic Oct 91 is subsequently expressed *de novo* in PGCs (Venkatarama et al., 2010). Oct 60 expression and function have primarily been investigated in somatic cells and its role in PGCs remains unknown. The maternally expressed transcription factor, Sox7, localizes to the vegetal pole and may have a role in PGC fate in addition to its known role in somatic cells (Zhang, Basta, Fawcett, & Klymkowsky, 2005; Zhang, Basta, & Klymkowsky, 2005). Interestingly, maternal expression of Oct 60 depends on an Octamer-Sox-binding motif in its promoter region and Oct 60 has been shown to bind this motif (Morichika, Sugimoto, Yasuda, & Kinoshita, 2014). Perhaps, Oct 60 and Sox proteins promote PGC transcription during the PCG-LGT. Consistent with this proposal, TRANSFAC promoter analysis identifies putative Oct and Sox binding motifs within the Oct 91 promoter region.

What gene programs are likely to be initiated at the PGC-LGT? Beyond keeping competency for totipotency, a zygotic differentiation program must be activated that confers the ability to migrate. Directed migration is one characteristic that PGCs acquire, which is missing from somatic endoderm. Because PGCs originate outside the gonads within embryonic endoderm, PGCs must migrate through somatic tissues and reach the developing gonads. Once inside the presumptive gonad, the PGCs divide, enter meiosis, and differentiate into definitive gametes.

Mutational analysis has identified three different signaling systems that affect PGC migration: CXCR4/SDF1, Notch/Delta2 (Morichika et al., 2010), and KIF13B-PIP3-GRIP2.1. PGCs begin their migration out of the endoderm and toward the presumptive gonads as early as tailbud (Stage 24), 10–12 h after PGC transcription begins (Terayama et al., 2013; Venkatarama et al., 2010). At this time, they are found clustered at the midline in the posterior endoderm. Early midline clustering is likely a passive step as CXCR4 is not expressed (RNA or protein) at this stage but it is during lateral and dorsal movements (Nishiumi, Komiya, & Ikenishi, 2005). Six hours later, PGCs are found spread more laterally and express CXCR4 on their surface (Takeuchi, Tanigawa, Minamide, Ikenishi, & Komiya, 2010) while its ligand, SDF-1 is expressed in the dorsal mesentery (Braun et al., 2002). Also around this time, a decreased attachment to ECM components like fibronectin, reduced adhesion to somatic endodermal cells, and downregulation of E-cadherin occurs in isolated PGCs but not in endodermal cells (Dzementsei et al., 2013).

The cellular movements that underpin these migration steps—blebbing and persistent blebbing at the leading edge—and which are required for directional migration, all occur in isolated PGCs cultured in a serum-free medium with a fibronectin substrate. As expected, these cellular behaviors require F-actin, myosin II activity, and RhoA/Rho-associated protein kinase (ROCK) signaling. ROCK was identified as an early zygotic transcript in PGCs (Venkatarama et al., 2010). Antisense depletion of either maternal *Xdazl* or *Dead-end* RNA prevents this step from occurring and PGCs are lost soon afterward (Horvay et al., 2006; Houston & King, 2000). Both of these RNA-binding proteins play a role in regulating translation, suggesting that maternal control of translation may extend to zygotically synthesized transcripts in the PGCs.

Finally, PGCs will give rise to the gametes and only they are competent to execute the reductional divisions during meiosis. How is this ability carried forward in PGCs through the MZT, but not in the soma? What maternal and/or zygotic factors are required to maintain competence for meiosis? These are questions that remain to be answered.

ACKNOWLEDGMENTS

The authors would like to thank Dr. Amanda Butler and Ms. Dawn Owens (Dept. of Cell Biology, University of Miami, Miller School of Medicine, Miami, FL, USA) who helped to assemble the data shown in Table 1. The authors would like to acknowledge support from NIH to M.L.K. (grant R01GM102397) and to J.Y. (grant R01GM111816).

REFERENCES

Ali-Murthy, Z., Lott, S. E., Eisen, M. B., & Kornberg, T. B. (2013). An essential role for zygotic expression in the pre-cellular Drosophila embryo. *PLoS Genetics*, *9*, e1003428.

Bashirullah, A., Halsell, S. R., Cooperstock, R. L., Kloc, M., Karaiskakis, A., Fisher, W. W., et al. (1999). Joint action of two RNA degradation pathways controls the timing of maternal transcript elimination at the midblastula transition in Drosophila melanogaster. *The EMBO Journal*, *18*, 2610–2620.

Benoit, B., He, C. H., Zhang, F., Votruba, S. M., Tadros, W., Westwood, J. T., et al. (2009). An essential role for the RNA-binding protein Smaug during the Drosophila maternal-to-zygotic transition. *Development*, *136*, 923–932.

Birsoy, B., Kofron, M., Schaible, K., Wylie, C., & Heasman, J. (2006). Vg 1 is an essential signaling molecule in Xenopus development. *Development*, *133*, 15–20.

Blythe, S. A., Cha, S. W., Tadjuidje, E., Heasman, J., & Klein, P. S. (2010). beta-Catenin primes organizer gene expression by recruiting a histone H3 arginine 8 methyltransferase, Prmt2. *Developmental Cell*, *19*, 220–231.

Bontems, F., Stein, A., Marlow, F., Lyautey, J., Gupta, T., Mullins, M. C., et al. (2009). Bucky ball organizes germ plasm assembly in zebrafish. *Current Biology*, *19*, 414–422.

Braun, M., Wunderlin, M., Spieth, K., Knochel, W., Gierschik, P., & Moepps, B. (2002). Xenopus laevis stromal cell-derived factor 1: Conservation of structure and function during vertebrate development. *Journal of Immunology*, *168*, 2340–2347.

Bushati, N., Stark, A., Brennecke, J., & Cohen, S. M. (2008). Temporal reciprocity of miRNAs and their targets during the maternal-to-zygotic transition in Drosophila. *Current Biology*, *18*, 501–506.

Cao, Y., Knochel, S., Donow, C., Miethe, J., Kaufmann, E., & Knochel, W. (2004). The POU factor Oct-25 regulates the Xvent-2B gene and counteracts terminal differentiation in Xenopus embryos. *The Journal of Biological Chemistry*, *279*, 43735–43743.

Cao, Y., Siegel, D., & Knochel, W. (2006). Xenopus POU factors of subclass V inhibit activin/nodal signaling during gastrulation. *Mechanisms of Development*, *123*, 614–625.

Cao, Y., Siegel, D., Oswald, F., & Knochel, W. (2008). Oct25 represses transcription of nodal/activin target genes by interaction with signal transducers during Xenopus gastrulation. *The Journal of Biological Chemistry*, *283*, 34168–34177.

Casey, E. S., Tada, M., Fairclough, L., Wylie, C. C., Heasman, J., & Smith, J. C. (1999). Bix4 is activated directly by VegT and mediates endoderm formation in Xenopus development. *Development*, *126*, 4193–4200.

Cha, S. W., Tadjuidje, E., Wylie, C., & Heasman, J. (2011). The roles of maternal Vangl2 and aPKC in Xenopus oocyte and embryo patterning. *Development*, *138*, 3989–4000.

Collart, C., Owens, N. D., Bhaw-Rosun, L., Cooper, B., De Domenico, E., Patrushev, I., et al. (2014). High-resolution analysis of gene activity during the Xenopus mid-blastula transition. *Development*, *141*, 1927–1939.

Collart, C., Ramis, J. M., Down, T. A., & Smith, J. C. (2009). Smicl is required for phosphorylation of RNA polymerase II and affects 3′-end processing of RNA at the midblastula transition in Xenopus. *Development*, *136*, 3451–3461.

Cuykendall, T. N., & Houston, D. W. (2009). Vegetally localized Xenopus trim36 regulates cortical rotation and dorsal axis formation. *Development*, *136*, 3057–3065.

Deshpande, G., Calhoun, G., Jinks, T. M., Polydorides, A. D., & Schedl, P. (2005). Nanos downregulates transcription and modulates CTD phosphorylation in the soma of early Drosophila embryos. *Mechanisms of Development*, *122*, 645–657.

Dzementsei, A., Schneider, D., Janshoff, A., & Pieler, T. (2013). Migratory and adhesive properties of Xenopus laevis primordial germ cells in vitro. *Biology Open*, *2*, 1279–1287.

Edgar, B. A., & Schubiger, G. (1986). Parameters controlling transcriptional activation during early Drosophila development. *Cell*, *44*, 871–877.

Eno, C., & Pelegri, F. (2013). Gradual recruitment and selective clearing generate germ plasm aggregates in the zebrafish embryo. *Bioarchitecture*, *3*, 125–132.
Ferg, M., Sanges, R., Gehrig, J., Kiss, J., Bauer, M., Lovas, A., et al. (2007). The TATA-binding protein regulates maternal mRNA degradation and differential zygotic transcription in zebrafish. *The EMBO Journal*, *26*, 3945–3956.
Ferrell, J. E., Jr. (1999). Xenopus oocyte maturation: new lessons from a good egg. *Bioessays*, *21*, 833–842.
Forristall, C., Pondel, M., Chen, L., & King, M. L. (1995). Patterns of localization and cytoskeletal association of two vegetally localized RNAs, Vg1 and Xcat-2. *Development*, *121*, 201–208.
Fox, M., Urano, J., & Reijo Pera, R. A. (2005). Identification and characterization of RNA sequences to which human PUMILIO-2 (PUM2) and deleted in Azoospermia-like (DAZL) bind. *Genomics*, *85*, 92–105.
Gard, D. L. (1991). Organization, nucleation, and acetylation of microtubules in Xenopus laevis oocytes: A study by confocal immunofluorescence microscopy. *Developmental Biology*, *143*, 346–362.
Gard, D. L. (1994). Gamma-tubulin is asymmetrically distributed in the cortex of Xenopus oocytes. *Developmental Biology*, *161*, 131–140.
Gard, D. L. (1999). Confocal microscopy and 3-D reconstruction of the cytoskeleton of Xenopus oocytes. *Microscopy Research and Technique*, *44*, 388–414.
Gard, D. L., Cha, B. J., & King, E. (1997). The organization and animal-vegetal asymmetry of cytokeratin filaments in stage VI Xenopus oocytes is dependent upon F-actin and microtubules. *Developmental Biology*, *184*, 95–114.
Ge, X., Grotjahn, D., Welch, E., Lyman-Gingerich, J., Holguin, C., Dimitrova, E., et al. (2014). Hecate/Grip2a acts to reorganize the cytoskeleton in the symmetry-breaking event of embryonic axis induction. *PLoS Genetics*, *10*, e1004422.
Giraldez, A. J., Cinalli, R. M., Glasner, M. E., Enright, A. J., Thomson, J. M., Baskerville, S., et al. (2005). MicroRNAs regulate brain morphogenesis in zebrafish. *Science*, *308*, 833–838.
Giraldez, A. J., Mishima, Y., Rihel, J., Grocock, R. J., Van Dongen, S., Inoue, K., et al. (2006). Zebrafish MiR-430 promotes deadenylation and clearance of maternal mRNAs. *Science*, *312*, 75–79.
Godsave, S. F., Anderton, B. H., Heasman, J., & Wylie, C. C. (1984). Oocytes and early embryos of Xenopus laevis contain intermediate filaments which react with anti-mammalian vimentin antibodies. *Journal of Embryology and Experimental Morphology*, *83*, 169–187.
Harrison, M. M., Botchan, M. R., & Cline, T. W. (2010). Grainyhead and Zelda compete for binding to the promoters of the earliest-expressed Drosophila genes. *Developmental Biology*, *345*, 248–255.
Harrison, M. M., Li, X. Y., Kaplan, T., Botchan, M. R., & Eisen, M. B. (2011). Zelda binding in the early Drosophila melanogaster embryo marks regions subsequently activated at the maternal-to-zygotic transition. *PLoS Genetics*, 7, e1002266.
Harvey, S. A., Sealy, I., Kettleborough, R., Fenyes, F., White, R., Stemple, D., et al. (2013). Identification of the zebrafish maternal and paternal transcriptomes. *Development*, *140*, 2703–2710.
Hashimoto, H., Hara, K., Hishiki, A., Kawaguchi, S., Shichijo, N., Nakamura, K., et al. (2010). Crystal structure of zinc-finger domain of Nanos and its functional implications. *EMBO Reports*, *11*, 848–853.
Heasman, J. (2006). Patterning the early Xenopus embryo. *Development*, *133*, 1205–1217.
Heasman, J., Quarmby, J., & Wylie, C. C. (1984). The mitochondrial cloud of Xenopus oocytes: The source of germinal granule material. *Developmental Biology*, *105*, 458–469.
Heasman, J., Wessely, O., Langland, R., Craig, E. J., & Kessler, D. S. (2001). Vegetal localization of maternal mRNAs is disrupted by VegT depletion. *Developmental Biology*, *240*, 377–386.

Heim, A. E., Hartung, O., Rothhamel, S., Ferreira, E., Jenny, A., & Marlow, F. L. (2014). Oocyte polarity requires a Bucky ball-dependent feedback amplification loop. *Development, 141*, 842–854.

Hinkley, C. S., Martin, J. F., Leibham, D., & Perry, M. (1992). Sequential expression of multiple POU proteins during amphibian early development. *Molecular and Cellular Biology, 12*, 638–649.

Horvay, K., Claussen, M., Katzer, M., Landgrebe, J., & Pieler, T. (2006). Xenopus Dead end mRNA is a localized maternal determinant that serves a conserved function in germ cell development. *Developmental Biology, 291*, 1–11.

Houliston, E., & Elinson, R. P. (1991). Patterns of microtubule polymerization relating to cortical rotation in Xenopus laevis eggs. *Development, 112*, 107–117.

Houston, D. W., & King, M. L. (2000). A critical role for Xdazl, a germ plasm-localized RNA, in the differentiation of primordial germ cells in Xenopus. *Development, 127*, 447–456.

Houston, D. W., Zhang, J., Maines, J. Z., Wasserman, S. A., & King, M. L. (1998). A Xenopus DAZ-like gene encodes an RNA component of germ plasm and is a functional homologue of Drosophila boule. *Development, 125*, 171–180.

Hudson, C., Clements, D., Friday, R. V., Stott, D., & Woodland, H. R. (1997). Xsox17alpha and -beta mediate endoderm formation in Xenopus. *Cell, 91*, 397–405.

Hudson, C., & Woodland, H. R. (1998). Xpat, a gene expressed specifically in germ plasm and primordial germ cells of Xenopus laevis. *Mechanisms of Development, 73*, 159–168.

Jaruzelska, J., Kotecki, M., Kusz, K., Spik, A., Firpo, M., & Reijo Pera, R. A. (2003). Conservation of a Pumilio-Nanos complex from Drosophila germ plasm to human germ cells. *Development Genes and Evolution, 213*, 120–126.

Jevtic, P., & Levy, D. L. (2015). Nuclear size scaling during Xenopus early development contributes to midblastula transition timing. *Current Biology, 25*, 45–52.

Kane, D. A., Hammerschmidt, M., Mullins, M. C., Maischein, H. M., Brand, M., van Eeden, F. J., et al. (1996). The zebrafish epiboly mutants. *Development, 123*, 47–55.

Karr, T. L., Weir, M. P., Ali, Z., & Kornberg, T. (1989). Patterns of engrailed protein in early Drosophila embryos. *Development, 105*, 605–612.

Kataoka, K., Yamaguchi, T., Orii, H., Tazaki, A., Watanabe, K., & Mochii, M. (2006). Visualization of the Xenopus primordial germ cells using a green fluorescent protein controlled by cis elements of the 3′ untranslated region of the DEADSouth gene. *Mechanisms of Development, 123*, 746–760.

Kedde, M., Strasser, M. J., Boldajipour, B., Oude Vrielink, J. A., Slanchev, K., le Sage, C., et al. (2007). RNA-binding protein Dnd1 inhibits microRNA access to target mRNA. *Cell, 131*, 1273–1286.

King, M. L. (2014). Germ cell specification in Xenopus. In M. Kloc & J. Z. Kubiak (Eds.), *Xenopus development* (pp. 75–100). Hoboken, New Jersey: John Wiley & Sons, Inc.

King, M. L., Messitt, T. J., & Mowry, K. L. (2005). Putting RNAs in the right place at the right time: RNA localization in the frog oocyte. *Biology of the Cell, 97*, 19–33.

Kloc, M., Bilinski, S., & Dougherty, M. T. (2007). Organization of cytokeratin cytoskeleton and germ plasm in the vegetal cortex of Xenopus laevis oocytes depends on coding and non-coding RNAs: Three-dimensional and ultrastructural analysis. *Experimental Cell Research, 313*, 1639–1651.

Kloc, M., Dougherty, M. T., Bilinski, S., Chan, A. P., Brey, E., King, M. L., et al. (2002). Three-dimensional ultrastructural analysis of RNA distribution within germinal granules of Xenopus. *Developmental Biology, 241*, 79–93.

Kloc, M., & Etkin, L. D. (1994). Delocalization of Vg1 mRNA from the vegetal cortex in Xenopus oocytes after destruction of Xlsirt RNA. *Science, 265*, 1101–1103.

Kloc, M., Wilk, K., Vargas, D., Shirato, Y., Bilinski, S., & Etkin, L. D. (2005). Potential structural role of non-coding and coding RNAs in the organization of the cytoskeleton at the vegetal cortex of Xenopus oocytes. *Development*, *132*, 3445–3457.

Koebernick, K., Loeber, J., Arthur, P. K., Tarbashevich, K., & Pieler, T. (2010). Elr-type proteins protect Xenopus Dead end mRNA from miR-18-mediated clearance in the soma. *Proceedings of the National Academy of Sciences of the United States of America*, *107*, 16148–16153.

Kotani, T., Yasuda, K., Ota, R., & Yamashita, M. (2013). Cyclin B1 mRNA translation is temporally controlled through formation and disassembly of RNA granules. *The Journal of Cell Biology*, *202*, 1041–1055.

Lai, F., & King, M. L. (2013). Repressive translational control in germ cells. *Molecular Reproduction and Development*, *80*, 665–676.

Lai, F., Singh, A., & King, M. L. (2012). Xenopus Nanos1 is required to prevent endoderm gene expression and apoptosis in primordial germ cells. *Development*, *139*, 1476–1486.

Lai, F., Zhou, Y., Luo, X., Fox, J., & King, M. L. (2011). Nanos1 functions as a translational repressor in the Xenopus germline. *Mechanisms of Development*, *128*, 153–163.

Langley, A. R., Smith, J. C., Stemple, D. L., & Harvey, S. A. (2014). New insights into the maternal to zygotic transition. *Development*, *141*, 3834–3841.

Leatherman, J. L., & Jongens, T. A. (2003). Transcriptional silencing and translational control: Key features of early germline development. *Bioessays*, *25*, 326–335.

Lee, M. T., Bonneau, A. R., Takacs, C. M., Bazzini, A. A., DiVito, K. R., Fleming, E. S., et al. (2013). Nanog, Pou5f1 and SoxB1 activate zygotic gene expression during the maternal-to-zygotic transition. *Nature*, *503*, 360–364.

Lerit, D. A., & Gavis, E. R. (2011). Transport of germ plasm on astral microtubules directs germ cell development in Drosophila. *Current Biology*, *21*, 439–448.

Liang, H. L., Nien, C. Y., Liu, H. Y., Metzstein, M. M., Kirov, N., & Rushlow, C. (2008). The zinc-finger protein Zelda is a key activator of the early zygotic genome in Drosophila. *Nature*, *456*, 400–403.

Lund, E., Liu, M., Hartley, R. S., Sheets, M. D., & Dahlberg, J. E. (2009). Deadenylation of maternal mRNAs mediated by miR-427 in Xenopus laevis embryos. *RNA*, *15*, 2351–2363.

Luo, X., Nerlick, S., An, W., & King, M. L. (2011). Xenopus germline nanos1 is translationally repressed by a novel structure-based mechanism. *Development*, *138*, 589–598.

MacArthur, H., Houston, D. W., Bubunenko, M., Mosquera, L., & King, M. L. (2000). DEADSouth is a germ plasm specific DEAD-box RNA helicase in Xenopus related to eIF4A. *Mechanisms of Development*, *95*, 291–295.

Marlow, F. L., & Mullins, M. C. (2008). Bucky ball functions in Balbiani body assembly and animal-vegetal polarity in the oocyte and follicle cell layer in zebrafish. *Developmental Biology*, *321*, 40–50.

Mathavan, S., Lee, S. G., Mak, A., Miller, L. D., Murthy, K. R., Govindarajan, K. R., et al. (2005). Transcriptome analysis of zebrafish embryogenesis using microarrays. *PLoS Genetics*, *1*, 260–276.

Mei, W., Jin, Z., Lai, F., Schwend, T., Houston, D. W., King, M. L., et al. (2013). Maternal Dead-End1 is required for vegetal cortical microtubule assembly during Xenopus axis specification. *Development*, *140*, 2334–2344.

Mendez, R., & Richter, J. D. (2001). Translational control by CPEB: A means to the end. *Nature Reviews. Molecular Cell Biology*, *2*, 521–529.

Morichika, K., Kataoka, K., Terayama, K., Tazaki, A., Kinoshita, T., Watanabe, K., et al. (2010). Perturbation of Notch/Suppressor of Hairless pathway disturbs migration of primordial germ cells in Xenopus embryo. *Development, Growth & Differentiation*, *52*, 235–244.

Morichika, K., Sugimoto, M., Yasuda, K., & Kinoshita, T. (2014). Possible regulation of Oct60 transcription by a positive feedback loop in Xenopus oocytes. *Zygote, 22*, 266–274.
Mosquera, L., Forristall, C., Zhou, Y., & King, M. L. (1993). A mRNA localized to the vegetal cortex of Xenopus oocytes encodes a protein with a nanos-like zinc finger domain. *Development, 117*, 377–386.
Murphy, C. R., Sabel, J. L., Sandler, A. D., & Dagle, J. M. (2002). Survivin mRNA is down-regulated during early Xenopus laevis embryogenesis. *Developmental Dynamics, 225*, 597–601.
Nair, S., Marlow, F., Abrams, E., Kapp, L., Mullins, M. C., & Pelegri, F. (2013). The chromosomal passenger protein birc5b organizes microfilaments and germ plasm in the zebrafish embryo. *PLoS Genetics, 9*, e1003448.
Nakakura, N., Miura, T., Yamana, K., Ito, A., & Shiokawa, K. (1987). Synthesis of heterogeneous mRNA-like RNA and low-molecular-weight RNA before the midblastula transition in embryos of Xenopus laevis. *Developmental Biology, 123*, 421–429.
Nakaya, M., Fukui, A., Izumi, Y., Akimoto, K., Asashima, M., & Ohno, S. (2000). Meiotic maturation induces animal-vegetal asymmetric distribution of aPKC and ASIP/PAR-3 in Xenopus oocytes. *Development, 127*, 5021–5031.
Newport, J., & Kirschner, M. (1982). A major developmental transition in early Xenopus embryos: II. Control of the onset of transcription. *Cell, 30*, 687–696.
Nieuwkoop, P. D., & Faber, J. (1956). *Normal Table of Xenopus laevis (Daudin): A systematical and chronological survey of the development from the fertilized egg till the end of metamorphosis.* Amsterdam: North-Holland Publ. Co.
Nieuwkoop, P. D., & Sutasurya, L. A. (1976). Embryological evidence for a possible polyphyletic origin of the recent amphibians. *Journal of Embryology and Experimental Morphology, 35*, 159–167.
Nishiumi, F., Komiya, T., & Ikenishi, K. (2005). The mode and molecular mechanisms of the migration of presumptive PGC in the endoderm cell mass of Xenopus embryos. *Development, Growth & Differentiation, 47*, 37–48.
Olson, D. J., Oh, D., & Houston, D. W. (2015). The dynamics of plus end polarization and microtubule assembly during Xenopus cortical rotation. *Developmental Biology, 401*, 249–263.
Ota, R., Kotani, T., & Yamashita, M. (2011). Biochemical characterization of Pumilio1 and Pumilio2 in Xenopus oocytes. *The Journal of Biological Chemistry, 286*, 2853–2863.
Padmanabhan, K., & Richter, J. D. (2006). Regulated Pumilio-2 binding controls RINGO/Spy mRNA translation and CPEB activation. *Genes & Development, 20*, 199–209.
Pfeiffer, D. C., & Gard, D. L. (1999). Microtubules in Xenopus oocytes are oriented with their minus-ends towards the cortex. *Cell Motility and the Cytoskeleton, 44*, 34–43.
Pondel, M. D., & King, M. L. (1988). Localized maternal mRNA related to transforming growth factor β mRNA is concentrated in a cytokeratin-enriched fraction from Xenopus oocytes. *Proceedings of the National Academy of Sciences of the United States of America, 85*, 7612–7616.
Pratt, C. A., & Mowry, K. L. (2013). Taking a cellular road-trip: mRNA transport and anchoring. *Current Opinion in Cell Biology, 25*, 99–106.
Prioleau, M. N., Huet, J., Sentenac, A., & Mechali, M. (1994). Competition between chromatin and transcription complex assembly regulates gene expression during early development. *Cell, 77*, 439–449.
Ressom, R. E., & Dixon, K. E. (1988). Relocation and reorganization of germ plasm in Xenopus embryos after fertilization. *Development, 103*, 507–518.
Robb, D. L., Heasman, J., Raats, J., & Wylie, C. (1996). A kinesin-like protein is required for germ plasm aggregation in Xenopus. *Cell, 87*, 823–831.

Saitou, M., & Yamaji, M. (2012). Primordial germ cells in mice. *Cold Spring Harbor Perspectives in Biology*, *4*, 1–20.

Savage, R. M., & Danilchik, M. V. (1993). Dynamics of germ plasm localization and its inhibition by ultraviolet irradiation in early cleavage Xenopus embryos. *Developmental Biology*, *157*, 371–382.

Schaner, C. E., Deshpande, G., Schedl, P. D., & Kelly, W. G. (2003). A conserved chromatin architecture marks and maintains the restricted germ cell lineage in worms and flies. *Developmental Cell*, *5*, 747–757.

Schroeder, M. M., & Gard, D. L. (1992). Organization and regulation of cortical microtubules during the first cell cycle of Xenopus eggs. *Development*, *114*, 699–709.

Seydoux, G., & Braun, R. E. (2006). Pathway to totipotency: Lessons from germ cells. *Cell*, *127*, 891–904.

Shiokawa, K., Misumi, Y., Tashiro, K., Nakakura, N., Yamana, K., & Oh-uchida, M. (1989). Changes in the patterns of RNA synthesis in early embryogenesis of Xenopus laevis. *Cell Differentiation and Development*, *28*, 17–25.

Sible, J. C., Anderson, J. A., Lewellyn, A. L., & Maller, J. L. (1997). Zygotic transcription is required to block a maternal program of apoptosis in Xenopus embryos. *Developmental Biology*, *189*, 335–346.

Skirkanich, J., Luxardi, G., Yang, J., Kodjabachian, L., & Klein, P. S. (2011). An essential role for transcription before the MBT in Xenopus laevis. *Developmental Biology*, *357*, 478–491.

Smith, L. D. (1966). The role of a "germinal plasm" in the formation of primordial germ cells in Rana pipiens. *Developmental Biology*, *14*, 330–347.

Song, H. W., Cauffman, K., Chan, A. P., Zhou, Y., King, M. L., Etkin, L. D., et al. (2007). Hermes RNA-binding protein targets RNAs-encoding proteins involved in meiotic maturation, early cleavage, and germline development. *Differentiation*, *75*, 519–528.

Stack, J. H., & Newport, J. W. (1997). Developmentally regulated activation of apoptosis early in Xenopus gastrulation results in cyclin A degradation during interphase of the cell cycle. *Development*, *124*, 3185–3195.

Stennard, F., Zorn, A. M., Ryan, K., Garrett, N., & Gurdon, J. B. (1999). Differential expression of VegT and Antipodean protein isoforms in Xenopus. *Mechanisms of Development*, *86*, 87–98.

Suzuki, A., Igarashi, K., Aisaki, K., Kanno, J., & Saga, Y. (2010). NANOS2 interacts with the CCR4-NOT deadenylation complex and leads to suppression of specific RNAs. *Proceedings of the National Academy of Sciences of the United States of America*, *107*, 3594–3599.

Suzuki, A., & Saga, Y. (2008). Nanos2 suppresses meiosis and promotes male germ cell differentiation. *Genes & Development*, *22*, 430–435.

Suzuki, A., Ueno, N., & Hemmati-Brivanlou, A. (1997). Xenopus msx1 mediates epidermal induction and neural inhibition by BMP4. *Development*, *124*, 3037–3044.

Tada, H., Mochii, M., Orii, H., & Watanabe, K. (2012). Ectopic formation of primordial germ cells by transplantation of the germ plasm: Direct evidence for germ cell determinant in Xenopus. *Developmental Biology*, *371*, 86–93.

Tadros, W., & Lipshitz, H. D. (2009). The maternal-to-zygotic transition: A play in two acts. *Development*, *136*, 3033–3042.

Takeuchi, T., Tanigawa, Y., Minamide, R., Ikenishi, K., & Komiya, T. (2010). Analysis of SDF-1/CXCR4 signaling in primordial germ cell migration and survival or differentiation in Xenopus laevis. *Mechanisms of Development*, *127*, 146–158.

Tan, M. H., Au, K. F., Yablonovitch, A. L., Wills, A. E., Chuang, J., Baker, J. C., et al. (2013). RNA sequencing reveals a diverse and dynamic repertoire of the Xenopus tropicalis transcriptome over development. *Genome Research*, *23*, 201–216.

Tao, Q., Yokota, C., Puck, H., Kofron, M., Birsoy, B., Yan, D., et al. (2005). Maternal wnt11 activates the canonical wnt signaling pathway required for axis formation in Xenopus embryos. *Cell*, *120*, 857–871.

Tarbashevich, K., Koebernick, K., & Pieler, T. (2007). XGRIP2.1 is encoded by a vegetally localizing, maternal mRNA and functions in germ cell development and anteroposterior PGC positioning in Xenopus laevis. *Developmental Biology*, *311*, 554–565.

ten Bosch, J. R., Benavides, J. A., & Cline, T. W. (2006). The TAGteam DNA motif controls the timing of Drosophila pre-blastoderm transcription. *Development*, *133*, 1967–1977.

Terayama, K., Kataoka, K., Morichika, K., Orii, H., Watanabe, K., & Mochii, M. (2013). Developmental regulation of locomotive activity in Xenopus primordial germ cells. *Development, Growth & Differentiation*, *55*, 217–228.

Theusch, E. V., Brown, K. J., & Pelegri, F. (2006). Separate pathways of RNA recruitment lead to the compartmentalization of the zebrafish germ plasm. *Developmental Biology*, *292*, 129–141.

Torpey, N. P., Heasman, J., & Wylie, C. C. (1990). Identification of vimentin and novel vimentin-related proteins in Xenopus oocytes and early embryos. *Development*, *110*, 1185–1195.

Torpey, N. P., Heasman, J., & Wylie, C. C. (1992). Distinct distribution of vimentin and cytokeratin in Xenopus oocytes and early embryos. *Journal of Cell Science*, *101*(Pt. 1), 151–160.

Venkatarama, T., Lai, F., Luo, X., Zhou, Y., Newman, K., & King, M. L. (2010). Repression of zygotic gene expression in the Xenopus germline. *Development*, *137*, 651–660.

Wakahara, M. (1977). Partial characterization of "primordial germ cell-forming activity" localized in vegetal pole cytoplasm in anuran eggs. *Journal of Embryology and Experimental Morphology*, *39*, 221–233.

Wakahara, M. (1978). Induction of supernumerary primordial germ cells by injecting vegetal pole cytoplasm into Xenopus eggs. *The Journal of Experimental Zoology*, *203*, 159–164.

Wang, Y., Opperman, L., Wickens, M., & Hall, T. M. (2009). Structural basis for specific recognition of multiple mRNA targets by a PUF regulatory protein. *Proceedings of the National Academy of Sciences of the United States of America*, *106*, 20186–20191.

Weidmann, C. A., & Goldstrohm, A. C. (2012). Drosophila Pumilio protein contains multiple autonomous repression domains that regulate mRNAs independently of Nanos and brain tumor. *Molecular and Cellular Biology*, *32*, 527–540.

Wharton, R. P., Sonoda, J., Lee, T., Patterson, M., & Murata, Y. (1998). The Pumilio RNA-binding domain is also a translational regulator. *Molecular Cell*, *1*, 863–872.

White, J. A., & Heasman, J. (2008). Maternal control of pattern formation in Xenopus laevis. *Journal of Experimental Zoology. Part B, Molecular and Developmental Evolution*, *310*, 73–84.

Whitfield, T., Heasman, J., & Wylie, C. (1993). XLPOU-60, a Xenopus POU-domain mRNA, is oocyte-specific from very early stages of oogenesis, and localised to presumptive mesoderm and ectoderm in the blastula. *Developmental Biology*, *155*, 361–370.

Wilk, K., Bilinski, S., Dougherty, M. T., & Kloc, M. (2005). Delivery of germinal granules and localized RNAs via the messenger transport organizer pathway to the vegetal cortex of Xenopus oocytes occurs through directional expansion of the mitochondrial cloud. *The International Journal of Developmental Biology*, *49*, 17–21.

Wylie, C. C., Brown, D., Godsave, S. F., Quarmby, J., & Heasman, J. (1985). The cytoskeleton of Xenopus oocytes and its role in development. *Journal of Embryology and Experimental Morphology*, *89*(Suppl.), 1–15.

Wylie, C. C., Heasman, J., Parke, J. M., Anderton, B., & Tang, P. (1986). Cytoskeletal changes during oogenesis and early development of Xenopus laevis. *Journal of Cell Science. Supplement*, *5*, 329–341.

Wylie, C. C., Snape, A., Heasman, J., & Smith, J. C. (1987). Vegetal pole cells and commitment to form endoderm in Xenopus laevis. *Developmental Biology*, *119*, 496–502.

Yabe, T., Ge, X., Lindeman, R., Nair, S., Runke, G., Mullins, M. C., et al. (2009). The maternal-effect gene cellular island encodes aurora B kinase and is essential for furrow formation in the early zebrafish embryo. *PLoS Genetics*, *5*, e1000518.

Yamaguchi, T., Kataoka, K., Watanabe, K., & Orii, H. (2014). Restriction of the Xenopus DEADSouth mRNA to the primordial germ cells is ensured by multiple mechanisms. *Mechanisms of Development*, *131*, 15–23.

Yamaguchi, T., Taguchi, A., Watanabe, K., & Orii, H. (2013). DEADSouth protein localizes to germ plasm and is required for the development of primordial germ cells in Xenopus laevis. *Biology Open*, *2*, 191–199.

Yan, X., Cao, L., Li, Q., Wu, Y., Zhang, H., Saiyin, H., et al. (2005). Aurora C is directly associated with Survivin and required for cytokinesis. *Genes to Cells*, *10*, 617–626.

Yang, J., Tan, C., Darken, R. S., Wilson, P. A., & Klein, P. S. (2002). Beta-catenin/Tcf-regulated transcription prior to the midblastula transition. *Development*, *129*, 5743–5752.

Zamore, P. D., Williamson, J. R., & Lehmann, R. (1997). The Pumilio protein binds RNA through a conserved domain that defines a new class of RNA-binding proteins. *RNA*, *3*, 1421–1433.

Zearfoss, N. R., Chan, A. P., Wu, C. F., Kloc, M., & Etkin, L. D. (2004). Hermes is a localized factor regulating cleavage of vegetal blastomeres in Xenopus laevis. *Developmental Biology*, *267*, 60–71.

Zhang, C., Basta, T., Fawcett, S. R., & Klymkowsky, M. W. (2005). SOX7 is an immediate-early target of VegT and regulates Nodal-related gene expression in Xenopus. *Developmental Biology*, *278*, 526–541.

Zhang, C., Basta, T., & Klymkowsky, M. W. (2005). SOX7 and SOX18 are essential for cardiogenesis in Xenopus. *Developmental Dynamics*, *234*, 878–891.

Zhang, B., Gallegos, M., Puoti, A., Durkin, E., Fields, S., Kimble, J., et al. (1997). A conserved RNA-binding protein that regulates sexual fates in the C. elegans hermaphrodite germ line. *Nature*, *390*, 477–484.

Zhang, J., Houston, D. W., King, M. L., Payne, C., Wylie, C., & Heasman, J. (1998). The role of maternal VegT in establishing the primary germ layers in Xenopus embryos. *Cell*, *94*, 515–524.

Zhang, J., & King, M. L. (1996). Xenopus VegT RNA is localized to the vegetal cortex during oogenesis and encodes a novel T-box transcription factor involved in mesodermal patterning. *Development*, *122*, 4119–4129.

Zhang, Y., Yan, L., Zhou, Z., Yang, P., Tian, E., Zhang, K., et al. (2009). SEPA-1 mediates the specific recognition and degradation of P granule components by autophagy in C. elegans. *Cell*, *136*, 308–321.

Zhao, Y., Tian, E., & Zhang, H. (2009). Selective autophagic degradation of maternally-loaded germline P granule components in somatic cells during C. elegans embryogenesis. *Autophagy*, *5*, 717–719.

Zust, B., & Dixon, K. E. (1975). The effect of u.v. irradiation of the vegetal pole of Xenopus laevis eggs on the presumptive primordial germ cells. *Journal of Embryology and Experimental Morphology*, *34*, 209–220.

CHAPTER NINE

Sculpting the Transcriptome During the Oocyte-to-Embryo Transition in Mouse

Petr Svoboda*,[1], Vedran Franke[†], Richard M. Schultz[‡,1]
*Institute of Molecular Genetics, Academy of Sciences of the Czech Republic, Prague, Czech Republic
[†]Bioinformatics Group, Division of Biology, Faculty of Science, Zagreb University, Zagreb, Croatia
[‡]Department of Biology, University of Pennsylvania, Philadelphia, Pennsylvania, USA
[1]Corresponding authors: e-mail address: svobodap@img.cas.cz; rschultz@sas.upenn.edu

Contents

Abstract

In mouse, the oocyte-to-embryo transition entails converting a highly differentiated oocyte to totipotent blastomeres. This transition is driven by degradation of maternal mRNAs, which results in loss of oocyte identity, and reprogramming of gene expression during the course of zygotic gene activation, which occurs primarily during the two-cell stage and confers blastomere totipotency. Full-grown oocytes are transcriptionally quiescent and mRNAs are remarkably stable in oocytes due to the RNA-binding protein MSY2, which stabilizes mRNAs, and low activity of the 5′ and 3′ RNA degradation machinery. Oocyte maturation initiates a transition from mRNA stability to instability due to phosphorylation of MSY2, which makes mRNAs more susceptible to the RNA degradation machinery, and recruitment of dormant maternal mRNAs that encode for critical components of the 5′ and 3′ RNA degradation machinery. Small RNAs (miRNA, siRNA, and piRNA) play little, if any, role in mRNA degradation that occurs during maturation. Many mRNAs are totally degraded but a substantial fraction is only

Current Topics in Developmental Biology, Volume 113
ISSN 0070-2153
http://dx.doi.org/10.1016/bs.ctdb.2015.06.004

partially degraded, their degradation completed by the end of the two-cell stage. Genome activation initiates during the one-cell stage, is promiscuous, low level, and genome wide (and includes both inter- and intragenic regions) and produces transcripts that are inefficiently spliced and polyadenylated. The major wave of genome activation in two-cell embryos involves expression of thousands of new genes. This unique pattern of gene expression is the product of maternal mRNAs recruited during maturation that encode for transcription factors and chromatin remodelers, as well as dramatic changes in chromatin structure due to incorporation of histone variants and modified histones.

1. INTRODUCTION

Oocytes are highly differentiated cells and their identity is defined by expression of oocyte-specific transcription factors (e.g., *Figla* (Soyal, Amleh, & Dean, 2000) and *Nobox* (Suzumori, Yan, Matzuk, & Rajkovic, 2002)), DNA-binding proteins (e.g., H1FOO (Tanaka, Hennebold, Macfarlane, & Adashi, 2001)), RNA-binding proteins (e.g., *Msy2* (Yu, Hecht, & Schultz, 2001)), genes that provide oocyte-specific functions (e.g., *Zp1–3* for sperm binding (Avella, Xiong, & Dean, 2013)), and their ability to undergo meiosis. The oocyte-to-embryo transition (OET) entails transforming a highly differentiated oocyte into totipotent blastomeres and represents one of the earliest obstacles that must be successfully hurdled for continued development. This loss of cell identity contrasts with later developmental events that are intimately linked with cell differentiation (e.g., formation of the primary germ layers, hematopoiesis, neurogenesis, etc.) in which expression of a set of transcription factors is inextricably coupled to establishing and maintaining cell identity.

Degradation of maternal mRNAs, which initiates during oocyte maturation and zygotic genome activation (ZGA), lies at the heart of OET (Fig. 1A). The current understanding of molecular basis underlying transcriptome remodeling during OET and its consequences is the primary focus of this review. We will use mouse as the main model system but will include other species when appropriate. Not discussed are changes in DNA methylation and incorporation of histone variants and modified histones that occur during oogenesis and preimplantation development as they relate to regulating gene expression because these topics have been reviewed recently (Beaujean, 2014; Dean, 2014).

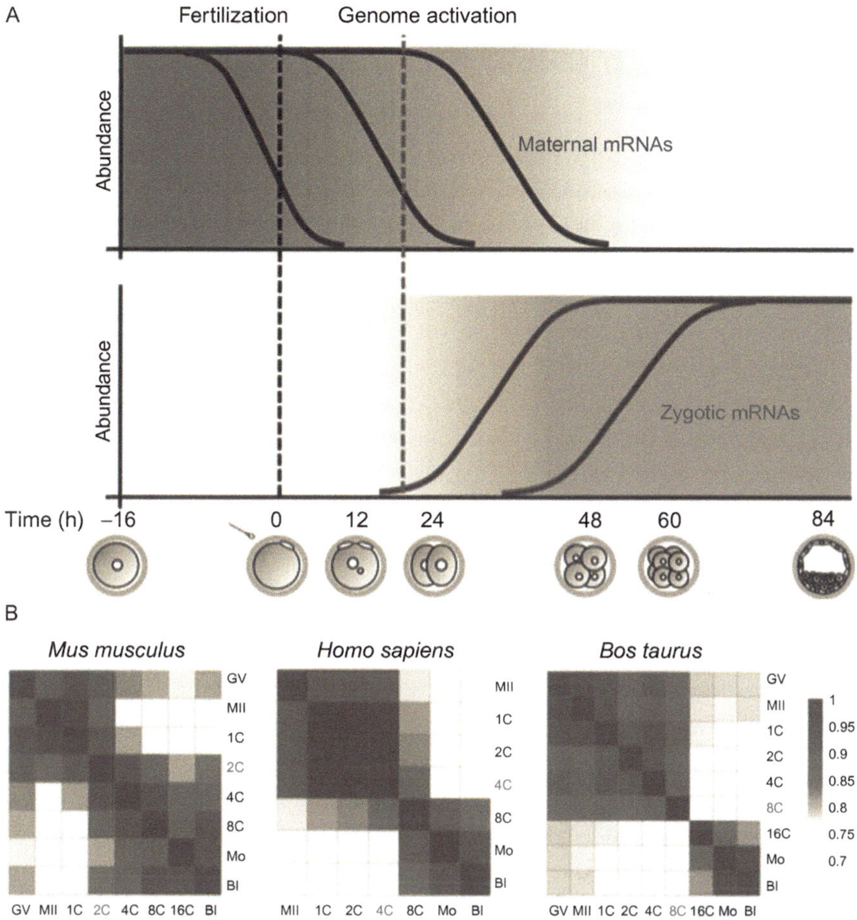

Figure 1 Transcriptome remodeling by maternal mRNA degradation and zygotic genome activation. (A) Schematic depiction of maternal mRNA degradation (upper graph) and ZGA (lower graph) during the OET in mice. The horizontal red (dark gray in the print version) gradient symbolizes disappearance of maternal mRNAs (mostly eliminated by the eight-cell stage). The blue (gray in the print version) gradient symbolizes synthesis of zygotic mRNAs, which first appear during minor ZGA at the one-cell stage but the main ZGA wave occurs during the two-cell stage. Individual curves represent the level of a specific mRNAs (or sets of mRNAs); for example, maternal mRNA degradation can be triggered by resumption of meiosis, fertilization, or ZGA; ZGA can be controlled by maternally provided transcription factors as well as by factors that need to be synthesized in the zygote early in ZGA. (B) Maternal-to-zygotic transcriptome remodeling in different mammals show a relatively fast switch from maternal-to-zygotic expression, which typically occurs within one developmental stage. Heatmaps depict correlation of pairwise comparisons of published microarray datasets from indicated species (Xie et al., 2010). Color intensity indicates a correlation coefficient according to the scale depicted on the right.

2. OVERVIEW OF SIGNIFICANT EVENTS DURING OET

Mice are born with their full complement of oocytes arrested in diplotene of the first meiotic prophase. These oocytes, which are ~10–15 μm in diameter, are contained in primordial follicles and surrounded by a single layer of flattened nonproliferating somatic cells (reviewed in Edson, Nagaraja, & Matzuk, 2009). A fraction of primordial follicles is periodically recruited into the growth phase in which oocyte diameter increases to a final diameter of ~80 μm and the surrounding somatic follicle cells are transformed into cuboidal granulosa cells that proliferate. Growing follicles at different stages of growth are classified as primary, secondary, and small and large antral based on morphological criteria, and oocytes within each type of follicle have distinct transcriptomes (Pan, O'Brien, Wigglesworth, Eppig, & Schultz, 2005).

The transcriptome changes coincide with a number of developmental changes that occur during oocyte growth. As oocytes grow they sequentially acquire both meiotic competence (i.e., the ability to resume meiosis and progress to and arrest at metaphase II) (Sorensen & Wassarman, 1976; Wickramasinghe & Albertini, 1992; Wickramasinghe, Ebert, & Albertini, 1991) and developmental competence (i.e., the ability to be fertilized and develop to term) (Eppig & Schroeder, 1989). Acquisition of meiotic competence occurs during the secondary to small antral follicle transition, whereas acquisition of developmental competence occurs during the small antral to large antral transition, with the ability of matured oocytes to develop to the blastocyst stage acquired during the last days of oocyte growth (Eppig & Schroeder, 1989). Also associated with acquisition of meiotic competence is acquisition of M-phase characteristics such as phosphorylation of centrosomal proteins (Wickramasinghe & Albertini, 1992; Wickramasinghe et al., 1991) and a change in DNA configuration from the nonsurrounded nucleolus (NSN) to surrounded nucleolus (SN) configuration, in which highly condensed DNA surrounds the nucleolus. A progressive decrease in transcription initiates around the time of antrum formation (Moore & Lintern-Moore, 1978), such that a full-grown oocyte is essentially transcriptionally quiescent. Typically, ~70–80% of full-grown oocytes display the SN configuration, the others remaining in the NSN configuration. Although it is not apparent why transcription ceases prior to completion of growth—it is observed in oocytes of all species examined to date—the developmental competence of full-grown NSN oocytes is markedly compromised with most arresting at the two-cell stage (Zuccotti, Piccinelli, Giorgi Rossi, Garagna, & Redi, 1995). This difference may be linked to an altered transcriptome (Ma, Li, et al., 2013) or changes in translation (Chalupnikova, Solc, Sulimenko, Sedlacek, & Svoboda, 2014).

Maternal mRNAs in mouse oocytes are unusually stable during the growth phase, which takes about 2.5 weeks, with an average half-life of ~10 days (as determined using an *in vitro* culture system that likely underestimates mRNA stability during oocyte growth *in vivo*) (Brower, Gizang, Boreen, & Schultz, 1981; De Leon, Johnson, & Bachvarova, 1983; Jahn, Baran, & Bachvarova, 1976) when compared to hours or minutes in somatic cells. Oocyte maturation is accompanied by a transition from mRNA stability to instability in which mRNAs are selectively degraded during the course of maturation, whereas an approximately equal number of mRNAs is recruited for translation during maturation (Chen et al., 2011; Su et al., 2007).

Fertilization triggers a set of events that comprise egg activation (e.g., cell cycle resumption, cortical granule exocytosis, and maternal mRNA recruitment) and that are temporally driven by oscillatory changes in intracellular calcium (Ducibella et al., 2002). Interestingly, experimentally manipulating the number of the calcium oscillations affects the reprogramming process (Ozil, Banrezes, Toth, Pan, & Schultz, 2006). The gamma isoform of CaMKII serves as the effector for cell cycle resumption and cortical granule exocytosis, whereas mRNA recruitment only requires an elevated intracellular calcium concentration (Backs et al., 2010).

ZGA entails extensive programming of gene expression as determined by microarray and RNA-seq analyses (Hamatani, Carter, Sharov, & Ko, 2004; Park et al., 2013; Wang et al., 2004; Zeng, Baldwin, & Schultz, 2004; Zeng & Schultz, 2005). Faithful reprogramming of gene expression is essential for continued development because perturbing reprogramming leads to early cleavage arrest (e.g., *Brg1* mutation Bultman et al., 2006). An ongoing challenge is to explain how a maternally inherited transcription machinery directs gene expression during the course of genome activation so that genes transcribed during ZGA distinctly differ from genes expressed in oocytes. Identifying the factors involved in the highly efficient and natural reprogramming process that is a hallmark of genome activation could improve the efficiency of generating iPS cells by first converting somatic cells to a more reprogrammable state.

The timing of OET observed in different taxons presumably reflects the environment in which ovulation, fertilization, and embryonic development occurs. Internal fertilization and development may remove constraints imposed on externally developing embryos. Oocytes supporting external development (e.g., *Xenopus* and zebrafish) are relatively large and contain large stores of yolk. Their large size permits accumulation of enough macromolecules and organelles (e.g., ribosomes) and sufficient endogenous energy sources (e.g., yolk) to support early development in a nutrient poor environment such as freshwater. In contrast, mammalian oocytes are small (typically 80–120 μm) and lack yolk. Because preimplantation embryos develop internally, they have access to nutrients that are present in oviductal and uterine fluids. Such access presumably has removed a selective pressure to accumulate yolk proteins as a nutrient reserve.

Another difference between embryos that develop internally and externally is the timing of genome activation. For example, mouse embryos initiate zygotic transcription during the one-cell stage with the major wave of genome activation occurring during the two-cell stage, whereas zebrafish and *Xenopus* embryos undergo multiple rounds of cleavage before genome activation, termed the mid-blastula transition (MBT). On a time basis, however, genome activation occurs much later in mammals (the major wave of genome activation occurring ~20 h postfertilization in mice), by which time a zebrafish embryo has a head, somites, and has gone through organogenesis, while a *Xenopus* embryo has completed gastrulation and formed a neural tube.

3. THE OET: THE MOTHER OF ALL TRANSCRIPTOME REMODELING

Transcriptome remodeling during the OET is the product of two processes, namely, maternal mRNA degradation and zygotic gene expression. Transcriptome profiling reveals that the transcriptome is remodeled during the OET through an erase-and-rewrite strategy where the maternal component of the transcriptome rapidly disappears during the course of ZGA (Fig. 1B). The "maternal" expression pattern persists until the stage when the major wave of ZGA occurs but it is essentially overwritten by the next cleavage stage as the embryonic transcriptome emerges. This developmental change in transcriptomes suggests that coexistence of maternal and zygotic "programs" is mainly restricted to the cleavage stage when the major wave of ZGA occurs. Loss of the maternal transcriptome, however, already initiates during resumption of meiosis.

3.1 Transcriptional Quiescence—When the End Is the Beginning

The striking correlation between the change in DNA configuration during the NSN-to-SN transition with the onset of transcription quiescence initially suggested a biochemical link between the two events, perhaps mediated by acquisition of M-phase traits in SN oocytes. Such a causal linkage, however, does not exist. For example, oocytes deficient in nucleoplasmin 2 do not undergo the NSN-to-SN transition but still become transcriptionally quiescent (De La Fuente et al., 2004). Reciprocally, inducing histone hyperacetylation with trichostatin A, which inhibits histone deacetylases, results in chromatin decondensation, but no increase in transcription is observed. Likewise, oocytes lacking MLL2, which methylates H3K4me1, undergo the NSN-to-SN transition but fail to become transcriptionally inactive (Andreu-Vieyra et al., 2010). Interestingly, inhibiting transcription in NSN oocytes with 5,6-dichlororibofuransyl-benzimidole robustly promotes their conversion to the SN configuration (Abe, Inoue, Suzuki, & Aoki, 2010; Liu, Nakamura, & Nakano, 2012), suggesting that transcriptional quiescence precedes or is concomitant with the transition. The histone H3K9me2-binding protein DPPA3 (also known as PGC7/Stella) is implicated in the transition because *Dpp3a*-null oocytes largely remain in the NSN configuration and expressing DPP3A in *Dpp3a*-null oocytes promotes their conversion to the SN configuration (Liu et al., 2012). Nevertheless, many *Dpp3a*-null SN oocytes remain transcriptionally active.

Changes in histone marks associated with transcriptional repression are unlikely to account for the absence of transcription in SN oocytes. Both activating (e.g., H3K4m2, H3K4me3, H3K9ac, H3K18ac, H4K5ac, and H4K12ac) and repressive (e.g., H3K9me2 and H3K9me3) histone marks increase during oocyte growth (Kageyama et al., 2007). It is formally possible that because only relative changes were measured, the molar amount of histones bearing repressive marks overwhelms and thereby trumps those possessing activating marks. Minimizing this possibility is that the 90% decrease in H3K9me3 (which is associated with transcription start sites; Barski et al., 2007) likely offsets the 50% increase in H4K12ac in *Mll2*-deficient oocytes that display no change in the repressive marks H3K9me2, H3K27me3, and H4K20me1 and remain transcriptionally active. Changes in modified histones that occur during oocyte growth, however, may have later effects on gene expression; for example, heterochromatin bearing acetylated histones in SN oocytes may mark and support a strand-specific burst of transcription in early embryos that is required for chromocenter formation (Probst et al., 2010).

The decrease in relative abundance of transcripts involved in transcription (e.g., *Polr2h* and *j*, *Sp1, Sp3, Sp4*, and *Tcf12)* could reduce RNA polymerase II (POLR2) activity, noting that the relative transcript abundance of *Sp4* and *Tcf12* decreases 33- and 12.5-fold, respectively, from the beginning to end of oocyte development (Pan et al., 2005). Although such changes could contribute to the decrease in global transcription, it seems unlikely that their decrease is the primary cause. Likewise, the increase in chromosome condensation that occurs during oocyte growth as oocytes acquire M-phase traits could be another contributing factor, but unlikely a primary factor. A key observation as to the molecular basis for transcriptional quiescence is that POLR2 is detected by immunocytochemistry in the nucleus of growing NSN oocytes but not SN oocytes when the oocytes are first permeabilized and then fixed (Abe et al., 2010). Moreover, this difference correlates with loss of phosphorylation of the carboxyl-terminal domain (CTD) of POLR2A, the largest POLR2 subunit in full-grown SN oocytes.

The CTD contains a repeating heptapeptide sequence ($Y_1S_2P_3T_4S_5P_6S_7$) that is phosphorylated by cyclin-dependent protein kinases (Phatnani & Greenleaf, 2006). POLR2 with a nonphosphorylated CTD initially binds to the promoter. CDK7/CCNH/MNAT1 (ménage-a-trois), which is regulated by CDK8/CCNC, phosphorylates S5 to convert the preinitiating POLR2 to an initiating form, which is further phosphorylated by CDK9/CCNT1/2 on S2 to form an elongating form. An S5 protein phosphatase converts POLR2 to a terminating form.

Analysis of microarray data (Pan et al., 2005) reveals that the transcripts encoding these CDKs and cyclins are expressed during oocyte development albeit with different patterns of expression but, strikingly, the relative abundance of *Ccnt1* mRNA (encoding Cyclin T2) declines between the secondary and small antral follicle stages and is not detected in full-grown oocytes. *Ccnt1*, however, is a product of zygotic gene activation in two-cell embryos. Furthermore, microarray data indicate that *Ccnt2* mRNA is likely recruited during oocyte maturation because it contains a cytoplasmic polyadenylation element (CPE) very close to the polyadenylation AATAAA signal sequence. Should the protein abundance of CCNT1 and CCNT2 be consistent with these developmental profiles in transcript abundance, the experimentally testable scenario that emerges is that loss of CCNT1 protein during oocyte growth would be the major factor for global transcriptional quiescence and synthesis of CCNT2 protein during maturation would "reactivate" POLR2 to support transcription in one-cell embryos (see below) with the burst of *Cnnt1* transcription in two-cell embryos then sustaining genome activation. Of note is that CDK9/CCNT1/2 appears critical for genome activation because flavopiridol-mediated inhibition of CDK9 results in arrest at the two-cell stage (Oqani, Kim, Diao, Park, & Jin, 2011).

A role for the poly(C)-binding protein PCBP1 has been proposed to maintain transcriptional quiescence in full-grown oocytes (Xia et al., 2012). Oocytes injected with siRNAs targeting *Pcbp1* mRNA were reported to undergo the NSN-to-SN transition and increase transcription in SN oocytes. These effects, however, were observed within 8 h following siRNA injection, a time that seems too short to promote not only mRNA degradation but also turnover of PCBP1 protein; loss of PCBP1 protein was not documented. Thus, a role for PCBP1 in maintaining transcriptional quiescence remains an open question.

It is now well recognized that oocytes and their companion follicle cells have an active dialog in which each influences the behavior of the other (Matzuk, Burns, Viveiros, & Eppig, 2002). Interestingly, development of the transcriptionally quiescent state requires communication with surrounding follicle cells. For example, when oocytes are cultured in the absence of granulosa cells (denuded oocytes), transcriptional activity remains high but declines when granulosa cell-enclosed oocytes are cultured (De La Fuente & Eppig, 2001). Similarly, full-grown NSN oocytes are approximately three times more transcriptionally active than their counterparts present in cumulus cell complexes (Liu & Aoki, 2002) and *Msy2*-null oocytes do not become transcriptionally quiescent—*Ccnt1* mRNA remains elevated in these oocytes—and are essentially devoid of attached cumulus cell following isolation (Medvedev, Pan, & Schultz, 2011). Although the signaling pathway is not known, a juxtacrine model seems more likely than one mediated by gap junction communication between the two cell types.

An interesting aspect of transcriptional quiescence is its potential direct impact on maternal mRNA decay. mRNA levels are defined by the balance between synthesis and turnover. As described above, mRNAs are in general highly stable in oocytes (Brower et al., 1981; De Leon et al., 1983). A small population of maternal mRNAs in full-grown oocytes, however, exhibit high turnover (Puschendorf et al., 2006), and their removal from the oocyte would be accelerated as a consequence of terminating transcription and thereby function as a timing mechanism (i.e., molecular clock). Decay of these mRNAs might be linked to developmental competence because experimentally extending the period of transcriptional quiescence in full-grown GV mouse oocytes does not reduce their ability to mature and be fertilized, but does reduce their competence to reach the blastocyst stage (De La Fuente & Eppig, 2001). Next-generation sequencing (NGS) of NSN and SN oocyte transcriptomes reveals differences in several metabolic pathways with more than 200 transcripts reduced greater than twofold in SN oocytes (Ma, Li, et al., 2013). Although there is minimal overlap between the unstable transcripts identified by Ma, Li, et al. (2013) and Puschendorf et al. (2006), the idea that mRNA turnover in transcriptionally quiescent preovulatory oocytes might contribute to the acquisition of developmental competence merits further investigation.

3.2 Dormant mRNAs—A Way to Synthesize a New Protein without Transcription

Oocytes utilize a class of mRNAs called dormant maternal mRNAs that are either not translated or poorly translated in the oocyte but are then recruited for translation following initiation of maturation. The molecular basis for recruitment of such maternal mRNAs is well defined and reviewed elsewhere (Groppo & Richter, 2009). What remains ill defined is the function of such mRNAs. Historically, discovering their identity, e.g., *Plat* (Huarte, Belin, Vassalli, Strickland, & Vassalli, 1987), was serendipitous. Results of microarray experiments in which oligo(dT) was used to generate cDNA (Zeng et al., 2004) and analysis of proteome changes and mRNAs recruited to polysomes during maturation (Chen et al., 2011; Wang et al., 2010), however, have provided a means to identify on a global scale these recruited maternal mRNAs and *cis*-acting 3′UTR regulatory sequences that govern the recruiting.

Recruitment of maternal mRNAs is associated with elongation of the poly(A) tail, which results in more efficient oligo(dT) priming. A consequence is that the measured relative abundance of the transcript in one-cell embryos is greater than that when compared to GV-intact oocytes (Zeng et al., 2004). In contrast, random priming does not reveal this increase and often reveals a decrease in mRNA abundance due to the coupling of mRNA recruitment with degradation (e.g., Ma, Flemr, Strnad, Svoboda, & Schultz, 2013). What emerges from analyzing transcripts that display an increase between the GV oocyte and one-cell stages is that they encode proteins central to processes that should be minimal in the oocyte but then are required by the one-cell stage; a similar conclusion is reached when examining mRNAs recruited to polysomes during maturation (Chen et al., 2011). For example, oocytes must not undergo a round of DNA replication between MI and MII, but need to replicate their DNA following fertilization. ORC6L, critical for assembly of a functional origin of replication complex, is encoded by a recruited maternal mRNA, thereby ensuring that DNA replication does not occur between MI and MII but that the one-cell embryo is capable of DNA synthesis (Murai, Stein, Buffone, Yamashita, & Schultz, 2010). The early events of egg activation are mediated by inositol 1,4,5-trisphosphate (IP_3), which releases calcium from intracellular stores by activating an IP_3 receptor (Ducibella, Schultz, & Ozil, 2006). Oocyte maturation is accompanied by an increase in synthesis of the IP_3 receptor 1 (ITPR1) that results in an increased sensitivity to IP_3-mediated release of intracellular calcium. Inhibiting the increase in ITPR using an RNAi approach results in impaired cortical granule exocytosis, which requires an increase in intracellular calcium, as well as a perturbed calcium oscillatory pattern following insemination (Xu, Williams, Kopf, & Schultz, 2003). The primary effector of the increase in intracellular calcium that mediates egg activation is CAMK2G, which is also encoded by a recruited maternal mRNA (Backs et al., 2010). The increases in both ITPR1 and CAMK2G ensure that eggs, but not oocytes, are successfully fertilized. Although the identities of mRNAs recruited following fertilization are poorly defined, inhibiting their recruitment following fertilization by preventing poly(A) elongation with 3′-deoxyadenosine markedly inhibits genome activation (Aoki, Hara, & Schultz, 2003), which normally initiates in the one-cell embryo (Aoki, Worrad, & Schultz, 1997; Bouniol, Nguyen, & Debey, 1995).

Other examples include the pluripotency factor LIN28, which regulates miRNA biogenesis during ZGA (Flemr, Moravec, Libova, Sedlacek, & Svoboda, 2014), as well as several mRNAs encoding E3 ubiquitin-protein ligases, which have the hallmark of a recruited maternal mRNA. E3 ligases recognize the targeted protein, catalyze the transfer of ubiquitin to that protein, and thereby provide specificity for selective protein degradation. Recruitment of mRNAs encoding E3 ligases likely results in the selective degradation of maternal proteins, thereby reinforcing the loss of oocyte identity due to mRNA degradation that initiates during maturation.

Recruitment of maternal mRNAs that encode transcriptional, posttranscriptional, and posttranslational regulatory factors provides an elegant solution for how gene expression can be extensively reprogrammed during genome activation using a maternally provided transcription apparatus, and thereby sever a Gordian knot. For example, *Sin3a* (chromatin remodeler), *Ezh2* (histone H3K27 methylase), and *Obox1/2* mRNAs are recruited during maturation (Jimenez, R., Stein, P., & Schultz, R.M., unpublished results). Inhibiting the increase in SIN3A using an siRNA/morpholino approach not only enhances the expression of a subset of genes that are normally zygotically expressed but also results in expression of a set of genes that are normally not expressed in the two-cell embryo; this unfaithful reprogramming likely accounts for the observed early cleavage arrest (Jimenez & Schultz, unpublished results).

A prominent feature of these recruited maternal transcripts is their relative high abundance. For example, for microarrays normalized to an average raw score of 150, the relative abundance of *Sin3a* and *Obox1/2* in GV oocytes is ~1000 and 7000, respectively. This high abundance in GV oocytes is not observed at later stages of preimplantation development and likely reflects a strategy for a large maturing oocyte to synthesize and accumulate a sufficient amount of the protein in a short period of time before its function is required.

3.3 Posttranscriptional Control by Small RNAs During OET

The OET also employs numerous mechanisms mediating posttranscriptional repression. One of the major posttranscriptional repressive mechanisms in most eukaryotes is RNA silencing, which employs small RNAs (20–30 nucleotides long) as guides that mediate sequence-specific silencing of gene expression. Three well-defined small RNA pathways operate in mammals: RNA interference (RNAi), microRNA (miRNA), and PIWI-associated RNA (piRNA) pathways. Oocytes are unique among mammalian cells because they contain significant amounts of small RNAs used by each of the three pathways (Garcia-Lopez, Hourcade Jde, Alonso, Cardenas, & del Mazo, 2014; Tam et al., 2008; Watanabe et al., 2008). Molecular mechanisms of RNAi, miRNA, and piRNA pathways have been reviewed in detail elsewhere (Chekulaeva & Filipowicz, 2009; Jinek & Doudna, 2009; Kim, Han, & Siomi, 2009; Siomi, Sato, Pezic, & Aravin, 2011); here, we summarize mainly information relevant for understanding the role of small RNAs during the OET.

3.3.1 *microRNA-Mediated Repression*

miRNAs are genome-encoded posttranscriptional regulators of gene expression. Nuclear miRNA precursor transcripts are processed, transported to the cytoplasm, and cleaved by Dicer, to produce ~22 nt long miRNAs, which are loaded on Argonaute (AGO) proteins that mediate the silencing effects (reviewed in Kim et al., 2009). Animal miRNAs typically base-pair imperfectly with the 3′UTR of target mRNAs, which results in translational repression and/or mRNA degradation. The silencing mode depends on the degree of base pair complementarity and the AGO isoform. AGO2 and perfect complementarity result in a direct endonucleolytic cleavage, which requires only AGO2 loaded with a small RNA. Other combinations result in translational repression followed by mRNA destabilization and require a full-formed effector complex including the translational repressor GW182 (reviewed in Braun, Huntzinger, & Izaurralde, 2013).

The miRNA pathway is the dominant mammalian small RNA pathway—miRNAs are found in all examined cell types and there is an ever-growing list of their essential roles in different cells and biological processes. NGS of small RNAs during the mouse OET reveals that maternal miRNAs are a relatively minor small RNA population in oocytes but become the dominant small RNA class toward the end of preimplantation development (Garcia-Lopez et al., 2014; Ohnishi et al., 2010). This pattern of expression is also consistent with zebrafish and *Xenopus* miRNA analyses (Chen et al., 2005; Tang & Maxwell, 2008; Yao et al., 2014). Thus, it appears that vertebrate oocytes generally do not accumulate maternal miRNAs during the growth phase but that miRNAs accumulate and remain abundant during early development.

Surprisingly, mouse oocytes lacking an essential miRNA biogenesis factor DGCR8 are able to give rise to viable offspring (Suh et al., 2010). Although such oocytes have slightly reduced developmental competence, they have an impressive tolerance to the absence of miRNAs. Remarkably, this tolerance extends into ZGA because *Dgcr8*$^{-/-}$ oocytes fertilized with *Dgcr8*$^{-/-}$ sperm develop as far as the blastocyst stage (Suh et al., 2010). Importantly, although miRNAs can be detected in the sperm and may enter the zygote, paternal miRNAs are extremely unlikely to contribute to the OET because their concentrations would be too low to have significant effects (Amanai, Brahmajosyula, & Perry, 2006).

miRNA-mediated repression is inefficient in mouse oocytes. First, the transcriptome of *Dgcr8*$^{-/-}$ oocytes does not differ from that of wild-type oocytes (Suh et al., 2010). Second, in contrast to *Dicer*$^{-/-}$ embryonic stem cells (ESCs), the transcriptome of *Dicer*$^{-/-}$ oocytes lacks a signature of relieved miRNA-mediated repression (Ma et al., 2010; Sinkkonen et al., 2008). Third, reporter assays suggest that miRNAs are effectively loaded on AGO2 and guide RNAi-like endonucleolytic cleavage of perfectly complementary targets but do not efficiently suppress partially complementary targets (Ma et al., 2010). Consistent with the notion of an incomplete effector complex, GW182-AGO2 colocalization is lost at the onset of the oocyte growth and is not restored until the eight-cell/morula stages (Flemr, Ma, Schultz, & Svoboda, 2010). The molecular basis for the minimal miRNA activity in oocytes remains unclear. In any case, uncoupling miRNAs from the repression of translation (and mRNA degradation) might be one of the earliest known reprogramming events in the mouse OET.

Why would mouse oocytes suppress miRNA function? One possibility is that the poor ability of miRNAs to promote mRNA degradation would contribute to mRNA stability and thereby facilitate mRNA accumulation during oocyte growth. Repression of miRNA function could also facilitate ZGA. The oocyte miRNA profile is somatic cell-like due to Let-7, which comprises a third of maternal miRNA (Tam et al., 2008). Let-7 marks differentiated somatic cells from *Caenorhabditis elegans* to humans and promotes differentiation and loss of totipotency in ESCs (reviewed in Svoboda, 2010). Abundant Let-7 in oocytes is unanticipated considering the reprogramming ability of oocyte cytoplasm and the need to express mRNAs encoding pluripotent transcription factors for ZGA. Thus, reduced miRNA activity during the OET would avoid undesirable effects of Let-7 and facilitate replacement of maternal miRNAs with zygotic miRNAs, particularly with the miR-290 cluster, which is highly abundant in pluripotent stem cells and counteracts Let-7 function (Melton, Judson, & Blelloch, 2010). The notion that Let-7 is detrimental to the OET is consistent with a Let-7-specific layer of miRNA repression by maternally provided LIN28, which prevents Let-7 biogenesis during ZGA (Flemr et al., 2014).

NGS of full-grown and meiotically mature porcine oocytes suggests that demise of maternal miRNAs starts during meiotic maturation (Yang et al., 2012). Global elimination of maternal miRNAs, however, also appears to be induced by fertilization (Garcia-Lopez et al., 2014; Tang et al., 2007; Tang & Maxwell, 2008). The global pattern of degradation likely reflects the fact that mature miRNAs do not carry any specific regulatory sequences, whereas their elimination must occur in a manner that permits accumulation of zygotic miRNAs. Although the molecular mechanism of maternal miRNA degradation in vertebrates has not been identified, it will likely involve RNA-specific ribonucleotidyl transferases (Martin & Keller, 2007). First, uridylation or adenylation of RNAs by these enzymes is well established in regulating miRNA stability (reviewed in Ha & Kim, 2014). Second, a noncanonical poly(A) polymerase, Wispy, was identified as the factor responsible for maternal miRNA adenylation and degradation in *Drosophila* (Lee et al., 2014).

3.3.2 RNA Interference by Short Interfering RNAs

RNAi is a sequence-specific mRNA degradation mechanism mediated by small RNAs generated by the RNase Dicer from long double-stranded RNA (dsRNA). Mammalian RNAi is an indolent pathway because, although it employs the same proteins as the miRNA pathway, endo-siRNAs (short interfering RNAs produced by Dicer from dsRNA) are usually present in negligible amounts in mammalian cells (e.g., Nejepinska et al., 2012). The full-length mammalian Dicer efficiently processes miRNA precursors (Chakravarthy, Sternberg, Kellenberger, & Doudna, 2010) but not long perfect duplexes because of the autoinhibitory role of the N-terminal helicase domains (Ma, MacRae, Kirsch, & Doudna, 2008). In addition, RNAi activity would normally be masked by the interferon response, which is a sequence-independent innate immune response to dsRNA (reviewed in de Veer, Sledz, & Williams, 2005; Gantier & Williams, 2007). Mouse oocytes, however, lack an interferon response to dsRNA (Stein, Zeng, Pan, & Schultz, 2005) and microinjected long dsRNA induces a robust RNAi response in oocytes (Svoboda, Stein, Hayashi, & Schultz, 2000; Wianny & Zernicka-Goetz, 2000).

Mouse oocytes contain abundant endo-siRNAs, including a unique class of endo-siRNAs derived from processed pseudogenes (Tam et al., 2008; Watanabe et al., 2008). In contrast to miRNAs and piRNAs, endo-siRNAs are essential for normal meiotic maturation because loss of *Dicer* and *Ago2* results in meiotic catastrophe, whereas loss of *Dgcr8* and canonical miRNAs has no obvious phenotype (Kaneda, Tang, O'Carroll, Lao, & Surani, 2009; Murchison et al., 2007; Suh et al., 2010; Tang et al., 2007). The essential role of endonucleolytic cleavage of cognate mRNAs was demonstrated using a catalytically inactive form of AGO2 (Stein et al., 2015). Interestingly, putative endo-siRNA targets are enriched in cell cycle regulators and genes involved in microtubule organization and dynamics (Tam et al., 2008), suggesting that the loss of gene-targeting endo-siRNAs might be responsible for the spindle defect phenotype.

Mouse oocytes express high levels of a unique Dicer isoform (denoted Dicer^{O}), which is adapted for endo-siRNA production. Dicer^{O} evolved in the *Muridae* rodent family as a consequence of a mouse transcript (MT) retrotransposon insertion into the *Dicer* gene, which resulted in oocyte-specific expression of an N-terminally truncated Dicer isoform (Flemr et al., 2013). Consistent with earlier studies (Ma et al., 2008), the Dicer^{O} isoform shows higher activity *in vitro* than the full-length version and its ectopic expression in cells increases endo-siRNA levels. Importantly, deletion of the Dicer^{O}-producing MT retrotransposon phenocopies the meiotic spindle defects observed in oocyte-specific *Ago2* and *Dicer* knockouts (Flemr et al., 2013).

Taken together, mouse oocytes represent a case where Dicer is adapted for endo-siRNA production and RNAi acquired an essential role to control endogenous gene expression. In contrast, bovine and human maternal transcriptomes (Graf et al., 2014; Xue, Cai, et al., 2013), and Western blot analysis of porcine oocytes (Svoboda et al., unpublished results) do not reveal any indication that other mammalian oocytes express a truncated Dicer isoform. This implies that RNAi pathway in other mammalian oocytes is weaker than that observed in mouse oocytes. Currently, the function and robustness of small RNA pathways in other mammalian species are not well understood. Although long dsRNA induces specific knockdown effects in bovine (Paradis, Vigneault, Robert, & Sirard, 2005), porcine (Anger et al., 2004), or ovine (Yan et al., 2010) oocytes, this only shows that mammalian oocytes do not have a strong interferon response and there is sufficient capacity to induce RNAi with a large excess of long dsRNA.

It is unlikely that RNAi controls meiotic progression in mammalian oocytes in general. The spindle defect phenotype in mouse oocytes is likely caused by the loss of pseudogene-derived endo-siRNAs. Thus, apart from rodent-restricted presence of Dicer^{O}, it is doubtful that pseudogene sequences giving rise to endo-siRNAs would be present in a common mammalian ancestor and remain conserved to retain specific endo-siRNA activities. Thus, we predict that the RNAi pathway in other mammalian oocytes is not masked by an interferon response but that, in contrast to mice, the RNAi activity in most mammals is relatively low and nonessential for meiotic progression and ZGA.

3.3.3 PIWI-Associated Small RNAs

piRNAs are longer than miRNAs and siRNAs and their complex biogenesis involves PIWI proteins (a different clade of the Argonaute protein family expressed during gametogenesis) but not Dicer (reviewed in Weick & Miska, 2014). piRNAs originate from distinct genomic regions and come in two flavors—retrotransposon-derived and those derived from other sequences. The first type helps to maintain genome integrity, whereas the role of the second type is being debated (Weick & Miska, 2014).

Three different mouse PIWI proteins participate in piRNA biogenesis and function in developing sperm; all three are essential for normal spermatogenesis but nonessential for oogenesis (Carmell et al., 2007;

Deng & Lin, 2002; Kuramochi-Miyagawa et al., 2004). A study of primordial follicles lacking piRNAs showed that maternal piRNAs also suppress retrotransposons. In contrast to developing sperm, however, this function is nonessential in oocytes (Lim et al., 2013). A possible explanation of nonessential function of piRNAs during mouse oocyte development is that loss of piRNAs is compensated by a highly active RNAi pathway. piRNAs are found in mouse, porcine, bovine, human, and macaque oocytes (Roovers et al., 2015; Tam et al., 2008; Watanabe et al., 2008; Yang et al., 2012). A comparative analysis of bovine and primate oocytes suggests that the piRNA pathway in mammalian oocytes is more active and functionally significant than the mouse data suggest (Roovers et al., 2015).

Postfertilization clearance of maternal piRNAs is observed in mouse zygotes (Garcia-Lopez et al., 2014; Ohnishi et al., 2010). Data from bovine early embryos suggest that piRNAs persist into early embryos, but there is no direct evidence for their functional significance (Roovers et al., 2015).

3.4 The Transition from mRNA Stability to Instability

Once transcription stops at the end of the growth phase, control of gene expression becomes posttranscriptional until ZGA and involves regulation of mRNA translation, exemplified by recruitment of maternal mRNAs discussed above, and regulation of mRNA stability. mRNA degradation (reviewed in Balagopal, Fluch, & Nissan, 2012; Houseley & Tollervey, 2009; Schoenberg & Maquat, 2012) usually involves deadenylation of the 3′-poly(A) tail and/or decapping. In mammalian somatic cells, deadenylation coupled with decapping is the main mRNA decay pathway (Yamashita et al., 2005). DCP1/DCP2-mediated decapping initiates degradation at the 5′ end of the mRNA and with further 5′-to-3′ degradation catalyzed by the XRN exonuclease. Degradation from the 3′ end is triggered by sequential PAN2/PAN3- and CCR4-NOT-mediated deadenylation that prepares the targeted mRNA for further degradation by the exosome complex. In somatic cells, deadenylation is often the first and rate-limiting step of mRNA degradation and, once degradation initiates, the mRNA is rapidly and totally degraded, typically within minutes. In specific cases, mRNAs can be cleaved by an endonucleolytic activity (e.g., by RNAi).

In terms of their role during the OET, maternal mRNAs fall into three categories: (1) mRNAs not needed beyond resumption of meiosis (e.g., oocyte-specific transcription factors), (2) continuously translated mRNAs needed beyond resumption of meiosis (e.g., housekeeping genes), and

(3) above discussed dormant mRNAs recruited for translation during the OET. Decay of maternal mRNAs does not occur according to these categories, but, rather, each developmental transition triggers a wave of mRNA degradation targeting a population of mRNAs, where representatives of all three categories might be found.

As mentioned above, maternal mRNA degradation starts prior to resumption of meiosis, which is usually considered the starting point for maternal mRNA clearance. The extent of mRNA degradation between the end of transcription and resumption of meiosis seems minor (Puschendorf et al., 2006) and its significance is unclear. Resumption of meiosis, however, induces a major change in the cytoplasmic environment from one that previously supported mRNA stability to one that fosters mRNA degradation. Some of the molecular mechanisms underlying the transition from mRNA stability to instability have recently been uncovered.

The germ cell-specific RNA-binding protein MSY2 plays a major role in the pronounced stability of mRNA in oocytes. MSY2, which displays little if any RNA sequence-binding specificity (Yu, Hecht, & Schultz, 2002), comprises ~2% of total oocyte protein and is essentially degraded by the two-cell stage (Yu et al., 2001), a time by which most maternal mRNAs are totally degraded. Female *Msy2*$^{-/-}$ mice are infertile (Yang et al., 2005) and oocytes lacking MSY2 protein contain ~25% less mRNA than their wild-type counterparts (Medvedev et al., 2011). Consistent with MSY2, stabilizing mRNAs is that an injected reporter mRNA is less stable in *Msy2*$^{-/-}$ oocytes, and coinjection of a cRNA-encoding wild-type MSY2, but not a cRNA encoding a mutant form unable to bind RNA, restores mRNA stability.

CDK1-mediated phosphorylation of MSY2 plays a critical role in triggering the transition from mRNA stability to instability that initiates following resumption of meiosis and nuclear envelope breakdown (Medvedev, Yang, Hecht, & Schultz, 2008). Inhibiting MSY2 phosphorylation by inhibiting CDK1 activity with roscovitine inhibits the maturation-associated decrease in mRNA degradation. Similarly, overexpressing nonphosphorylatable forms of MSY2 reduces the degree of mRNA degradation, whereas overexpressing a putative constitutively active form of MSY2 in oocytes that are prevented from resuming meiosis induces mRNA degradation. MSY2 phosphorylation apparently renders mRNAs more susceptible to the RNA degradation machinery because treating permeabilized GV-intact oocytes with RNase under conditions that lead to little mRNA degradation results in extensive mRNA degradation when the putative constitutively active MSY2 is expressed prior to permeabilization (Medvedev et al., 2008).

mRNA stability in oocytes also stems from an environment low in RNA degradation activity in which an increase in the activity of 5′ and 3′ RNA degradation machinery would promote the transition to mRNA instability. An environment of intrinsically low RNA degradation activity would support accumulation of mRNA during the protracted period of oocyte growth that occurs without cell division, and in particular during the late stages when transcription is absent or reduced.

Oocyte maturation triggers recruitment of mRNAs encoding critical components of both the 5′ and 3′ mRNA degradation machinery. Both *Dcp1a* and *Dcp2* are recruited during maturation—both have well-defined CPEs—and inhibiting their recruitment using siRNAs to target *Dcp1a* and *Dcp2* mRNAs retards degradation of mRNAs that are normally degraded during the course of maturation (Ma, Flemr, et al., 2013). mRNAs encoding PAN2, the catalytic subunit, and CNOT6L and CNOT7, the two catalytic subunits of the CCR4-NOT complex, are also recruited during maturation. Inhibiting recruitment of *Cnot7* mRNA using an siRNA approach inhibits deadenylation as assessed by poly(A) tail (PAT) assays, whereas there is little effect when recruitment of *Pan2* mRNA is inhibited (Ma & Schultz, unpublished results). This difference likely reflects the fact that the average length of the poly(A) tail in oocyte is ~50–100 nt and PAN2 prefers longer poly(A) tails, with the initial degradation mediated by PAN2/PAN3 generating a substrate for the CCR4–NOT complex. Although inhibiting either 5′ or 3′ mRNA degradation does not inhibit oocyte maturation, genome activation, (i.e., transcription) in two-cell embryos is inhibited ~50% for both cases (Ma, Flemr, et al., 2013). The role of other deadenylases such as PARN in degradation of maternal mRNA in mouse oocytes is not known.

A striking feature of mRNA degradation during oocyte maturation is that initiation of mRNA degradation does not lead to total mRNA degradation but, rather, the partially degraded transcripts are stable until ZGA. In fact, inhibiting genome activation stabilizes many transcripts that are normally degraded. In mouse, failure to degrade mRNA completely was deduced more than 25 years ago by assessing changes in the amount of actin mRNA by Northern blot analysis and determining that the deadenylated mRNA was stable and only degraded following genome activation (Paynton, Rempel, & Bachvarova, 1988). More recent studies using RT-qPCR and microarrays led to the conclusion that such is the fate of many mRNAs. For example, RT-qPCR using random primers to generate cDNA reveals a range of mRNA degradation from ~30% to almost complete degradation (Medvedev et al., 2008). A probe-level analysis of transcript abundance of microarray data is also consistent with the generation of stable deadenylated transcripts during maturation (Salisbury, Hutchison, Wigglesworth, Eppig, & Graber, 2009). This analysis also revealed differential degradation of transcript isoforms.

Results of RNA-seq experiments examining the time course of mRNA degradation during maturation are also illuminating (Ma, Gregory, Vandivier, & Schultz, unpublished results). Different patterns of degradation emerge with some transcripts being degraded to completion from the 5′ end (*Zp2*) or 3′ end (*Zp3*), whereas others are nibbled at both ends with the body of the transcript remaining (*Foxj3*). A role for MSY2 in conferring stability to deadenylated mRNAs is not known, but the correlation between loss of MSY2 by the two-cell stage and completion of degradation of maternal mRNAs by this time is striking, and studies using *Msy2*$^{-/-}$ oocytes should be useful to establish whether a causal relationship exists. Degradation of maternal mRNAs initiating during maturation and resulting in stable deadenylated mRNAs is likely a conserved process in vertebrates. *Xenopus* oocytes express FRGY2, an ortholog of MSY2, and stable deadenylated mRNAs are present in *Xenopus* following oocyte maturation and early cleavage, with degradation only occurring following activation of the embryonic genome (Tafuri & Wolffe, 1993).

One consequence of degradation of maternal mRNAs, with many mRNAs being partially degraded and nontranslatable, is to initiate the irreversible process of erasing oocyte identity. More intriguing is why many deadenylated mRNAs are not degraded until the time of genome activation. A tantalizing proposition is that resumption of degradation of these partially degraded mRNAs in two-cell embryos serves as a source of nucleotides to support transcription during genome activation. Such a strategy would circumvent the likely rapid catabolism of nucleotides if mRNAs were totally degraded during maturation as well as the poor ability of one- and two-cell embryos to transport ribonucleosides (Daentl & Epstein, 1971; Epstein, Daentl, Smith, & Kwok, 1971).

3.5 Maternal RNA Degradation—Yin of OET Transcriptome Remodeling

As noted above, RNA degradation is a major driving force for remodeling RNA content during the OET. A remarkable feature of mammalian oocyte maturation is the significant elimination of ribosomal RNA (rRNA) and ribosomes. In zebrafish and *Xenopus* models, the total RNA (bulk of which is rRNA) remains relatively constant during the OET (Brown & Littna, 1964; de Jong et al., 2010). In mammalian oocytes, however, total RNA is markedly reduced (by 50% or more) between resumption of meiosis and ZGA (Gilbert et al., 2009; Piko & Clegg, 1982), which means that maternal ribosomes are eliminated faster during the OET than their zygotic production. In addition, studies in the mouse model show that POL1R-mediated transcription and pre-rRNA processing machineries that produce pre-rRNA transcripts and mature rRNAs are largely eliminated by metaphase II (Fulka & Langerova, 2014; Zatsepina, Bouniol-Baly, Amirand, & Debey, 2000). Furthermore, microarray data suggest degradation of mRNAs encoding ribosomal protein-encoding genes during meiotic maturation and after fertilization (Su et al., 2007; Zeng et al., 2004). Although the zygotic rRNA expression is initiated at the two-cell stage, the main increase in rRNA synthesis in mouse zygotes takes place at the four-cell stage (Abe et al., 2015). Taken together, the protein synthetic machinery is not stockpiled for the mammalian OET but, rather, is extensively eliminated and rapidly replaced during the course of genome activation, which is in stark contrast with stockpiling observed in externally developing embryos of other vertebrates.

Why oocyte maturation is accompanied by destruction of a significant fraction of the ribosome pool is not apparent. The rDNA array is not composed of multiple copies of identical transcription units but, rather, by at least seven rDNA variant subtypes (Tseng et al., 2008). This finding raises the intriguing possibility that maternal ribosomes are replaced by zygotic ribosomes containing different rRNA variants with different translational capacities as proposed in the ribosome filter hypothesis (Mauro & Edelman, 2002). Such is unlikely to be the case, however, because the pattern of zygotically expressed variant rRNA is similar to that of oocytes (Ihara, Tseng, & Schultz, 2011).

As described above, maternal mRNA is degraded during the mammalian OET. In mice, the poly(A) RNA content of full-grown GV oocytes (~80–90 pg) is reduced severalfold by the two-cell stage (Fig. 2A). Such extensive elimination of maternal mRNA suggests that any dependence on maternal mRNAs does not extend far beyond ZGA. This situation contrasts with zebrafish and *Xenopus*, which activate their genomes within a few hours after fertilization and zygotic RNAs replenish the maternal RNA pool without a significant decrease in RNA content (Brown & Littna, 1964;

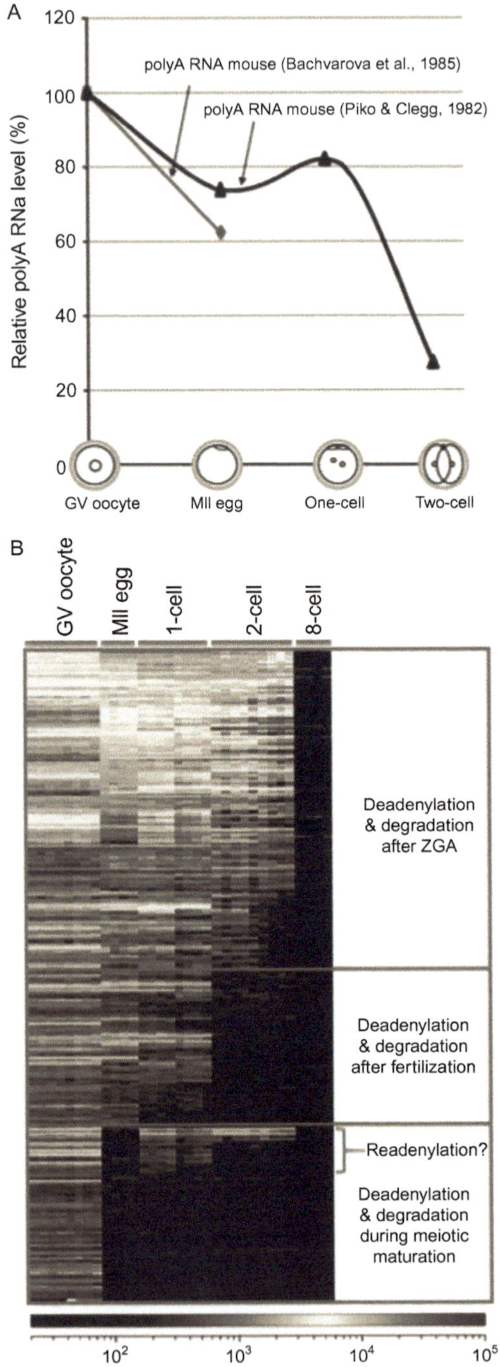

Figure 2 See legend on opposite page.

de Jong et al., 2010). In zebrafish and *Xenopus*, maternal mRNAs are still present during organogenesis and they control germline development (Giraldez et al., 2006; Pelegri, 2003; Wylie, 2000).

Transcriptome profiling of mammalian oocytes and early embryos reveals that maternal mRNA degradation occurs in waves associated with major OETs—resumption of meiosis, fertilization, and ZGA (Hamatani et al., 2004; Park et al., 2013; Su et al., 2007; Wang et al., 2004; Xie et al., 2010; Xue, Cai, et al., 2013; Yan et al., 2013; Zeng & Schultz, 2005; Zeng et al., 2004). The first major wave of maternal mRNA degradation is triggered by resumption of meiosis and, as described above, is driven in large part by recruitment of dormant maternal mRNAs encoding critical components of the mRNA degradation machinery. Because *Dcp2* and *Cnot7* translation occur after MI (Ma, Flemr, et al., 2013), the bulk of mRNA degradation during meiotic maturation likely occurs after polar body extrusion. A puzzling feature of this first wave of maternal mRNA degradation is why a large number of degraded mRNAs encode proteins with housekeeping and basic metabolic functions (Su et al., 2007).

It is not clear what controls selectivity during the first wave of mRNA degradation, i.e., why some mRNAs are degraded and others are not although, as noted above, transcript isoforms are differentially degraded (Salisbury et al., 2009). Bioinformatic analysis of data by Su et al. (2007) suggests that transcripts degraded during meiosis tend to have shorter 3′UTRs and degradation is not associated with any specific 3′UTR motif. In contrast, stable mRNAs have longer 3′UTRs and carry AU-rich motifs (Ma, Flemr, et al., 2013; Svoboda et al., unpublished results). Thus, when considering selective degradation of maternal mRNAs, selective stabilization of mRNAs must also be considered. Further research is needed to estimate the extent to which selectivity of maternal mRNA degradation during oocyte maturation is driven by sequence-specific recruitment of the degradation machinery and

Figure 2 Maternal mRNA degradation during the OET in mice. (A) Analysis of relative poly(A) RNA amounts in different stages shows that the majority of maternal poly(A) RNA is degraded by the time the embryo undergoes ZGA. (B) Microarray profiling reveals phased maternal mRNA degradation where each developmental transition (resumption of meiosis, fertilization, and ZGA) triggers a wave of mRNA degradation. Shown is microarray profiling (Zeng et al., 2004; Zeng & Schultz, 2005) of ~300 oocyte-specific genes (expressed in oocytes but not expressed in somatic tissues; Su et al., 2007). Colors indicate mRNA abundance approximated by fluorescence intensity (arbitrary units). The median expression value for expressed genes across the entire dataset was set to 500. (See the color plate.)

to what extent it is driven by the loss of protection of a large portion of the transcriptome against degradation (e.g., protection conferred by RNA-binding proteins such as MSY2).

The second wave of maternal RNA degradation is triggered by fertilization. Although the exact mechanism is unclear, it might involve terminal ribonucleotidyl transferases (reviewed in Martin & Keller, 2007). First, there is an increase in the amount of poly(A) RNA upon fertilization (Piko & Clegg, 1982), which likely reflects fertilization-induced cytoplasmic polyadenylation. Second, analysis of NGS data suggests increased turnover of a large number of mRNAs polyadenylated after fertilization (Abe et al., 2015; Xue, Huang, et al., 2013). Third, degradation of miRNAs, which requires RNA-specific ribonucleotidyl transferases, is also triggered by fertilization and, last but not least, transcripts from several terminal ribonucleotidyl transferases are present in relatively high abundance in mouse oocytes (Su et al., 2002). The final wave of degradation of maternal mRNA is triggered by ZGA and fits a model where *cis*-factors in maternal mRNAs are degraded by *trans*-acting factors expressed upon ZGA (Alonso, 2012). Murine maternal mRNAs are essentially gone by the eight-cell stage (Fig. 2B).

3.6 Turning Chaos into Order—Zygotic Genome Activation

A poorly understood aspect of mammalian ZGA is how its onset is regulated. Except for placental mammals, the initial cleavages are typically characterized by rapid cell cycles without transcriptional activation of the genome (reviewed in O'Farrell, Stumpff, & Su, 2004) where different mechanisms may regulate the delayed onset of ZGA (Schier, 2007). One of the classical models suggests that ZGA timing is controlled by nucleocytoplasmic ratio changes during cleavage where the geometrically increasing amount of genomic DNA titrates transcriptional repression (Newport & Kirschner, 1982).

It has been proposed that early mammalian development should be aligned with early development of other animals by gastrulation and not fertilization (O'Farrell et al., 2004). From this perspective, fertilization in mammals occurs at a comparatively earlier stage in oocyte development and the rapid cleavage cycles of other metazoans are homologous to rapid perigastrulation cycles (O'Farrell et al., 2004). Thus, mammalian preimplantation development would be a shared-derived character, which evolved as a solution for viviparous development in terrestrial ancestors of placental mammals. The early cleavage divisions during preimplantation development are not rapid (e.g., the first cleavage of mouse zygotes occurs by the time development in other animal embryos is considerably advanced). Thus, cleavage of a mammalian zygote may not be an homologous process to the cleavage in nonmammalian species because developing mammalian embryos first separate the extraembryonic and embryonic lineages before undergoing gastrulation.

Mammalian ZGA is traditionally divided into two phases: (i) minor ZGA, during which zygotic transcription initiates but the contribution of zygotic transcripts to overall gene expression is minimal, if any and (ii) major ZGA, which is characterized by a global reprogramming of gene expression that is essential for continued development. The concept of a minor and major ZGA is nicely illustrated on the mouse ZGA where the minor ZGA occurs already at the one-cell stage and the major ZGA at the two-cell stage is conveniently separated by the first cleavage (Fig. 3A). Other mammalian embryos, however, undergo ZGA several cleavage divisions later (4–16 cell), although recent NGS data suggest that the very first transcription occurs earlier than previously thought (Graf et al., 2014; Xue, Cai, et al., 2013). Importantly, minor and major ZGA differ not only in the amount of transcription but also in what is transcribed; that is, the major wave of ZGA is not simply more of the same transcripts expressed during the minor phase of ZGA but rather constitutes a constellation of new genes that are expressed.

Chromatin dynamics are unquestionably a major factor regulating ZGA and reprogramming gene expression. Such reprogramming is ultimately a matter of RNA polymerase's ability to access a promoter, a process driven by chromatin structure and transcription factors. It should be noted that, in contrast to somatic cells, where transcription factors retained on mitotic chromosomes mark promoters for transcription following entry into interphase, such is apparently not the case for condensed metaphase II chromosomes (Nagashima et al., 2007).

Plasmid-borne reporter genes have provided insights regarding mechanisms that govern transcription in one- and two-cell mouse embryos (Majumder & DePamphilis, 1995; Nothias, Majumder, Kaneko, & DePamphilis, 1995). For example, one-cell embryos do not require an enhancer for efficient expression, whereas an enhancer is required for efficient expression in two-cell embryos. The requirement for an enhancer reflects formation of a transcriptionally repressive state in two-cell embryos. Thus, one-cell embryos are transcriptionally permissive but development to

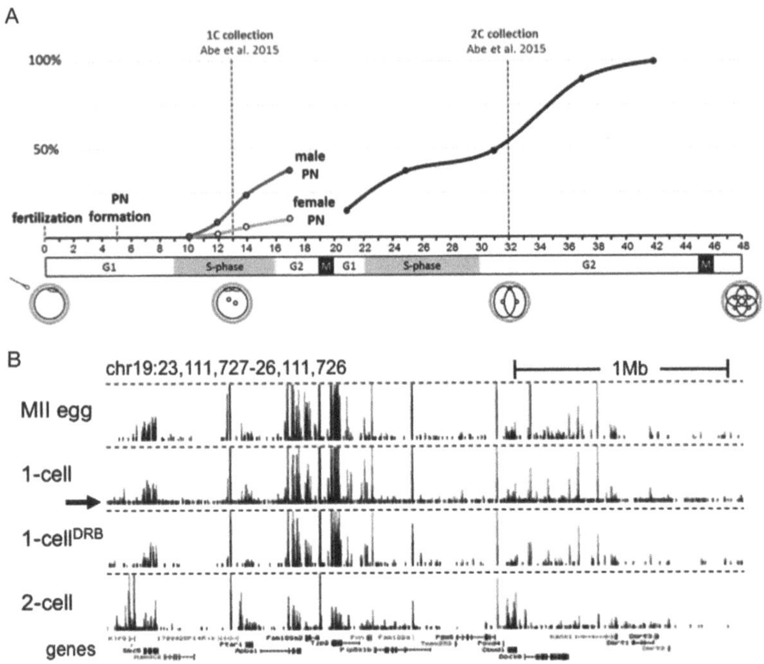

Figure 3 Minor and major ZGA in mice. (A) Minor ZGA as seen in a classical experiment measuring BrUTP incorporation (Aoki et al., 1997). Vertical blue (light gray in the print version) lines indicate time points for samples shown in (B). (B) Minor ZGA observed in NGS analysis of total RNA. Shown are HTS data (a customized screenshot from the UCSC Genome browser (Karolchik, Hinrichs, & Kent, 2012)) from different stages from a randomly selected 3 Mb region of the genome. The vertical scale was trimmed at 0.5 counts per million (CPM); trimming is indicated by horizontal dashed lines. In this view of the genome, the minor ZGA is manifested as a "grass in the forest" pattern, which likely reflects genome-wide and promiscuous nature of the initial transcription (Abe et al., 2015).

the two-cell stage is accompanied by formation of a transcriptionally repressive state and genome-wide accumulation of repressive histone modification marks (Santos, Peters, Otte, Reik, & Dean, 2005). Consistent with a role of enhancers to relieve chromatin-mediated repression of transcription is that increasing histone acetylation relieves this requirement (Majumder & DePamphilis, 1995; Nothias et al., 1995). In addition, DNA replication is also needed for such relief of repression and presumably reflects replication-dependent changes in chromatin structure. Although these conclusions were drawn using viral promoters, similar conclusions were reached using expression of the two-cell transiently expressed endogenous *Eif1a* gene (Davis, De Sousa, & Schultz, 1996; Davis & Schultz, 1997) and a plasmid-borne reporter gene driven by an endogenous promoter (*Tktls*) (Hamamoto, Suzuki, Suzuki, & Aoki, 2014).

One of the well-recognized features of epigenetic reprogramming during the OET in mouse is elimination of parental heterochromatin prior to ZGA. All paternal histone modifications are almost completely erased during spermatogenesis (reviewed in Braun, 2001; Kimmins & Sassone-Corsi, 2005). Maternal heterochromatin histone marks are also reduced after fertilization. In mice, H4K20me3 and H3K64me3 are rapidly lost by the two-cell stage (Daujat et al., 2009; Kourmouli et al., 2004) and H3K9me3 is reduced in pericentric heterochromatin until the eight-cell stage after which time it again becomes highly enriched (Puschendorf et al., 2008). This pattern suggests that erasure of parental constitutive heterochromatin marks is a part of the reprogramming process that results in totipotent blastomeres. The pattern is also consistent with an open chromatin structure observed during early development (Ahmed et al., 2010; Cho, Sakai, Nagata, & Aoki, 2002) and reminiscent of the chromatin structure observed in pluripotent ESCs (Ahmed et al., 2010; Martens et al., 2005).

Reprogramming that results in totipotent blastomeres may involve eliminating maternal histone variants associated with transcriptional repression, such as the maternal histone variant macroH2A, which is eliminated shortly after fertilization (Chang et al., 2005) and can act as a barrier to induced pluripotency (Gaspar-Maia et al., 2013). Another example is an oocyte-specific linker histone variant H1FOO. H1FOO is highly expressed in GV oocytes, but it is gradually lost after fertilization (McGraw, Vigneault, Tremblay, & Sirard, 2006; Tanaka et al., 2001), being replaced with the somatic linker histone H1 (Gao et al., 2004). The linker histone B4 (H1FOO homologue) in *Xenopus* also accumulates during oogenesis and is eliminated around the MBT when it is replaced by somatic type linker histones (Dimitrov, Almouzni, Dasso, & Wolffe, 1993). Interestingly, elimination of oocyte-specific linker histones correlates with the time of genome activation in mouse, bovine, and *Xenopus* embryos. This timing implies that elimination of H1FOO might be a prerequisite for successful embryonic genome activation and reprogramming of gene expression, a notion supported by impaired pluripotency upon ectopic expression of H1FOO in ESCs (Hayakawa, Ohgane, Tanaka, Yagi, & Shiota, 2012).

Taken together, in mouse zygotes, chromatin dynamics suggests that reduced heterochromatin, "immature" paternal chromatin, and a lack of an enhancer requirement provide a permissive environment for global genome activation, which would subsequently be shaped by "maturing" chromatin and formation of a transcriptionally repressive environment.

3.6.1 Minor ZGA—Selectivity Hidden in Promiscuous Expression

Mice exhibit the earliest known onset of ZGA among mammals with transcription initiating ~7 h following pronucleus formation (Aoki et al., 1997; Bouniol et al., 1995; Matsumoto et al., 1994). Interestingly, the male pronucleus supports a higher level of transcription than the female pronucleus (Aoki et al., 1997). This difference has been ascribed to the extensive chromatin remodeling in the male pronucleus due to the protamine–histone exchange (Nonchev & Tsanev, 1990) that would provide a window of opportunity for maternal transcription factors to associate preferentially with the male pronucleus (Aoki et al., 1997).

Transcription in the one-cell embryo could also be promiscuous; e.g., repetitive elements such as B2-containing sequences (Vasseur, Condamine, & Duprey, 1985) and MuERV-L (Kigami, Minami, Takayama, & Imai, 2003) are expressed in one-cell mouse embryos and plasmid-borne reporter gene expression in one-cell embryos does not require an enhancer for efficient expression, whereas an enhancer is required for efficient expression in two-cell embryos (Majumder & DePamphilis, 1995). Microarray studies, however, failed to detect zygotic transcripts generated by one-cell embryos (Hamatani et al., 2004; Zeng & Schultz, 2005). The reason for this failure could be that, if the zygotic transcript is also expressed in the oocyte, the increase in abundance may not be sufficient to detect as being α-amanitin sensitive given the sensitivity of microarray data. Similarly, if the transcript is expressed for the first time (i.e., not present in the oocyte), it may not accumulate to a level that sufficient to be detected by microarray analysis.

Recent studies using RNA-seq, which is more sensitive than microarrays for detection of low abundance transcripts, identified hundreds of potential minor ZGA mRNAs (Deng, Ramskold, Reinius, & Sandberg, 2014; Park et al., 2013; Xue, Cai, et al., 2013). These results, however, are confounded by the increase in mRNA adenylation that occurs following fertilization in one-cell embryos (Piko & Clegg, 1982) because poly(A) RNA was isolated and sequenced. Therefore, changes in transcript abundance could reflect changes polyadenylation and not one-cell transcription *per se*.

The best insight into the products of the minor wave of ZGA is provided by two recent studies, which used relatively large amounts of total RNA for NGS (Abe et al., 2015; Park et al., 2013). Park et al. (2013) used total RNA 50 nt single-end SOLiD sequencing after rRNA depletion and identified ~600 genes that are upregulated >1.5-fold between oocyte and one-cell stage. Abe et al. (2015) used 76 paired-end Illumina sequencing and observed transcription within ~4000 genes. The first wave of zygotic transcription has characteristics of being promiscuous, low level, and genome wide (Fig. 3B) and produces transcripts that are inefficiently spliced and polyadenylated (Abe et al., 2015). The low level of genome-wide transcription that is both inter- and intragenic may reflect that transcription in one-cell embryos only requires minimal promoter features (Hamamoto et al., 2014). The relaxed control of splicing and polyadenylation could be linked to the absence of nuclear speckles in one-cell embryos (Abe et al., 2015); disassembly of nuclear speckles is associated with reduced pre-mRNA splicing (Sacco-Bubulya & Spector, 2002).

Complex interacting relationships exist between chromatin structure, transcription, and mRNA processing. Chromatin structure in one-cell embryos is likely the key feature that gives rise to the unique features of minor ZGA. Chromatin is extensively remodeled between the one- and two-cell stages (reviewed in Burton & Torres-Padilla, 2010, 2014). Briefly, in addition to the protamine–histone exchange, 5-methyl cytosine residues in DNA become hydroxymethylated (Iqbal, Jin, Pfeifer, & Szabo, 2011; Wossidlo et al., 2011). 5-Hydroxymethyl cytosine can be further oxidized to 5-formylcytosine, which leads to DNA unwinding (Raiber et al., 2015). Should 5-formylcytosine be generated, even transiently, in the male pronucleus, a more relaxed chromatin may be formed. Furthermore, the extensive exchange of modified histones and histone variants that occurs after fertilization (reviewed in Beaujean, 2014) could also contribute to a "relaxed" chromatin structure that would support promiscuous transcription. In fact, genomic DNA is more sensitive to DNase I at the one- than at the two-cell stage (Cho et al., 2002), a finding consistent with the lack of a requirement for an enhancer for efficient gene expression in one-cell embryos (Majumder & DePamphilis, 1995); enhancers stimulate gene expression by relieving chromatin-mediated repression of transcription.

The genome-wide promiscuous transcription that occurs in one-cell embryos, including a broad list of retrotransposons, results in a large array of inappropriately expressed genes, which could compromise developmental competence. The mRNAs transcribed in one-cell embryos, however, generally do not produce functional proteins due to their inefficient post-transcriptional processing. The inability of these nascent transcripts to mature and generate functional proteins could confer a protective mechanism against such promiscuous gene expression. This inability may also explain, in part, the uncoupling of transcription and translation observed for plasmid-borne reporter genes in one-cell embryos (Nothias, Miranda, & DePamphilis, 1996).

The significance of the minor ZGA remains unclear. It is possible that some transcripts from the broad spectrum of genes transcribed in one-cell embryos are spliced and translated to generate functional proteins important for continued development. Transcription from intergenic regions in one-cell embryos might also encode long noncoding RNAs (lncRNA), which are transcribed from intergenic regions. lncRNAs, some of which do not require splicing (Hutchinson et al., 2007), can play essential roles in various cellular processes, including regulation of chromatin structure (Marchese & Huarte, 2014; Wu et al., 2013; Yoon, Abdelmohsen, & Gorospe, 2013). In light of recent findings that transcription *per se* is required for certain cellular processes, such as genomic imprinting (Tomizawa, Nowacka-Woszuk, & Kelsey, 2012), it is possible that minor ZGA transcription (i.e., not its products) drives chromatin remodeling, which will set the stage for reprogramming of gene expression during major ZGA.

3.6.2 Major ZGA—Yang of Transcriptome Remodeling During the OET

The second wave of OET occurs during the two-cell stage and results in the observed major reprogramming of gene expression. Maternal mRNA degradation can be viewed as an oocyte identity-erasing mechanism and ZGA as a zygotic (totipotent) identity-generating mechanism. Transcriptome profiling shows an oocyte-like identity of the first ZGA stage, which suggests that the amount of remaining maternal mRNAs still dominates over novel zygotic mRNAs (Fig. 1B). The continued degradation of maternal mRNA coupled with ongoing zygotic transcription generates a distinct zygotic transcriptome identity in the following developmental stage (Fig. 1B).

Analysis of appearance of nascent zygotic transcripts (represented by intronic sequences of nascent transcripts Fig. 4A) in mouse early embryos suggests that (i) minor ZGA transcription occurs at many genes but generates insignificant amounts of mature mRNAs in the transcriptome, (ii) a genome-wide transcriptional reprogramming takes place already during minor ZGA, and (iii) ZGA represents a transcriptional program completely distinct from that observed in oocyte. It should be noted that the dynamics of transcription and transcriptomes varies among mammals. For example, populations of nascent transcripts in bovine early embryos show very similar trends as mature mRNAs (Fig. 4A). This observation could reflect that there may not be a genome-wide low-level transcription uncoupled from post-transcriptional processing during the first wave of transcription in bovine embryos or that a bovine equivalent of murine minor ZGA is not separated from the major ZGA by a cleavage event. Consequently, the first wave of

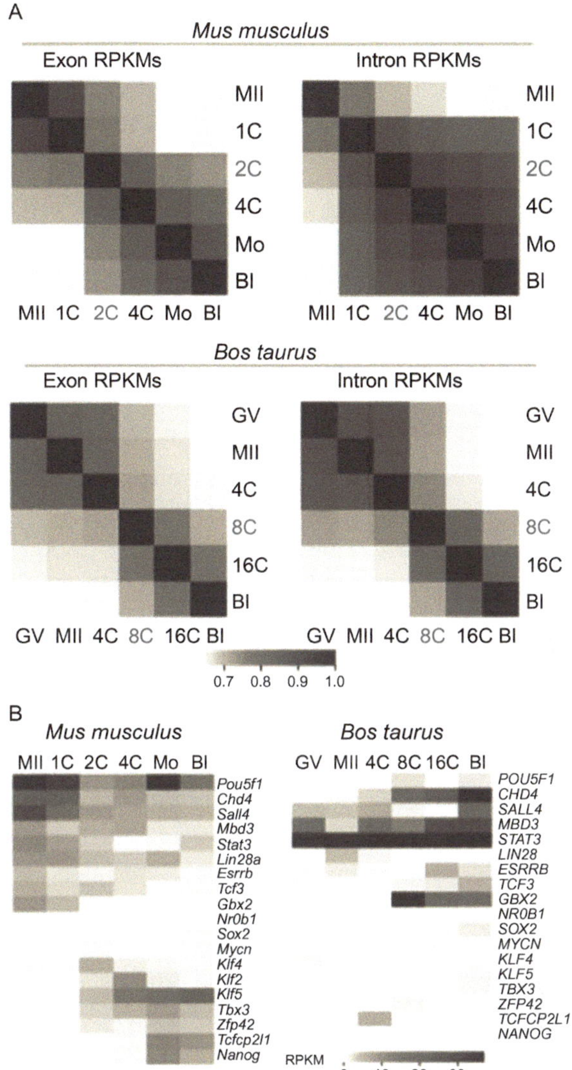

Figure 4 ZGA transcriptome remodeling and the core pluripotency network. (A) Exon- and intron-derived reads reveal that minor ZGA in mice does not produce enough mRNAs to cause a significant transcriptome remodeling, whereas bovine ZGA is more gradual and the appearance of nascent transcripts (i.e., intron-derived reads) corresponds well to the observed transcriptome remodeling. Heatmaps depict correlation of pairwise comparisons of published NGS datasets from indicated species (Abe et al., 2015; Graf et al., 2014). Color intensity indicates a correlation coefficient according to the scale depicted on the right. (B) Expression of the core pluripotency network factors during the OET in mice and cows. Heatmaps depict reads per million per kilobase (RPKMs) for the indicated genes in published datasets from the indicated species (Abe et al., 2015; Graf et al., 2014). Both heatmaps use the same color (gray in the print version) scale indicating RPKMS. STAT3 expression is far above the displayed maximum, with the mean expression across stages = 267.7 RPKMs.

RNA synthesis produces enough mature mRNAs in a single developmental stage to make a significant impact on the mRNA population.

Reprogramming of gene expression during major ZGA is selective, although estimates of genes transcribed during ZGA vary as a result of different experimental filtering conditions. For example, microarray profiling of mouse early embryos identified ~3000 α-amanitin-sensitive genes in two-cell embryos (Zeng & Schultz, 2005). NGS, however, shows that microarrays underestimate the number of expressed genes because arrays lack many nonannotated genes and high sequencing depth is more sensitive to detect low expression levels (Park et al., 2013). NGS profiling of early bovine development identified ~100 transcripts upregulated at the four-cell stage and ~1000 and ~1300 transcripts upregulated at the 8- and 16-cell stages, respectively (Graf et al., 2014).

Profiling of human early embryos identified ~2500 genes, whose expression increased between the four- and eight-cell stages. (Yan et al., 2013). Another study suggested that human ZGA occurs in three waves, with ~2000 genes upregulated between two- and four-cell embryos (early risers), ~1500 genes upregulated between four- and eight-cell embryos (middle risers), ~1400 genes upregulated after eight-cell (late risers) (Vassena et al., 2011). These data suggest that transcriptional activation in other mammals progressively increases during the early cleavage divisions. In fact, Xue, Cai, et al. (2013) reported minor ZGA in human one-cell embryos, although to what extent these results are biased by cytoplasmic polyadenylation is not clear. Taken together, ZGA and reprogramming of gene expression entails thousands—not hundreds—of genes.

To date, the set of transcription factors and binding sites in the regulated genes that constitute mammalian ZGA is poorly defined. Apart from ubiquitously expressed genes controlled by ubiquitous transcription factors, NGS data suggest that germ cell-specific transcription factors, which control expression during oocyte growth (such as FIGLA, NOBOX, LHX8, SOHLH1/2, and FOXO3A), do not contribute to ZGA in mouse embryos (Abe et al., 2015). As described above, recruitment of mRNAs encoding transcription factors and chromatin remodelers during maturation (or following fertilization) may be the driving force responsible for reprogramming gene expression. It has also been proposed that minor ZGA in mice is "primed" by sperm because >300 α-amanitin-sensitive minor ZGA transcripts are also expressed in sperm but not in parthenogenetic embryos (Park et al., 2013). Minimizing this proposal is that only a few highly expressed testis-specific genes are expressed during minor ZGA and a bioinformatic analysis of nascent transcripts reveals that minor ZGA appears to prime major ZGA (Abe et al., 2015; Fig. 4A).

Cloning experiments indicate that oocytes have the ability to reprogram gene expression of a somatic nucleus into a totipotent state. Analysis of the transcription factor network in pluripotent ESCs and experimental induction of pluripotency converged on a set of factors denoted core pluripotency factors (reviewed in Dunn, Martello, Yordanov, Emmott, & Smith, 2014); a subset of these, such as the famous "Yamanaka Factor" combination of *Pou5f1*, *Sox2*, *Klf4*, and c-*Myc*, are sufficient to reprogram gene expression and induce pluripotency in differentiated somatic cells (Takahashi & Yamanaka, 2006). These findings raise the question to what extent ZGA is related to induction of the core pluripotency network. The answer is that, although major ZGA and induction of the core pluripotency network have some overlap, the two events are distinct and the contribution of core pluripotency factors to the selectivity of mammalian major ZGA is relatively minor (Fig. 4B). This situation contrasts with zebrafish in which the conserved core network components *Nanog*, *Pou5f1*, and *SoxB1* family activate the first wave of zygotic expression in the zebrafish (Lee et al., 2013).

The behavior of the mammalian core pluripotency network during early development is illustrated by a small set of pluripotency factors expressed during murine and bovine early development (Dunn et al., 2014). Most of these pluripotency factors are expressed in mouse blastocysts, which is consistent with the origin of ESCs. About one-half of the pluripotency factors are maternally provided and the other half expressed upon ZGA in mice (Fig. 4B). When looking at "Yamanaka Factors," whose homologs constitute the maternal foundation for ZGA in the zebrafish (Lee et al., 2013), only POU5F1 is expressed maternally, whereas SOX2 is transiently expressed at the two-cell stage and then again at the blastocyst stage. Interestingly, *Klf4* is one of the few pluripotency factors that are strongly upregulated during major ZGA in mouse. *Nanog* is expressed from the two-cell stage and its transcripts accumulate until the blastocyst stage. Taken together, the mouse model suggests that a fraction of the core pluripotency factors is provided maternally to initiate a reprogramming cascade to establish a pluripotent state in inner mass cells.

When performing a similar analysis of expression of pluripotency factors in bovine embryos, striking differences in terms of temporal/stage expression of pluripotency factors emerge (Fig. 4B). These results suggest high variability in wiring the core pluripotency network during early mammalian development and are consistent with genome-wide analysis of transcriptome dynamics in early development of mice, cows, and humans (Xie et al., 2010) and a comparative ZGA analysis in fish, fly, and mice (Heyn et al., 2014). The notion that the core pluripotency factors have a minor contribution to ZGA and the associated massive reprogramming of gene expression in mammals is further supported by a simple analysis of annotated transcription factor binding sites (Chen et al., 2008) upstream of genes transcribed during major ZGA (Zeng & Schultz, 2005), which shows that only a minority of ZGA genes carry such binding sites (Franke & Svoboda, unpublished observations). This situation contrasts with the zebrafish where *Nanog*, *Pou5f1*, and *SoxB1* affect expression of >75% of zygotic genes (Lee et al., 2013). Taken together, these data suggest that the mammalian pluripotency circuitry is deeply rooted in vertebrate ZGA but, in mammals, it was adapted into a pluripotency program that is established in the inner cell mass upon its separation from the extraembryonic lineages.

Acquisition and loss of transcription control elements may evolve relatively fast even among tissue-specific genes expressed in terminally differentiated cells, such as liver cells (Wilson et al., 2008). An even higher frequency of transcriptional rewiring of ZGA during evolution might be expected because such rewiring would likely be under selective pressure during speciation and strongly influenced by retrotransposons. Retrotransposon expansion in the genome occurs through retrotransposition in the germline. Different retrotransposons adopt different strategies and are expressed during different stages of the germline cycle. For example, mouse L1 retrotransposons are expressed in germ cells and then later during early development (Branciforte & Martin, 1994; Trelogan & Martin, 1995). Intracisternal A particle retroviruses have a maternal–zygotic pattern of gene expression and accumulate from ZGA to the blastocyst stage (Svoboda et al., 2004), and MT retroelements are transcribed maternally where they provide promoter platforms for dozens of maternally expressed genes (Peaston et al., 2004). Interestingly, mouse endogenous retrovirus-type L (MuERV-L) retrotransposons are exclusively expressed during ZGA where they comprise a significant fraction of mRNAs (Svoboda et al., 2004) and thus represent a potential platform for rewiring ZGA. In fact, detailed analysis of ESC populations shows that a small fraction (~1%) of cells expresses MuERV-L and has unique characteristics in terms of gene expression and totipotent differentiation potential normally found in cells of two-cell embryos (Macfarlan et al., 2012). This finding suggests that the developmental potential of ESCs, which is typically restricted to pluripotency (i.e., formation of embryonic lineages), may fluctuate and sporadically revert to a totipotent state. Whether MuERV-L is an active player in this fluctuation and whether it is a unique feature of the mouse model is an open question. Nevertheless, retrotransposons provide a mechanism to generate great plasticity to rewire gene expression.

ACKNOWLEDGMENTS

We thank Fugaku Aoki for sharing his unpublished data and Renee Ketting for providing their manuscript prior to publication. P.S.'s research on the OET was supported by a Czech Science Foundation Grant P305/12/G034, V.F. was supported through the European Commission Seventh Framework Program (Integra-Life; Grant 315997), and Croatian Ministry of Science, Education and Sports Grant 119-0982913-1211 awarded to Kristian Vlahovicek and R.M.S. by a grant from the NIH (HD022681). The collaborative work was supported by an Academy of Sciences of the Czech Republic project M200521202 (P.S. and V.F.) and a Czech Ministry of Education Grant KONTAKT II LH13084 (P.S. and R.M.S.).

REFERENCES

Abe, K., Inoue, A., Suzuki, M. G., & Aoki, F. (2010). Global gene silencing is caused by the dissociation of RNA polymerase II from DNA in mouse oocytes. *The Journal of Reproduction and Development, 56*, 502–507.

Abe, K., Yamamoto, R., Franke, V., Cao, M., Suzuki, Y., Vlahovicek, K., et al. (2015). The first murine zygotic transcription is promiscuous and uncoupled from splicing and 3' processing. *The EMBO Journal, 34*(11), 1523–1537.

Ahmed, K., Dehghani, H., Rugg-Gunn, P., Fussner, E., Rossant, J., & Bazett-Jones, D. P. (2010). Global chromatin architecture reflects pluripotency and lineage commitment in the early mouse embryo. *PLoS One, 5*, e10531.

Alonso, C. R. (2012). A complex 'mRNA degradation code' controls gene expression during animal development. *Trends in Genetics: TIG, 28*, 78–88.

Amanai, M., Brahmajosyula, M., & Perry, A. C. (2006). A restricted role for sperm-borne microRNAs in mammalian fertilization. *Biology of Reproduction, 75*, 877–884.

Andreu-Vieyra, C. V., Chen, R., Agno, J. E., Glaser, S., Anastassiadis, K., Stewart, A. F., et al. (2010). MLL2 is required in oocytes for bulk histone 3 lysine 4 trimethylation and transcriptional silencing. *PLoS Biology, 8*, e1000453.

Anger, M., Klima, J., Kubelka, M., Prochazka, R., Motlik, J., & Schultz, R. M. (2004). Timing of Plk1 and MPF activation during porcine oocyte maturation. *Molecular Reproduction and Development, 69*, 11–16.

Aoki, F., Hara, K. T., & Schultz, R. M. (2003). Acquisition of transcriptional competence in the 1-cell mouse embryo: Requirement for recruitment of maternal mRNAs. *Molecular Reproduction and Development, 64*, 270–274.

Aoki, F., Worrad, D. M., & Schultz, R. M. (1997). Regulation of transcriptional activity during the first and second cell cycles in the preimplantation mouse embryo. *Developmental Biology, 181*, 296–307.

Avella, M. A., Xiong, B., & Dean, J. (2013). The molecular basis of gamete recognition in mice and humans. *Molecular Human Reproduction, 19*, 279–289.

Bachvarova, R., De Leon, V., Johnson, A., Kaplan, G., & Paynton, B. V. (1985). Changes in total RNA, polyadenylated RNA, and actin mRNA during meiotic maturation of mouse oocytes. *Developmental Biology, 108*, 325–331.

Backs, J., Stein, P., Backs, T., Duncan, F. E., Grueter, C. E., McAnally, J., et al. (2010). The gamma isoform of CaM kinase II controls mouse egg activation by regulating cell cycle resumption. *Proceedings of the National Academy of Sciences of the United States of America, 107*, 81–86.

Balagopal, V., Fluch, L., & Nissan, T. (2012). Ways and means of eukaryotic mRNA decay. *Biochimica et Biophysica Acta, 1819*, 593–603.

Barski, A., Cuddapah, S., Cui, K., Roh, T. Y., Schones, D. E., Wang, Z., et al. (2007). High-resolution profiling of histone methylations in the human genome. *Cell, 129*, 823–837.

Beaujean, N. (2014). Histone post-translational modifications in preimplantation mouse embryos and their role in nuclear architecture. *Molecular Reproduction and Development, 81*, 100–112.

Bouniol, C., Nguyen, E., & Debey, P. (1995). Endogenous transcription occurs at the 1-cell stage in the mouse embryo. *Experimental Cell Research, 218*, 57–62.

Branciforte, D., & Martin, S. L. (1994). Developmental and cell type specificity of LINE-1 expression in mouse testis: Implications for transposition. *Molecular and Cellular Biology, 14*, 2584–2592.

Braun, R. E. (2001). Packaging paternal chromosomes with protamine. *Nature Genetics, 28*, 10–12.

Braun, J. E., Huntzinger, E., & Izaurralde, E. (2013). The role of GW182 proteins in miRNA-mediated gene silencing. *Advances in Experimental Medicine and Biology, 768*, 147–163.

Brower, P. T., Gizang, E., Boreen, S. M., & Schultz, R. M. (1981). Biochemical studies of mammalian oogenesis: Synthesis and stability of various classes of RNA during growth of the mouse oocyte in vitro. *Developmental Biology, 86*, 373–383.

Brown, D. D., & Littna, E. (1964). Variations in the synthesis of stable RNA's during oogenesis and development of Xenopus laevis. *Journal of Molecular Biology, 8*, 688–695.

Bultman, S. J., Gebuhr, T. C., Pan, H., Svoboda, P., Schultz, R. M., & Magnuson, T. (2006). Maternal BRG1 regulates zygotic genome activation in the mouse. *Genes & Development, 20*, 1744–1754.

Burton, A., & Torres-Padilla, M. E. (2010). Epigenetic reprogramming and development: A unique heterochromatin organization in the preimplantation mouse embryo. *Briefings in Functional Genomics, 9*, 444–454.

Burton, A., & Torres-Padilla, M. E. (2014). Chromatin dynamics in the regulation of cell fate allocation during early embryogenesis. *Nature Reviews. Molecular Cell Biology, 15*, 723–734.

Carmell, M. A., Girard, A., van de Kant, H. J., Bourc'his, D., Bestor, T. H., de Rooij, D. G., et al. (2007). MIWI2 is essential for spermatogenesis and repression of transposons in the mouse male germline. *Developmental Cell, 12*, 503–514.

Chakravarthy, S., Sternberg, S. H., Kellenberger, C. A., & Doudna, J. A. (2010). Substrate-specific kinetics of Dicer-catalyzed RNA processing. *Journal of Molecular Biology, 404*, 392–402.

Chalupnikova, K., Solc, P., Sulimenko, V., Sedlacek, R., & Svoboda, P. (2014). An oocyte-specific ELAVL2 isoform is a translational repressor ablated from meiotically competent antral oocytes. *Cell Cycle, 13*, 1187–1200.

Chang, C. C., Ma, Y., Jacobs, S., Tian, X. C., Yang, X., & Rasmussen, T. P. (2005). A maternal store of macroH2A is removed from pronuclei prior to onset of somatic macroH2A expression in preimplantation embryos. *Developmental Biology, 278*, 367–380.

Chekulaeva, M., & Filipowicz, W. (2009). Mechanisms of miRNA-mediated post-transcriptional regulation in animal cells. *Current Opinion in Cell Biology, 21*, 452–460.

Chen, P. Y., Manninga, H., Slanchev, K., Chien, M., Russo, J. J., Ju, J., et al. (2005). The developmental miRNA profiles of zebrafish as determined by small RNA cloning. *Genes & Development, 19*, 1288–1293.

Chen, J., Melton, C., Suh, N., Oh, J. S., Horner, K., Xie, F., et al. (2011). Genome-wide analysis of translation reveals a critical role for deleted in azoospermia-like (Dazl) at the oocyte-to-zygote transition. *Genes & Development, 25*, 755–766.

Chen, X., Xu, H., Yuan, P., Fang, F., Huss, M., Vega, V. B., et al. (2008). Integration of external signaling pathways with the core transcriptional network in embryonic stem cells. *Cell, 133*, 1106–1117.

Cho, T., Sakai, S., Nagata, M. a., & Aoki, F. (2002). Involvement of chromatin structure in the regulation of mouse zygotic gene activation. *Animal Science Journal, 73*, 113–122.

Daentl, D. L., & Epstein, C. J. (1971). Developmental interrelationships of uridine uptake. Nucleotide formation and incorporation into RNA by early mammalian embryos. *Developmental Biology, 24*, 428–442.

Daujat, S., Weiss, T., Mohn, F., Lange, U. C., Ziegler-Birling, C., Zeissler, U., et al. (2009). H3K64 trimethylation marks heterochromatin and is dynamically remodeled during developmental reprogramming. *Nature Structural & Molecular Biology, 16*, 777–781.

Davis, W., Jr., De Sousa, P. A., & Schultz, R. M. (1996). Transient expression of translation initiation factor eIF-4C during the 2-cell stage of the preimplantation mouse embryo: Identification by mRNA differential display and the role of DNA replication in zygotic gene activation. *Developmental Biology, 174*, 190–201.

Davis, W., Jr., & Schultz, R. M. (1997). Role of the first round of DNA replication in reprogramming gene expression in the preimplantation mouse embryo. *Molecular Reproduction and Development, 47*, 430–434.

de Jong, M., Rauwerda, H., Bruning, O., Verkooijen, J., Spaink, H. P., & Breit, T. M. (2010). RNA isolation method for single embryo transcriptome analysis in zebrafish. *BMC Research Notes, 3*, 73.

De La Fuente, R., & Eppig, J. J. (2001). Transcriptional activity of the mouse oocyte genome: Companion granulosa cells modulate transcription and chromatin remodeling. *Developmental Biology, 229*, 224–236.

De La Fuente, R., Viveiros, M. M., Burns, K. H., Adashi, E. Y., Matzuk, M. M., & Eppig, J. J. (2004). Major chromatin remodeling in the germinal vesicle (GV) of mammalian oocytes is dispensable for global transcriptional silencing but required for centromeric heterochromatin function. *Developmental Biology, 275*, 447–458.

De Leon, V., Johnson, A., & Bachvarova, R. (1983). Half-lives and relative amounts of stored and polysomal ribosomes and poly(A) + RNA in mouse oocytes. *Developmental Biology, 98*, 400–408.

de Veer, M. J., Sledz, C. A., & Williams, B. R. (2005). Detection of foreign RNA: Implications for RNAi. *Immunology and Cell Biology, 83*, 224–228.

Dean, W. (2014). DNA methylation and demethylation: A pathway to gametogenesis and development. *Molecular Reproduction and Development, 81*, 113–125.

Deng, W., & Lin, H. (2002). miwi, a murine homolog of piwi, encodes a cytoplasmic protein essential for spermatogenesis. *Developmental Cell, 2*, 819–830.

Deng, Q., Ramskold, D., Reinius, B., & Sandberg, R. (2014). Single-cell RNA-seq reveals dynamic, random monoallelic gene expression in mammalian cells. *Science (New York, NY), 343*, 193–196.

Dimitrov, S., Almouzni, G., Dasso, M., & Wolffe, A. P. (1993). Chromatin transitions during early Xenopus embryogenesis: Changes in histone H4 acetylation and in linker histone type. *Developmental Biology, 160*, 214–227.

Ducibella, T., Huneau, D., Angelichio, E., Xu, Z., Schultz, R. M., Kopf, G. S., et al. (2002). Egg-to-embryo transition is driven by differential responses to Ca(2+) oscillation number. *Developmental Biology, 250*, 280–291.

Ducibella, T., Schultz, R. M., & Ozil, J. P. (2006). Role of calcium signals in early development. *Seminars in Cell & Developmental Biology, 17*, 324–332.

Dunn, S. J., Martello, G., Yordanov, B., Emmott, S., & Smith, A. G. (2014). Defining an essential transcription factor program for naive pluripotency. *Science, 344*, 1156–1160.

Edson, M. A., Nagaraja, A. K., & Matzuk, M. M. (2009). The mammalian ovary from genesis to revelation. *Endocrine Reviews, 30*, 624–712.
Eppig, J. J., & Schroeder, A. C. (1989). Capacity of mouse oocytes from preantral follicles to undergo embryogenesis and development to live young after growth, maturation, and fertilization in vitro. *Biology of Reproduction, 41*, 268–276.
Epstein, C. J., Daentl, D. L., Smith, S. A., & Kwok, L. W. (1971). Guanine metabolism in preimplantation mouse embryo. *Biology of Reproduction, 5*, 308–313.
Flemr, M., Ma, J., Schultz, R. M., & Svoboda, P. (2010). P-body loss is concomitant with formation of a messenger RNA storage domain in mouse oocytes. *Biology of Reproduction, 82*, 1008–1017.
Flemr, M., Malik, R., Franke, V., Nejepinska, J., Sedlacek, R., Vlahovicek, K., et al. (2013). A retrotransposon-driven dicer isoform directs endogenous small interfering RNA production in mouse oocytes. *Cell, 155*, 807–816.
Flemr, M., Moravec, M., Libova, V., Sedlacek, R., & Svoboda, P. (2014). Lin28a is dormant, functional, and dispensable during mouse oocyte-to-embryo transition. *Biology of Reproduction, 90*, 131.
Fulka, H., & Langerova, A. (2014). The maternal nucleolus plays a key role in centromere satellite maintenance during the oocyte to embryo transition. *Development (Cambridge, England), 141*, 1694–1704.
Gantier, M. P., & Williams, B. R. (2007). The response of mammalian cells to double-stranded RNA. *Cytokine & Growth Factor Reviews, 18*, 363–371.
Gao, S., Chung, Y. G., Parseghian, M. H., King, G. J., Adashi, E. Y., & Latham, K. E. (2004). Rapid H1 linker histone transitions following fertilization or somatic cell nuclear transfer: Evidence for a uniform developmental program in mice. *Developmental Biology, 266*, 62–75.
Garcia-Lopez, J., Hourcade Jde, D., Alonso, L., Cardenas, D. B., & del Mazo, J. (2014). Global characterization and target identification of piRNAs and endo-siRNAs in mouse gametes and zygotes. *Biochimica et Biophysica Acta, 1839*, 463–475.
Gaspar-Maia, A., Qadeer, Z. A., Hasson, D., Ratnakumar, K., Leu, N. A., Leroy, G., et al. (2013). MacroH2A histone variants act as a barrier upon reprogramming towards pluripotency. *Nature Communications, 4*, 1565.
Gilbert, I., Scantland, S., Sylvestre, E. L., Gravel, C., Laflamme, I., Sirard, M. A., et al. (2009). The dynamics of gene products fluctuation during bovine pre-hatching development. *Molecular Reproduction and Development, 76*, 762–772.
Giraldez, A. J., Mishima, Y., Rihel, J., Grocock, R. J., Van Dongen, S., Inoue, K., et al. (2006). Zebrafish MiR-430 promotes deadenylation and clearance of maternal mRNAs. *Science, 312*, 75–79.
Graf, A., Krebs, S., Zakhartchenko, V., Schwalb, B., Blum, H., & Wolf, E. (2014). Fine mapping of genome activation in bovine embryos by RNA sequencing. *Proceedings of the National Academy of Sciences of the United States of America, 111*, 4139–4144.
Groppo, R., & Richter, J. D. (2009). Translational control from head to tail. *Current Opinion in Cell Biology, 21*, 444–451.
Ha, M., & Kim, V. N. (2014). Regulation of microRNA biogenesis. *Nature Reviews. Molecular Cell Biology, 15*, 509–524.
Hamamoto, G., Suzuki, T., Suzuki, M. G., & Aoki, F. (2014). Regulation of transketolase like 1 gene expression in the murine one-cell stage embryos. *PLoS One, 9*, e82087.
Hamatani, T., Carter, M. G., Sharov, A. A., & Ko, M. S. (2004). Dynamics of global gene expression changes during mouse preimplantation development. *Developmental Cell, 6*, 117–131.
Hayakawa, K., Ohgane, J., Tanaka, S., Yagi, S., & Shiota, K. (2012). Oocyte-specific linker histone H1foo is an epigenomic modulator that decondenses chromatin and

impairs pluripotency. *Epigenetics: Official Journal of the DNA Methylation Society, 7*, 1029–1036.

Heyn, P., Kircher, M., Dahl, A., Kelso, J., Tomancak, P., Kalinka, A. T., et al. (2014). The earliest transcribed zygotic genes are short, newly evolved, and different across species. *Cell Reports, 6*, 285–292.

Houseley, J., & Tollervey, D. (2009). The many pathways of RNA degradation. *Cell, 136*, 763–776.

Huarte, J., Belin, D., Vassalli, A., Strickland, S., & Vassalli, J. D. (1987). Meiotic maturation of mouse oocytes triggers the translation and polyadenylation of dormant tissue-type plasminogen activator mRNA. *Genes & Development, 1*, 1201–1211.

Hutchinson, J. N., Ensminger, A. W., Clemson, C. M., Lynch, C. R., Lawrence, J. B., & Chess, A. (2007). A screen for nuclear transcripts identifies two linked noncoding RNAs associated with SC35 splicing domains. *BMC Genomics, 8*, 39.

Ihara, M., Tseng, H., & Schultz, R. M. (2011). Expression of variant ribosomal RNA genes in mouse oocytes and preimplantation embryos. *Biology of Reproduction, 84*, 944–946.

Iqbal, K., Jin, S. G., Pfeifer, G. P., & Szabo, P. E. (2011). Reprogramming of the paternal genome upon fertilization involves genome-wide oxidation of 5-methylcytosine. *Proceedings of the National Academy of Sciences of the United States of America, 108*, 3642–3647.

Jahn, C. L., Baran, M. M., & Bachvarova, R. (1976). Stability of RNA synthesized by the mouse oocyte during its major growth phase. *The Journal of Experimental Zoology, 197*, 161–171.

Jinek, M., & Doudna, J. A. (2009). A three-dimensional view of the molecular machinery of RNA interference. *Nature, 457*, 405–412.

Kageyama, S., Liu, H., Kaneko, N., Ooga, M., Nagata, M., & Aoki, F. (2007). Alterations in epigenetic modifications during oocyte growth in mice. *Reproduction (Cambridge, England), 133*, 85–94.

Kaneda, M., Tang, F., O'Carroll, D., Lao, K., & Surani, M. A. (2009). Essential role for Argonaute2 protein in mouse oogenesis. *Epigenetics & Chromatin, 2*, 9.

Karolchik, D., Hinrichs, A. S., & Kent, W. J. (2012). The UCSC genome browser. *Current Protocols in Bioinformatics*. editorial board, Andreas D. Baxevanis [et al.] Chapter 1, Unit1 4.

Kigami, D., Minami, N., Takayama, H., & Imai, H. (2003). MuERV-L is one of the earliest transcribed genes in mouse one-cell embryos. *Biology of Reproduction, 68*, 651–654.

Kim, V. N., Han, J., & Siomi, M. C. (2009). Biogenesis of small RNAs in animals. *Nature Reviews Molecular Cell Biology, 10*, 126–139.

Kimmins, S., & Sassone-Corsi, P. (2005). Chromatin remodelling and epigenetic features of germ cells. *Nature, 434*, 583–589.

Kourmouli, N., Jeppesen, P., Mahadevhaiah, S., Burgoyne, P., Wu, R., Gilbert, D. M., et al. (2004). Heterochromatin and tri-methylated lysine 20 of histone H4 in animals. *Journal of Cell Science, 117*, 2491–2501.

Kuramochi-Miyagawa, S., Kimura, T., Ijiri, T. W., Isobe, T., Asada, N., Fujita, Y., et al. (2004). Mili, a mammalian member of piwi family gene, is essential for spermatogenesis. *Development (Cambridge, England), 131*, 839–849.

Lee, M. T., Bonneau, A. R., Takacs, C. M., Bazzini, A. A., DiVito, K. R., Fleming, E. S., et al. (2013). Nanog, Pou5f1 and SoxB1 activate zygotic gene expression during the maternal-to-zygotic transition. *Nature, 503*, 360–364.

Lee, M., Choi, Y., Kim, K., Jin, H., Lim, J., Nguyen, T. A., et al. (2014). Adenylation of maternally inherited microRNAs by Wispy. *Molecular Cell, 56*, 696–707.

Lim, A. K., Lorthongpanich, C., Chew, T. G., Tan, C. W., Shue, Y. T., Balu, S., et al. (2013). The nuage mediates retrotransposon silencing in mouse primordial ovarian follicles. *Development (Cambridge, England), 140*, 3819–3825.

Liu, H., & Aoki, F. (2002). Transcriptional activity associated with meiotic competence in fully grown mouse GV oocytes. *Zygote, 10*, 327–332.
Liu, Y. J., Nakamura, T., & Nakano, T. (2012). Essential role of DPPA3 for chromatin condensation in mouse oocytogenesis. *Biology of Reproduction, 86*, 40.
Ma, J., Flemr, M., Stein, P., Berninger, P., Malik, R., Zavolan, M., et al. (2010). MicroRNA activity is suppressed in mouse oocytes. *Current Biology, 20*, 265–270.
Ma, J., Flemr, M., Strnad, H., Svoboda, P., & Schultz, R. M. (2013). Maternally recruited DCP1A and DCP2 contribute to messenger RNA degradation during oocyte maturation and genome activation in mouse. *Biology of Reproduction, 88*, 11.
Ma, J. Y., Li, M., Luo, Y. B., Song, S., Tian, D., Yang, J., et al. (2013). Maternal factors required for oocyte developmental competence in mice: Transcriptome analysis of non-surrounded nucleolus (NSN) and surrounded nucleolus (SN) oocytes. *Cell Cycle, 12*, 1928–1938.
Ma, E., MacRae, I. J., Kirsch, J. F., & Doudna, J. A. (2008). Autoinhibition of human dicer by its internal helicase domain. *Journal of Molecular Biology, 380*, 237–243.
Macfarlan, T. S., Gifford, W. D., Driscoll, S., Lettieri, K., Rowe, H. M., Bonanomi, D., et al. (2012). Embryonic stem cell potency fluctuates with endogenous retrovirus activity. *Nature, 487*, 57–63.
Majumder, S., & DePamphilis, M. L. (1995). A unique role for enhancers is revealed during early mouse development. *BioEssays: News and Reviews in Molecular, Cellular and Developmental Biology, 17*, 879–889.
Marchese, F. P., & Huarte, M. (2014). Long non-coding RNAs and chromatin modifiers: Their place in the epigenetic code. *Epigenetics: Official Journal of the DNA Methylation Society, 9*, 21–26.
Martens, J. H., O'Sullivan, R. J., Braunschweig, U., Opravil, S., Radolf, M., Steinlein, P., et al. (2005). The profile of repeat-associated histone lysine methylation states in the mouse epigenome. *The EMBO Journal, 24*, 800–812.
Martin, G., & Keller, W. (2007). RNA-specific ribonucleotidyl transferases. *RNA (New York, NY), 13*, 1834–1849.
Matsumoto, K., Anzai, M., Nakagata, N., Takahashi, A., Takahashi, Y., & Miyata, K. (1994). Onset of paternal gene activation in early mouse embryos fertilized with transgenic mouse sperm. *Molecular Reproduction and Development, 39*, 136–140.
Matzuk, M. M., Burns, K. H., Viveiros, M. M., & Eppig, J. J. (2002). Intercellular communication in the mammalian ovary: Oocytes carry the conversation. *Science, 296*, 2178–2180.
Mauro, V. P., & Edelman, G. M. (2002). The ribosome filter hypothesis. *Proceedings of the National Academy of Sciences of the United States of America, 99*, 12031–12036.
McGraw, S., Vigneault, C., Tremblay, K., & Sirard, M. A. (2006). Characterization of linker histone H1FOO during bovine in vitro embryo development. *Molecular Reproduction and Development, 73*, 692–699.
Medvedev, S., Pan, H., & Schultz, R. M. (2011). Absence of MSY2 in mouse oocytes perturbs oocyte growth and maturation, RNA stability, and the transcriptome. *Biology of Reproduction, 85*, 575–583.
Medvedev, S., Yang, J., Hecht, N. B., & Schultz, R. M. (2008). CDC2A (CDK1)-mediated phosphorylation of MSY2 triggers maternal mRNA degradation during mouse oocyte maturation. *Developmental Biology, 321*, 205–215.
Melton, C., Judson, R. L., & Blelloch, R. (2010). Opposing microRNA families regulate self-renewal in mouse embryonic stem cells. *Nature, 463*, 621–626.
Moore, G. P., & Lintern-Moore, S. (1978). Transcription of the mouse oocyte genome. *Biology of Reproduction, 18*, 865–870.
Murai, S., Stein, P., Buffone, M. G., Yamashita, S., & Schultz, R. M. (2010). Recruitment of Orc6l, a dormant maternal mRNA in mouse oocytes, is essential for DNA replication in 1-cell embryos. *Developmental Biology, 341*, 205–212.

Murchison, E. P., Stein, P., Xuan, Z., Pan, H., Zhang, M. Q., Schultz, R. M., et al. (2007). Critical roles for Dicer in the female germline. *Genes & Development*, *21*, 682–693.
Nagashima, T., Maruyama, T., Furuya, M., Kajitani, T., Uchida, H., Masuda, H., et al. (2007). Histone acetylation and subcellular localization of chromosomal protein BRD4 during mouse oocyte meiosis and mitosis. *Molecular Human Reproduction*, *13*, 141–148.
Nejepinska, J., Malik, R., Filkowski, J., Flemr, M., Filipowicz, W., & Svoboda, P. (2012). dsRNA expression in the mouse elicits RNAi in oocytes and low adenosine deamination in somatic cells. *Nucleic Acids Research*, *40*, 399–413.
Newport, J., & Kirschner, M. (1982). A major developmental transition in early Xenopus embryos: II. Control of the onset of transcription. *Cell*, *30*, 687–696.
Nonchev, S., & Tsanev, R. (1990). Protamine-histone replacement and DNA replication in the male mouse pronucleus. *Molecular Reproduction and Development*, *25*, 72–76.
Nothias, J. Y., Majumder, S., Kaneko, K. J., & DePamphilis, M. L. (1995). Regulation of gene expression at the beginning of mammalian development. *The Journal of Biological Chemistry*, *270*, 22077–22080.
Nothias, J. Y., Miranda, M., & DePamphilis, M. L. (1996). Uncoupling of transcription and translation during zygotic gene activation in the mouse. *The EMBO Journal*, *15*, 5715–5725.
O'Farrell, P. H., Stumpff, J., & Su, T. T. (2004). Embryonic cleavage cycles: How is a mouse like a fly? *Current Biology: CB*, *14*, R35–45.
Ohnishi, Y., Totoki, Y., Toyoda, A., Watanabe, T., Yamamoto, Y., Tokunaga, K., et al. (2010). Small RNA class transition from siRNA/piRNA to miRNA during pre-implantation mouse development. *Nucleic Acids Research*, *38*, 5141–5151.
Oqani, R. K., Kim, H. R., Diao, Y. F., Park, C. S., & Jin, D. I. (2011). The CDK9/cyclin T1 subunits of P-TEFb in mouse oocytes and preimplantation embryos: A possible role in embryonic genome activation. *BMC Developmental Biology*, *11*, 33.
Ozil, J. P., Banrezes, B., Toth, S., Pan, H., & Schultz, R. M. (2006). Ca2+ oscillatory pattern in fertilized mouse eggs affects gene expression and development to term. *Developmental Biology*, *300*, 534–544.
Pan, H., O'Brien, M,. J., Wigglesworth, K., Eppig, J. J., & Schultz, R. M. (2005). Transcript profiling during mouse oocyte development and the effect of gonadotropin priming and development in vitro. *Developmental Biology*, *286*, 493–506.
Paradis, F., Vigneault, C., Robert, C., & Sirard, M. A. (2005). RNA interference as a tool to study gene function in bovine oocytes. *Molecular Reproduction and Development*, *70*, 111–121.
Park, S. J., Komata, M., Inoue, F., Yamada, K., Nakai, K., Ohsugi, M., et al. (2013). Inferring the choreography of parental genomes during fertilization from ultralarge-scale whole-transcriptome analysis. *Genes & Development*, *27*, 2736–2748.
Paynton, B. V., Rempel, R., & Bachvarova, R. (1988). Changes in state of adenylation and time course of degradation of maternal mRNAs during oocyte maturation and early embryonic development in the mouse. *Developmental Biology*, *129*, 304–314.
Peaston, A. E., Evsikov, A. V., Graber, J. H., de Vries, W. N., Holbrook, A. E., Solter, D., et al. (2004). Retrotransposons regulate host genes in mouse oocytes and preimplantation embryos. *Developmental Cell*, *7*, 597–606.
Pelegri, F. (2003). Maternal factors in zebrafish development. *Developmental Dynamics: An Official Publication of the American Association of Anatomists*, *228*, 535–554.
Phatnani, H. P., & Greenleaf, A. L. (2006). Phosphorylation and functions of the RNA polymerase II CTD. *Genes & Development*, *20*, 2922–2936.
Piko, L., & Clegg, K. B. (1982). Quantitative changes in total RNA, total poly(A), and ribosomes in early mouse embryos. *Developmental Biology*, *89*, 362–378.
Probst, A. V., Okamoto, I., Casanova, M., El Marjou, F., Le Baccon, P., & Almouzni, G. (2010). A strand-specific burst in transcription of pericentric satellites is required for

chromocenter formation and early mouse development. *Developmental Cell, 19*, 625–638.

Puschendorf, M., Stein, P., Oakeley, E. J., Schultz, R. M., Peters, A. H., & Svoboda, P. (2006). Abundant transcripts from retrotransposons are unstable in fully grown mouse oocytes. *Biochemical and Biophysical Research Communications, 347*(1), 36–43.

Puschendorf, M., Terranova, R., Boutsma, E., Mao, X., Isono, K., Brykczynska, U., et al. (2008). PRC1 and Suv39h specify parental asymmetry at constitutive heterochromatin in early mouse embryos. *Nature Genetics, 40*, 411–420.

Raiber, E. A., Murat, P., Chirgadze, D. Y., Beraldi, D., Luisi, B. F., & Balasubramanian, S. (2015). 5-Formylcytosine alters the structure of the DNA double helix. *Nature Structural & Molecular Biology, 22*, 44–49.

Roovers, E., Rosenkranz, D., Mahdipour, M., Han, C., He, N., de Sousa Lopes, S. C., et al. (2015). PIWI proteins and piRNAs in mammalian oocytes and early embryos. *Cell Reports, 10*(12), 2069–2082.

Sacco-Bubulya, P., & Spector, D. L. (2002). Disassembly of interchromatin granule clusters alters the coordination of transcription and pre-mRNA splicing. *The Journal of Cell Biology, 156*, 425–436.

Salisbury, J., Hutchison, K. W., Wigglesworth, K., Eppig, J. J., & Graber, J. H. (2009). Probe-level analysis of expression microarrays characterizes isoform-specific degradation during mouse oocyte maturation. *PLoS One, 4*, e7479.

Santos, F., Peters, A. H., Otte, A. P., Reik, W., & Dean, W. (2005). Dynamic chromatin modifications characterise the first cell cycle in mouse embryos. *Developmental Biology, 280*, 225–236.

Schier, A. F. (2007). The maternal-zygotic transition: Death and birth of RNAs. *Science, 316*, 406–407.

Schoenberg, D. R., & Maquat, L. E. (2012). Regulation of cytoplasmic mRNA decay. *Nature Reviews Genetics, 13*, 246–259.

Sinkkonen, L., Hugenschmidt, T., Berninger, P., Gaidatzis, D., Mohn, F., Artus-Revel,- C. G., et al. (2008). MicroRNAs control de novo DNA methylation through regulation of transcriptional repressors in mouse embryonic stem cells. *Nature Structural & Molecular Biology, 15*, 259–267.

Siomi, M. C., Sato, K., Pezic, D., & Aravin, A. A. (2011). PIWI-interacting small RNAs: The vanguard of genome defence. *Nature Reviews. Molecular Cell Biology, 12*, 246–258.

Sorensen, R. A., & Wassarman, P. M. (1976). Relationship between growth and meiotic maturation of the mouse oocyte. *Developmental Biology, 50*, 531–536.

Soyal, S. M., Amleh, A., & Dean, J. (2000). FIGalpha, a germ cell-specific transcription factor required for ovarian follicle formation. *Development, 127*, 4645–4654.

Stein, P., Rozhkov, N. V., Li, F., Cardenas, F. L., Davydenk, O., Vandivier, L. E., et al. (2015). Essential role for endogenous siRNAs during meiosis in mouse oocytes. *PLoS Genetics, 11*, e1005013.

Stein, P., Zeng, F., Pan, H., & Schultz, R. M. (2005). Absence of non-specific effects of RNA interference triggered by long double-stranded RNA in mouse oocytes. *Developmental Biology, 286*, 464–471.

Su, A. I., Cooke, M. P., Ching, K. A., Hakak, Y., Walker, J. R., Wiltshire, T., et al. (2002). Large-scale analysis of the human and mouse transcriptomes. *Proceedings of the National Academy of Sciences of the United States of America, 99*, 4465–4470.

Su, Y. Q., Sugiura, K., Woo, Y., Wigglesworth, K., Kamdar, S., Affourtit, J., et al. (2007). Selective degradation of transcripts during meiotic maturation of mouse oocytes. *Developmental Biology, 302*, 104–117.

Suh, N., Baehner, L., Moltzahn, F., Melton, C., Shenoy, A., Chen, J., et al. (2010). MicroRNA function is globally suppressed in mouse oocytes and early embryos. *Current Biology, 20*, 271–277.

Suzumori, N., Yan, C., Matzuk, M. M., & Rajkovic, A. (2002). Nobox is a homeobox-encoding gene preferentially expressed in primordial and growing oocytes. *Mechanisms of Development*, *111*, 137–141.

Svoboda, P. (2010). Why mouse oocytes and early embryos ignore miRNAs? *RNA Biology*, 7, 559–563.

Svoboda, P., Stein, P., Anger, M., Bernstein, E., Hannon, G. J., & Schultz, R. M. (2004). RNAi and expression of retrotransposons MuERV-L and IAP in preimplantation mouse embryos. *Developmental Biology*, *269*, 276–285.

Svoboda, P., Stein, P., Hayashi, H., & Schultz, R. M. (2000). Selective reduction of dormant maternal mRNAs in mouse oocytes by RNA interference. *Development (Cambridge, England)*, *127*, 4147–4156.

Tafuri, S. R., & Wolffe, A. P. (1993). Selective recruitment of masked maternal mRNA from messenger ribonucleoprotein particles containing FRGY2 (mRNP4). *The Journal of Biological Chemistry*, *268*, 24255–24261.

Takahashi, K., & Yamanaka, S. (2006). Induction of pluripotent stem cells from mouse embryonic and adult fibroblast cultures by defined factors. *Cell*, *126*, 663–676.

Tam, O. H., Aravin, A. A., Stein, P., Girard, A., Murchison, E. P., Cheloufi, S., et al. (2008). Pseudogene-derived small interfering RNAs regulate gene expression in mouse oocytes. *Nature*, *453*, 534–538.

Tanaka, M., Hennebold, J. D., Macfarlane, J., & Adashi, E. Y. (2001). A mammalian oocyte-specific linker histone gene H1oo: Homology with the genes for the oocyte-specific cleavage stage histone (cs-H1) of sea urchin and the B4/H1M histone of the frog. *Development (Cambridge, England)*, *128*, 655–664.

Tang, F., Kaneda, M., O'Carroll, D., Hajkova, P., Barton, S. C., Sun, Y. A., et al. (2007). Maternal microRNAs are essential for mouse zygotic development. *Genes & Development*, *21*, 644–648.

Tang, G. Q., & Maxwell, E. S. (2008). Xenopus microRNA genes are predominantly located within introns and are differentially expressed in adult frog tissues via post-transcriptional regulation. *Genome Research*, *18*, 104–112.

Tomizawa, S., Nowacka-Woszuk, J., & Kelsey, G. (2012). DNA methylation establishment during oocyte growth: Mechanisms and significance. *The International Journal of Developmental Biology*, *56*, 867–875.

Trelogan, S. A., & Martin, S. L. (1995). Tightly regulated, developmentally specific expression of the first open reading frame from LINE-1 during mouse embryogenesis. *Proceedings of the National Academy of Sciences of the United States of America*, *92*, 1520–1524.

Tseng, H., Chou, W., Wang, J., Zhang, X., Zhang, S., & Schultz, R. M. (2008). Mouse ribosomal RNA genes contain multiple differentially regulated variants. *PLoS One*, *3*, e1843.

Vassena, R., Boue, S., Gonzalez-Roca, E., Aran, B., Auer, H., Veiga, A., et al. (2011). Waves of early transcriptional activation and pluripotency program initiation during human preimplantation development. *Development*, *138*, 3699–3709.

Vasseur, M., Condamine, H., & Duprey, P. (1985). RNAs containing B2 repeated sequences are transcribed in the early stages of mouse embryogenesis. *The EMBO Journal*, *4*, 1749–1753.

Wang, S., Kou, Z., Jing, Z., Zhang, Y., Guo, X., Dong, M., et al. (2010). Proteome of mouse oocytes at different developmental stages. *Proceedings of the National Academy of Sciences of the United States of America*, *107*, 17639–17644.

Wang, Q. T., Piotrowska, K., Ciemerych, M. A., Milenkovic, L., Scott, M. P., Davis, R. W., et al. (2004). A genome-wide study of gene activity reveals developmental signaling pathways in the preimplantation mouse embryo. *Developmental Cell*, *6*, 133–144.

Watanabe, T., Totoki, Y., Toyoda, A., Kaneda, M., Kuramochi-Miyagawa, S., Obata, Y., et al. (2008). Endogenous siRNAs from naturally formed dsRNAs regulate transcripts in mouse oocytes. *Nature*, *453*, 539–543.

Weick, E. M., & Miska, E. A. (2014). piRNAs: From biogenesis to function. *Development (Cambridge, England)*, *141*, 3458–3471.

Wianny, F., & Zernicka-Goetz, M. (2000). Specific interference with gene function by double-stranded RNA in early mouse development. *Nature Cell Biology*, *2*, 70–75.

Wickramasinghe, D., & Albertini, D. F. (1992). Centrosome phosphorylation and the developmental expression of meiotic competence in mouse oocytes. *Developmental Biology*, *152*, 62–74.

Wickramasinghe, D., Ebert, K. M., & Albertini, D. F. (1991). Meiotic competence acquisition is associated with the appearance of M-phase characteristics in growing mouse oocytes. *Developmental Biology*, *143*, 162–172.

Wilson, M. D., Barbosa-Morais, N. L., Schmidt, D., Conboy, C. M., Vanes, L., Tybulewicz, V. L., et al. (2008). Species-specific transcription in mice carrying human chromosome 21. *Science*, *322*, 434–438.

Wossidlo, M., Nakamura, T., Lepikhov, K., Marques, C. J., Zakhartchenko, V., Boiani, M., et al. (2011). 5-Hydroxymethylcytosine in the mammalian zygote is linked with epigenetic reprogramming. *Nature Communications*, *2*, 241.

Wu, P., Zuo, X., Deng, H., Liu, X., Liu, L., & Ji, A. (2013). Roles of long noncoding RNAs in brain development, functional diversification and neurodegenerative diseases. *Brain Research Bulletin*, *97*, 69–80.

Wylie, C. (2000). Germ cells. *Current Opinion in Genetics & Development*, *10*, 410–413.

Xia, M., He, H., Wang, Y., Liu, M., Zhou, T., Lin, M., et al. (2012). PCBP1 is required for maintenance of the transcriptionally silent state in fully grown mouse oocytes. *Cell Cycle*, *11*, 2833–2842.

Xie, D., Chen, C. C., Ptaszek, L. M., Xiao, S., Cao, X., Fang, F., et al. (2010). Rewirable gene regulatory networks in the preimplantation embryonic development of three mammalian species. *Genome Research*, *20*, 804–815.

Xu, Z., Williams, C. J., Kopf, G. S., & Schultz, R. M. (2003). Maturation-associated increase in IP3 receptor type 1: Role in conferring increased IP3 sensitivity and Ca2+ oscillatory behavior in mouse eggs. *Developmental Biology*, *254*, 163–171.

Xue, L., Cai, J. Y., Ma, J., Huang, Z., Guo, M. X., Fu, L. Z., et al. (2013). Global expression profiling reveals genetic programs underlying the developmental divergence between mouse and human embryogenesis. *BMC Genomics*, *14*, 568.

Xue, Z., Huang, K., Cai, C., Cai, L., Jiang, C. Y., Feng, Y., et al. (2013). Genetic programs in human and mouse early embryos revealed by single-cell RNA sequencing. *Nature*, *500*, 593–597.

Yamashita, A., Chang, T. C., Yamashita, Y., Zhu, W., Zhong, Z., Chen, C. Y., et al. (2005). Concerted action of poly(A) nucleases and decapping enzyme in mammalian mRNA turnover. *Nature Structural & Molecular Biology*, *12*, 1054–1063.

Yan, Z., Ma, Y. Z., Liu, D. J., Cang, M., Wang, R., & Bao, S. (2010). Targeted suppression of connexin 43 in ovine preimplantation embryos by RNA interference using long double-stranded RNA. *Asian-Australasian Journal of Animal Sciences*, *23*, 456–464.

Yan, L., Yang, M., Guo, H., Yang, L., Wu, J., Li, R., et al. (2013). Single-cell RNA-Seq profiling of human preimplantation embryos and embryonic stem cells. *Nature Structural & Molecular Biology*, *20*, 1131–1139.

Yang, C. X., Du, Z. Q., Wright, E. C., Rothschild, M. F., Prather, R. S., & Ross, J. W. (2012). Small RNA profile of the cumulus-oocyte complex and early embryos in the pig. *Biology of Reproduction*, *87*, 117.

Yang, J., Medvedev, S., Yu, J., Tang, L. C., Agno, J. E., Matzuk, M. M., et al. (2005). Absence of the DNA-/RNA-binding protein MSY2 results in male and female

infertility. *Proceedings of the National Academy of Sciences of the United States of America, 102*, 5755–5760.
Yao, Y., Ma, L., Jia, Q., Deng, W., Liu, Z., Zhang, Y., et al. (2014). Systematic characterization of small RNAome during zebrafish early developmental stages. *BMC Genomics, 15*, 117.
Yoon, J. H., Abdelmohsen, K., & Gorospe, M. (2013). Posttranscriptional gene regulation by long noncoding RNA. *Journal of Molecular Biology, 425*, 3723–3730.
Yu, J., Hecht, N. B., & Schultz, R. M. (2001). Expression of MSY2 in mouse oocytes and preimplantation embryos. *Biology of Reproduction, 65*, 1260–1270.
Yu, J., Hecht, N. B., & Schultz, R. M. (2002). RNA-binding properties and translation repression in vitro by germ cell-specific MSY2 protein. *Biology of Reproduction, 67*, 1093–1098.
Zatsepina, O. V., Bouniol-Baly, C., Amirand, C., & Debey, P. (2000). Functional and molecular reorganization of the nucleolar apparatus in maturing mouse oocytes. *Developmental Biology, 223*, 354–370.
Zeng, F., Baldwin, D. A., & Schultz, R. M. (2004). Transcript profiling during preimplantation mouse development. *Developmental Biology, 272*, 483–496.
Zeng, F., & Schultz, R. M. (2005). RNA transcript profiling during zygotic gene activation in the preimplantation mouse embryo. *Developmental Biology, 283*, 40–57.
Zuccotti, M., Piccinelli, A., Giorgi Rossi, P., Garagna, S., & Redi, C. A. (1995). Chromatin organization during mouse oocyte growth. *Molecular Reproduction and Development, 41*, 479–485.

CHAPTER TEN

The Maternal-to-Zygotic Transition in Flowering Plants: Evidence, Mechanisms, and Plasticity

Célia Baroux, Ueli Grossniklaus[1]
Institute of Plant Biology & Zurich-Basel Plant Science Center, University of Zurich, Zurich, Switzerland
[1]Corresponding author: e-mail address: grossnik@botinst.uzh.ch

Contents

Abstract

The maternal-to-zygotic transition (MZT) defines a developmental phase during which the embryo progressively emancipates itself from a developmental control relying largely on maternal information. The MZT is a functional readout of two processes: the clearance of maternally derived information and the *de novo* expression of the inherited, parental alleles enabled by zygotic genome activation (ZGA). In plants, for many years the debate about whether the MZT exists at all focused on the ZGA alone. However, several recent studies provide evidence for a progressive alleviation of the maternal control over embryogenesis that is correlated with a gradual ZGA, a process that is itself maternally controlled. Yet, several examples of zygotic genes that are expressed and/or functionally required early in embryogenesis demonstrate a certain flexibility in the dynamics and kinetics of the MZT among plant species and also intraspecific hybrids.

Current Topics in Developmental Biology, Volume 113
ISSN 0070-2153
http://dx.doi.org/10.1016/bs.ctdb.2015.06.005

1. SEED DEVELOPMENT: A LARGELY MATERNAL AFFAIR

Sexual reproduction of flowering plants culminates in the formation of seeds within which the embryo develops and matures. Seeds are derived from ovules, which harbor the female gametophytes (embryo sacs), each consisting of two female gametes and several accessory cells (Drews & Koltunow, 2011; Grossniklaus & Schneitz, 1998; Sprunck & Gross-Hardt, 2011). During double fertilization, the egg and central cell (Fig. 1A) are fertilized by one sperm cell each and give rise to the embryo and endosperm, respectively. These two products of fertilization develop coordinately with the maternal ovule integuments, which give rise to the seed coat. Genetic screens have uncovered that seed development is influenced by the maternal parent (reviewed in Sun, 2014) and that maternal effects can be mediated via different, interconnected routes.

In brief, maternal effects can be conveyed by

- **(i)** *the egg cell*: as in animals, the egg cell of flowering plants has a large cytoplasm containing a highly complex transcriptome that is thought to influence zygotic development (Anderson et al., 2013; Johnston et al., 2007; Le et al., 2005; Ning et al., 2006; Wuest et al., 2010);
- **(ii)** *genomic imprinting*: several genes are specifically expressed from the maternal allele only during early embryogenesis of *Arabidopsis*, maize and, likely, rice (Grossniklaus, Vielle-Calzada, Hoeppner, & Gagliano, 1998; Jahnke & Scholten, 2009; Luo et al., 2011; Raissig, Bemer, Baroux, & Grossniklaus, 2013);
- **(iii)** *the endosperm*: the endosperm provides embryo-nurturing, protective functions (Costa et al., 2012; Hehenberger, Kradolfer, & Köhler, 2012; Lopes & Larkins, 1993) as well as developmental cues regulating embryogenesis (Costa et al., 2014);
- **(iv)** *the ovule integuments*: they provide physical, metabolic, and developmental constraints on embryo and endosperm development (reviewed in Li & Berger, 2012; Li & Li, 2015; Radchuk & Borisjuk, 2014); and
- **(v)** *the maternal plant*: nutrient allocation from mother to offspring results in a maternal effect on seed performance and thus the next generation. The effects of maternal provisioning depend on the health and vigor of the mother and, thus, reflect to some extent environmental conditions (He et al., 2014).

The endosperm plays an important role for maternal effects through its maternally dominated genetic constitution: the endosperm inherits two maternal genomes and one paternal genome from the central cell and a

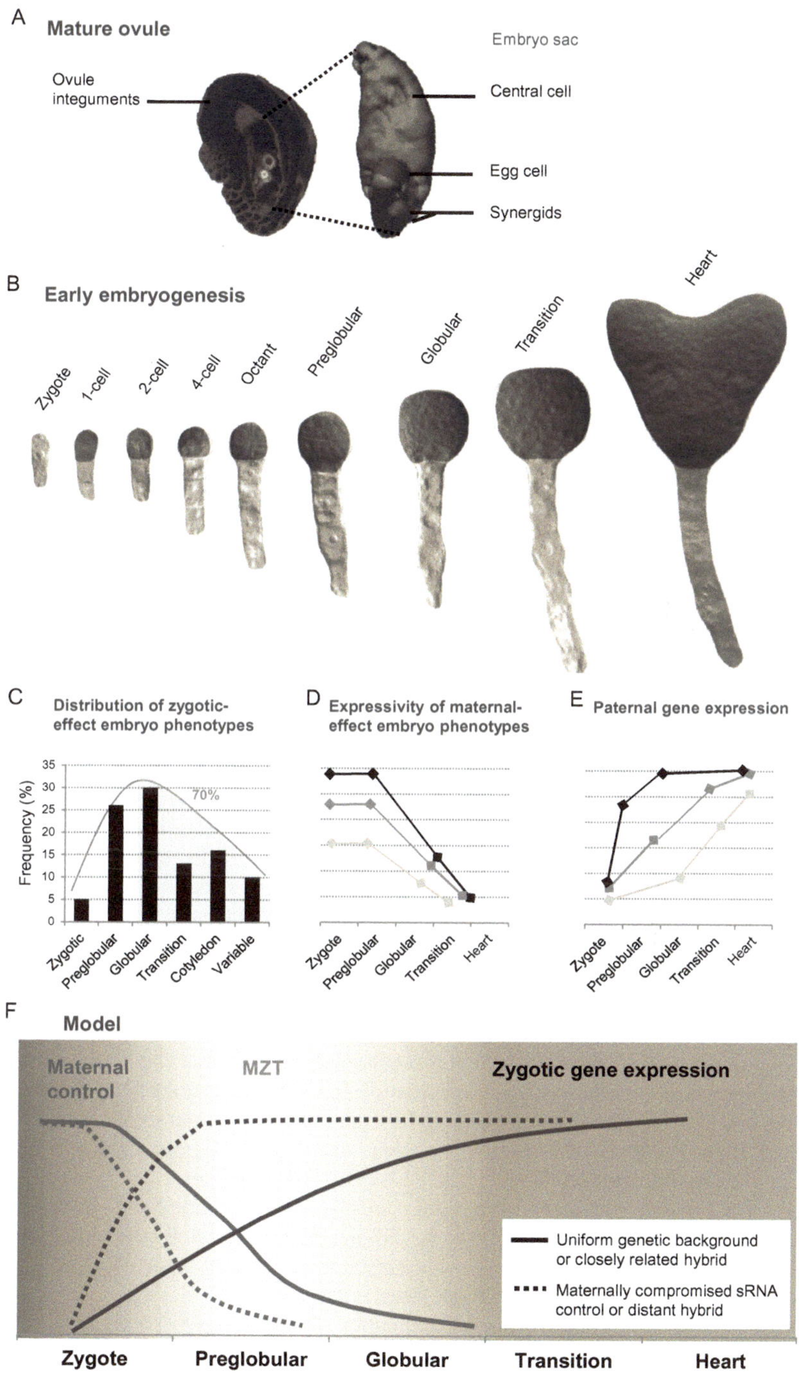

Figure 1 See figure legend on next page.

sperm cell, respectively, such that the vast majority of transcripts in the endosperm of *Arabidopsis*, maize, rice, and castor bean show a 2:1 maternal:paternal stoichiometry (Gehring, Missirian, & Henikoff, 2011; Luo et al., 2011; Waters et al., 2011; Wolff et al., 2011; Xu, Dai, Li, & Liu, 2014). Nevertheless, the endosperm does not seem to critically influence early

Figure 1 Embryogenesis, maternal effects, and progressive ZGA in *Arabidopsis*. (A) Mature *Arabidopsis* ovule (left, 3D reconstruction of a confocal scanning light microscopy image series of an ovule with cell wall staining; cells of the embryo sac are pseudocolored: central cell in blue, egg cell in yellow, synergids in green; ovule integuments are in red) and isolated embryo sac (right, 3D rendering of embryo sac cells extracted from the image series shown on the left, powered by Imaris, Bitplane AG, CH). (B) *Arabidopsis* early embryogenesis. Consecutive embryonic stages are indicated. The embryo proper is highlighted in yellow while the suspensor is displayed in gray. The surrounding endosperm and seed integuments are not shown. Embryo images were extracted using Adobe Photoshop from whole, cleared seed. (C) Distribution of zygotic-effect embryo phenotypes based on frequencies described among 1501 *emb* mutants curated by the Meinke laboratory (McElver et al., 2001). The graph is based on published class frequencies with slight modifications for simplicity: the classes "zygotic/preglobular," "preglobular/globular," and "other" were pooled together in the class "zygotic," "preglobular," and "variable", respectively. (D) Schematic interpretation of the expressivity of several independent maternal-effect embryo phenotypes reported by the Gillmor laboratory (Del Toro-De Leon, Garcia-Aguilar, & Gillmor, 2014). In all cases, the zygotic rescue of maternal-effect embryo defects mediated by a wild-type paternal allele is gradual, leading to a decreasing proportion (*y* axis) of mutant phenotypes as development proceeds. The rescue kinetics reflecting the maternal effect depends on the *emb* locus but also the genetic background. The greater the genetic divergence between the parents, the shorter the maternal effects last. (E) Schematic representation of paternal gene expression kinetics obtained in reporter and profiling experiments (see main text for references), indicating that ZGA is a gradual rather than an all-or-none process, targeting progressively more and more loci as development progresses. (F) Integrative model of the MZT in flowering plants reconciling different reports (see main text for references). The MZT is a long transition initiated in the zygote and largely completed at the globular stage. The MZT is the combination of the maternal clearance of mRNAs, about which we have currently no information in plants, and gradual ZGA. Maternal control (red lines) has been determined in functional studies of maternal-effect embryo mutants or transcriptionally disabled zygotes, and likely relies on maternally stored products inherited through the egg cell. ZGA (blue lines) initiates in the zygote, where several loci are activated, becoming more robust as development proceeds and more and more loci are transcribed. However, based on cytological staining of active RNA Pol II, the zygote remains in a relatively quiescent, transcriptional state. Maternal effects have a briefer impact while ZGA is more rapid when embryos inherit genetically divergent parental genomes or are maternally impaired in ZGA repressive mechanisms, such as mutants of the RdDM pathway (dashed lines). *Panel (B): Pictures courtesy of Quy A. Ngo.* (See the color plate.)

embryogenesis because embryos can develop to the late globular stage in the absence of endosperm (Ngo et al., 2012; Pillot et al., 2010). In addition, endosperm-derived small RNAs have been proposed to safeguard the genome's integrity in the embryo; however, this model awaits direct experimental evidence (reviewed in Lafon-Placette & Köhler, 2014).

In this intricate maternal affair, the question arises whether, when, and how the embryo emancipates itself from maternal control. In a traditional approach, the MZT is looked at from the viewpoint of the embryo.

2. EMANCIPATION OF THE PLANT EMBRYO

As in animals, plant embryogenesis involves proliferation, morphogenesis, and organogenesis. The mature embryo is a miniature organism that anticipates the basic body plan of the adult plant organized along radial, lateral, and apical–basal axes but essentially consists of an axis with a shoot meristem at one end and a root meristem at the other. Shoot and root organogenesis takes place postembryonically throughout the plant's growth phase through activity of the apically located meristems comprising the stem cell populations specified during embryogenesis. For a detailed overview of the genetic and molecular control of embryo patterning and development in the model plant *Arabidopsis,* the reader is referred to recent reviews (e.g., Jenik, Gillmor, & Lukowitz, 2007; Wendrich & Weijers, 2013); a brief description of early embryogenesis is given below and illustrated in Fig. 1. After fertilization, the zygote rapidly elongates and divides asymmetrically to produce an apical and a basal cell, which give rise to the embryo proper and the suspensor, respectively. Oriented (anticlinal) divisions of the basal cell produce a file of 8–12 cells forming the suspensor. The apical cell divides symmetrically three times to produce an octant-stage embryo. From there on, distinctly oriented cell divisions shape the radial pattern: at the globular stage protoderm, ground tissue, and provasculature elements are formed from the periphery to the center of the embryo. Morphogenesis, with oriented divisions and anisotropic cell elongation, forms first a heart- and then a torpedo-shaped embryo that comprises the basic tissue patterns and meristematic stem cells that generate the future root and shoot organs.

Because plant embryos can also form from somatic cells or microspores (Mordhorst, Toonen, de Vries, & Meinke, 1997), the egg cytoplasm was traditionally not considered to play an essential role during early embryogenesis. This view was supported by the fact that some zygotic recessive embryonic lethal (*emb*) mutations affected the first, asymmetric division of the zygote when homozygous (e.g., Mayer, Buttner, & Jürgens, 1993), indicating activity of both parental alleles already in the zygote. This notion was challenged when several genes were found to be active during early embryogenesis only when maternally inherited (Baroux, Blanvillain, & Gallois, 2001; Vielle-Calzada, Baskar, & Grossniklaus, 2000). Since then, the analysis of a large number of *emb* mutants has confirmed that embryogenesis does not entirely rely on zygotic resources immediately after fertilization. In *Arabidopsis*, among 1501 *emb* mutant lines, close to 70% showed developmental arrest or patterning defects at or after the globular stage (McElver et al., 2001) (Fig. 1C). This suggests that early embryogenesis may, at least partially, rely on information inherited from either parent. However, the existence of other *emb* mutations leading to embryonic phenotypes at the preglobular or zygotic stage (Mayer et al., 1993; McElver et al., 2001; Ronceret et al., 2005, 2008) clearly indicates that (maternally or paternally) inherited information does not cover all cellular functions necessary for normal development. A similar situation is found in maize and rice where a large proportion of zygotic *emb* mutants arrest at postglobular stages, although these studies involved fewer mutants than the *Arabidopsis* ones (Clark & Sheridan, 1991; Hong, Aoki, Kitano, Satoh, & Nagato, 1995).

An elegant approach to elucidate the extent of maternal control over early embryogenesis in *Arabidopsis* is to look for transient developmental phenotypes generated by purely maternal inheritance of *emb* mutations, i.e., in crosses of a mutant mother with a wild-type father. Originally, this approach proved successful to uncover previously unrecognized maternal effects for a small number of loci (Baroux, Autran, Gillmor, Grimanelli, & Grossniklaus, 2008; Vielle-Calzada et al., 2000) and, more recently, for a wide range of genes with diverse cellular functions (transcription, replication, translation, cytoskeletal function, organelle biology, signaling components, basic cellular metabolism, etc.) (Del Toro-De Leon et al., 2014). Interestingly, maternal effects in hemizygous embryos are progressively alleviated as development procedes, a phenomenon largely completed by the globular stage (Fig. 1D). The timing of this paternal rescue varies among embryo siblings (from the 2–4 cell to the early heart stage), mutant loci, and genetic backgrounds (Del Toro-De Leon et al., 2014). The observation that the corresponding paternal, wild-type allele gradually increases expression before the phenotypic rescue indicates that a zygotically provided function is responsible for this. Consistent with such a period of maternal control over early embryogenesis is the isolation of many female gametophytic mutants that have normal embryo sacs and only show developmental defects during embryogenesis. About half of the female gametophytic mutants isolated in genetic screens show such maternal effects on seed development (Moore, 2002; Ngo et al., 2012; Pagnussat et al., 2005), indicating that many factors usually provided through the female gametes are required for normal embryo and/or endosperm development.

Collectively, these studies provide unequivocal evidence for an extensive maternal control during early embryogenesis in *Arabidopsis*. Specifically, the analysis of zygotic embryo mutants provides both a starting point and an end point for the MZT. The observation that some loci are functionally required for zygote elongation, the first division of the zygote, cell fate decisions in the apical and basal cells, or early mitoses of the embryo proper (Mayer et al., 1993; McElver et al., 2001; Ronceret et al., 2005, 2008) indicates that the MZT initiates already in the zygote (reviewed in Sun, 2014). However, globally, and for a large number of processes, paternal rescue assays have shown that the *Arabidopsis* embryo progressively emancipates itself to rely almost entirely on zygotically expressed cellular functions only by the globular stage (Del Toro-De Leon et al., 2014), thereby defining the end of the MZT. This model is consistent with the observation that *Arabidopsis* zygotes that are genetically deprived of transcriptional activity can develop until about the 16-cell stage. This developmental progression presumably rely on maternally inherited information (Pillot et al., 2010), although paternally contributed transcripts also provide developmental instructions (Bayer et al., 2009).

Thus, at least in *Arabidopsis*, the MZT appears to be a long process initiated in the zygote and ending at the globular stage, which marks the transition to morphogenetic events. A similar situation may exist in maize, considering the major restructuring of the transcriptome at the globular stage (Grimanelli, Perotti, Ramirez, & Leblanc, 2005), while the situation may differ in other plant species, such as tobacco (Zhao et al., 2011) or hybrid embryos from genetically divergent parents (Autran et al., 2011; Baroux, Autran, Raissig, Grimanelli, & Grossniklaus, 2013; Del Toro-De Leon et al., 2014; Nodine & Bartel, 2012).

3. PROGRESSIVE ZYGOTIC GENE ACTIVATION

Several studies have reported on the timing of zygotic gene expression, either on a gene-by-gene basis or using genome-wide transcriptome profiling approaches. Measuring the onset of zygotic gene activation (ZGA) requires the detection of transcripts produced *de novo* from the zygotic genome. However, maternally stored transcripts have confounding effects in transcriptome profiling or RT-PCR analyses that cannot easily be resolved. A possible solution is to perform differential profiling to identify transcripts that are present in the early embryo but undetectable in the egg cell as has been done in maize, wheat, rice, tobacco, and *Arabidopsis* (Abiko, Maeda, Tamura, Hara-Nishimura, & Okamoto, 2013; Autran et al., 2011; Grimanelli et al., 2005; Meyer & Scholten, 2007; Ning et al., 2006; Nodine & Bartel, 2012; Sprunck, Baumann, Edwards, Langridge, & Dresselhaus, 2005). Another solution is to approximate ZGA with the detection of paternal transcripts only (Autran et al., 2011; Del Toro-De Leon et al., 2014; Vielle-Calzada et al., 2000), but this relies on the assumption that the sperm cytoplasm does not contribute a significant amount of transcripts to the developing embryo.

ZGA can be characterized by the time point at which the first zygotic gene expression is detected or by the developmental stage at which many zygotic genes are strongly expressed. It was earlier proposed that ZGA is a gradual process that may be characterized by minor and major activation waves (Baroux et al., 2001, 2008). This scenario is consistent with the observation of a gradual rescue of maternal-effect embryo phenotypes by paternal alleles (Baroux et al., 2008; Del Toro-De Leon et al., 2014). Importantly, this model predicts that the onset of ZGA is not uniform throughout the genome and that the number of loci becoming transcriptionally permissive increases as development proceeds, thereby conciliating seemingly conflicting reports. Indeed, analyses of transgenic or endogenous gene expression has identified certain loci with detectable zygotic transcription soon after fertilization in *Arabidopsis*, tobacco, and wheat (Abdalla, Yoshizawa, & Hochi, 2009; Autran et al., 2011; Baroux et al., 2001; Dresselhaus, Hagel, Lorz, & Kranz, 1996; Ning et al., 2006; Nodine & Bartel, 2012; Sprunck et al., 2005; Weijers, Geldner, Offringa, & Jürgens, 2001) while, for other loci, transcripts from paternal alleles are only detected later, around the globular stage in maize and *Arabidopsis* (Autran et al., 2011; Baroux et al., 2001; Grimanelli et al., 2005; Vielle-Calzada et al., 2000).

Time-resolved expression analyses of paternally inherited reporter transgenes indicate that zygotic expression is not fully synchronized among sibling embryos and that the fraction of embryos showing detectable reporter gene expression increases as embryogenesis progresses (Autran et al., 2011; Baroux et al., 2001; Grimanelli et al., 2005; Vielle-Calzada et al., 2000). This indicates a gradual ZGA that gains robustness toward the globular stage. This model was corroborated by allele-specific transcriptome profiling in *Arabidopsis* hybrid embryos originating from the common laboratory strains Landsberg *erecta* (L*er*) and Columbia-0 (Col) (Autran et al., 2011). With the use of early next-generation sequencing methods, the relative parental contributions of 3000–4000 genes (depending on strain and stage) were assessed, providing several important insights (Autran et al., 2011):

(i) at the 2–4 cell embryo stage, the transcriptome is maternally dominated with 88% of the reads being maternal—a snapshot that, however, does not distinguish between maternally inherited and *de novo* expressed transcripts;

(ii) despite this maternal dominance, 66% of genes show biallelic expression, indicating an early onset of ZGA;

(iii) more than half of the biparentally expressed genes show a marked maternal dominance (>75% maternal reads) that decreases to one-third at the globular stage; and

(iv) over 500 of the 3000–4000 genes (about 15%) showed *de novo* paternal expression (the only class that can be unambiguously attributed to ZGA) at the globular stage.

In summary, quantitative reporter gene analyses as well as transcriptome profiling studies in *Arabidopsis* indicate that ZGA occurs early and is a gradual rather than an all-or-none process (Fig. 1E). Similarly, in maize, a major restructuring of the transcriptome occurs in kernels containing embryos at the late globular stage, suggesting that ZGA is near-complete at this stage (Grimanelli et al., 2005). Nevertheless, also in maize, several genes have undergone zygotic activation at earlier stages (Meyer & Scholten, 2007). Together, these data indicate that in the majority of plant species analyzed so far (*Arabidopsis*, tobacco, wheat, maize, rice) ZGA starts soon after fertilization, in the (elongated) zygote, but affects a progressively higher number of loci as development proceeds.

4. MECHANISMS OF ZYGOTIC GENOME ACTIVATION

As in animals, the chromatin of plant sperm cells is highly condensed (Baroux, Pien, & Grossniklaus, 2007; Southworth, 1996). Even though flowering plants do not possess the protamines typical of animals, arginine-rich chromatin-associated proteins and sperm-specific histone variants, such as those isolated in *Lilium longiflorum* (Ueda et al., 2000; Ueda & Tanaka, 1995), may contribute to tight packaging. After karyogamy, structural changes are necessary to enable access of the transcription and replication machinery to the paternal and, possibly, also to the maternal chromatin. In animals, this process is largely controlled by maternal factors (Jenkins & Carrell, 2012; Loppin, Berger, & Couble, 2001). In the mouse, paternal and maternal chromatin remains topologically distinct for a few cell divisions (Mayer, Smith, Fundele, & Haaf, 2000), while the onset of ZGA occurs at the 2-cell stage (Nothias, Miranda, & DePamphilis, 1996; Tadros & Lipshitz, 2009). This does not seem to be the case in plants, where zygotic chromatin undergoes restructuring soon after fertilization, including the active replacement of inherited histone H3 variants as shown *in planta* for *Arabidopsis* (Ingouff, Hamamura, Gourgues, Higashiyama, & Berger, 2007), and the rapid decondensation of the paternal chromatin a few hours after karyogamy in *in vitro* fertilized zygotes of rice and maize (Ohnishi, Hoshino, & Okamoto, 2014; Scholten, Lörz, & Kranz, 2002). While there is evidence that active, global chromatin remodeling may rapidly restructure the zygotic genome, the exact molecular events, their regulation, and their effect on the spreading of ZGA in the genome remain to be elucidated.

Even though the zygote is competent to sustain *de novo* gene expression, global levels of active RNA polymerase II (Pol II) remain barely detectable in immunostaining experiments (Autran et al., 2011). Together with the low proportion of active loci in the zygote compared to later stages of embryogenesis, this suggests that active transcription may be restricted to a small number of loci. Yet, higher levels of active Pol II are observed in zygotes lacking Pol IV activity (Autran et al., 2011). Pol IV is involved in the biogenesis of 24-nt small-interfering RNAs (siRNAs) that mediate RNA-dependent DNA methylation (RdDM), a posttranscriptional gene silencing pathway that leads to DNA methylation and the deposition of repressive H3K9me2 marks at target loci (reviewed in Haag & Pikaard, 2011; Law & Jacobsen, 2010; Matzke & Mosher, 2014). Impairment of siRNA-mediated silencing in the zygote has consequences for the kinetics of ZGA consistent with the cytological observations: embryos inheriting maternal mutations affecting the RdDM pathway show precocious expression of paternally inherited reporter transgenes and endogenous genes (Autran et al., 2011). In particular, genome-wide transcriptome profiling in 2–4 cell embryos lacking maternal activity of the histone methyltransferase KRYPTONITE (Jackson, Lindroth, Cao, & Jacobsen, 2002), a downstream effector of the RdDM pathway, revealed parental contributions—globally and at the single gene level—similar to those in wild-type embryos at the globular stage, while only marginally affecting the overall composition of the transcriptome (Autran et al., 2011). A scenario implicating maternally derived siRNAs in ZGA is supported by the presence of 24 nt siRNAs in ovules, which contain unfertilized egg cells. While 24 nt siRNAs usually target transposable elements and repeats (reviewed in Haag & Pikaard, 2011; Law & Jacobsen, 2010; Matzke & Mosher, 2014), this distinct population maps to coding sequences and promoter regions (Autran et al., 2011).

In contrast to the repressing activity of the RdDM pathway, the CHROMATIN ASSEMBLY FACTOR1 (CAF1, Ramirez-Parra & Gutierrez, 2007) is required for efficient ZGA: embryos maternally inheriting mutations affecting this histone chaperone complex show a delay in the activation of the paternal alleles at several loci (Autran et al., 2011). However, the genome-wide consequences of such a maternal deficiency remain to be established.

Finally, microRNAs (miRNAs) might contribute to the regulation of ZGA: embryos deficient for DICER-LIKE1 (DCL1; Schauer, Jacobsen, Meinke, & Ray, 2002), required for miRNA biogenesis, have severe patterning defects (Golden et al., 2002; Seefried, Willmann, Clausen, & Jenik, 2014) and express transcripts that normally accumulate only during embryo maturation at earlier stages (Nodine & Bartel, 2010). This is because miR156 downregulates transcripts of the SPL10 and SPL11 transcription factors, which activate these maturation genes, until the late globular stage (Nodine & Bartel, 2010). However, while miRNAs indirectly regulate the abundance of some embryonic mRNAs, they have not been shown to be directly involved in transcriptional regulation, the classical focus of ZGA.

Based on these data, the following plausible model for the regulation of ZGA in *Arabidopsis* can be proposed:

(i) rapid, structural chromatin changes enable a minor wave of zygotic expression at selected loci;

(ii) wide-spread—but not global—transcriptional repression is mediated by maternally derived siRNAs (hence the apparent transcriptional quiescence in immunocytochemical experiments detecting active Pol II);

(iii) progressive release of silencing enabling a major wave of ZGA as embryogenesis proceeds, either through passive or active mechanisms, conferring a locus-specific dynamic to ZGA that may be somewhat stochastic, explaining the nonuniform ZGA pattern among siblings;

(iv) concomitant with the release of siRNA-based repression, CAF1-mediated chromatin assembly or remodeling may establish a robust transcriptional competence, thereby completing ZGA.

This working model motivates future investigations of silencing processes, as well as chromatin assembly, modification, and remodeling in the developing embryo at higher temporal and spatial resolution.

5. THE DYNAMICS OF THE MATERNAL-TO-ZYGOTIC TRANSITION VARY DEPENDING ON PLANT SPECIES AND GENETIC BACKGROUND

Recent studies have made it clear that it is difficult to define the MZT based on individual loci, with some loci being expressed and/or functionally required shortly after fertilization but others only at the globular stage. Thus, this developmental transition has to be described by the collective behavior of the zygotic genome. A model accommodating these observations thus proposes a long-lasting MZT that starts at the zygote stage with a few loci being active and is complete as the embryo reaches the globular stage when the majority of loci are active from both maternal and paternal alleles. This model offers many possibilities for a flexible tuning of the number, timing, and dynamics of zygotically active genes. The model would also explain the discrepancy in MZT measurements between plant species and within the same species with embryos originating from genetically divergent parents. In tobacco, for instance, the inhibition of zygotic transcription leads to a failure of the embryo to develop beyond the elongated 1-cell stage (Zhao et al., 2011). Consistent with this finding, a vast number of transcripts are detected *de novo* in the tobacco zygote (Ning et al., 2006; Zhao et al., 2011). This is in contrast to *Arabidopsis,* where transcriptionally compromised zygotes can develop until the preglobular to globular stage (Pillot et al., 2010). These discrepancies likely reflect differences in ZGA and the MZT between these two species. However, they may also, at least in part, be attributable to differences in the experimental approaches used: toxicological inhibition studies (Zhao et al., 2011) often have additional, undesired effects (Xin, Zhao, & Sun, 2012; Zhao et al., 2011) and transgenic downregulation of Pol II (Pillot et al., 2010) may not be entirely complete. In addition to interspecific variation, several studies have illustrated intraspecific variation in the dynamics of the MZT. In *Arabidopsis*, allele-specific transcriptome profiling in hybrid embryos showed discrepancies depending on the strains (accessions) used as parents. Hybrid embryos derived from the L*er* and Col accessions showed a maternally dominated transcriptome at the 2–4 cell stage (Autran et al., 2011). In contrast, hybrid embryos derived from the *Cape Verde Island* (Cvi) and Col-0 accessions showed equal parental contributions (Nodine & Bartel, 2012). While there were also several technical differences in transcriptome profiling (embryo isolation, RNA population sampling), it has been argued that the main factor for this discrepancy is likely the different parental accessions used (Baroux et al., 2013). Indeed, the maternal genotype strongly influences both the kinetics of ZGA, as measured by paternal reporter gene expression (Autran et al., 2011), and the MZT, as measured by functional rescue of maternal-effect embryonic phenotypes (Del Toro-De Leon et al., 2014). The effect of genetic background is strongest for the C24 and Cvi accessions. While the effect of Cvi may be related to an unusual epigenetic feature (discussed in Baroux et al., 2013), more generally it appears to reflect the genetic relatedness between the parental genomes. Indeed, in a SNP-marker analysis Cvi was found to be the most genetically divergent from Col-0 among 12 accessions (including *Ler*) studied (Schmid et al., 2003). A larger, genome-wide survey of polymorphisms further confirmed that Cvi greatly differs from Col-0 and most other accessions studied (Nordborg et al., 2005).

In maize, different ZGA readouts were obtained by profiling hybrid embryos from different parental inbred lines. When generating hybrids using the CML216 and CML72 parental inbred lines, 16 out of 16 genes tested showed delayed paternal expression where only maternal transcripts were detected in kernels until 3 days after pollination (Grimanelli et al., 2005). In contrast, in hybrids derived from the UH005 and UH301 inbred lines, 24 out of the 25 genes tested showed detectable levels of paternal transcripts in isolated zygotes (Meyer & Scholten, 2007). However, also in the latter study, 10 genes showed a bias toward maternally derived transcripts in the zygote (reviewed in Baroux et al., 2008).

The question remains, however, how the degree of genetic relatedness between parental genomes is detected and how this controls the timing of ZGA in the hybrid zygote. Maternal siRNAs targeting paternal alleles provide plausible sensors for genetic divergence as their silencing effect depends on sequence homology between the siRNA derived from the maternal genome and their target sites, for instance on the paternally inherited genome.

6. BIOLOGICAL FUNCTIONS OF THE MATERNAL-TO-ZYGOTIC TRANSITION IN PLANTS

There have been few experimental investigations that allow conclusions about the functional role of the MZT in plants. However, it is likely that similar selective pressures have led to the evolution of maternal control over the early phase of embryogenesis in both plants and animals. It is possible that the evolution of anisogamy, leading to the formation of large egg cells and small sperm, resulted in the deposition of the factors required for early embryogenesis to allow rapid development after fertilization. The MZT can thus be viewed as a consequence of anisogamy, making it necessary that the zygotic genome gets activated and maternally stored products get cleared. The evolution of the timing and dynamics of ZGA may have been modulated by parent–offspring conflicts, which will depend on life-history traits and mating systems, explaining the large diversity observed in both animals and plants.

In *Drosophila*, the clearance of maternally derived mRNA can have permissive functions, e.g., the replacement of ubiquitously distributed, maternal mRNAs by zygotic mRNAs that are expressed in a spatially or temporally restricted pattern (De Renzis, Elemento, Tavazoie, & Wieschaus, 2007). A similar function can be envisioned during plant embryogenesis, although it awaits experimental support because it has so far not been possible to distinguish maternally and zygotically produced *de novo* transcripts, except for those derived from the paternal genome only. However, maternal mutants in the RdDM pathway, such as *kryptonite*, affect ZGA and have embryonic patterning defects (Autran et al., 2011), indicating that the spatially restricted expression of zygotic mRNAs may play a role.

On the other hand, the removal of maternal mRNAs can also have instructive functions. The increase or decrease of maternal mRNAs of the cell cycle regulator encoded by *string*, the *Drosophila* homolog of *cdc25*, results in a longer or shorter period of nuclear division cycles before cellularization occurs, respectively (Edgar & Datar, 1996). To our knowledge, an instructive function for the clearing of maternal mRNAs in the plant embryo has not yet been demonstrated but it is likely that this is also the case in plants. Clearly, the activation of specific loci at distinct stages during embryogenesis is expected to have instructive functions. However, precisely timed misexpression experiments have not been performed to test effects on the plant MZT.

Apart from shared functions of the MZT in animals and plants, there may also be a role that is unique to plants and intimately related to the plant life cycle, which alternates between a haploid, gametophytic and a diploid, sporophytic generation. Although the extent of the haploid phase has been greatly reduced during land plant evolution, it has not been eliminated in flowering plants. It has been speculated that the retention of a haploid phase is linked to the fact that somatic mutations in plants have the potential to be inherited, as the germline is set aside very late during flower development

(Grossniklaus, 2011; Walbot & Evans, 2003). The haploid phase would then serve as a filter against deleterious mutations that are required for basic cellular functions. It has been speculated that a delayed ZGA would prolong this filter and make it more effective, thus preventing the spread of deleterious mutations in the population (Vielle-Calzada et al., 2000).

7. UNRESOLVED QUESTIONS AND OUTLOOK

In summary, even though the occurrence of an MZT in flowering plants has remained a topic of controversy for more than a decade, we have shown here that all published observations can be reconciled into a reasonable scenario.

First, the MZT cannot be defined based on a single or a few genes, as some functions are zygotically required for progression beyond the 1-cell embryo stage, while many others are required only after several cell divisions, as it is also the case in animals. Instead, the MZT in flowering plants is a gradual transition that initiates in the zygote with some genes becoming zygotically expressed soon after fertilization, with the extent of this early activation likely differing between species and depending on the genetic distance between parents of hybrid zygotes. The MZT becomes more robust as ZGA encompasses a larger fraction of the genome as embryogenesis progresses. Functional rescue experiments suggest that the MZT is largely complete at the onset of morphogenesis although ZGA might continue beyond this point.

Second, the MZT in flowering plants shows a remarkable flexibility. On the one hand, this is true between species—a situation similar to that in animals where the onset of the MZT varies from one to hundreds of cell divisions after fertilization (Andéol, 1994). On the other hand this flexibility is also seen within species, depending on the genetic background, an effect that may be mediated by maternal siRNAs that directly detect the relatedness between the parental genomes and modulate the kinetics of ZGA.

This model has the virtue of being integrative by reconciling various studies but it also provokes questions for future investigations, particularly in species like *Arabidopsis* or maize, which are amenable to molecular genetic approaches. In animals, the silencing of the zygotic genome during early development relies on several mechanisms, including chromatin-mediated repression and transcriptional quiescence that result from both deficiencies in the transcription machinery and active transcriptional repression (Schier, 2007; Tadros & Lipshitz, 2009). The regulation of ZGA in flowering plants has some mechanistic similarities with that of animals as chromatin-remodeling events in the zygote (Ingouff et al., 2007; Ohnishi et al., 2014) and (partial) transcriptional repression by siRNAs that modulate zygotic gene expression (Autran et al., 2011) play a role. Yet, a much higher temporal resolution of epigenome profiling including DNA methylation, chromatin modifications, and siRNAs in the developing embryo is needed to get a better understanding of the dynamics and mechanisms underlying the MZT. While the plant embryo is amenable to transcriptome profiling (Autran et al., 2011; Nodine & Bartel, 2012) and cytogenetic chromatin analyses (Gernand et al., 2005; Raissig, Gagliardini, Jaenisch, Grossniklaus, & Baroux, 2013), epigenome profiling remains challenging due to the inaccessibility and small size of the embryo. However, progress in affinity purification of cell-specific, tagged nuclei (Deal & Henikoff, 2011; Palovaara, Saiga, & Weijers, 2013) or flow sorting of GFP-tagged embryonic nuclei (Slane et al., 2014) may overcome this technical challenge in the near future.

In addition, nothing is currently known about the clearance of stored, maternal mRNA in plants. It is unclear whether maternally inherited RNAs are passively diluted, actively degraded, or both. First, profiling of nascent nuclear RNAs in developing embryos, from the zygote stage onward, is necessary to quantify the relative proportions of zygotically produced *versus* maternally inherited mRNAs, which cannot be distinguished by traditional transcriptome profiling approaches. The recent development of techniques that allow efficient isolation of embryonic nuclei should also help to answer this question. Second, genetic screens will be instrumental for the identification of regulators of maternal mRNA clearance. Such screens could be designed, for instance, to find maternal modifiers that influence the developmental life-time of an egg-cell-produced visual marker. In animals, maternal mRNA degradation is induced by the binding of proteins and miRNAs to the 3′-untranslated region of target RNAs (Schier, 2007; Tadros & Lipshitz, 2009). In plants, the molecular basis of mRNA decay is beginning to be unveiled (Seefried et al., 2014) and known regulators of the mRNA decapping complex could serve as candidates to screen for modifiers of maternal-marker accumulation in the developing embryo. Furthermore, whether miRNAs also have a function in maternal mRNA clearance beside their role in embryonic gene repression (Nodine & Bartel, 2010) remains to be determined. miRNAs have the ability to trigger both translational inhibition and mRNA decay in animals and flowering plants (Hu & Coller, 2012) where investigations into the breadth of this functional relationship are in their infancy (Jiao & Meyerowitz, 2010).

ACKNOWLEDGMENTS

Our work related to the MZT and ZGA is supported by the University of Zürich, the Swiss National Science Foundation, and the European Research Council. We thank Nuno Pires for discussions on the genetic relatedness among *Arabidopsis* strains and Quy A. Ngo for sharing the images of cleared *Arabidopsis* seeds shown in Fig. 1.

REFERENCES

Abdalla, H., Yoshizawa, Y., & Hochi, S. (2009). Active demethylation of paternal genome in mammalian zygotes. *The Journal of Reproduction and Development*, *55*(4), 356–360.

Abiko, M., Maeda, H., Tamura, K., Hara-Nishimura, I., & Okamoto, T. (2013). Gene expression profiles in rice gametes and zygotes: Identification of gamete-enriched genes and up- or down-regulated genes in zygotes after fertilization. *Journal of Experimental Botany*, *64*(7), 1927–1940.

Andéol, Y. (1994). Early transcription in different animal species: Implication for transition from maternal to zygotic control in development. *Roux's Archives of Developmental Biology*, *204*, 3–10.

Anderson, S. N., Johnson, C. S., Jones, D. S., Conrad, L. J., Gou, X., Russell, S. D., et al. (2013). Transcriptomes of isolated *Oryza sativa* gametes characterized by deep sequencing: Evidence for distinct sex-dependent chromatin and epigenetic states before fertilization. *The Plant Journal*, *76*(5), 729–741.

Autran, D., Baroux, C., Raissig, M. T., Lenormand, T., Wittig, M., Grob, S., et al. (2011). Maternal epigenetic pathways control parental contributions to *Arabidopsis* early embryogenesis. *Cell*, *145*(5), 707–719.

Baroux, C., Autran, D., Gillmor, C. S., Grimanelli, D., & Grossniklaus, U. (2008). The maternal to zygotic transition in animals and plants. *Cold Spring Harbor Symposia on Quantitative Biology*, *73*, 89–100.

Baroux, C., Autran, D., Raissig, M. T., Grimanelli, D., & Grossniklaus, U. (2013). Parental contributions to the transcriptome of early plant embryos. *Current Opinion in Genetics & Development*, *23*(1), 72–74.

Baroux, C., Blanvillain, R., & Gallois, P. (2001). Paternally inherited transgenes are down-regulated but retain low activity during early embryogenesis in *Arabidopsis*. *FEBS Letters*, *509*(1), 11–16.

Baroux, C., Pien, S., & Grossniklaus, U. (2007). Chromatin modification and remodeling during early seed development. *Current Opinion in Genetics & Development*, *17*(6), 473–479.

Bayer, M., Nawy, T., Giglione, C., Galli, M., Meinnel, T., & Lukowitz, W. (2009). Paternal control of embryonic patterning in *Arabidopsis thaliana*. *Science*, *323*(5920), 1485–1488.

Clark, J. K., & Sheridan, W. F. (1991). Isolation and characterization of 51 embryo-specific mutations of maize. *Plant Cell*, *3*(9), 935–951.

Costa, L. M., Marshall, E., Tesfaye, M., Silverstein, K. A., Mori, M., Umetsu, Y., et al. (2014). Central cell-derived peptides regulate early embryo patterning in flowering plants. *Science*, *344*(6180), 168–172.

Costa, L. M., Yuan, J., Rouster, J., Paul, W., Dickinson, H., & Gutierrez-Marcos, J. F. (2012). Maternal control of nutrient allocation in plant seeds by genomic imprinting. *Current Biology*, *22*(2), 160–165.

De Renzis, S., Elemento, O., Tavazoie, S., & Wieschaus, E. F. (2007). Unmasking activation of the zygotic genome using chromosomal deletions in the *Drosophila* embryo. *PLoS Biology*, *5*(5), e117.

Deal, R. B., & Henikoff, S. (2011). The INTACT method for cell type-specific gene expression and chromatin profiling in *Arabidopsis thaliana*. *Nature Protocols*, *6*(1), 56–68.

Del Toro-De Leon, G., Garcia-Aguilar, M., & Gillmor, C. S. (2014). Non-equivalent contributions of maternal and paternal genomes to early plant embryogenesis. *Nature*, *514*(7524), 624–627.

Dresselhaus, T., Hagel, C., Lörz, H., & Kranz, E. (1996). Isolation of a full-length cDNA encoding calreticulin from a PCR library of *in vitro* zygotes of maize. *Plant Molecular Biology*, *31*(1), 23–34.

Drews, G. N., & Koltunow, A. M. G. (2011). The female gametophyte. *The Arabidopsis Book*, *9*, e0155. http://dx.doi.org/10.1199/tab.0155.

Edgar, B. A., & Datar, S. A. (1996). Zygotic degradation of two maternal Cdc25 mRNAs terminates *Drosophila*'s early cell cycle program. *Genes & Development*, *10*(15), 1966–1977.

Gehring, M., Missirian, V., & Henikoff, S. (2011). Genomic analysis of parent-of-origin allelic expression in *Arabidopsis thaliana* seeds. *PLoS One*, *6*(8), e23687.

Gernand, D., Rutten, T., Varshney, A., Rubtsova, M., Prodanovic, S., Bruss, C., et al. (2005). Uniparental chromosome elimination at mitosis and interphase in wheat and pearl millet crosses involves micronucleus formation, progressive heterochromatinization, and DNA fragmentation. *Plant Cell*, *17*(9), 2431–2438.

Golden, T. A., Schauer, S. E., Lang, J. D., Pien, S., Mushegian, A. R., Grossniklaus, U., et al. (2002). *SHORT INTEGUMENTS1/SUSPENSOR1/CARPEL FACTORY*, a dicer homolog, is a maternal effect gene required for embryo development in *Arabidopsis*. *Plant Physiology*, *130*(2), 808–822.

Grimanelli, D., Perotti, E., Ramirez, J., & Leblanc, O. (2005). Timing of the maternal-to-zygotic transition during early seed development in maize. *Plant Cell*, *17*(4), 1061–1072.

Grossniklaus, U. (2011). Plant germline development: A tale of cross-talk, signaling, and cellular interactions. *Sexual Plant Reproduction*, *24*(2), 91–95.

Grossniklaus, U., & Schneitz, K. (1998). The molecular and genetic basis of ovule and megagametophyte development. *Seminars in Cell & Developmental Biology*, *9*(2), 227–238.

Grossniklaus, U., Vielle-Calzada, J. P., Hoeppner, M. A., & Gagliano, W. B. (1998). Maternal control of embryogenesis by *MEDEA*, a *Polycomb* group gene in *Arabidopsis*. *Science*, *280*(5362), 446–450.

Haag, J. R., & Pikaard, C. S. (2011). Multisubunit RNA polymerases IV and V: Purveyors of non-coding RNA for plant gene silencing. *Nature Reviews Molecular Cell Biology*, *12*(8), 483–492.

He, H., de Souza Vidigal, D., Snoek, L. B., Schnabel, S., Nijveen, H., Hilhorst, H., et al. (2014). Interaction between parental environment and genotype affects plant and seed performance in *Arabidopsis*. *Journal of Experimental Botany*, *65*(22), 6603–6615.

Hehenberger, E., Kradolfer, D., & Köhler, C. (2012). Endosperm cellularization defines an important developmental transition for embryo development. *Development*, *139*(11), 2031–2039.

Hong, S.-K., Aoki, T., Kitano, H., Satoh, H., & Nagato, Y. (1995). Phenotypic diversity of 188 rice embryo mutants. *Developmental Genetics*, *16*(4), 298–310.

Hu, W., & Coller, J. (2012). What comes first: Translational repression or mRNA degradation? The deepening mystery of microRNA function. *Cell Research*, *22*(9), 1322–1324.

Ingouff, M., Hamamura, Y., Gourgues, M., Higashiyama, T., & Berger, F. (2007). Distinct dynamics of HISTONE3 variants between the two fertilization products in plants. *Current Biology*, *17*(12), 1032–1037.

Jackson, J. P., Lindroth, A. M., Cao, X., & Jacobsen, S. E. (2002). Control of CpNpG DNA methylation by the KRYPTONITE histone H3 methyltransferase. *Nature*, *416*(6880), 556–560.

Jahnke, S., & Scholten, S. (2009). Epigenetic resetting of a gene imprinted in plant embryos. *Current Biology*, *19*(19), 1677–1681.

Jenik, P. D., Gillmor, C. S., & Lukowitz, W. (2007). Embryonic patterning in *Arabidopsis thaliana*. *Annual Review of Cell and Developmental Biology*, *23*, 207–236.

Jenkins, T. G., & Carrell, D. T. (2012). Dynamic alterations in the paternal epigenetic landscape following fertilization. *Frontiers in Genetics*, *3*, 143.

Jiao, Y., & Meyerowitz, E. M. (2010). Cell-type specific analysis of translating RNAs in developing flowers reveals new levels of control. *Molecular Systems Biology*, *6*, 419.

Johnston, A. J., Meier, P., Gheyselinck, J., Wuest, S. E., Federer, M., Schlagenhauf, E., et al. (2007). Genetic subtraction profiling identifies genes essential for *Arabidopsis* reproduction and reveals interactions between the female gametophyte and the maternal sporophyte. *Genome Biology*, *8*(10), R204.

Lafon-Placette, C., & Köhler, C. (2014). Embryo and endosperm, partners in seed development. *Current Opinion in Plant Biology*, *17*, 64–69.

Law, J. A., & Jacobsen, S. E. (2010). Establishing, maintaining and modifying DNA methylation patterns in plants and animals. *Nature Reviews Genetics*, *11*(3), 204–220.

Le, Q., Gutierrez-Marcos, J. F., Costa, L. M., Meyer, S., Dickinson, H. G., Lörz, H., et al. (2005). Construction and screening of subtracted cDNA libraries from limited populations of plant cells: A comparative analysis of gene expression between maize egg cells and central cells. *The Plant Journal*, *44*(1), 167–178.

Li, J., & Berger, F. (2012). Endosperm: Food for humankind and fodder for scientific discoveries. *The New Phytologist*, *195*(2), 290–305.

Li, N., & Li, Y. (2015). Maternal control of seed size in plants. *Journal of Experimental Botany*, *66*(4), 1087–1097.

Lopes, M. A., & Larkins, B. A. (1993). Endosperm origin, development, and function. *Plant Cell*, *5*(10), 1383–1399.

Loppin, B., Berger, F., & Couble, P. (2001). The *Drosophila* maternal gene *sesame* is required for sperm chromatin remodeling at fertilization. *Chromosoma*, *110*(6), 430–440.

Luo, M., Taylor, J. M., Spriggs, A., Zhang, H., Wu, X., Russell, S., et al. (2011). A genome-wide survey of imprinted genes in rice seeds reveals imprinting primarily occurs in the endosperm. *PLoS Genetics*, 7(6), e1002125.

Matzke, M. A., & Mosher, R. A. (2014). RNA-directed DNA methylation: An epigenetic pathway of increasing complexity. *Nature Reviews Genetics*, *15*(6), 394–408.

Mayer, U., Buttner, G., & Jürgens, G. (1993). Apical-basal pattern formation in the *Arabidopsis* embryo: Studies on the role of the *gnom* gene. *Development*, *117*(1), 149–162.

Mayer, W., Smith, A., Fundele, R., & Haaf, T. (2000). Spatial separation of parental genomes in preimplantation mouse embryos. *The Journal of Cell Biology*, *148*(4), 629–634.

McElver, J., Tzafrir, I., Aux, G., Rogers, R., Ashby, C., Smith, K., et al. (2001). Insertional mutagenesis of genes required for seed development in *Arabidopsis thaliana*. *Genetics*, *159*(4), 1751–1763.

Meyer, S., & Scholten, S. (2007). Equivalent parental contribution to early plant zygotic development. *Current Biology*, *17*(19), 1686–1691.

Moore, J. M. (2002). *Isolation and Characterization of Gametophytic Mutants in* Arabidopsis thaliana. State University of New York, Stony Brook (PhD thesis).

Mordhorst, A. P., Toonen, M. A. J., de Vries, S. C., & Meinke, D. (1997). Plant embryogenesis. *Critical Reviews in Plant Sciences*, *16*(6), 535–576.

Ngo, Q. A., Baroux, C., Guthorl, D., Mozerov, P., Collinge, M. A., Sundaresan, V., et al. (2012). The armadillo repeat gene *ZAK IXIK* promotes *Arabidopsis* early embryo and endosperm development through a distinctive gametophytic maternal effect. *Plant Cell*, *24*(10), 4026–4043.

Ning, J., Peng, X. B., Qu, L. H., Xin, H. P., Yan, T. T., & Sun, M. (2006). Differential gene expression in egg cells and zygotes suggests that the transcriptome is restructed before the first zygotic division in tobacco. *FEBS Letters*, *580*(7), 1747–1752.

Nodine, M. D., & Bartel, D. P. (2010). MicroRNAs prevent precocious gene expression and enable pattern formation during plant embryogenesis. *Genes & Development, 24*(23), 2678–2692.

Nodine, M. D., & Bartel, D. P. (2012). Maternal and paternal genomes contribute equally to the transcriptome of early plant embryos. *Nature, 482*(7383), 94–97.

Nordborg, M., Hu, T. T., Ishino, Y., Jhaveri, J., Toomajian, C., Zheng, H., et al. (2005). The pattern of polymorphism in *Arabidopsis thaliana. PLoS Biology, 3*(7), e196.

Nothias, J. Y., Miranda, M., & DePamphilis, M. L. (1996). Uncoupling of transcription and translation during zygotic gene activation in the mouse. *The EMBO Journal, 15*(20), 5715–5725.

Ohnishi, Y., Hoshino, R., & Okamoto, T. (2014). Dynamics of male and female chromatin during karyogamy in rice zygotes. *Plant Physiology, 165*(4), 1533–1543.

Pagnussat, G. C., Yu, H. J., Ngo, Q. A., Rajani, S., Mayalagu, S., Johnson, C. S., et al. (2005). Genetic and molecular identification of genes required for female gametophyte development and function in *Arabidopsis. Development, 132*(3), 603–614.

Palovaara, J., Saiga, S., & Weijers, D. (2013). Transcriptomics approaches in the early *Arabidopsis* embryo. *Trends in Plant Science, 18*(9), 514–521.

Pillot, M., Baroux, C., Vazquez, M. A., Autran, D., Leblanc, O., Vielle-Calzada, J. P., et al. (2010). Embryo and endosperm inherit distinct chromatin and transcriptional states from the female gametes in *Arabidopsis. Plant Cell, 22*(2), 307–320.

Radchuk, V., & Borisjuk, L. (2014). Physical, metabolic and developmental functions of the seed coat. *Frontiers in Plant Science, 5*, 510.

Raissig, M. T., Bemer, M., Baroux, C., & Grossniklaus, U. (2013). Genomic imprinting in the *Arabidopsis* embryo is partly regulated by PRC2. *PLoS Genetics, 9*(12), e1003862.

Raissig, M. T., Gagliardini, V., Jaenisch, J., Grossniklaus, U., & Baroux, C. (2013). Efficient and rapid isolation of early-stage embryos from *Arabidopsis thaliana* seeds. *Journal of Visualized Experiments, 76*, e50371.

Ramirez-Parra, E., & Gutierrez, C. (2007). The many faces of chromatin assembly factor 1. *Trends in Plant Science, 12*(12), 570–576.

Ronceret, A., Gadea-Vacas, J., Guilleminot, J., Lincker, F., Delorme, V., Lahmy, S., et al. (2008). The first zygotic division in *Arabidopsis* requires *de novo* transcription of thymidylate kinase. *The Plant Journal, 53*(5), 776–789.

Ronceret, A., Guilleminot, J., Lincker, F., Gadea-Vacas, J., Delorme, V., Bechtold, N., et al. (2005). Genetic analysis of two *Arabidopsis* DNA polymerase epsilon subunits during early embryogenesis. *The Plant Journal, 44*(2), 223–236.

Schauer, S. E., Jacobsen, S. E., Meinke, D. W., & Ray, A. (2002). *DICER-LIKE1*: Blind men and elephants in *Arabidopsis* development. *Trends in Plant Science*, 7(11), 487–491.

Schier, A. F. (2007). The maternal-zygotic transition: Death and birth of RNAs. *Science, 316*(5823), 406–407.

Schmid, K. J., Sorensen, T. R., Stracke, R., Torjek, O., Altmann, T., Mitchell-Olds, T., et al. (2003). Large-scale identification and analysis of genome-wide single-nucleotide polymorphisms for mapping in *Arabidopsis thaliana. Genome Research, 13*(6A), 1250–1257.

Scholten, S., Lörz, H., & Kranz, E. (2002). Paternal mRNA and protein synthesis coincides with male chromatin decondensation in maize zygotes. *The Plant Journal, 32*(2), 221–231.

Seefried, W. F., Willmann, M. R., Clausen, R. L., & Jenik, P. D. (2014). Global regulation of embryonic patterning in *Arabidopsis* by microRNAs. *Plant Physiology, 165*(2), 670–687.

Slane, D., Kong, J., Berendzen, K. W., Kilian, J., Henschen, A., Kolb, M., et al. (2014). Cell type-specific transcriptome analysis in the early *Arabidopsis thaliana* embryo. *Development, 141*(24), 4831–4840.

Southworth, D. (1996). Gametes and fertilization in flowering plants. *Current Topics in Developmental Biology, 34*, 259–279.

Sprunck, S., Baumann, U., Edwards, K., Langridge, P., & Dresselhaus, T. (2005). The transcript composition of egg cells changes significantly following fertilization in wheat (*Triticum aestivum* L.). *The Plant Journal*, *41*(5), 660–672.

Sprunck, S., & Gross-Hardt, R. (2011). Nuclear behavior, cell polarity, and cell specification in the female gametophyte. *Sexual Plant Reproduction*, *24*(2), 123–136.

Sun, M.-X. (2014). The expression and roles of parent-of-origin genes in early embryogenesis of angiosperms. *Frontiers in Plant Science*, *5*.

Tadros, W., & Lipshitz, H. D. (2009). The maternal-to-zygotic transition: A play in two acts. *Development*, *136*(18), 3033–3042.

Ueda, K., Kinoshita, Y., Xu, Z. J., Ide, N., Ono, M., Akahori, Y., et al. (2000). Unusual core histones specifically expressed in male gametic cells of *Lilium longiflorum*. *Chromosoma*, *108*(8), 491–500.

Ueda, K., & Tanaka, I. (1995). The appearance of male gamete-specific histones gH2B and gH3 during pollen development in *Lilium longiflorum*. *Developmental Biology*, *169*(1), 210–217.

Vielle-Calzada, J. P., Baskar, R., & Grossniklaus, U. (2000). Delayed activation of the paternal genome during seed development. *Nature*, *404*(6773), 91–94.

Walbot, V., & Evans, M. M. (2003). Unique features of the plant life cycle and their consequences. *Nature Reviews Genetics*, *4*(5), 369–379.

Waters, A. J., Makarevitch, I., Eichten, S. R., Swanson-Wagner, R. A., Yeh, C. T., Xu, W., et al. (2011). Parent-of-origin effects on gene expression and DNA methylation in the maize endosperm. *Plant Cell*, *23*(12), 4221–4233.

Weijers, D., Geldner, N., Offringa, R., & Jürgens, G. (2001). Seed development: Early paternal gene activity in Arabidopsis. *Nature*, *414*(6865), 709–710.

Wendrich, J. R., & Weijers, D. (2013). The *Arabidopsis* embryo as a miniature morphogenesis model. *The New Phytologist*, *199*(1), 14–25.

Wolff, P., Weinhofer, I., Seguin, J., Roszak, P., Beisel, C., Donoghue, M. T., et al. (2011). High-resolution analysis of parent-of-origin allelic expression in the *Arabidopsis* endosperm. *PLoS Genetics*, 7(6), e1002126.

Wuest, S. E., Vijverberg, K., Schmidt, A., Weiss, M., Gheyselinck, J., Lohr, M., et al. (2010). *Arabidopsis* female gametophyte gene expression map reveals similarities between plant and animal gametes. *Current Biology*, *20*(6), 506–512.

Xin, H. P., Zhao, J., & Sun, M. X. (2012). The maternal-to-zygotic transition in higher plants. *Journal of Integrative Plant Biology*, *54*(9), 610–615.

Xu, W., Dai, M., Li, F., & Liu, A. (2014). Genomic imprinting, methylation and parent-of-origin effects in reciprocal hybrid endosperm of castor bean. *Nucleic Acids Research*, *42*(11), 6987–6998.

Zhao, J., Xin, H., Qu, L., Ning, J., Peng, X., Yan, T., et al. (2011). Dynamic changes of transcript profiles after fertilization are associated with *de novo* transcription and maternal elimination in tobacco zygote, and mark the onset of the maternal-to-zygotic transition. *The Plant Journal*, *65*(1), 131–145.

CHAPTER ELEVEN

The Maternal-to-Zygotic Transition in Higher Plants: Available Approaches, Critical Limitations, and Technical Requirements

Peng Zhao, Meng-Xiang Sun[1]
Department of Cell and Developmental Biology, College of Life Sciences, State Key Laboratory of Hybrid Rice, Wuhan University, Wuhan, China
[1]Corresponding author: e-mail address: mxsun@whu.edu.cn

Contents

Abstract

Fertilization marks the turnover from the gametophyte to sporophyte generation in higher plants. After fertilization, sporophytic development undergoes genetic turnover from maternal to zygotic control: the maternal-to-zygotic transition (MZT). The MZT is thought to be critical for early embryogenesis; however, little is known about the time course or developmental impact of the MZT in higher plants. Here, we discuss what is known in the field and focus on techniques used in relevant studies and their limitations. Some significant questions and technical requirements for further investigations are also discussed.

Current Topics in Developmental Biology, Volume 113
ISSN 0070-2153
http://dx.doi.org/10.1016/bs.ctdb.2015.06.006

1. ZYGOTE DEVELOPMENT AND EMBRYOGENESIS IN ANGIOSPERMS

The angiosperm life cycle alternates between sporophytic and gametophytic generations. Male and female sporophytes produce microspores and megaspores, which develop into male and female gametophytes, respectively. The early male gametophyte, or pollen, germinates and produces a pollen tube that grows through the female pistil to deliver a pair of sperm cells to the female gametophyte, which usually consists of seven cells of four different cell types: three antipodal cells, two synergid cells, one egg cell, and one central cell (Fig. 1). One of the sperm cells fuses with the egg cell to produce the zygote, which develops into the embryo, and another sperm cell fuses with the central cell to produce the triploid endosperm, a nutritive tissue supporting early embryonic development. Fertilization marks the turnover from gametophyte to sporophyte development. Thus, the zygote is the starting point for embryogenesis.

Embryogenesis is a key process in higher plant development, and early morphological events during this period have been well described in a number of species, such as *Arabidopsis thaliana* and *Nicotiana tabacum* (Goldberg, de Paiva, & Yadegari, 1994; He, He, Qu, Sun, & Yang, 2007; Zhao et al., 2013). The large vacuole in the egg cell disappears shortly after fertilization, and the fused nucleus moves to the center of the fertilized egg, giving rise to the symmetric state of the fertilized egg cell, without obvious polarity (Fig. 1). Then, the fertilized egg cell begins to actively synthesize a cell wall and starts to elongate. This process is followed by the appearance of obvious zygote polarity. Another large vacuole is generated, and the nucleus moves to the chalazal end of the zygote. The elongated zygote usually divides asymmetrically, giving rise to a larger basal cell and a smaller apical cell with distinct developmental fates. The small apical cell develops into the main body of the embryo proper and exhibits an isotropic growth pattern. The apical cell and the daughter cell thereof usually divide twice longitudinally and once transversely to form a spherical eight-cell proembryo, whereas the larger basal cell continues to expand longitudinally and divides transversely to form a suspensor composed of several cells (Fig. 1). The uppermost suspensor cell, termed the hypophysis, eventually becomes part of the primary root meristem, whereas the other cells form an extra-embryonic suspensor, which degrades at later stages of embryogenesis. Comparable to the morphological difference between the apical cell and basal cell lineages, some

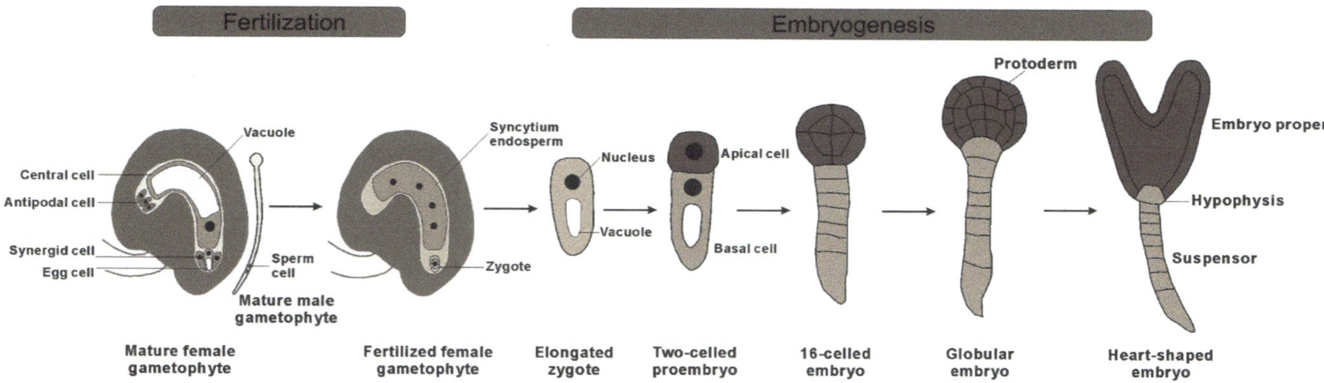

Figure 1 The process of fertilization and embryogenesis in higher plants. In angiosperms, the early male gametophyte, or pollen, germinates and produces a pollen tube that grows through the pistil to deliver a pair of sperm cells to the female gametophyte. The female gametophyte, or embryo sac, usually consists of seven cells: three antipodal cells, two synergid cells, one egg cell, and one central cell. During fertilization, one of the sperm cells fuses with the egg cell to produce a diploid zygote, which develops into an embryo, and the other sperm cell fuses with the central cell to produce a triploid primary endosperm cell. During embryogenesis, the elongated zygote divides asymmetrically, giving rise to a larger basal cell and a smaller apical cell with distinct developmental fates. The small apical cell develops into the main body of the embryo proper, whereas the larger basal cell continues to expand longitudinally and divides transversely to form a suspensor. The uppermost suspensor cell, termed the hypophysis, eventually becomes part of the primary root meristem. The pale green color labels egg and fertilized egg cell, the orange color labels central cell and fertilized central cell, the dark green color labels apical cell lineage, and the green-yellow color labels basal cell linage. (See the color plate.)

apical/basal cell-specific genes have been identified (Haecker et al., 2004; Ma et al., 2011). However, whether they exert maternal effects relevant to the determination of cell fate remains largely unknown.

Early embryogenesis is a critical period during plant development; a series of fundamental and complex developmental events occur. Although the question of which critical processes are controlled by maternal factors, or by *de novo*-synthesized zygotic transcripts, is interesting, this remains a mystery. Hence, an understanding of the fundamental roles played by maternal and *de novo*-synthesized transcripts (from the zygotic genome) during the maternal-to-zygotic transition (MZT) is essential to reveal the molecular mechanisms of embryogenesis.

2. THE MZT DURING EARLY ANGIOSPERM EMBRYOGENESIS

The MZT is the first major transition in the life of higher plants and animals and has been divided into two processes: clearance of maternal transcripts and proteins deposited in the egg cell and gradual *de novo* activation of zygotic genome transcription. Therefore, the MZT is a period, not a specific time point. Although different definitions of the MZT have been proposed, the more popular concept is that the MZT is the embryonic developmental period beginning with clearance of maternal transcripts, continuing with *de novo* transcription of zygotic transcripts, and ending with the first major morphological requirement of the *de novo*-synthesized transcripts (Tadros & Lipshitz, 2009). In other words, early embryogenesis proceeds from a maternally controlled to a zygotically controlled process. Three critical time points are evident during the MZT; these are initiation of degradation of maternal transcripts after fertilization, the onset of zygotic genome transcription and, ultimately, the replacement of parental transcripts with zygotic transcripts.

Remarkable progress has been made in our understanding of the MZT in different animal species, particularly the timing and scale of the MZT, the molecular mechanisms regulating degradation of maternal mRNAs and zygotic genome activation (ZGA), and the roles played by maternal factors activating the zygotic genome (Lee, Bonneau, & Giraldez, 2014; Li, Lu, & Dean, 2013; Tadros & Lipshitz, 2009). However, studies on the MZT of higher plants are still at an early stage. Several key questions, such as conservation (or not) of the molecular mechanisms underlying the MZTs of plants and animals, timing of the MZT in higher plants, the roles played by maternal factors and *de novo*-synthesized transcripts in specific developmental events during early embryogenesis, the molecular mechanisms regulating clearance of maternal factors, and activation of the zygotic genome, remain poorly understood. Well-established methods used to isolate egg cells and early embryos, as well as technical advances in single-cell cDNA library construction and next-generation transcript sequencing, have facilitated MZT studies in plants. *De novo* transcripts can be detected in the zygote shortly after fertilization and before zygotic cell division. However, it remains difficult to draw a clear picture based on our current knowledge of the MZT in higher plants, particularly in terms of the duration of the MZT and mechanisms regulating the degradation of maternal transcripts and activation of the zygotic genome. This is primarily attributable to critical limitations in techniques.

3. STUDIES ON THE MZT DURING EMBRYOGENESIS AND TECHNICAL LIMITATIONS

Two major coordinated molecular events occur during early embryogenesis, as mentioned above. The first is clearance of maternal gene products, and the second is *de novo* transcription of the zygotic genome, which is a hallmark of the MZT. Remarkable progress has been made in our understanding of the time courses of these two processes, and the molecular mechanisms thereof, in different animal species. These studies have revealed that maternal transcripts originally stored in egg cells drive early embryonic development and, upon gradual elimination of such transcripts, replacement thereof by *de novo* zygotic transcripts occurs (Lee et al., 2014; Li et al., 2013; Tadros & Lipshitz, 2009; Walser & Lipshitz, 2011). However, only a few pioneering studies on the MZTs of higher plants have been conducted; these have addressed the duration of the MZT, the activation stage of the zygotic genome, and the contributions of parental transcripts to early embryogenesis. Here, we provide a brief overview on what is known about the MZT in higher plants with particular reference to technical aspects of such studies.

3.1 Timing of the MZT in Higher Plants

The timing of the MZT has been studied intensively in animals. The MZT occurs at different time points ranging from a few hours to several days after fertilization, in different animal species. The midblastula transition (MBT) in animals represents the first developmental event that requires transcription of the zygotic genome, often being defined by blocking zygotic genome transcriptional activity with pharmacological inhibitors. Of such inhibitors, α-amanitin has been widely used as a transcriptional inhibitor in animal experiments. α-Amanitin is an eight-amino acid cyclic peptide that inhibits RNA polymerase II, slightly inhibits RNA polymerase III, but does not inhibit RNA polymerase I. Addition of α-amanitin to fertilized eggs or later embryonic stages blocks zygotic transcription, and the developmental arrest that ensues marks the time of occurrence of the MBT. Global gene transcription expression profiling after addition of pharmacological inhibitors has also been used to explore MBT timing in animals. Mouse embryos of different stages, including unfertilized eggs, fertilized eggs, two-celled embryos, four-celled embryos, eight-celled embryos, morulas, and blastocysts, have been so evaluated. Global expression profiles and single-gene time-course analyses have revealed distinct patterns of maternal RNA degradation and activation of zygotic genes during mouse embryonic development (Hamatani, Carter, Sharov, & Ko, 2004). Furthermore, the expression profiles of pronuclei to late two-celled embryos treated with α-amanitin (to inhibit transcription), cycloheximide (to inhibit translation), and aphidicolin (to inhibit DNA replication) have been compared (Hamatani et al., 2004). The results showed that translation of maternal RNAs deposited in early embryos is essential to initiate the ZGA. This type of work illustrates the utility of inhibitors in relevant MZT studies. However, these experiments also depend on the availability of egg cells/early embryos and appropriate *in vitro* culture systems for developing zygotes.

No similar research strategy has been widely used to study the timing of the MZT in higher plants. MZT stages have been explored using pharmacological experimental techniques based on well-established embryo isolation methods, and an *in vitro* zygote culture system, in *N. tabacum* (Zhao et al., 2011). The cited authors confirmed that ZGA initiated during the zygote stage, was necessary for zygotic cell division, and triggered embryogenesis. However, α-amanitin has not been successfully used in plant MZT experiments, possibly because the drug lacks useful effects on plant embryos. Early pioneering work in *Daucus carota* showed that carrot cells inactivate α-amanitin under embryonic conditions (Pitto, Schiavo, & Terzi, 1985). Two other inhibitors, actinomycin D and cordycepin, have been used in *N. tabacum* MZT experiments. Actinomycin D binds to the DNA transcription initiation complex and interferes with elongation of the RNA chain by RNA polymerase (Sobell, 1985). Cordycepin, a derivative of adenosine, is incorporated into RNA and inhibits transcription elongation because of the absence of a hydroxyl moiety at the 3′ position (Holbein et al., 2009). In combination with a well-established embryonic isolation method and a culture system, 20 μg/mL actinomycin D and 40 μg/mL cordycepin were added to a semi-*in vivo* zygote culture system, and the inhibitory effects of the drugs on RNA transcription confirmed by reverse transcription-polymerase chain reaction (RT-PCR) analysis. Zygotic division and elongation were strongly affected by both actinomycin D and cordycepin, suggesting that the onset of *de novo* transcription of the genome occurs before or during the zygote elongation stage and that *de novo* zygotic transcripts are likely required for elongation and division of the zygote (Zhao et al., 2011). These results show that the ZGA is initiated before the first zygotic cell division in *N. tabacum* (Fig. 2). However, actinomycin D and cordycepin may have other effects apart from inhibiting polymerases. Actinomycin D can stimulate DNA topoisomerase-induced DNA cleavage and impair DNA replication (Guy & Taylor, 1978; Wassermann et al., 1990). Moreover, cordycepin not only inhibits chain termination, but also exerts a very strong effect on mRNA polyadenylation in certain strains (Holbein et al., 2009). Although these side effects are known, the inhibitors have been successfully used in many experiments. It is important to use a suitable inhibitor at an appropriate concentration to reduce side effects and ensure the reliability of results.

Certain other approaches, including silencing of RNA *polymerase II* using an RNA interference (RNAi) technique, have been used to study the timing of the MZT in higher plants. *Polymerase II* was downregulated using an RNAi driven by the *pNG* promoter of the embryo and endosperm to block transcription; the RNAi was present in central and egg cells immediately prior to fertilization and maternally contributed to the early embryo and endosperm (Pillot et al., 2010). RNA polymerase II activity was detected in *polymerase II* RNAi lines using an antibody (H5) directed against the active isoform of the polymerase II main subunit. It was found that embryos developed to the globular stage (16–32 cells) in *polymerase II* RNAi lines. These results indicate that egg cell-deposited products are sufficient to sustain early embryonic development in *A. thaliana* (Fig. 2).

The use of RNAi *in vivo* avoids the possible side effects of pharmacological treatments; the efficiency of any RNAi in suppressing gene expression should be considered to make sure whether the targeted gene transcripts were totally silenced via the RNAi technique. Transcription activator-like effector nucleases and clustered regularly interspaced short palindromic repeats (CRISPR)-based genome editing technology have been developed to knock out genes. These techniques have been widely used to perform

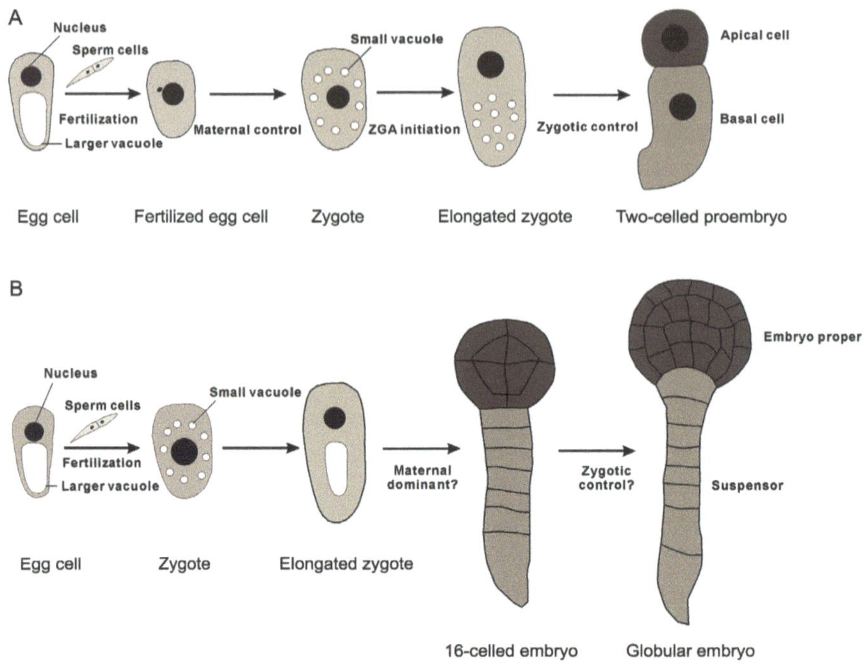

Figure 2 The timing of MZT in higher plants. (A) In *N. tabacum*, fertilization occurs about 48 h after pollination (HAP). The zygote becomes elongated 96 HAP and asymmetrically divides 108 HAP. Right after fertilization, about 60 HAP, male transcripts can be detected in zygotes. The onset of *de novo* transcription of the zygotic genome is during the zygote elongation stage and the *de novo* zygotic transcripts are likely required for zygotic elongation and division (Ning et al., 2006; Zhao et al., 2011). (B) In *A. thaliana*, a similar process of fertilization and embryogenesis is observed. Fertilization occurs about 8 HAP and the zygote starts to divide 24 HAP. The male transcripts could also be detected at the early zygote stage, but the exact timing of the MZT in early embryogenesis cannot be determined from our current knowledge. There is evidence indicating that ZGA has likely initiated in the zygote (Kawashima & Berger, 2014) and that early embryogenesis (from the zygote to the 16–32 cell embryo) is predominantly controlled by maternal factors (Pillot et al., 2010).

reverse genetic studies in plants. In particular, the CRISPR-based method of genome engineering can be used to efficiently and conveniently edit the plant genome. In a recent report, the CRISPR system was modified to allow cell- and tissue-specific genome editing, in which gRNA transcription was independent of the U6 and U3 snRNA promoters and gRNA was generated from pre-gRNA via a self-catalyzed process based on the ribozyme nuclease activity (Gao & Zhao, 2014). Editing of the polymerase II genome sequence in fertilized egg cells using this modified CRISPR system would enhance our understanding of the timing of the MZT in higher plants.

3.2 ZGA in Higher Plants

The MZT requires clearance of maternal factors and activation of the zygotic genome. Previous studies of the MZT in higher plants have primarily investigated ZGA. One early study on the MZT of *A. thaliana* showed that no paternally inherited allele from 19 tested loci was detectable in early embryos using β-glucuronidase (GUS) assays (Vielle-Calzada, Baskar, & Grossniklaus, 2000). In addition, transcriptional profile analyses of embryos produced via apomixis from a hybrid between *Zea mays* and the wild relative thereof, *Tripsacum*, compared to those developing from fertilized ovules of sexually reproducing *Z. mays*, indicated that early embryos have transcription profiles similar to those of ovules prior to fertilization (Grimanelli, Perotti, Ramirez, & Leblanc, 2005). The results of these two pioneering studies suggest that early embryogenesis in higher plants is controlled by maternal factors and that ZGA is delayed. Additional *de novo* transcripts have been identified as early as the zygote stage, shortly after fertilization. In addition, transcriptome profile analyses of embryos at successive stages from the zygote to the mature embryo have been performed in *A. thaliana* (Xiang et al., 2011) and several stage-specific transcripts have also been identified. However, whether these are transcribed *de novo* in embryos and the exact stage of production was not comprehensively analyzed, possibly because of a lack of transcriptome information from gametes.

Loss-of-function mutants defective in early embryonic development have helped to define the MZT in animals. Two different classes of mutants are useful in this context. The first is maternal-effect mutants exhibiting embryonic developmental defects; these can be used to explore the roles played by maternal factors. The second class is zygotic recessive mutants, which are useful to detect contributions made by *de novo* zygotic transcripts. As the numbers of available *A. thaliana* mutants increase, similar genetic approaches may potentially be used to study the MZT of higher plants. Some mutants exhibit early embryonic defects and die at different stages of embryonic development; these mutants are useful to identify maternal or paternal influences on early embryogenesis. However, to date, few plant mutants have been useful for analysis of the MZT. Plant embryos are usually surrounded by endosperm and deeply imbedded in the seed, which increases the practical difficulties. A genome-wide analysis of *A. thaliana* embryo-defective mutations showed that a substantial number of mutations were recessive, suggesting that expression of either parentally derived allele in the early embryo could support complete embryogenesis (Muralla, Lloyd, &

Meinke, 2011). Furthermore, 49 of the *A. thaliana* embryo-defective genes were evaluated in terms of relative parental contributions to early embryogenesis (Del Toro-De Leon, Garcia-Aguilar, & Gillmor, 2014). The 49 heterozygous embryo-defective mutants were crossed with wild-type pollen, and embryos exhibiting mutant phenotypes at approximately four-celled embryo, globular embryo, heart-shaped embryo, and mature embryo stages were scored, respectively. Wild-type paternal alleles allowed recovery of embryonic defects in nine mutants within the first 2 days, whereas paternal alleles only gradually allowed recovery of the mutant phenotypes of the remaining 40 mutants, and only at later stages (Del Toro-De Leon et al., 2014). This suggests that zygotic genome transcription begins within the first 2 days after pollination, and that transcription is not synchronous. This important feature should be considered when analyzing the timing of ZGA. Such information may help to improve estimates of the timing and extent of ZGA (Fig. 2).

3.3 The Contribution of Parental Transcripts to Early Embryogenesis

As discussed above, pharmacological, genetic, and molecular experiments have shown that maternal factors deposited in egg cells play major roles during early animal embryogenesis (e.g., BRG1; Bultman et al., 2006). However, the contribution of parental transcripts to early embryogenesis in plants remains largely unknown. A common perspective is that parental transcripts are involved in early embryogenesis and may play specific roles. Two recent transcriptome analyses indicated that both paternal and maternal transcripts are present in early *A. thaliana* embryos (Autran et al., 2011; Nodine & Bartel, 2012). In the first study, two- to four-cell and globular stage hybrid embryos, derived from a cross between Landsberg erecta (Ler) and Columbia (Col), were isolated, and their transcriptomes sequenced. Reads covering single-nucleotide polymorphisms (SNPs) were extracted and used to determine the origins of the transcripts (Autran et al., 2011). It was found that the transcriptome of two- to four-cell embryos, derived from the cross between Ler and Col, was dominated by maternal reads (88.4%), which decreased gradually in proportion from the two- to four-cell stage to the globular-embryo stage (64.1% maternal reads) (Autran et al., 2011). In the second study, the hybrid embryo transcriptome from a reciprocal cross between Col and Cape Verde Islands (Cvi) was sequenced, and the ratios of RNA-seq reads covering maternal and paternal SNPs within the transcribed regions of annotated genes scored. It was found that the 1-2-, 8- and ~32-cell embryo transcriptomes from a cross between Col and Cvi had equal numbers of reads from maternal and paternal alleles (Nodine & Bartel, 2012). Such differences in the parental transcript ratio might be explained by variations in the materials used (Del Toro-De Leon et al., 2014) but this matter requires further exploration.

Parental genes that are candidates for playing important roles in early embryos have been identified in using different methods. The maternally expressed maize imprinting gene, *MEE 1*, was identified in early embryos via analyses of cleaved amplified polymorphic sequences of hybrid embryo cDNA, and allele-specific expression thereof correlated with differences in allelic methylation levels (Jahnke & Scholten, 2009). However, the exact role played by *MEE 1* in zygotic embryogenesis remains unknown.

Another attractive possibility is that gamete-delivered mRNAs and proteins from sperm and egg cells may be contributed to the zygote, via fertilization. Early electron microscopic observations suggested that mitochondria in the sperm nucleus are transmitted to the egg and central cells during fertilization (Yu & Russell, 1994), indicating that cytoplasmic components of the male gamete could be brought into the zygote. Unfortunately, research on this subject has been hindered by difficulties associated with distinguishing factors delivered by gametes from those synthesized *de novo* after ZGA. The first example of a gamete-delivered transcript exerting a known function during early embryonic development is the SHORT SUSPENSOR (*SSP*). *SSP* transcripts have been detected in mature pollen via *in situ* hybridization, whereas no yellow fluorescent protein (YFP)-tagged SSP fluorescence was evident in germinated pollen. Only transient YFP-tagged SSP fluorescence was visible in the zygote (Bayer et al., 2009). Thus, *SSP* transcripts are thought to be delivered to the zygote by the sperm cell during fertilization. After transmission to the zygote, the *SSP* transcripts are translated, and transiently accumulate to regulate asymmetric zygotic division by activating the downstream YODA pathway (Bayer et al., 2009). However, whether other paternal transcripts are transported into the zygote during fertilization to exert a parent-of-origin effect on later embryo development has not been explored. An earlier transcriptome analysis showed that some transcripts were detectable in sperm cells and zygotes, but not in egg cells or embryos at later stages (Ning et al., 2006). Interestingly, two sperm cell transcripts, *Ntsp0002* and *Ntsp0023*, were detected in zygotes 60 h after pollination (HAP) (about 10 h after fertilization) using RT-PCR, but were degraded rapidly in zygotes at 96 HAP, indicating that the two paternal transcripts may play critical roles immediately after fertilization (Xin et al., 2011).

These sperm-delivered transcripts must be considered when *de novo* transcription is studied by examining differences in transcripts between egg cells and zygotes.

In addition, some transcripts have been detected in egg cells and zygotes, but not in sperm cells or embryos at later stages, and may be transmitted to the zygote from egg cells (Zhao et al., 2011). However, it is difficult to confirm whether such transcripts are actually gamete-delivered; technical limitations are in play.

4. CURRENT TECHNIQUES USED TO STUDY THE MZT IN HIGHER PLANTS

As discussed above, great progress has been made in our understanding of the MZT in higher plants, attributable to the development of specific techniques, such as single-cell sequencing, *in vitro* culture systems for early embryo, and construction of loss-of-function mutant libraries. Here, we provide an overview of available approaches that have been applied successfully to studies of the MZT in higher plants. These include methods for collecting gamete cells and early embryos, of single-cell sequencing, of detecting *de novo* zygotic genome transcripts, and of distinguishing paternal from maternal transcripts.

4.1 Current Methods for Isolating Gamete Cells and Early Embryos

The inaccessibility of egg cells and early embryos, which are usually embedded in maternal tissues, has impeded MZT research in higher plants. A critical step in performance of transcriptome analysis and allowing *de novo* transcripts synthesized in early embryos to be distinguished is the isolation and collection of individual gamete cells and early embryos. Techniques used to successfully isolate gametes and early embryos have been derived for different species (Table 1).

4.1.1 Microdissection

Microdissection is a traditional method used to isolate tissue-specific cells; a fine glass needle is manipulated under an inverted microscope to assist in dissection. Such methods, either direct dissection or dissection after brief enzyme treatment, are well established for isolation of sperm cells, egg cells, and early embryos from *N. tabacum* (Fu, Sun, Zhou, & Yang, 1996; He et al., 2007; Sun, Yang, & Zhou, 1993; Xin et al., 2011), *Oryza sativa*

Table 1 Overview of Available Methods Used in the Isolation and Analysis of Gamete Cells and Early Embryos

Method	Cell Type	Species	Amplification Method	Application	References
Microdissection	Sperm cell	*N. tabacum*	PCR	cDNA library construction	Xin et al. (2011)
	Egg cell, zygote	*N. tabacum*	PCR	Suppression subtractive hybridization	Ning et al. (2006)
	Egg cell, zygote, two-celled proembryo	*N. tabacum*	PCR	cDNA library construction	Zhao et al. (2011)
	Two- to four-celled embryo, globular embryo	*A. thaliana*	SPIA amplification	RNA-seq	Autran et al. (2011)
	One to two-celled embryo, about 32-celled embryo	*A. thaliana*	*In vitro* transcription	RNA-seq	Nodine and Bartel (2012)
	Zygote to mature embryo	*A. thaliana*	*In vitro* transcription	Microarray	Xiang et al. (2011)
	Egg cell, central cell	*Z. mays*	PCR	Suppression subtractive hybridization	Li, Cui, and Kim (2006)
	Sperm cell, central cell	*Z. mays*	–	Fusion	Kranz, von Wiegen, Quader, and Lorz (1998)
	Egg cells, synergid cell, central cells	*Z. mays*	–	Electrofusion	Kranz, Bautor, and Lorz (1991)

Continued

Table 1 Overview of Available Methods Used in the Isolation and Analysis of Gamete Cells and Early Embryos—cont'd

Method	Cell Type	Species	Amplification Method	Application	References
	Egg cell	*Z. mays*	PCR	cDNA library construction	Dresselhaus, Lorz, and Kranz (1994)
	Early embryo	*Z. mays*	PCR	Suppression subtractive hybridization	Meyer, Pospisil, and Scholten (2007)
	Early embryo	*Z. mays*	–	–	Jahnke and Scholten (2009) and Meyer and Scholten (2007)
	Egg cell, sperm cells	*O. sativa*	–	*In vitro* fusion	Khalequzzaman and Haq (2005)
	Egg cell, synergid cell	*O. sativa*	–	Microarray	Ohnishi et al. (2011)
FACS	Sperm cell	*A. thaliana*	*In vitro* transcription	Microarray	Borges et al. (2008)
LCM	Egg cell, central cell, synergid cell	*A. thaliana*	*In vitro* transcription	Microarray	Wuest et al. (2010)
	Apical cell, basal cell	*N. tabacum*	PCR	cDNA library construction	Ma et al. (2011)

(Khalequzzaman & Haq, 2005; Ohnishi et al., 2011; Takanashi et al., 2010; Zhang, Dong, Potrykus, & Galli, 1999), and *Z. mays* (Kranz et al., 1991). Living paired *N. tabacum* sperm cells can be released after osmotic shock of a pollen tube, and the paired sperm cells washed and collected under an inverted microscope, using a special micropipette, for further analysis (Xin et al., 2011). Egg cells can be isolated from ovules via manual microdissection after brief enzymatic treatment (Sun et al., 1993). In addition, fertilized embryo sacs can be isolated by enzymatic maceration followed by gentle grinding with a small glass pestle (Fu et al., 1996). Zygotes can be mechanically isolated from embryo sacs after a second enzymatic treatment (He et al., 2007). Various protocols for isolating *O. sativa* egg cells and zygotes are well established. Of these, nonenzymatic manual dissection methods are most convenient and efficient. After a transverse incision is made in the middle region of an ovule placed in mannitol solution, the egg cell or zygote is released by gentle pushing on the ovule base (Ohnishi et al., 2011; Takanashi et al., 2010). Methods for isolating *Z. mays* gamete cells and embryos are also established. Living sperm cells are released automatically from pollen grains after rupture via osmotic shock (Kranz et al., 1991). Female gamete cells can be obtained by mechanical micromanipulation of embryo sacs after enzymatic treatment (Kranz et al., 1991). The relevant techniques continue to improve; avoidance of enzyme treatment and shortening the duration of dissection, are being combined with the use of transcriptional inhibitors, to reduce possible stress-induced gene expression.

4.1.2 Fluorescence-Activated Cell Sorting

Fluorescence-activated cell sorting (FACS) is a specialized type of flow cytometry that rapidly separates labeled living cells of particular types from a heterogeneous mixture of different cell types, based on the specific fluorescence characteristics and light-scattering properties of the target cells. Several different labeling methods are available, including fluorescent labeling, quantum dots, and isotope labeling. Of these, fluorescent labeling has been widely used to isolate cells of interest employing specifically targeted fluorescence. In plants, FACS has been used to isolate enriched cell populations expressing a fluorescent protein such as green fluorescent protein (GFP) driven by a cell- or tissue-specific promoter. Different types of root cells have been isolated via FACS (Birnbaum et al., 2003, 2005). However, several potential limitations are apparent. The first is the requirement for a relatively large initial volume of cell-containing solution, which hampers the isolation of cells from ovules, particularly egg cells and early embryos. It is relatively difficult to collect high numbers of such cells from maternal tissues after enzymatic treatment. The second is sensitivity; FACS is unsuitable for isolation of relatively rare cells from a large cell population; false-positive signals may occur. FACS has not been used successfully to collect egg cells or early embryos of higher plants because of this technical limitation, but has been valuable for collecting living sperm cells of *A. thaliana* for transcriptome analysis (Borges et al., 2008). Transgenic plants expressing *pAtGEX2::eGFP*, a sperm-specific gene, were generated. Sperm cell populations labeled with *pAtGEX2::eGFP* were prepared from inflorescences, and a series of preliminary purification steps was performed prior to FACS. Finally, living sperm cell fractions were sorted for RNA extraction and further gene chip analysis of the sperm cell transcriptome (Borges et al., 2008).

In view of the difficulties associated with preparing early embryo protoplasts, a modified sorting method termed fluorescence-activated nuclear sorting (FANS) has been developed for transcriptome analysis where fluorescence-labeled nuclear material serves as the starting material. This type of analysis has been applied successfully to generate proembryo and suspensor-nuclei expression profiles. Proembryo and suspensor nuclei were labeled with nucleus-localized GFP driven by proembryo or suspensor-specific promoters, and RNA was extracted via FANS for microarray analysis (Slane et al., 2014).

4.1.3 Laser Capture Microdissection

Laser capture microdissection (LCM) is a high-resolution method used to isolate cells from their surrounding tissues, with the aid of a laser beam, under direct microscopic visualization. Specific cells can be isolated directly by cutting target cells away from unwanted cells, to obtain enriched cell populations. Such populations can be used to generate cDNA libraries or to conduct transcriptome or proteome analyses. The advantage of LCM is that specific cell types can be harvested from tissue sections regardless of whether the sample is labeled with a fluorescent marker. However, plant tissue must be frozen or paraffin-embedded prior to LCM, which usually therefore yields only small numbers of the cells of interest, of poor quality, with chemically cross-linked RNA. Nevertheless, LCM has been applied successfully to capture egg cells, synergids, and central cells, for cell type-specific expression profile analysis (Wuest et al., 2010). This method requires preparation of fixed samples and sections. It is important to guard against possible contamination during actual LCM manipulation.

4.2 Current Approaches Toward Single-Cell Sequencing

Construction of a single-cell transcriptome is helpful when analyzing transcriptome changes associated with the MZT of plants. Traditional methods, such as RT-PCR, *in situ* hybridization, and quantitative real-time PCR-based methods, use single cells as starting materials, but can only assay a limited number of genes at any one time. Hence, global transcriptome analysis of egg cells and early embryos, using high-throughput technology, is required to study the MZT of higher plants. High-throughput technologies such as microarrays and RNA-seq are powerful when an overview of gene expression is required, but are usually limited by the large amounts of RNA needed for analysis. Thus, templates extracted from single cells must be amplified either via PCR or linear *in vitro* transcription prior to microarray and RNA-seq analyses. RNA-seq has emerged as a powerful tool for transcriptome analyses, because of falling costs, and yields more comprehensive gene expression information than does microarray analysis. Here, we overview current methods of single-cell RNA-seq.

4.2.1 Single-Cell mRNA Extraction

Diverse RNAs in sperm, egg cells, and early embryos (before the eight-cell stage) are released automatically following osmotic shock with a buffer, and can be reverse transcribed. However, the major limitation of this method is that RNA solutions cannot be stored for long period since that RNA in the shock buffer are easily degraded. Another mRNA isolation method based on a poly(A) selection strategy has been developed, using a limited number of plant cells. Poly(A) is bound to the surface of superparamagnetic beads, allowing mRNA to be efficiently and rapidly isolated This method has the advantage of capturing and isolating pure intact poly(A) mRNA at the microscale, while excluding undesirable rRNAs and tRNAs in a single step. However, nonpolyadenylated yet informative RNAs, including small RNAs generated by RNA polymerase III and nonpolyadenylated lncRNAs, are lost using this method (Yang, Duff, Graveley, Carmichael, & Chen, 2011).

4.2.2 Construction of Single-Cell cDNA Libraries

PCR-based amplification is very efficient when used to construct single-cell cDNA libraries and has been widely employed for RNA-seq, in combination with different sequencing technologies. This method uses high-efficiency PCR to amplify low-level cDNA templates after reverse transcription, employing universal primers binding to the 5′ and 3′ ends. Several different methods are available for generating 5′ universal PCR primers. Of these, the SMARTer cDNA synthesis methods efficiently generate cDNAs representing full-length transcripts (Chenchik et al., 1998).

In vitro transcription for linear mRNA amplification can also be efficiently used to amplify initial low-level templates from a limited number of starting cells. First-strand cDNA is synthesized with the aid of an oligo(dT) primer bearing an *in vitro* transcription promoter, and the RNA transcripts are next linearly amplified by RNA polymerase from a linear cDNA template containing the corresponding promoter sequence (Nygaard & Hovig, 2006).

4.3 Current Techniques Used to Detect *De Novo* Synthesized Transcripts

Initiation of the MZT requires both clearance of maternal factors and the ZGA. The onset of zygotic genome transcription is the hallmark of the MZT. *De novo*-synthesized transcripts are detectable when gamete cells and early embryos are compared via suppression subtractive hybridization (SSH), fluorescence *in situ* hybridization (RNA-FISH), and transcriptome analysis.

4.3.1 Suppression Subtractive Hybridization

SSH is a valuable tool used to distinguish two closely related cDNA samples based on a suppression polymerase chain reaction technique. However, a challenging drawback of the SSH technique is the appearance of "false positive" cDNAs, attributable to unequal enrichment of low-abundance cDNAs, usually associated with positive results during screening but which cannot be confirmed in later experiments. Mirror orientation selection (MOS) is used to substantially decrease the number of "false positive" background clones in SSH-generated libraries. Egg cell and zygote cDNA libraries have been constructed and subtracted using MOS (Ning et al., 2006). Several *de novo*-synthesized transcripts have been detected in zygotes compared to gametes, indicating the reliability of the method.

4.3.2 RNA Fluorescence In Situ *Hybridization*

RNA-FISH is another useful method employed to detect transcriptional changes and has been applied to identify nascent transcripts within the animal nucleus (Ohno, Aoki, & Sasaki, 2001; Ronshaugen & Levine, 2004). However, in plants, RNA-FISH has been applied only to detect *de novo* transcripts in *A. thaliana* endosperm (Vielle-Calzada et al., 1999), but not in the early embryos. The maternally expressed imprinting gene MEDEA was confirmed to be actively transcribed after fertilization, in the early endosperm, using the RNA-FISH technique.

4.3.3 Transcription Profile Analysis of Gamete Cells and Embryos

Comprehensive transcription profile analyses of sperm cells, egg cells, and early embryos, will help identify *de novo*-synthesized transcripts in early embryos. Expression profiles of sperm cells, egg cells, and early embryos have been obtained for some species, including *A. thaliana*, *N. tabacum*, and *Z. mays*. However, *de novo*-synthesized transcripts have not been analyzed comprehensively in sperm cells, egg cells, or early embryos. Only a few *de novo* tobacco zygotic transcripts have been identified, based on transcription profile differences between egg cells and zygotes; the differences were further confirmed by RT-PCR (Ning et al., 2006; Zhao et al., 2011).

4.3.4 Time-Resolved Expression Profile Analysis

Time-resolved expression profile analysis of embryos at different stages will help identify *de novo*-synthesized embryonic transcripts. This method requires isolation of embryos from successive stages (e.g., from the zygote to the late globular-embryo stage), and performance of genome-wide gene expression analyses at successive stages. A transcriptome profile analysis of embryos at stages from the zygote to the mature embryo has been performed only for *A. thaliana* (Xiang et al., 2011). A comprehensive analysis of *de novo*-transcribed embryonic transcripts at different stages may follow in the near future.

4.4 Current Approaches to Distinguish Paternal from Maternal Transcripts

During plant sexual reproduction, a pair of sperm cells fuses with the egg and central cells, which form the diploid zygote and the triploid endosperm, respectively. Both parents contribute their transcripts (delivered directly or via *de novo* synthesis from parental alleles in the zygotic genome) to early embryonic development. Several parental-origin genes have been identified in early embryos of different species (Autran et al., 2011; Meyer & Scholten, 2007; Nodine & Bartel, 2012; Zhang et al., 2011). Work on the contributions made by parental transcripts to embryonic development in higher plants is just beginning; thus, no clear picture of the contributions of parental-origin genes to embryonic development can be drawn at this stage.

Establishment of efficient methods allowing paternal and maternal transcripts to be distinguished is essential for study of the ZGA in higher plants. Below, we overview current methods used to distinguish paternal from maternal transcripts; these include allele-specific RT-PCR, allele-specific restriction enzyme digestion, SSR-based RT-PCR, and SNP-based RNA-seq (usually based on SNP or SSR differences between paternal and maternal alleles) (Table 2).

4.4.1 Allele-Specific RT-PCR

Allele-specific RT-PCR is a simple method used to distinguish paternal from maternal transcripts based on SNP differences between the two alleles. PCR reactions are performed using allele-specific primers with 3′-end-bases complementary to the SNP site. The two advantages of this method are the relatively low cost and the short time needed to detect allele-specific transcripts. This method has been applied to *A. thaliana*, and several allele-specific transcripts have been detected (Vielle-Calzada et al., 2000). However, a single nucleotide change at the 3′-end of a primer is not always sufficient to ensure reliable discrimination between the two parents because of mismatches between the allele-specific primers used for PCR. Thus, a modified allele-specific RT-PCR method using 3′-end-locked nucleic acid primers has been developed to improve the reliability of allele-specific amplification (Latorra, Campbell, Wolter, & Hurley, 2003; Nakitandwe, Trognitz, & Trognitz, 2007).

Table 2 Overview of Available Methods Used to Distinguish Parental Transcripts in Early Embryos

Method	Material	Species	References
Allele-specific restriction enzyme reaction	Seed	*A. thaliana*	Bayer et al. (2009)
Relative gene expression analysis	Early embryo	*Z. mays*	Meyer and Scholten (2007)
Allele-specific RT-PCR	Seed	*A. thaliana*	Vielle-Calzada et al. (2000)
GFP/GUS report analysis	Seed	*A. thaliana*	Vielle-Calzada et al. (2000)
SNP based RNA-seq	Early embryo	*A. thaliana*	Autran et al. (2011) and Nodine and Bartel (2012)
SSR based RT-PCR	Zygote, eight-celled embryo	*N. tabacum* *N. rustica*	Zhang et al. (2011)

4.4.2 Simple Sequence Repeat-Based RT-PCR

SSRs are useful molecular markers distinguish paternal from maternal transcripts. SSR markers usually feature repetitive di- or tri-nucleotide repeats, which can be used in conjunction with PCR and gel electrophoresis to detect different transcripts identified by the lengths of the PCR products. However, the principal limitation of this method is that SSR sites are fewer in number than SNPs. Nevertheless, the method was applied successfully to study activation of parental genomes in interspecific tobacco hybrids (Zhang et al., 2011). Four paternal-origin transcripts and one maternal-origin transcript were identified in hybrid zygotes (Hamayan × SR1) using SSR-based RT-PCR.

4.4.3 GFP/GUS Reporter-Based Analysis

Reciprocal crosses between nontransgenic and transgenic plants carrying the GFP/GUS reporter gene can be efficiently used to detect early embryonic transcription. However, the contributions of parental transcripts may be influenced by the T-DNA chromosomal insertion site. In early pioneering work, no GUS signal was detected in 19 tested early embryonic loci when plants carrying the GUS reporter gene served as male parents. In contrast, when plants carrying the reporter served as female parents, the GUS signal was detected in embryos, similar to what is seen in self-pollinating plants carrying the GUS reporter gene (Vielle-Calzada et al., 2000). The GUS signal was detected as early as the two-cell embryo stage when wild-type egg cells were fertilized with sperm cells containing AtRPS5A::GUS (Weijers, Geldner, Offringa, & Jurgens, 2001). A similar result was obtained in a reciprocal cross between plants carrying CYC (*A. thaliana* cyclin B)::GUS and wild-type plants.

4.4.4 SNP-Based RNA-seq

In the time since development of single-cell sequencing technologies, RNA-seq has become a powerful tool used to study changes in the transcriptome during the MZT of plants. Early embryos from hybrids derived from crosses between two polymorphic inbred lines can be collected for transcriptome analysis using the single-cell RNA-seq technique described above. The relative parental contributions can be quantified based on analysis of SNPs that distinguish the two parental genomes. As mentioned earlier, transcriptome profiles of two- to four-cell and globular-stage hybrid embryos derived from a cross between Ler and Col, has been analyzed using RNA-seq; 88.4% and 64.1% of reads were maternal at the two- to four-cell stage and in globular hybrid embryos, respectively (Autran et al., 2011). A different study that sequenced hybrid embryos derived from a cross between Cvi and Col, found that the transcriptome contributions from the paternal and maternal genomes were equal (Nodine & Bartel, 2012). It is difficult to reconcile these conflicting results based on our present knowledge of the contributions of parental transcripts to early plant embryos.

5. CONCLUSIONS AND PERSPECTIVES

Studies of the MZT and the molecular basis thereof in plants, as well as technical advances, have allowed us to begin to answer certain questions, such as when the MZT commences, when zygotic genome transcription commences, and the contributions of parental transcripts to early embryogenesis; all seem to differ from those of animals. It is common in plants for both maternal and paternal genetic information to be involved in early embryogenesis, whereas early animal embryogenesis is maternally controlled. In addition, in plants but not in animals, sperm-delivered transcripts probably play important roles during zygotic cell division. Novel directions of investigation into the possible roles played by parental transcripts in the MZT may become apparent when maternal and *de novo* transcripts are successfully distinguished. Depending on the outcome of such global analyses, it may indeed become appropriate, for angiosperm plants, to describe the transfer of genetic control during early embryogenesis as the "parental-to-zygotic transition."

The inaccessibility of the egg cell and the absence of an *in vitro* culture system for early plant embryos mean that several techniques commonly used in animal MZT research have not been applied successfully to plant MZT research. Such technical limitations are the major barrier to study of the MZT in higher plants. Thus, no clear picture of the plant MZT has emerged. Many basic questions are yet to be answered. What is the exact timing of the MZT in different plant species? What maternal factors deposited in egg cells, if any, are needed for fertilization and zygote development? Are these cleared during early embryogenesis? What are the key molecular components triggering degradation of maternal factors and activation of zygotic genome transcription? Are any transcriptional waves evident during early embryogenesis, as seen in animals? What are the exact roles played by *de novo*-synthesized transcripts during early embryogenesis? More powerful technologies, particularly for analysis of single cells, are required to answer these questions. The poly(A) test used to analyze poly(A) tail length, ribosome profiling used to detect actively translated mRNAs, morpholino microinjections used to block translation, and BrUTP microinjections used to trace *de novo*-synthesized transcripts, all of which are commonly used in animal MZT research, have not been successfully applied to plant MZT research. These techniques will need to be modified, and their sensitivity improved, for use in studies on the MZT of higher plants.

More reliable *in vitro* culture systems and microinjection techniques for fertilized egg cells should be established for plants. These would help in studies on the timing of the MZT, transcriptome and epigenetic changes during the MZT, pharmacological treatments, and the use of single-cell sequencing technology for transcriptome and epigenetic analyses. A detailed understanding of the mechanisms of transcription and transcript degradation would allow interruption of such processes by blocking of essential pathways *in vivo*. This would allow the fates and roles of such transcripts to be recognized. Hopefully, such techniques will be available in the near future. Specific gene knockout technology, using the modified CRISPR system, will need to be developed to explore the roles played by cell type-specific genes in egg cells and early embryos during the MZT. Such technological advances will help us discover more details of the molecular mechanisms of the plant MZT.

ACKNOWLEDGMENTS

This project was supported by National Natural Science Foundation of China key project (31430007), the "973" Project (2013CB126900), and National Natural Science Foundation of China project (31170297, 31400171).

REFERENCES

Autran, D., Baroux, C., Raissig, M. T., Lenormand, T., Wittig, M., Grob, S., et al. (2011). Maternal epigenetic pathways control parental contributions to Arabidopsis early embryogenesis. *Cell*, *145*, 707–719.

Bayer, M., Nawy, T., Giglione, C., Galli, M., Meinnel, T., & Lukowitz, W. (2009). Paternal control of embryonic patterning in Arabidopsis thaliana. *Science*, *323*, 1485–1488.

Birnbaum, K., Jung, J. W., Wang, J. Y., Lambert, G. M., Hirst, J. A., Galbraith, D. W., et al. (2005). Cell type-specific expression profiling in plants via cell sorting of protoplasts from fluorescent reporter lines. *Nature Methods*, *2*, 615–619.

Birnbaum, K., Shasha, D. E., Wang, J. Y., Jung, J. W., Lambert, G. M., Galbraith, D. W., et al. (2003). A gene expression map of the Arabidopsis root. *Science*, *302*, 1956–1960.

Borges, F., Gomes, G., Gardner, R., Moreno, N., McCormick, S., Feijo, J. A., et al. (2008). Comparative transcriptomics of Arabidopsis sperm cells. *Plant Physiology*, *148*, 1168–1181.

Bultman, S. J., Gebuhr, T. C., Pan, H., Svoboda, P., Schultz, R. M., & Magnuson, T. (2006). Maternal BRG1 regulates zygotic genome activation in the mouse. *Genes & Development, 20*, 1744–1754.

Chenchik, A., Zhu, Y. Y., Diatchenko, L., Li, R., Hill, J., & Siebert, P. D. (1998). Generation and use of high-quality cDNA from small amounts of total RNA by SMARTPCR. In P. D. Siebert & J. W. Larrick (Eds.), *Gene cloning and analysis by RT-PCR* (pp. 305–319). Natick, MA: BioTechniques Books.

Del Toro-De Leon, G., Garcia-Aguilar, M., & Gillmor, C. S. (2014). Non-equivalent contributions of maternal and paternal genomes to early plant embryogenesis. *Nature, 514*, 624–627.

Dresselhaus, T., Lorz, H., & Kranz, E. (1994). Representative cDNA libraries from few plant cells. *The Plant Journal, 5*, 605–610.

Fu, C. M., Sun, M. X., Zhou, C., & Yang, H. Y. (1996). Isolation of fertilized embryo sacs and zygotes and triggering of zygote division in vitro in Nicotiana tabacum. *Acta Botanica Sinica, 38*, 262–267.

Gao, Y., & Zhao, Y. (2014). Self-processing of ribozyme-flanked RNAs into guide RNAs in vitro and in vivo for CRISPR-mediated genome editing. *Journal of Integrative Plant Biology, 56*, 343–349.

Goldberg, R. B., de Paiva, G., & Yadegari, R. (1994). Plant embryogenesis: Zygote to seed. *Science, 266*, 605–614.

Grimanelli, D., Perotti, E., Ramirez, J., & Leblanc, O. (2005). Timing of the maternal-to-zygotic transition during early seed development in maize. *Plant Cell, 17*, 1061–1072.

Guy, A. L., & Taylor, J. H. (1978). Actinomycin D inhibits initiation of DNA replication in mammalian cells. *Proceedings of the National Academy of Sciences of the United States of America, 75*, 6088–6092.

Haecker, A., Gross-Hardt, R., Geiges, B., Sarkar, A., Breuninger, H., Herrmann, M., et al. (2004). Expression dynamics of WOX genes mark cell fate decisions during early embryonic patterning in Arabidopsis thaliana. *Development, 131*, 657–668.

Hamatani, T., Carter, M. G., Sharov, A. A., & Ko, M. S. (2004). Dynamics of global gene expression changes during mouse preimplantation development. *Developmental Cell, 6*, 117–131.

He, Y. C., He, Y. Q., Qu, L. H., Sun, M. X., & Yang, H. Y. (2007). Tobacco zygotic embryogenesis in vitro: The original cell wall of the zygote is essential for maintenance of cell polarity, the apical-basal axis and typical suspensor formation. *The Plant Journal, 49*, 515–527.

Holbein, S., Wengi, A., Decourty, L., Freimoser, F. M., Jacquier, A., & Dichtl, B. (2009). Cordycepin interferes with 3' end formation in yeast independently of its potential to terminate RNA chain elongation. *RNA, 15*, 837–849.

Jahnke, S., & Scholten, S. (2009). Epigenetic resetting of a gene imprinted in plant embryos. *Current Biology, 19*, 1677–1681.

Kawashima, T., & Berger, F. (2014). Epigenetic reprogramming in plant sexual reproduction. *Nature Reviews. Genetics, 15*, 613–624.

Khalequzzaman, M., & Haq, N. (2005). Isolation and in vitro fusion of egg and sperm cells in Oryza sativa. *Plant Physiology and Biochemistry: PPB, 43*, 69–75.

Kranz, E., Bautor, J., & Lorz, H. (1991). In vitro fertilization of single, isolated gametes of maize mediated by electrofusion. *Sexual Plant Reproduction, 4*, 12–16.

Kranz, E., von Wiegen, P., Quader, H., & Lorz, H. (1998). Endosperm development after fusion of isolated, single maize sperm and central cells in vitro. *Plant Cell, 10*, 511–524.

Latorra, D., Campbell, K., Wolter, A., & Hurley, J. M. (2003). Enhanced allele-specific PCR discrimination in SNP genotyping using 3′ locked nucleic acid (LNA) primers. *Human Mutation, 22*, 79–85.

Lee, M. T., Bonneau, A. R., & Giraldez, A. J. (2014). Zygotic genome activation during the maternal-to-zygotic transition. *Annual Review of Cell and Developmental Biology*, *30*, 581–613.

Li, X. Y., Cui, X. S., & Kim, N. H. (2006). Transcription profile during maternal to zygotic transition in the mouse embryo. *Reproduction, Fertility, and Development*, *18*, 635–645.

Li, L., Lu, X., & Dean, J. (2013). The maternal to zygotic transition in mammals. *Molecular Aspects of Medicine*, *34*, 919–938.

Ma, L., Xin, H., Qu, L., Zhao, J., Yang, L., Zhao, P., et al. (2011). Transcription profile analysis reveals that zygotic division results in uneven distribution of specific transcripts in apical/basal cells of tobacco. *PLoS One*, *6*, e15971.

Meyer, S., Pospisil, H., & Scholten, S. (2007). Heterosis associated gene expression in maize embryos 6 days after fertilization exhibits additive, dominant and overdominant pattern. *Plant Molecular Biology*, *63*, 381–391.

Meyer, S., & Scholten, S. (2007). Equivalent parental contribution to early plant zygotic development. *Current Biology*, *17*, 1686–1691.

Muralla, R., Lloyd, J., & Meinke, D. (2011). Molecular foundations of reproductive lethality in Arabidopsis thaliana. *PLoS One*, *6*, e28398.

Nakitandwe, J., Trognitz, F., & Trognitz, B. (2007). Reliable allele detection using SNP-based PCR primers containing Locked Nucleic Acid: Application in genetic mapping. *Plant Methods*, *3*, 2.

Ning, J., Peng, X. B., Qu, L. H., Xin, H. P., Yan, T. T., & Sun, M. (2006). Differential gene expression in egg cells and zygotes suggests that the transcriptome is restructed before the first zygotic division in tobacco. *FEBS Letters*, *580*, 1747–1752.

Nodine, M. D., & Bartel, D. P. (2012). Maternal and paternal genomes contribute equally to the transcriptome of early plant embryos. *Nature*, *482*, 94–97.

Nygaard, V., & Hovig, E. (2006). Options available for profiling small samples: A review of sample amplification technology when combined with microarray profiling. *Nucleic Acids Research*, *34*, 996–1014.

Ohnishi, T., Takanashi, H., Mogi, M., Takahashi, H., Kikuchi, S., Yano, K., et al. (2011). Distinct gene expression profiles in egg and synergid cells of rice as revealed by cell type-specific microarrays. *Plant Physiology*, *155*, 881–891.

Ohno, M., Aoki, N., & Sasaki, H. (2001). Allele-specific detection of nascent transcripts by fluorescence in situ hybridization reveals temporal and culture-induced changes in Igf2 imprinting during pre-implantation mouse development. *Genes to Cells*, *6*, 249–259.

Pillot, M., Baroux, C., Vazquez, M. A., Autran, D., Leblanc, O., Vielle-Calzada, J. P., et al. (2010). Embryo and endosperm inherit distinct chromatin and transcriptional states from the female gametes in Arabidopsis. *Plant Cell*, *22*, 307–320.

Pitto, L., Schiavo, F. L., & Terzi, M. (1985). alpha-Amanitin resistance is developmentally regulated in carrot. *Proceedings of the National Academy of Sciences of the United States of America*, *82*, 2799–2803.

Ronshaugen, M., & Levine, M. (2004). Visualization of trans-homolog enhancer-promoter interactions at the Abd-B Hox locus in the Drosophila embryo. *Developmental Cell*, *7*, 925–932.

Slane, D., Kong, J., Berendzen, K. W., Kilian, J., Henschen, A., Kolb, M., et al. (2014). Cell type-specific transcriptome analysis in the early Arabidopsis thaliana embryo. *Development*, *141*, 4831–4840. http://dx.doi.org/10.1242/dev.116459.

Sobell, H. M. (1985). Actinomycin and DNA transcription. *Proceedings of the National Academy of Sciences of the United States of America*, *82*, 5328–5331.

Sun, M. X., Yang, H. Y., & Zhou, C. (1993). A new method for embryo sac isolation and in situ fusion of egg and synergid protoplasts in Nicotiana tabacum. *Acta Botanica Sinica*, *35*, 893–900.

Tadros, W., & Lipshitz, H. D. (2009). The maternal-to-zygotic transition: A play in two acts. *Development, 136*, 3033–3042.

Takanashi, H., Ohnishi, T., Mogi, M., Okamoto, T., Arimura, S., & Tsutsumi, N. (2010). Studies of mitochondrial morphology and DNA amount in the rice egg cell. *Current Genetics, 56*, 33–41.

Vielle-Calzada, J. P., Baskar, R., & Grossniklaus, U. (2000). Delayed activation of the paternal genome during seed development. *Nature, 404*, 91–94.

Vielle-Calzada, J. P., Thomas, J., Spillane, C., Coluccio, A., Hoeppner, M. A., & Grossniklaus, U. (1999). Maintenance of genomic imprinting at the Arabidopsis medea locus requires zygotic DDM1 activity. *Genes & Development, 13*, 2971–2982.

Walser, C. B., & Lipshitz, H. D. (2011). Transcript clearance during the maternal-to-zygotic transition. *Current Opinion in Genetics & Development, 21*, 431–443.

Wassermann, K., Markovits, J., Jaxel, C., Capranico, G., Kohn, K. W., & Pommier, Y. (1990). Effects of morpholinyl doxorubicins, doxorubicin, and actinomycin D on mammalian DNA topoisomerases I and II. *Molecular Pharmacology, 38*, 38–45.

Weijers, D., Geldner, N., Offringa, R., & Jurgens, G. (2001). Seed development: Early paternal gene activity in Arabidopsis. *Nature, 414*, 709–710.

Wuest, S. E., Vijverberg, K., Schmidt, A., Weiss, M., Gheyselinck, J., Lohr, M., et al. (2010). Arabidopsis female gametophyte gene expression map reveals similarities between plant and animal gametes. *Current Biology, 20*, 506–512.

Xiang, D., Venglat, P., Tibiche, C., Yang, H., Risseeuw, E., Cao, Y., et al. (2011). Genome-wide analysis reveals gene expression and metabolic network dynamics during embryo development in Arabidopsis. *Plant Physiology, 156*, 346–356.

Xin, H. P., Peng, X. B., Ning, J., Yan, T. T., Ma, L. G., & Sun, M. X. (2011). Expressed sequence-tag analysis of tobacco sperm cells reveals a unique transcriptional profile and selective persistence of paternal transcripts after fertilization. *Sexual Plant Reproduction, 24*, 37–46.

Yang, L., Duff, M. O., Graveley, B. R., Carmichael, G. G., & Chen, L. L. (2011). Genomewide characterization of non-polyadenylated RNAs. *Genome Biology, 12*, R16.

Yu, H. S., & Russell, S. D. (1994). Occurrence of mitochondria in the nuclei of tobacco sperm cells. *Plant Cell, 6*, 1477–1484.

Zhang, J., Dong, W. H., Potrykus, I., & Galli, A. (1999). Regeneration of fertile plants from isolated zygotes of rice. *Plant Cell Reports, 19*, 128–132.

Zhang, J. E., Luo, A., Xin, H. P., Zhao, J., Li, S. S., Qu, L. H., et al. (2011). Genes of both parental origins are differentially involved in early embryogenesis of a tobacco interspecies hybrid. *PLoS One, 6*, e23153.

Zhao, J., Xin, H., Qu, L., Ning, J., Peng, X., Yan, T., et al. (2011). Dynamic changes of transcript profiles after fertilization are associated with de novo transcription and maternal elimination in tobacco zygote, and mark the onset of the maternal-to-zygotic transition. *The Plant Journal, 65*, 131–145.

Zhao, P., Zhou, X. M., Zhang, L. Y., Wang, W., Ma, L. G., Yang, L. B., et al. (2013). A bipartite molecular module controls cell death activation in the Basal cell lineage of plant embryos. *PLoS Biology, 11*, e1001655.

INDEX

Note: Page numbers followed by "*f*" indicate figures and "*t*" indicate tables.

D

H

P

R

S

T

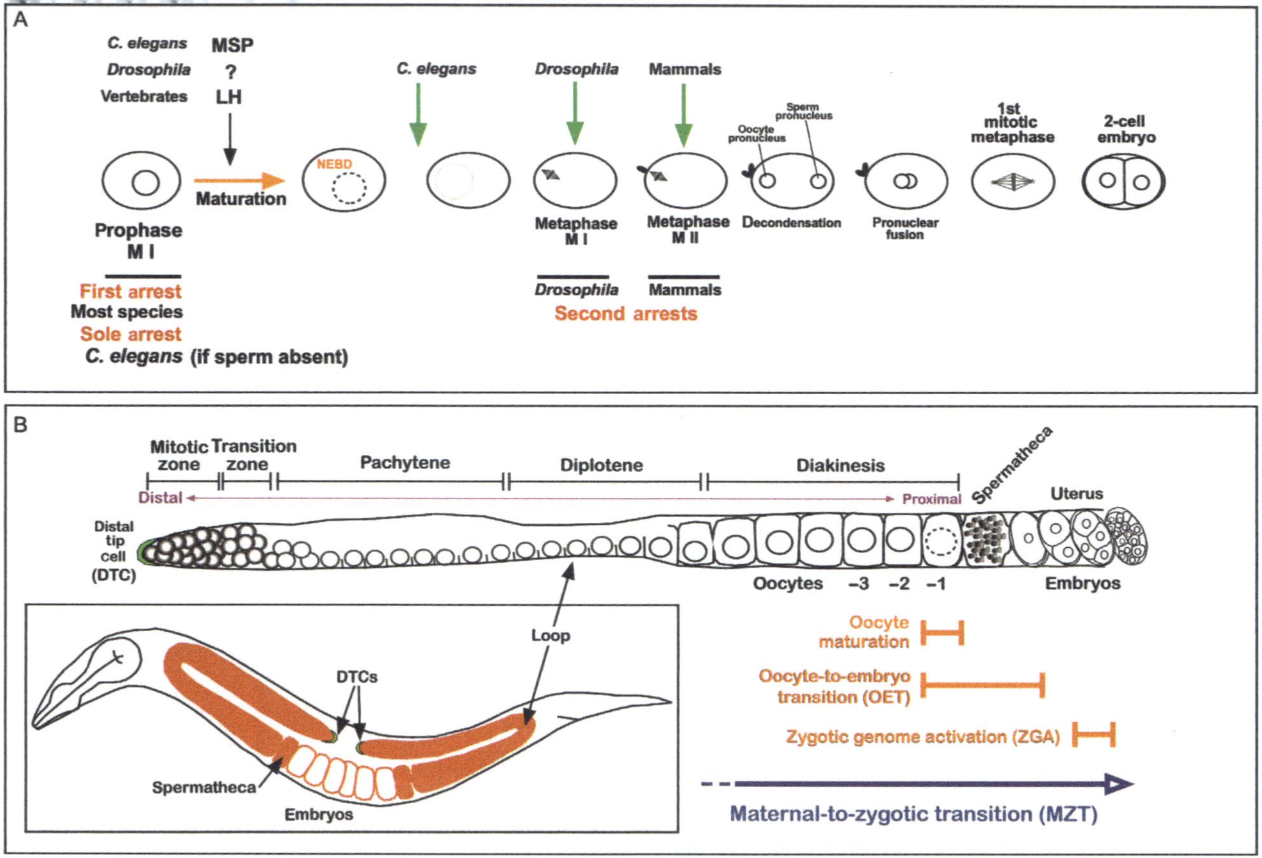

Scott Robertson and Rueyling Lin, Figure 1 (A) Oocyte development and fertilization. General schematic of oocyte maturation and fertilization, highlighting the meiotic arrest(s) (red text), inducers of oocyte maturation (orange arrow), and timing of fertilization (green arrows) for *C. elegans*, *Drosophila*, and mammals. MI, meiosis I; MII, meiosis II; NEBD, nuclear envelope breakdown; MSP, major sperm protein; LH, luteinizing hormone. (B) Schematic of *C. elegans* gonad. One arm of the adult *C. elegans* hermaphrodite gonad is represented. Specific mitotic/meiotic stages are indicated above the gonad by the bars. Below the gonad, bars and arrow indicate oocyte maturation, OET, ZGA, and MZT, respectively. Inset: Schematic of an adult hermaphrodite, highlighting the gonad (red fill) and developing embryos (red outline). DTCs, distal tip cells.

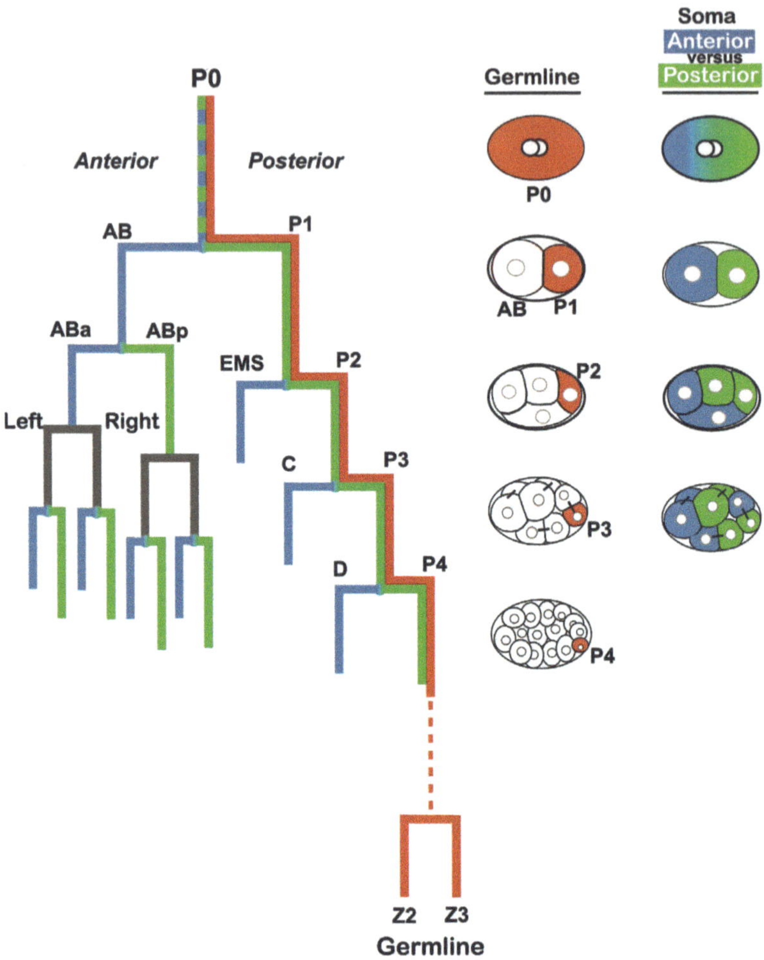

Scott Robertson and Rueyling Lin, Figure 4 Segregation of germline versus anterior or posterior developmental potential in the early *C. elegans* embryo. The cytoplasm of the 1-cell embryo (P0) contains multiple maternal factors, both protein and mRNA, that direct lineage-specific differentiation. Segregation of these developmental determinants begins with the first mitosis and occurs at each following mitosis to generate the correct blastomere-specific constellation of proteins and maternal transcripts to effect both early development in the absence of zygotic transcription as well as an orderly, blastomere-specific transition from maternal-to-zygotic control of embryogenesis.

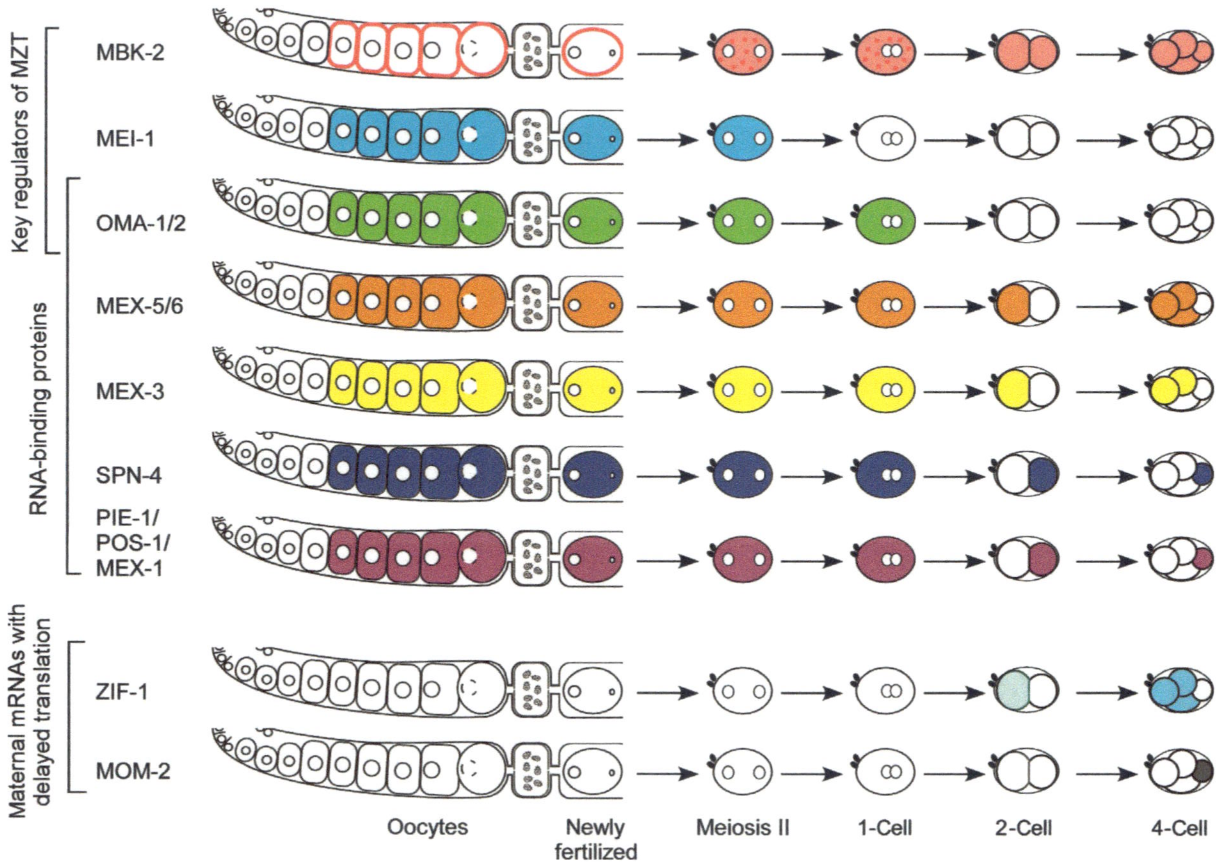

Figure 6 See the legend on next page.

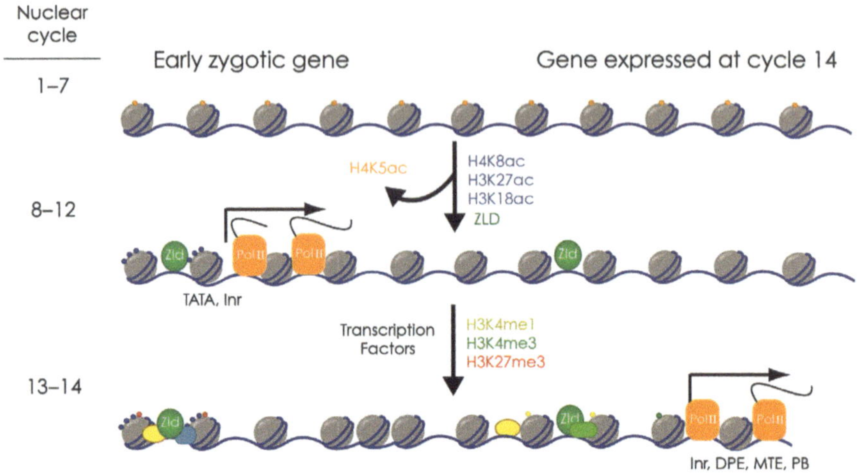

Melissa M. Harrison and Michael B. Eisen, Figure 3 Model of the multiple processes that govern gene expression during the MZT. During nuclear cycles 1–7, the genome is likely reprogrammed to a relatively naïve state. Due to the rapid DNA replication, there are high levels of acetylated histone H4 lysine 5. At cycles 8–12, a number of early zygotic genes are activated. These genes have TATA box-containing promoters. At the *cis*-regulatory regions driving gene expression, ZLD is bound and histones are acetylated. As the embryo transitions through cycle 14, histone methylation is first detected, and ZLD facilitates the binding of additional transcription factors. Widespread RNA polymerase II pausing is established, especially at promoters containing DPE, MTE, and PB elements.

Scott Robertson and Rueyling Lin, Figure 6 Distribution of key maternal proteins in early *C. elegans* embryos. The top 7 rows indicate the spatiotemporal localization patterns for 11 maternally supplied proteins, including two redundant pairs (OMA-1/2 and MEX-5/6) and one pattern shown for the three germline-localized proteins—PIE-1, POS-1, and MEX-1. Note that, although not indicated, PIE-1, unlike POS-1 and MEX-1, concentrates in the nuclei of the germline blastomeres. Many of these patterns result from asymmetric protein segregation at mitotic divisions (asymmetry usually precedes division) and/or selective degradation. The bottom two rows are examples of regulated translation of maternally supplied mRNAs, primarily through the highly spatiotemporally regulated binding of a cohort of positive- and negative-acting maternally supplied RNA-binding proteins that produce a blastomere-specific functional readout for what appears to be a 3′UTR "code" of binding sites (see text for details).

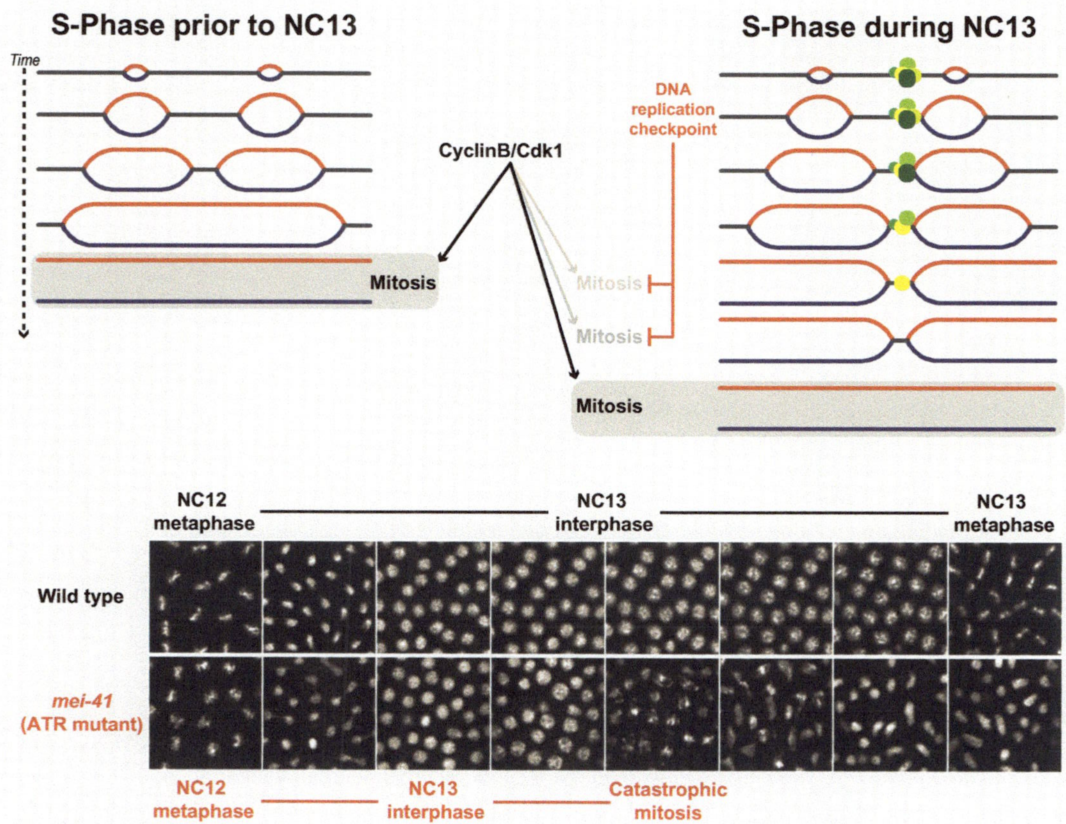

Figure 2 See the legend on next page.

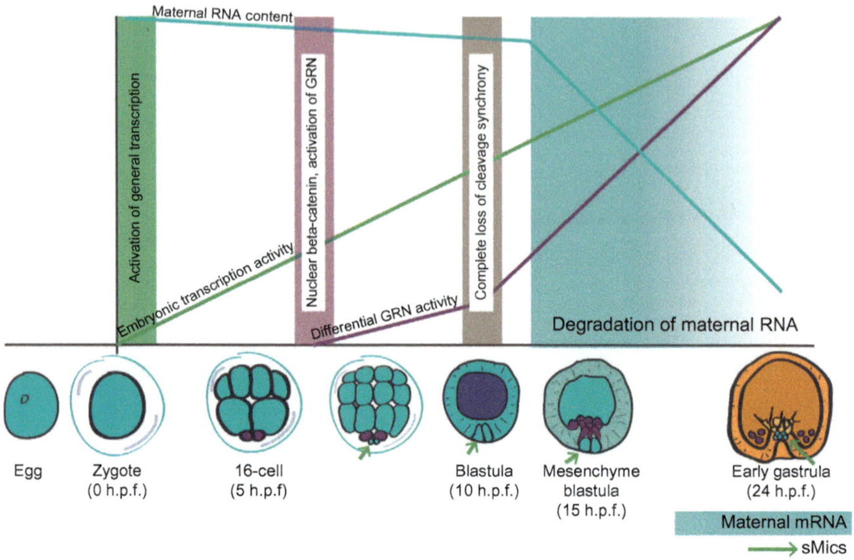

S. Zachary Swartz and Gary M. Wessel, Figure 4 Summary of METs in sea urchin embryogenesis. Important embryonic milestones are denoted with embryo illustrations. General transcription based on incorporation of radioactive nucleotides initiates at fertilization and steadily increases through early development (green line). The regionally specific gene regulatory network and fate specification first occur at the 16-cell stage, when β-catenin is nuclearized in the micromeres (purple line). We hypothesize that the differential GRN further diversifies in activity when the cell cycle is remodeled at approximately 10 h.p.f. and cleavage synchrony is lost. Based on mRNA expression dynamics, we suggest that there is a major degradation of maternal RNA that occurs as the embryo prepares to gastrulate. Maternal mRNA is represented in blue in embryo descriptions, and is lost in somatic cells in the gastrula, but retained in the sMics (green arrows).

Shelby A. Blythe and Eric F. Wieschaus, Figure 2 Cell cycle lengthening by the DNA damage checkpoint. Top panels: prior to NC13 (top left), S-phase proceeds at a maximal rate. During NC13 (top right), chromatin remodeling results in increased promoter occupancy (yellow-green spots) that causes reduced replication fork progression. Such stalling of DNA replication triggers the DNA replication checkpoint, which serves to inhibit transiently mitotic entry as signaled by cyclin/Cdk1 (center). Upon completion of DNA replication, the checkpoint is inactivated and mitosis can proceed. Bottom panels: frames from time-lapse micrographs of wild-type or *mei-41* mutant embryos expressing *Histone H2Av-GFP* demonstrate the relative contribution of DNA replication checkpoint activity on cell cycle slowing during NC13. Note the premature mitotic entry for *mei-41* mutant embryos and the subsequent catastrophic mitosis. Image sequences represent 3-min interval between frames of a 2500 μm^2 area on the embryo surface.

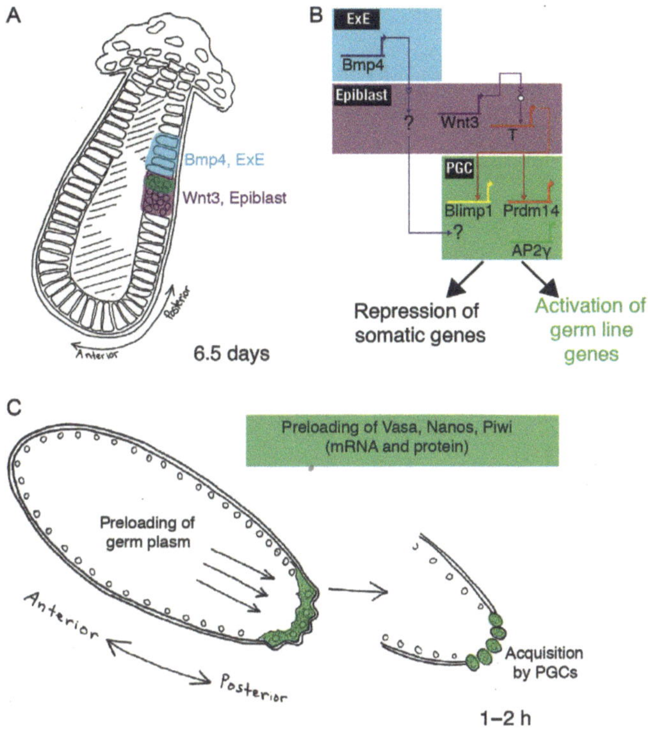

S. Zachary Swartz and Gary M. Wessel, Figure 5 Strategies for germ line segregation by induction or preloading. (A) Germ line specification by induction. Shown is a simplified mouse embryo 6.5 days postfertilization. Bmp4 signaling from the extraembryonic ectoderm (ExE, blue) and Wnt3 expressed in the proximal epiblast (purple) are synergistically required to activate germ line transcriptional regulators in the PGCs (green). (B) Regulatory diagram depicting Wnt3 activating the mesodermal transcription factor T, which in turn activates Blimp1 and Prdm14, which along with AP2γ, form a tripartite network that activates germ line gene expression and represses somatic fate. The activating role of Bmp4 has not been directly delineated, but is required to license the PGCs to respond to the Wnt3 signal. (C) Germ line segregation by preloading. Shown is a syncytial blastoderm-staged *Drosophila* embryo, with maternally localized germ plasm in green. The PGCs are the first to cellularize in this embryo, and acquisition of the germ plasm directs them toward germ line fate.

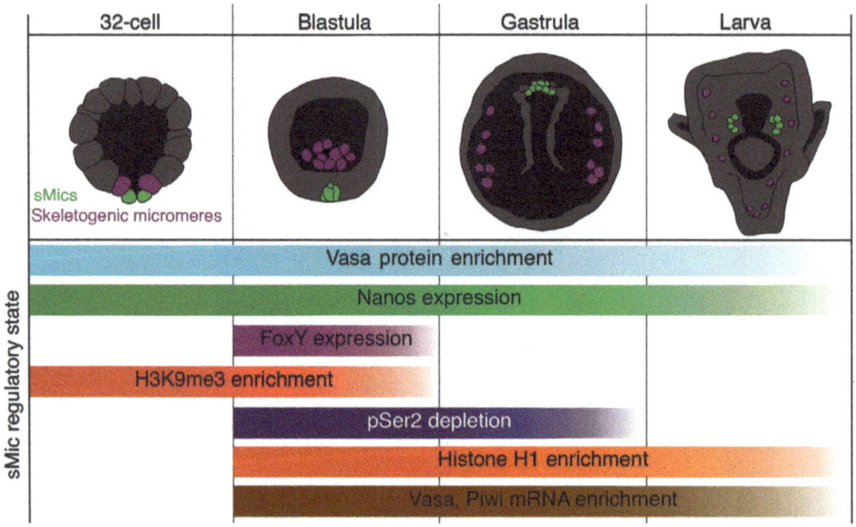

S. Zachary Swartz and Gary M. Wessel, Figure 6 Temporal dynamics of sMic regulatory state. Location of the sMic lineage is indicated in green, from 32-cell stage into the larva. Colored bars indicate the presence or absence of gene expression, or enrichment or depletion for specific markers. *Adapted from Wessel et al. (2014).*

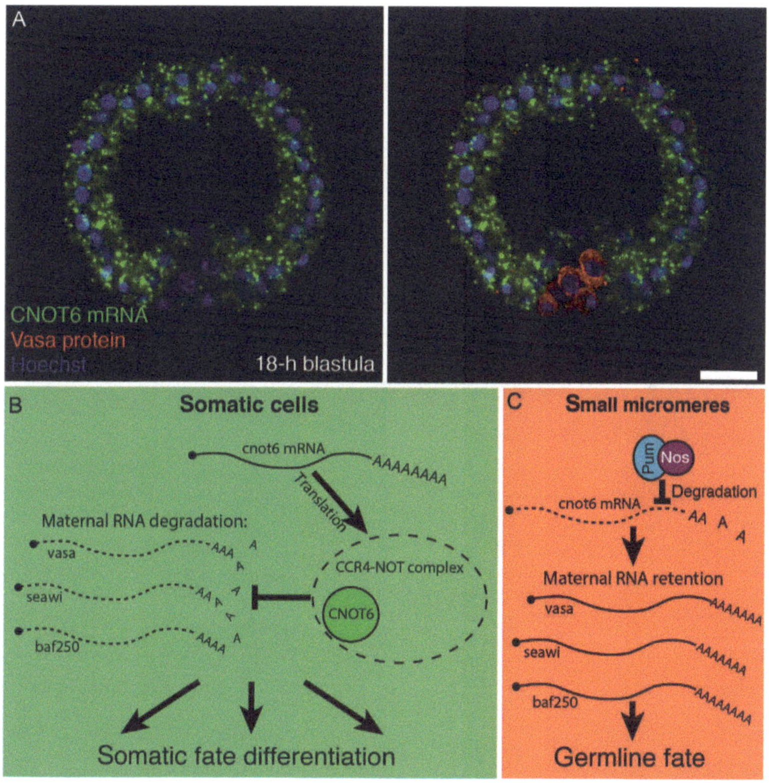

S. Zachary Swartz and Gary M. Wessel, Figure 7 Time Capsule model for (A) CNOT6 mRNA (green) is present in all somatic cells but depleted in the sMics (red, Vasa protein) in the 18 h.p.f. blastula. (B and C) Model for differential RNA stability between the somatic cells and sMics. (B) CNOT6 deadenylase is broadly present, and destabilizes maternal transcripts such as Vasa, Seawi, and Baf250 by shortening their poly-A tails. (B) These RNAs are protected in the sMics, because Nanos (Nos) and its partner Pumilio (Pum) repress the accumulation of CNOT6. Scale bar = 20 μm. *Reproduced from Wessel et al. (2014).*

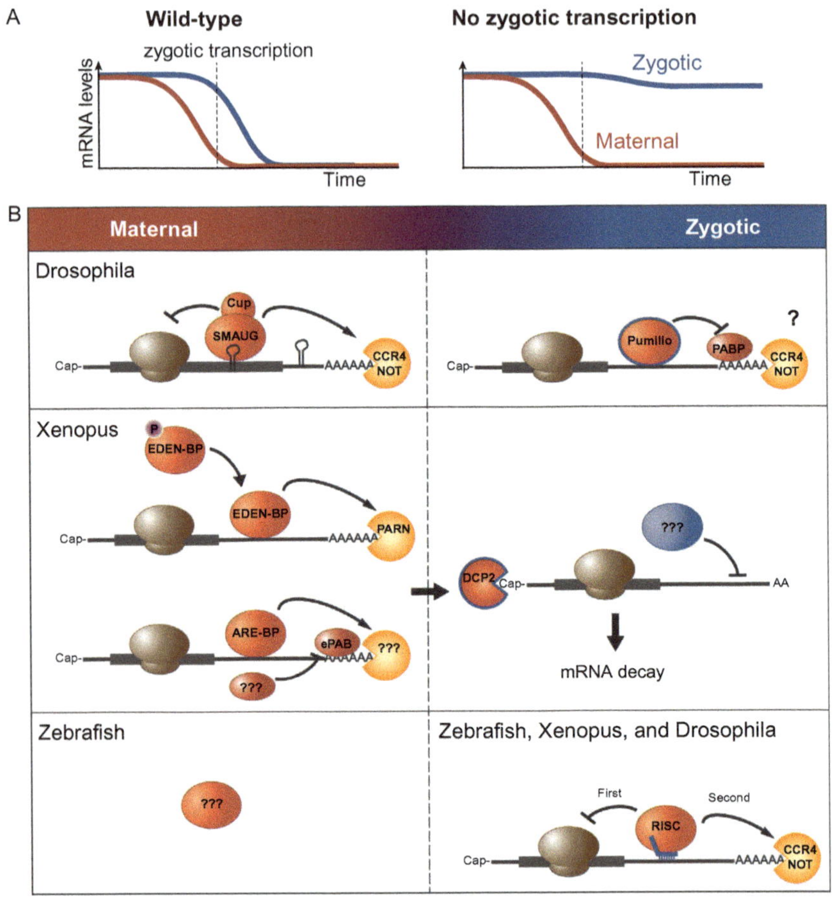

Valeria Yartseva and Antonio J. Giraldez, Figure 2 Maternal and zygotic mechanisms of maternal mRNA clearance. (A) Maternal mRNAs under the regulation of the "maternal mode" mechanisms (red) will be destabilized independently of zygotic transcription, while mRNAs under the "zygotic mode" mechanisms (blue) will be stable in the absence of zygotic transcription. (B) Examples of characterized pathways of maternal and zygotic mode mechanisms across species. Maternal mode factors (red), zygotic mode factors (blue), and maternal factors activated after zygotic transcription (red with blue outline).

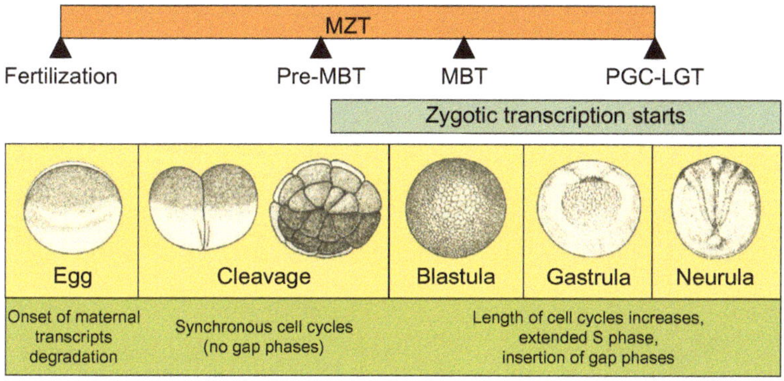

Jing Yang *et al.*, Figure 1 The maternal-to-zygotic transition (MZT). MZT is a period that occurs very early during embryonic development, when degradation of maternal transcripts is initiated and is complete with the first morphological change caused by zygotic transcription, gastrulation. After the 12 mitotic divisions, the synchronous cell cycles are followed by asynchronous cleavages, with the introduction of gap phases during a developmental event called the "mid-blastula transition" (MBT). Most zygotic genes are silent until large-scale zygotic gene activation (ZGA) that, in *Xenopus*, coincides with the MBT. Several genes required for embryo patterning are transcribed before the MBT (pre-MBT) starting at the 32-cell stage. PGC transcription begins later, during the transition between gastrula and neurula (*p*rimordial *g*erm *c*ell *l*ate-*g*astrula *t*ransition, PGC-LGT). Images of *Xenopus* embryos are shown to highlight the different developmental stages.

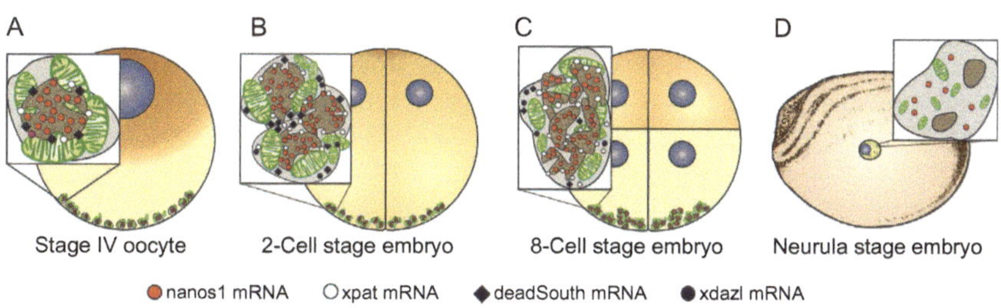

Jing Yang *et al.*, Figure 2 Schemes of germ-plasm structure during *Xenopus laevis* embryogenesis. At oocyte Stage VI, germ-plasm islands are associated with the vegetal cortex (A). Between fertilization and the 8-cell stage, germinal granules (GGs) undergo ultrastructural changes thought to represent a permissive state for translation. After fertilization, the germ plasm contains a large number of small (250 nm) spherical GGs (B). During early cleavage, GGs coalesce into progressively more complex aggregates at the apex of vegetal pole blastomeres (C). Germ plasm at the 8-cell stage contains a few irregularly shaped large GG (about 2000 nm) that, in cross section, show complex and variable morphology (C). Between the gastrula and neurula stages, large GGs disappear and primordial germ cells from the neurula stage contain only a few small (300 nm) GGs (D).

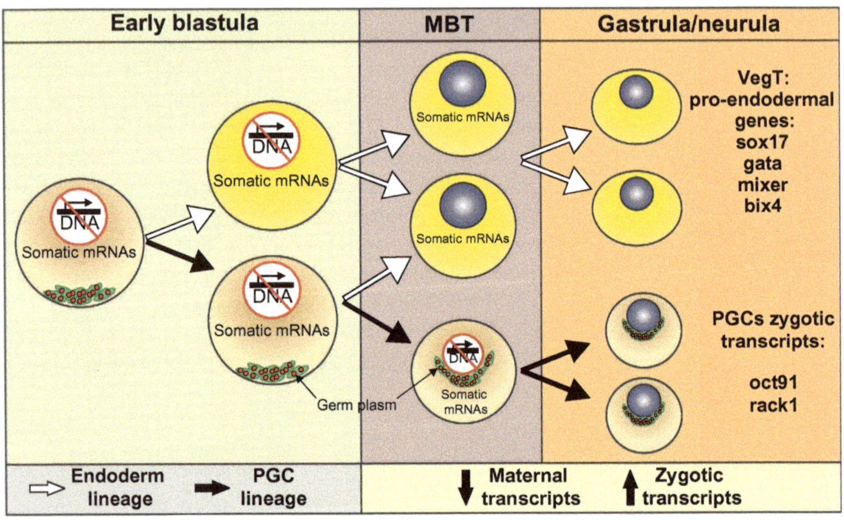

Jing Yang *et al.*, Figure 3 Scheme of primordial germ cell (PGC) segregation from soma. In *Xenopus*, the germline is specified through the inheritance of germ plasm formed during oogenesis and asymmetrically segregated into the future germ-cell lineage (black arrows). Endodermal somatic fate is established by the inheritance of maternal somatic transcripts, including VegT (white arrows). Germ cells repress somatic maternal mRNAs. PGCs delay starting mRNA transcription to ensure that somatic differentiation programs remain inactive at the time that zygotic transcription is initiated in the rest of the embryo.

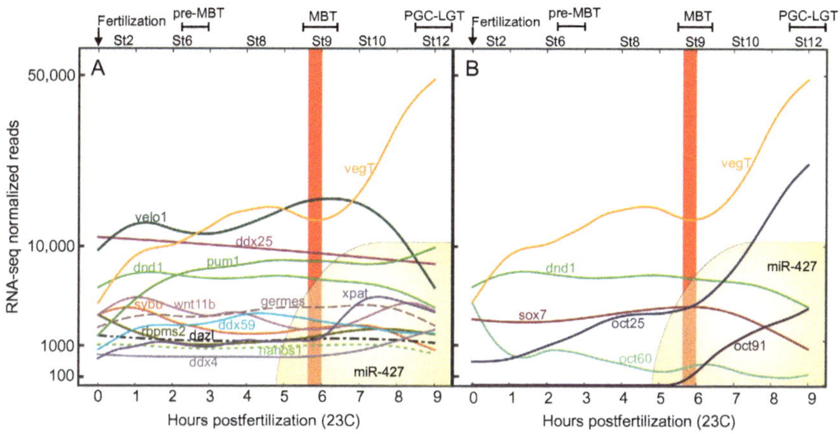

Jing Yang *et al.*, Figure 4 Transcriptional profiling in *Xenopus* early development. Representation of gene-expression patterns of (A) PGC-related genes and (B) pluripotency factors between fertilization and the onset of gastrulation. Before the MBT, positive/negative changes in reads are because of polyadenylation and deadenylation rather than degradation while, after the MBT, miR-427-dependent degradation occurs (Collart et al., 2014). Polyadenylation of maternally deposited transcripts occurs immediately after fertilization and is required for zygotic transcription. Two waves of zygotic genome activation occur: At the 32/64-cell stage (pre-MBT), ~150 new transcripts are synthesized while the bulk of transcription occurs hours later, at the MBT (Tan et al., 2013). The maternal determinant *VegT* is degraded in PGCs while it is polyadenlyated and translated in the soma and zygotically transcribed at the MBT (included in both graphics as a reference). miR-427 is essential for deadenylation of maternal mRNAs during the MZT (yellow curves in A and B). (A) Germline RNAs appear to be polyadenylated soon after fertilization and remain stable through the MBT. (B) Pluripotency factors generally decline after the MBT except for Oct 91, which is newly transcribed in the PGCs. *Modified from Collart et al. (2014).*

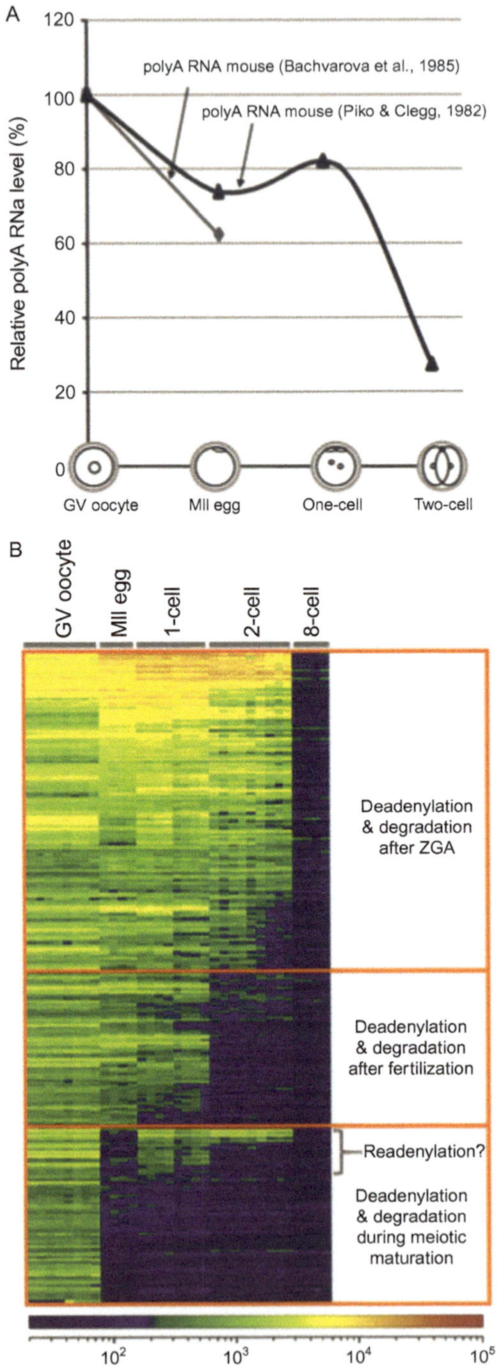

Figure 2 See the legend on opposite page.

Petr Svoboda *et al.*, Figure 2 Maternal mRNA degradation during the OET in mice. (A) Analysis of relative poly(A) RNA amounts in different stages shows that the majority of maternal poly(A) RNA is degraded by the time the embryo undergoes ZGA. (B) Microarray profiling reveals phased maternal mRNA degradation where each developmental transition (resumption of meiosis, fertilization, and ZGA) triggers a wave of mRNA degradation. Shown is microarray profiling (Zeng et al., 2004; Zeng & Schultz, 2005) of ~300 oocyte-specific genes (expressed in oocytes but not expressed in somatic tissues; Su et al., 2007). Colors indicate mRNA abundance approximated by fluorescence intensity (arbitrary units). The median expression value for expressed genes across the entire dataset was set to 500.

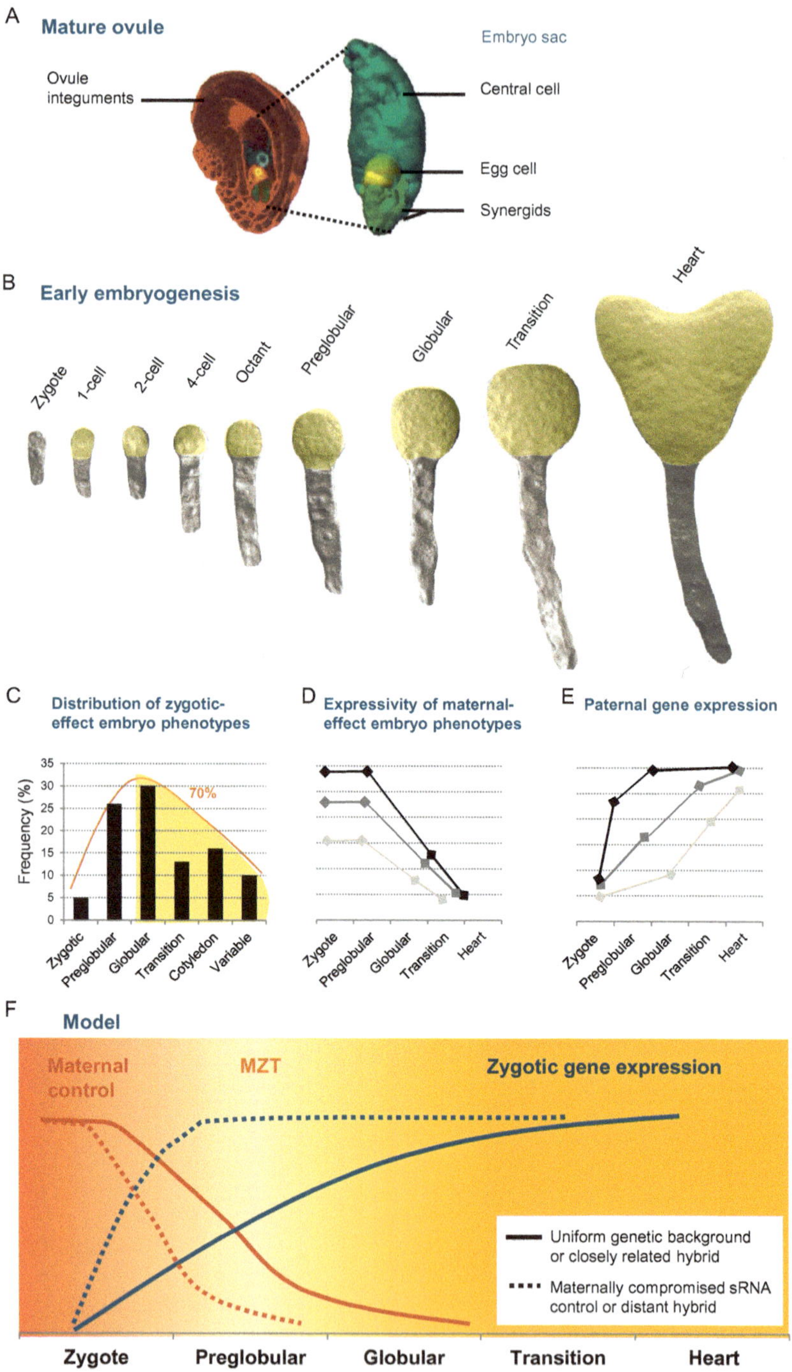

Figure 1 See the legend on opposite page.

Célia Baroux and Ueli Grossniklaus, Figure 1 Embryogenesis, maternal effects, and progressive ZGA in *Arabidopsis.* (A) Mature *Arabidopsis* ovule (left, 3D reconstruction of a confocal scanning light microscopy image series of an ovule with cell wall staining; cells of the embryo sac are pseudo-colored: central cell in blue, egg cell in yellow, synergids in green; ovule integuments are in red) and isolated embryo sac (right, 3D rendering of embryo sac cells extracted from the image series shown on the left, powered by Imaris, Bitplane AG, CH). (B) *Arabidopsis* early embryogenesis. Consecutive embryonic stages are indicated. The embryo proper is highlighted in yellow while the suspensor is displayed in gray. The surrounding endosperm and seed integuments are not shown. Embryo images were extracted using Adobe Photoshop from whole, cleared seed. (C) Distribution of zygotic-effect embryo phenotypes based on frequencies described among 1501 *emb* mutants curated by the Meinke laboratory (McElver et al., 2001). The graph is based on published class frequencies with slight modifications for simplicity: the classes "zygotic/preglobular," "preglobular/globular," and "other" were pooled together in the class "zygotic," "preglobular," and "variable", respectively. (D) Schematic interpretation of the expressivity of several independent maternal-effect embryo phenotypes reported by the Gillmor laboratory (Del Toro-De Leon, Garcia-Aguilar, & Gillmor, 2014). In all cases, the zygotic rescue of maternal-effect embryo defects mediated by a wild-type paternal allele is gradual, leading to a decreasing proportion (*y* axis) of mutant phenotypes as development proceeds. The rescue kinetics reflecting the maternal effect depends on the *emb* locus but also the genetic background. The greater the genetic divergence between the parents, the shorter the maternal effects last. (E) Schematic representation of paternal gene expression kinetics obtained in reporter and profiling experiments (see main text for references), indicating that ZGA is a gradual rather than an all-or-none process, targeting progressively more and more loci as development progresses. (F) Integrative model of the MZT in flowering plants reconciling different reports (see main text for references). The MZT is a long transition initiated in the zygote and largely completed at the globular stage. The MZT is the combination of the maternal clearance of mRNAs, about which we have currently no information in plants, and gradual ZGA. Maternal control (red lines) has been determined in functional studies of maternal-effect embryo mutants or transcriptionally disabled zygotes, and likely relies on maternally stored products inherited through the egg cell. ZGA (blue lines) initiates in the zygote, where several loci are activated, becoming more robust as development proceeds and more and more loci are transcribed. However, based on cytological staining of active RNA Pol II, the zygote remains in a relatively quiescent, transcriptional state. Maternal effects have a briefer impact while ZGA is more rapid when embryos inherit genetically divergent parental genomes or are maternally impaired in ZGA repressive mechanisms, such as mutants of the RdDM pathway (dashed lines). *Panel (B): Pictures courtesy of Quy A. Ngo.*

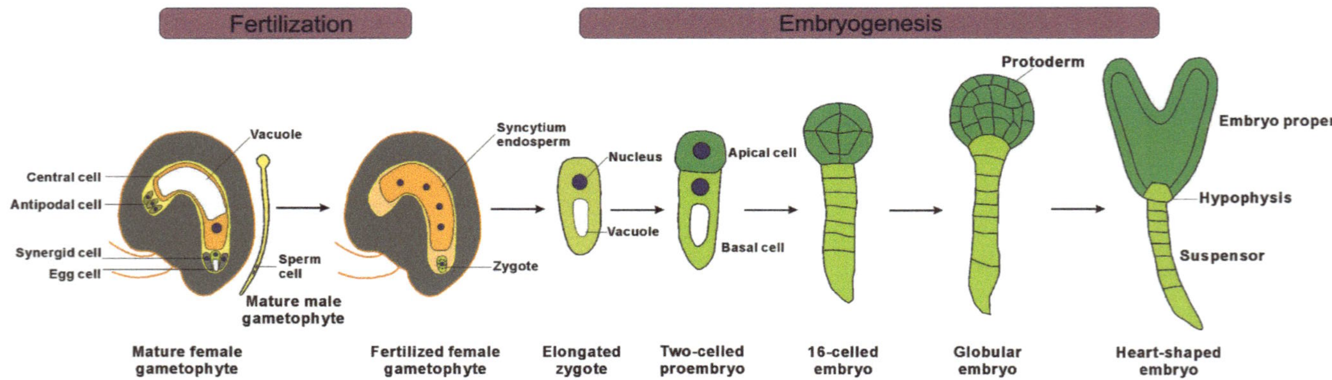

Peng Zhao and Meng-Xiang Sun, Figure 1 The process of fertilization and embryogenesis in higher plants. In angiosperms, the early male gametophyte, or pollen, germinates and produces a pollen tube that grows through the pistil to deliver a pair of sperm cells to the female gametophyte. The female gametophyte, or embryo sac, usually consists of seven cells: three antipodal cells, two synergid cells, one egg cell, and one central cell. During fertilization, one of the sperm cells fuses with the egg cell to produce a diploid zygote, which develops into an embryo, and the other sperm cell fuses with the central cell to produce a triploid primary endosperm cell. During embryogenesis, the elongated zygote divides asymmetrically, giving rise to a larger basal cell and a smaller apical cell with distinct developmental fates. The small apical cell develops into the main body of the embryo proper, whereas the larger basal cell continues to expand longitudinally and divides transversely to form a suspensor. The uppermost suspensor cell, termed the hypophysis, eventually becomes part of the primary root meristem. The pale green color labels egg and fertilized egg cell, the orange color labels central cell and fertilized central cell, the dark green color labels apical cell lineage, and the green-yellow color labels basal cell linage.

FSC
www.fsc.org
MIX
Papier | Fördert gute Waldnutzung
FSC® C083411

Kolibri 360 GmbH
Ferdinand-Jühlke-Straße 7
99095 Erfurt, Deutschland
produktsicherheit@kolibri360.de